TUTORIALS IN EVENT RELATED RESEARCH: ENDOGENOUS COMPONENTS

ADVANCES IN PSYCHOLOGY
10

Editors

G. E. STELMACH

P. A. VROON

NORTH-HOLLAND PUBLISHING COMPANY
AMSTERDAM • NEW YORK • OXFORD

TUTORIALS IN EVENT RELATED POTENTIAL RESEARCH: ENDOGENOUS COMPONENTS

Edited by

Anthony W. K. GAILLARD
Institute for Perception TNO
Soesterberg, The Netherlands

and

Walter RITTER
Albert Einstein College of Medicine
and
Lehman College, City University of New York
New York, U.S.A.

1983

NORTH-HOLLAND PUBLISHING COMPANY
AMSTERDAM • NEW YORK • OXFORD

ISBN: 0 444 86551 9

Publishers:
NORTH-HOLLAND PUBLISHING COMPANY
AMSTERDAM • NEW YORK • OXFORD

Sole distributors for the U.S.A. and Canada:
ELSEVIER SCIENCE PUBLISHING COMPANY, INC.
52 VANDERBILT AVENUE
NEW YORK, N.Y. 10017

PRINTED IN THE NETHERLANDS

PREFACE

This book is an attempt to cover important topics in research on what are referred to as the "endogenous" ERP components. These components are related to cognitive processing of stimulus information or the organization of behavior, rather than evoked by the presentation of the stimulus or the execution of a motor response as such.

The chapters in this book are extended versions of talks given at an international symposium on Event-Related Potential (ERP) correlates of information processing. The symposium was held at the Woudschoten Conference Center in Zeist (The Netherlands; December 7-12, 1980) and was organized by Anthony Gaillard, Albert Kok and Walter Ritter.

The idea of organizing such a symposium arose during the EPIC V conference that was held in June, 1979 in Ulm, West-Germany. The EPIC Conferences, which constitute the major meetings for investigators who work in all areas of ERP research, have of necessity become increasingly diversified. Sessions or panels are devoted to a large variety of topics, such as various clinical endeavors, developmental studies, attempts to delineate the generator sources of ERP components, cognitive processes, etc. A consequence of the growth and diversification of ERP research is that less time can be devoted to particular topic areas. Considering that research which examines the relationship of endogenous ERP components to information processing has itself expanded, it was thought that there is a place for conferences focused on ERPs and cognitive processes which would supplement the EPIC meetings.

The symposium held in Zeist was the first of such conferences and had a different format than that of EPIC. About eighteen investigators were invited to give one hour long talks, consisting of either a review of a line of research conducted in their laboratories or a review of some aspect of the literature. In this way, each speaker had an opportunity to develop, in some depth, central issues of a theoretical or methodological nature.

An initial draft of each chapter was commented on by the two editors, as well as by at least one other person chosen by the authors of the various papers. The comments received by the authors were in the nature of suggestions, but were mostly incorporated into the final published versions.

The chapters are ordered according to the latency range of the ERP components reviewed, except for the first two and the last four chapters which have a more general character. Chapter 1 discusses the relationship between ERP components and their intracranial generator sources. The second chapter deals with some theoretical and methodological issues regarding the relationship between ERP components and cognitive processes. In the next two chapters, the effects of selective attention on short-latency ERP components are reviewed. Chapter 5 through 7 focus on the N2 components with regard to the orienting reflex and the process of discrimination. The following five chapters are concerned with the P300 component. In chapters 8 and 9 this component is discussed in terms of the orienting reflex; chapter 10 reviews intracranial recordings obtained with clinical populations; chapter 11 discusses the relationship between P300 and the "Slow Wave"; and chapter 12 deals with the utilization of P300 in the chronometric analysis of information processing. Chapter 13 reviews the very slow potentials which comprise the CNV and chapter 14 is on the ERP components related to the organization of motor responses.

The last four chapters concentrate on general areas of research in ERPs. Chapter 15 addresses the relationship between human development and ERP components related to cognitive processing. Chapter 16 deals with the relationship between ERPs and the processing of language; chapter 17 reviews hemispheric asymmetries in ERPs; and chapter 18 reviews the background EEG and related hemispheric asymmetries.

In addition to the invited talks, papers were presented in poster sessions. The latter have been published as a special issue of Biological Psychology (Volume 13, 1981).

Since the conference was judged to be very successful by the participants, a second conference will be held in the fall of 1982 in Kingston (Canada), and a third in England in 1984. The intent is to hold these conferences every other year, alternating between Europe and North America and the EPIC conferences.

The symposium was sponsored jointly by the Free University of Amsterdam, the Netherlands Psychonomic Foundation, Beckman Instruments and Medilog. We are deeply indebted to these organizations, and also gratefully acknowledge the assistance of Elly Plooij-van Gorsel and Huib Looren de Jong who contributed much effort to administrative matters.

Anthony W.K. Gaillard
Walter Ritter

CONTENTS

Tutorials in ERP Research: Endogenous Components
A.W.K. Gaillard and W. Ritter (eds.)

1

NEUROPHYSIOLOGICAL CONSIDERATIONS IN EVENT-RELATED POTENTIAL RESEARCH

Herbert G. Vaughan, Jr., Walter Ritter and Richard Simson

Departments of Neuroscience and Neurology
Rose F. Kennedy Center
Albert Einstein College of Medicine
Bronx, New York
U.S.A.

The experiments described in this book provide evidence of relationships between averaged event-related potentials (ERPs) and mental phenomena. The delineation of these relationships provides a basis for understanding how the mind works. In order to relate psychological phenomena to brain functioning it is necessary to elucidate the neurophysiological mechanisms that give rise to ERPs. Although the anatomical and biophysical substrates are complex, the integration of methods from neuroscience and cognitive psychology provides promising approaches to this problem.

In relating scalp-recorded ERPs to underlying brain activity on the one hand and psychological processes on the other, a key issue is the definition of ERP components. It is often assumed that the voltage peaks of scalp-recorded ERPs reflect distinct physiological processes within the brain, and that measuring the latency, amplitude and scalp distribution of each peak provides direct information on the timing, magnitude and spatial extent of the intracranial neural processes that generated it. However, ERP peaks often represent composites of electrical activity arising from complex configurations of transmembrane current sources and sinks within more than one brain structure. The fusion of field potentials generated by different cellular elements within several concurrently active structures makes it difficult if not impossible to resolve the contribution of a particular structure or cellular group from the analysis of scalp recordings alone.

Intracranial mapping of ERPs in non-human primates has demonstrated that the generators of surface-recorded potentials are of two principal kinds: (1) synchronous volleys of action potentials within subcortical afferent fiber tracts, such as those that give rise to the short-latency auditory and somatosensory components (Arezzo, Legatt and Vaughan, 1979; Legatt, 1981; Legatt, Arezzo and Vaughan, in preparation), and (2) graded, postsynaptic potentials of neurons within both surface and deep cortical structures (Arezzo, Vaughan and Legatt, 1981; Steinschneider, Arezzo and Vaughan, 1982). Although studies of cortically generated ERPs in experimental primates have concentrated on sensory and movement-related potentials (e.g., Arezzo, Pickoff and Vaughan, 1975; Arezzo and Vaughan, 1975, 1980), it is not unlikely that the longer-latency task-related ERP components are generated by graded postsynaptic activity as well.

The basic principles governing field potential generation by groups of

*This work was supported by grants HD 10804 and MH 06723 from the USPHS.
The second author is also at the Department of Psychology, Lehman College, City University of New York.

neurons which these experimental data confirm, were first advanced by Lorente de No (1947). In this formulation, "open-field" configurations capable of generating potentials that can be recorded at a distance comprise arrays of neurons with a consistent axial orientation, such as the apical dendrites of pyramidal cells within the cerebral cortex. When such cells are activated, each neuron generates a characteristic dipolar field associated with current flow along the apical dendrite. The basal dendrites, by virtue of their radial symmetry, do not contribute to the fields recorded at a distance from the active membranes. The regular orientation of the pyramidal cells with respect to the cortical surface causes the potential field generated by a group of synchronously activated cortical neurons to appear as if it were generated by a time-varying dipole layer coextensive with the active region.

If the current sources and sinks within a region remain stationary over time, the polarity and scalp distribution of potentials will be constant, determined jointly by the configuration and strength of the dipolar layer, and the shape and electrical properties of the brain and its coverings. Thus, an ERP component can be defined at a given point in time as the sum of field potentials generated by a set of uniformly active cellular elements within a specific brain region. Both the topography and polarity of the field produced by this stable neural activity will be invariant and characteristic of the geometry of the active structure. Inasmuch as patterns of synaptic activity and the laminar distribution of current sources and sinks may vary, during the course of activation of a brain region, the polarity and amplitude of ERP components may change. However, the scalp distribution of the components generated within a specific structure will remain constant. Thus, potentials generated within a particular brain region may change in sign or magnitude according to the neural activity within the structure at each point in time but the topography of a given component is determined by the gross anatomy of the structure and is invariant.

A comprehensive understanding of the genesis of ERP components will include detailed descriptions of the thalamocortical, cortico-cortical and transcallosal synaptic contacts on the various types of intracortical neurons. Data bearing on these matters have become increasingly available from antero- and retrograde transport studies and from Golgi-electron microscope studies (Winfield et al., 1982). Furthermore, current source density (CSD) analysis (Nicholson and Freeman, 1975), employing the one-dimensional transitional approach of Mitzdorf and Singer (1979), provides a method for estimating the intracortical current sources and sinks during the actual course of an ERP (Vaughan, 1982). While it is not possible to dissect out the contributions of specific cellular events from observations of extracranial ERPs alone, the information obtained from intracranial investigations in experimental animals may ultimately make it possible to interpret the physiological processes that underlie each surface-recorded ERP component.

Exemplifying the importance of such information is a possible interpretation of the origins of negative potentials such as N2 that arise when stimulus discriminations are required (Ritter et al., 1979). Topographic studies (e.g., Simson et al., 1977a) have demonstrated extensive spatial overlap between N2 and sensory ERP components in both the visual and auditory modalities. The initial surface-positive sensory ERP components are presumably generated by depolarization of cells receiving afferent termination in lamina IV, with more superficial current sources. Recent studies in monkeys have demonstrated recurrent connections between several visual cortical areas which terminate mainly in layer I (Rockland and Pandya, 1979), the activation of which would generate a surface negative potential. Thus, we can speculate that the surface negative potentials seen during discrimination may represent activation of recurrent cortico-cortical connections. This would provide a physiological mechanism whereby stored information about an anticipated stimulus could be brought directly into comparison with the incoming stimulus information.

STRATEGIES FOR ERP COMPONENT RESOLUTION AND IDENTIFICATION OF INTRACRANIAL GENERATORS

Several approaches can be used to resolve ERPs into discrete components generated within various intracranial structures. These techniques share the use of topographic analysis to identify the contribution of each structure to the ERP waveform.

Generator Modelling

The effective use of topographic data depends upon a quantitative determination of the relationship between the electrical and geometrical characteristics of the intracranial generators and the field potential distributions they set up within the head. This can be done if the generator parameters and the impedance characteristics of the brain and its coverings are known. For practical purposes, the generators are assumed to be current dipoles arrayed to represent surface and sulcal cortex (Vaughan, 1974). The brain is considered to be spherical and covered by shells of uniform thickness that represent the scalp, skull and subarachnoid space. The volume impedance of the brain and its coverings is essentially resistive and constant throughout each kind of tissue. Although the actual physical situation departs somewhat from these assumptions, it can be shown that the computed fields at the scalp are relatively insensitive to variations in conductivity and relative thickness of the brain's coverings, as well as to inhomogeneities within the brain itself (Witwer et al., 1972).

There are two methods for using empirical scalp potential distributions to estimate intracranial generators. In the 'forward' method, hypothetical potential fields are calculated from assumed generators with specified electrical and geometric characteristics. The computed theoretical surface distributions are compared to the observed topography to determine the predictive adequacy of the assumed generators. Alternative generator configurations can be compared with one another to determine the best fit to the empirical data. By contrast, with the 'inverse' method intracranial generator configurations are directly computed from the scalp potential distribution (viz., Wood, 1982). The estimation of intracranial generators by the inverse method is convenient but it is limited by the fact that it cannot effectively deal with the problem of multiple anatomically extended generators without introducing constraints on the computations based upon anatomical knowledge.

Since the forward method provides for the computation of unique intracranial and surface potential distributions from any specified generator properties, this seems the more useful approach.

As more empirical data on the electrical properties of intracranial generators become available from detailed intracerebral recordings, increasingly accurate models of the consequences of specific patterns of neural activation within specific structures can be constructed. Klee and Rall (1977) provided an example of this approach, in which simple model neurons, arranged as pyramidal cells within surface cerebral cortex, were used as the basis for computing the external field distribution. This model differed from earlier applications of the forward method in that transmembrane currents produced by a specific neuronal model were used to estimate the strength of the generator, rather than assuming an equivalent dipolar layer of arbitrary strength. The rough validity of this pyramidal cell generator model is attested by the fact that the computed voltage across the modelled cortex of about 400 uV is similar to the observed transcortical evoked potential amplitude in the monkey (Arezzo et al., 1981).

The analysis of scalp topography is complicated by the presence of multiple, concurrently active intracranial generators. In the auditory and visual modalities the primary projection cortex is bilaterally activated. Responses within secondary areas and corpus callosum are generated within a few milliseconds after initiation of the primary cortical response. Inasmuch as activity generated in connected cortical regions temporally overlaps and is often generated in contiguous areas, it is difficult to resolve their relative contributions. Thus, although it is convenient to map the topography of scalp ERP at a few points in time, the usual practice of identifying components by the presence of peaks and mapping them at the times corresponding to these peaks, risks the loss of important topographic information and errors due to temporally overlapping but independently generated components. A more detailed chronotopographic analysis (viz., Wood, 1982; Renault, this volume) is of great value. A field topography that is constant over time suggests origins in an anatomically stable set of generators, whereas a shifting topography implies a changing configuration of intracranial sources.

From a series of topographic studies of human ERPs (Vaughan, 1968; Simson et al., 1976), we have concluded that the obligatory components of cortical origin arise from modality specific primary and secondary cortical areas. Studies of movement-related potentials (Vaughan et al., 1969) indicate principal origins in the pre- and post-central cortex; potentials preceding saccadic eye movements arise in the posterior frontal and parietal regions (Kurtzberg and Vaughan, 1977, 1982) and potentials associated with auditory and visual discriminative tasks are generated in the secondary cortex of each modality as well as in modality unspecific loci believed to involve the parietal association cortex (Simson et al., 1977a, 1977b). The spatiotemporal overlap of the various ERP components observed in these studies was often manifested by complex peaks, suggesting that the intracranial generators we inferred from measurements of individual peaks may not be homogeneous. Only when ERP components are substantially different in their timing and spatial distribution can they be readily differentiated and mapped from surface recordings. In most circumstances the experimental segregation of ERP components will undoubtedly require several complementary approaches.

Principal Components Analysis (PCA)

The scalp potentials produced by temporally overlapping but anatomically distinct intracranial generators represent the linear summation of the fields generated by each structure. This raises the possibility that PCA might be useful for segregating the potentials associated with different generators. In most applications of PCA to ERP, variance associated with both experimental manipulations and electrode locations has been included in the data matrix (Donchin and Hefley, 1978). However, if the source of variance is restricted to the ERP topography within a single experimental condition, it may be possible to identify factors that correspond to the fields set up by specific intracranial structures. While PCA of ERP across scalp recording sites can produce misleading results when spatial and temporal overlap of components is substantial (Rotkin, Vaughan and Ritter, unpublished simulations), success in resolving several components, whose timing and topography are stable across experimental conditions (Friedman et al., 1981) suggests that PCA may yield physiologically meaningful components when clear topographic differences are present. The application of PCA to animal experiments where the intracranial generators and the potentials they produce can be directly examined remains to be exploited as a test of its utility in this regard.

Stimulus and Task Variables

The experimental manipulation of stimulus and task variables to selectively activate specific brain processes is a central experimental approach in ERP investigations. For example, restricted portions of the visual field can be stimulated to elicit responses within different regions of visual cortex. The introduction of specific task requirements in association with ERP recording has provided an opportunity to investigate the electrophysiological correlates of cognitive processes, an endeavor that forms the focus of this volume. The additional components that emerge with various cognitive processing requirements may represent activation of specific brain structures, or more likely, systems that include more than a single anatomical structure. In these experiments it is often useful to define the topography of components observed in "difference waveforms" which are obtained by subtraction of the ERPs derived from conditions that are intended to selectively manipulate a particular perceptual or cognitive process, such as pattern recognition or stimulus selection (viz., Ritter et al., this volume). Although it is difficult to be sure that the imposition of a specific processing requirement affects only a single stage in the sequence of information processing, topographic analysis of difference ERPs can provide clues to the brain structure or structures activated by the particular task demand. A great strength of human ERP studies is their capacity for disclosing brain potentials associated with various conveniently manipulated aspects of cognitive processing. The existence and localization of these cognition-related neural processes would, where possible, require extremely laborious and often impractical behavioral manipulations in experimental animals. Although animal investigations are required to elucidate the neural mechanisms that underlie perception, cognition and behavior, human ERP studies form the guiding and in some cases the only basis for the neurophysiological investigation of cognitive processes.

Effects of Brain Lesions

The study of ERPs in patients with well defined intracranial lesions may provide valuable information on the sources of ERP components. There are, however, several difficulties in the interpretation of such data. Early studies of the impact of local brain lesions on ERPs were impeded by the lack of generally applicable methods for defining the location and extent of intracranial pathology. This difficulty has to some extent been alleviated by the introduction of CT scan technology which provides a non-invasive visualization of cerebral structures, that unlike post-mortem anatomical evidence, can be obtained at the time of ERP recording. Despite the undoubted value of the CT scan, these data can be misleading, as shown by two recent case reports (Wood et al., 1982) on the effect of bitemporal lesions on the auditory ERP. In both cases the CT scans were considered to demonstrate bilateral destruction of auditory cortex. In one, no auditory ERP was recorded, whereas in the other case, it was normal. There is no clear explanation for this striking discrepancy inasmuch as the location and extent of the lesions disclosed by the CT scans were quite similar. However, the substantial variation in the sulcal patterns and cytoarchitectonic extent of human cortex (e.g., Galaburda and Sanides, 1980), must be taken into account in the interpretation of the impact of cerebral lesions on ERPs. Radiographic evidence should be regarded with caution for the definitive localization of intracranial pathology to ERP generators. Furthermore, given the complexity of the connections among cortical and subcortical structures, as well as the capricious and often uncertain extent of human cerebral pathology, it is unlikely that lesion effects on scalp ERPs will be simple. Possible lesion-related alterations in the patterns of excitatory and inhibitory interactions make it difficult to interpret changes in ERP waveforms in the presence of brain damage without detailed information on both the anatomy and physiological interactions of the affected structures. Much additional work, utilizing recordings in animals will be required to define the ERP alterations that are associated with destruction of specific intracranial generators and interruption of their interconnections.

Intracranial Recording

It is evident that direct intracranial recordings will be required to definitively establish the anatomical location of the structures that generate surface-recorded ERPs and to delineate patterns of neuronal activity within these generators. Inasmuch as activity within deeper brain structures may not produce field potentials that are volume conducted to the scalp, intracranial data disclose a more complete picture of brain processes than can be achieved from surface recordings alone. Data obtained from human patients during the performance of discriminative tasks has yielded evidence of a complex intracerebral distribution of late task-related potentials (Wood et al., 1982) and the region including the hippocampus has been identified as a generator of some of these potentials (Squires et al., this volume). Although important evidence on intracerebral ERP distribution can be obtained from human studies, these are necessarily restricted to clinically indicated exploration so opportunities for detailed intracranial mapping and histological verification of recording sites are rare and limited in scope. Accordingly, the establishment of homologies between ERPs recorded in humans and experimental animals is of considerable importance for furthering our knowledge of the physiological basis of human ERPs.

Selection of an appropriate animal model for the comparative analysis of ERP generators and surface topography involves several important considerations. First, the gross anatomy of brain structures and their relationship to the cranial surface should be similar to the human so that the macroscopic relationships between source geometry and surface potential distributions are comparable. Further, the organization of afferent pathways and their central projections, as well as the cellular architecture of cortical regions and the patterns of their intrinsic and extrinsic connections must be comparable so as to maximize similarities in physiological activity across species. Finally, both the human and experimental animal should exhibit similar psychophysical responses and discriminative behavior so that correspondence of ERP components can be assessed on functional as well as on anatomical and physiological grounds. Although small mammals such as cats and rats have been extensively employed in neurophysiological investigations, only primates adequately meet the above requirements for comparative studies. It is evident that there are important differences in the gross configuration and cellular organization of sensorimotor systems between primates and the non-primate experimental animals, rendering the identification of equivalent cross-species ERP components problematic. Among available primate species, the old world monkeys provide the closest practical experimental model for human ERP generation.

We have examined the surface topography and detailed intracranial distribution of somatosensory and auditory evoked potentials, as well as of movement related potentials in alert monkeys (Arezzo et al., 1975; Arezzo et al., 1979; Arezzo and Vaughan, 1980). These data, while not yet establishing a definitive cross species equivalence of ERP components, generally confirm the previously noted inferences about sources derived from our studies of human scalp topography. These findings encourage extension of intracranial analyses of ERPs in the monkey to discriminative and task conditions that are similar to those employed in humans. It will be necessary in these studies to take into account anatomical differences in the extent of the cortical regions related to cognitive processes in the two species. Nevertheless, the study of the neurophysiological basis of task-related components represents an important frontier of ERP investigation that can begin to bridge the gap between cognitive psychology and cellular neurophysiology.

Tutorials in ERP Research: Endogenous Components
A.W.K. Gaillard and W. Ritter (eds.)

2

ENDOGENOUS ER"Ps" AND COGNITION: PROBES, PROSPECTS, AND PITFALLS IN MATCHING PIECES OF THE MIND-BODY PUZZLE

Frank Rösler

Institut für Psychologie
Christian-Albrechts-Universität
Kiel
Federal Republic of Germany

The first part of this chapter deals with some epistemological problems of ERP research and tackles the question of how the functional state of the brain, which manifests itself in an endogenous ERP component, can be defined exactly. In the second part some methodological issues are discussed which are critical for the definition and interpretation of endogenous ERP components. Both themes, which are closely interrelated, are elaborated by reviewing material accumulated on "P300".

INTRODUCTION

The intent of the present paper is to elucidate somewhat the sins of commission and omission of researchers who want to establish psychophysiological relationships by means of event-related potentials (ERPs). In particular, I want to think about the situations in which intervening variables are defined and related to variations of ERP phenomena (e.g., "expectency" and "P300"), or in which hypothetical constructs are introduced and accepted as causes or generators of endogenous ERPs (e.g. a "context updating operation"). That means I will try to explicate some of the assumptions made implicitly by researchers who want to pinpoint the functional significance of ERP phenomena, and I will examine the sources of information researchers can rely on to define intervening variables or hypothetical constructs.[1]

The discussion will be restricted to endogenous ERP phenomena. The term endogenous component was coined by Sutton, Braren, Zubin, and John in 1965, who discovered that the now well known P300 was not related to physical aspects of the evoking stimuli, but rather to intrinsic characteristics of the subject. The distinction between exogenous and endogenous ERP components was elaborated in detail by Donchin, Ritter & McCallum (1978). According to their definition the main characteristics of the two ERP types are: Exogenous components are evoked by events extrinsic to the nervous system and their variance is accounted for primarily by a variation of physical stimulus parameters, such as intensity, quality, modality etc. Endogenous components, on the other hand, are also triggered by external events, however, they are only partially related to the physical parameters of these. Instead, their variance is primarily determined by the particular tasks and instructions assigned to the eliciting events. Thus, the very same stimulus may evoke an endogenous component in one experimental condition but not in another, or physically very different stimuli may evoke the same endogenous component when they have the same meaning to the subject.

The chapter clusters around two themes which are closely interrelated. The first part deals with epistemological issues of ERP research and could be titled "On how to define the functional state of the brain which is indexed by an endogenous ERP component". Apart from some brief abstract statements this discussion will be closely linked to empirical studies, in particular to P300 studies. Optionally, this part of the chapter can also be taken as a review of P300 research. The second theme refers to some methodological problems which are critical for research that deals with endogenous components. With that I want to make obvious some of the fallacies one can run into with the methods used to detect and define endogenous ERPs. In detail, I will address the following questions: (1) What are the basic implicit assumptions made in an ERP experiment? (2) What kind of information is available and how can this information be used to define the psychological state prevailing during the registration of an endogenous component? (3) Is it a reasonable assumption that the psychological state or process related to an ERP peak is invariant during all epochs of EEG activity sampled for one averaged potential? To what extent are gradual changes of peak amplitudes, which are observed in averaged potentials, only due to an artifact of averaging? (4) What are the explicit and implicit definitions of an ERP-component?

IMPLICIT ASSUMPTIONS MADE IN ERP RESEARCH

Let us begin with what might be called the "core" of each experimental paradigm used to study the behavior of ERPs. If we strip ERP paradigms to the skin we will find a situation such as depicted in figure 1. The continuous electroencephalogram is monitored throughout the total course of an experimental session and somewhere in time a physical stimulus is presented to the organism or an overt response of the organism is registered. In some cases both stimulus and response are considered, especially when the response follows shortly after presentation of a stimulus. Anyway, at least one of these events is defined exactly in time. This means either the onset or offset of these events can be fixed in the range of milliseconds with high precision. This exactly defined point in time is used to segment the EEG so that one or another signal extraction method can be applied (averaging, Woody-filtering, etc., see e.g. Glaser & Ruchkin, 1976, Rösler, 1980a). With the signal successfully extracted from the background "noise" of the spontaneous EEG, we have either a <u>stimulus-locked</u> or <u>response-locked</u> ERP.

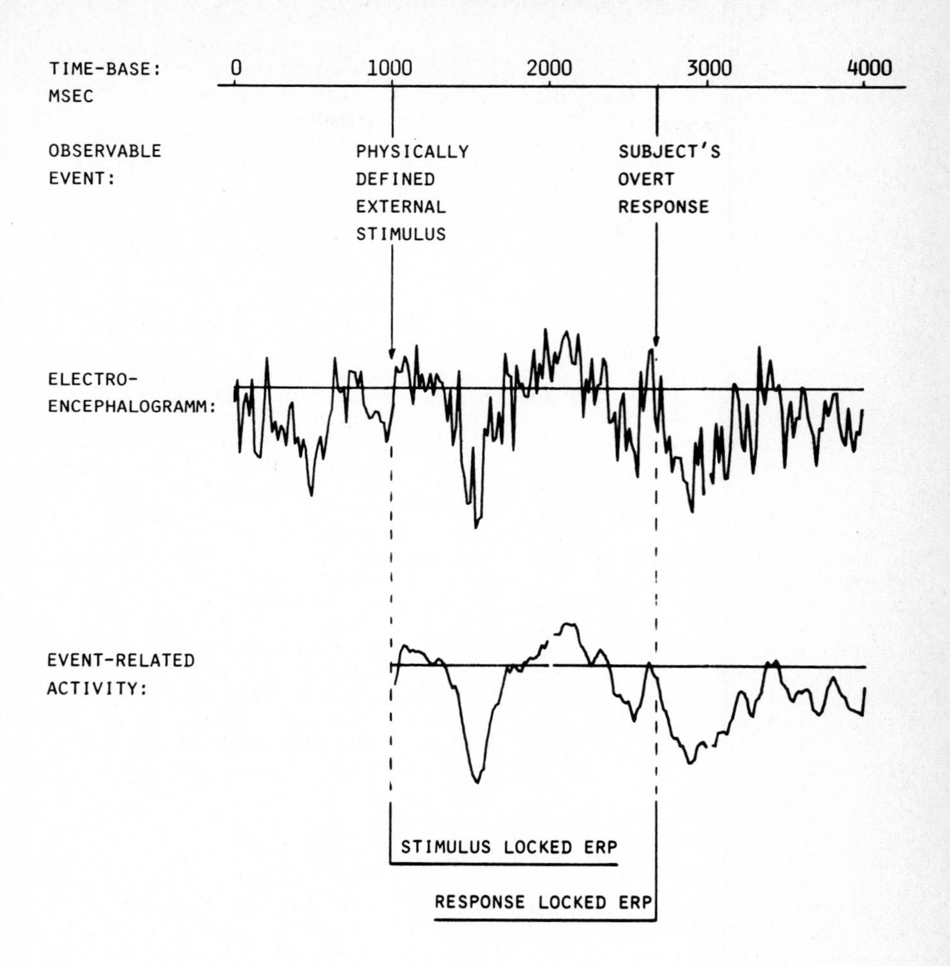

Figure 1
Basics of ERP research (see text for explanation)

Because of the temporal coincidence, the reasonable implicit assumption is made that there exists a causality relationship between the triggering event (the stimulus or the response) and the specific brain wave pattern that follows or precedes that event. However, this causality assumption is only an indirect one. The primary and direct causality relation is drawn between the triggering event and some internal information processing activity which follows stimulus-input or precedes

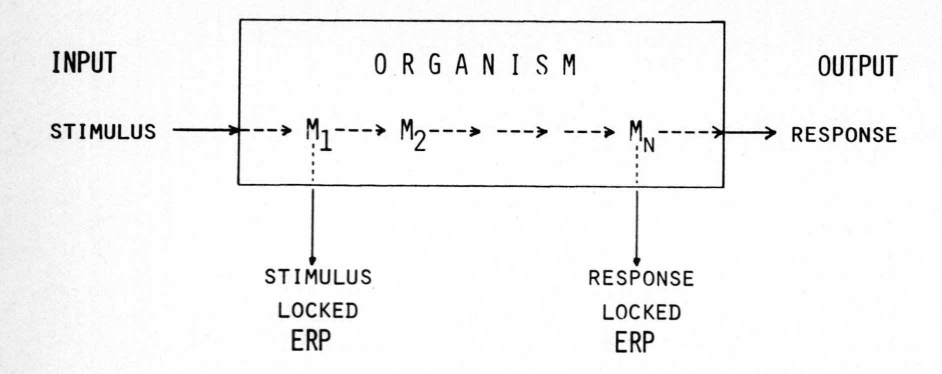

Figure 2
Basic assumptions made in the study of endogenous ERPs. Stimulus or response locked ERPs are seen as manifestations (epiphenomena) of mediating processes of the organism. It is assumed that these processes (functional states of the brain, cognitive events, subroutines) are either triggered by an input (stimulus) or that they themselves trigger an output (response).

response-output. The implication of this is that the ERP is accepted only as a manifestation or as an epiphenomenon by the central nervous system (CNS). The information processing activities themselves might be defined either in <u>physiological</u> terms (e.g. transmission characteristics of particular parts of the brain, inhibitory or excitatory states of cell populations) or in <u>psychological</u> terms (e.g. sensations, perceptions, decisions, preparations, etc.).

With the epiphenomenon assumption it is implied that there exists a one-to-one relationship between variations of an assumed internal process (a hypothetical construct) and variations of a given ERP phenomenon. In practice, the assumption is specified as follows: an ERP phenomenon, defined by polarity, scalp topography, and latency range, usually addressed as "peak" or "component", is related to one and only one internal process; the amplitude of this peak is then taken as a manifestation of the strength, and the latency as a manifestation of the timing of that internal process.[2] I will discuss the problems linked to this definition of a peak as an unified ERP phenomenon later. For now, let us assume that a peak defined as above, is indeed a meaningful entity and an epiphenomenon of one particular functional state of the brain. The question then is, how can we define this state, how can we uncover its significance for information processing activities performed by the CNS?

SOURCES OF INFORMATION RELEVANT TO EXPLAIN THE VARIATION OF ENDOGENOUS ERP COMPONENTS

The different sources of information, which, in principle, can be used to pinpoint the psychophysiological meaning of a specified ERP-phenomenon, are collected in figure 3. To make things a bit more concrete in that what follows I will refer only to the iridescent phenomenon of P300 (for a summarized definition see e.g. Donchin, Ritter, & McCallum, 1978, p. 376, Pritchard, 1981, Rösler, 1980b, Tueting, 1978). I think it is an example *par excellence* to use in discussing issues involving investigations of endogenous ERP components.

EXPERIMENTAL VARIABLES AND OPERATIONS

Physical characteristics of triggering events. Remembering the skeleton of an ERP experiment the most obvious variables which might be considered as bearing information about internal processes are, of course, the physical characteristics of the triggering events (e.g. modality, quality, and intensity of the time-locked stimuli, or the body region, quality, and intensity of the time-locked responses). As far as P300 is concerned, it is a well known fact that these characteristics give no proper clue about the antecedents of this ERP phenomenon.

The careful studies of Simson, Vaughan & Ritter (1976, 1977a, b) and Snyder, Hillyard & Galambos (1980) have shown that P300 is modality unspecific, i.e. scalp topography and waveshape are the same whether acoustic, visual, or somatosensory stimuli are used as triggering events. Moreover, even if the triggering events are stimulus omissions in a regularly presented series of stimuli, the positivity is emitted after the stimulus omission with a latency and a scalp distribution similar to that evoked by a real physical stimulus (Ruchkin & Sutton, 1979, Ruchkin, Sutton & Tueting, 1975, Simson et al., 1976). These facts are generally taken as evidence that there is a common neural generating system for this component. As it seems, the latency and the amplitude differences, which can be observed between evoked and emitted late positive components (LPCs) (Ruchkin & Sutton, 1979) or between visually and acoustically triggered LPCs (Snyder et al., 1980) are not due to the physical differences *per se*. It is much more likely that other, intrinsic properties of the nervous system are the bases for these differences (see below and for details Ruchkin & Sutton, 1979, Snyder et al., 1980, Squires, N., Donchin, Squires & Grossberg, 1977).

Similarly, the physical characteristics of the response do not tell much about the functional significance of P300. From various paradigms it is known that P300 appears with similar

BACKGROUND VARIABLES:

- STATE VARIABLES (TIME OF DAY, DRUGS, ETC.)
- SUBJECT CHARACTERISTICS (SEX, QUESTIONNAIRE DATA, ETC.)

CONTEXT VARIABLES:

- DEFINITION OF TASK (COUNT STIMULI OF TYPE X)
- SPECIFIC INSTRUCTIONS (STIMULI OF TYPE Y ARE IRRELEVANT)
- INCENTIVES (WIN ONE DOLLAR FOR A CORRECT GUESS OF EVENT Z)
- A PRIORI PROBABILITY OF EVENTS, OVERALL AND LOCAL

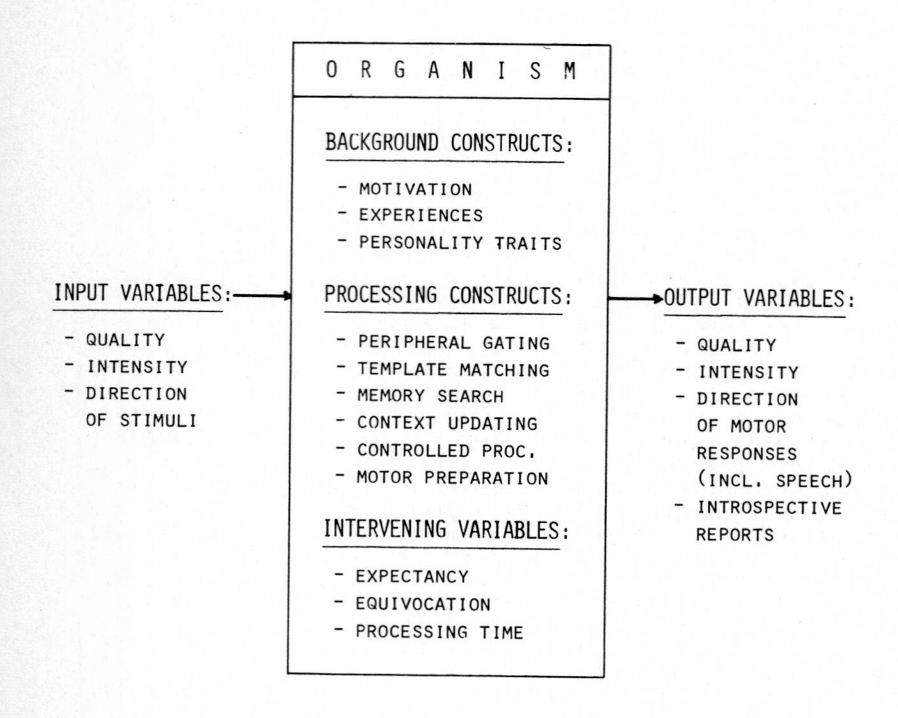

Figure 3
Classes of variables and constructs available to define functional states of the brain prevailing during the recording of endogenous components. For each class of variables and constructs only some examples are listed.

characteristic whether an overt response (button press, etc.) is committed or not and whether such a response is given orally or manually (e.g., Courchesne, Hillyard & Courchesne, 1977). The only response related variable that might explain some of the variability of P300 is response latency (Friedman, Vaughan & Erlenmeyer-Kimling, 1978, Kutas, McCarthy & Donchin, 1977, McCarthy, this volume). However, response latency is not a simple physical characteristic of a motor response, rather, it is a much more complex entity. It already states a relationship between stimulus and response, and therefore it should be treated as a context- or even as an intervening-variable (see below).

<u>Context variables</u>. The second class of variables and possible candidates to explain P300 variability can be derived from a more extended behavioristic analysis of the experimental situation.[3] On naming these variables one has to take into account more of the experiment than just the information available during an individual trial sketched in figure 1. This includes the instructions given, proportions (probabilities) of different types of stimuli, sequences of stimuli, etc..

For example, stimuli declared as task relevant by the experimenter, such that they have to be counted or responded to by a key press, evoke a P300 of considerable amplitude, while stimuli declared as task irrelevant, such that they can be ignored, do not (Corby & Kopell, 1973, Courchesne et al., 1977, Ford, Roth, Dirks & Kopell, 1973, Ford, Roth & Kopell, 1976a, Harter & Salmon, 1972, Hillyard, Hink, Schwent & Picton, 1973, Simson et al., 1977a, Squires, K., Donchin, Herning & McCarthy, 1977). In the past, the <u>a priori</u> probability of the stimuli was considered to be of similar importance. Infrequent stimuli, whether task-relevant or not, evoked P300s with higher amplitude than did frequent stimuli (whether task relevant or not) (Duncan-Johnson & Donchin, 1977, Squires, K. et al., 1977, Squires, N., Squires, K. & Hillyard, 1975). As far as <u>a priori</u> probability is concerned, it was not settled for some time whether the critical variable for P300-amplitude was the <u>a priori</u> probability of the stimuli <u>per se</u> or rather the <u>a priori</u> probability of the responses related to these stimuli. In the classic "odd-ball" experiments these two types of <u>a priori</u> probability were confounded because one type of stimulus was always mapped to only one response type, e.g., <u>count</u> the infrequent <u>high</u> pitched tones but <u>do not count</u> the frequent <u>low</u> pitched tones. Recent data presented by Johnson & Donchin (1980) confirm earlier results of Karlin & Martz (1973) that it might be more the <u>a priori</u> probability of the response category than that of the stimulus category which is responsible for the strong probability effects observed with P300 amplitude (cf. also Courchesne et al., 1977, Courchesne, Courchesne & Hillyard, 1978).

Another variable is that of stimulus sequence, the effects of which have been studied extensively by K. Squires and colleagues (Squires, Wickens, Squires & Donchin, 1976, Squires, Petuchowski, Wickens & Donchin, 1977) and by Johnson (Johnson

& Donchin, 1981, Johnson, 1980). Squires and colleagues (1976, 1977) analysed the ERPs seperately for different sequences of two events of a Bernoulli series. ERPs from sequences up to the fifth order were considered, the ERP trials from the 2^5 possible fifth order sequences were treated seperately. For example, a task relevant event, a signal of type A, was sorted into one bin for averaging if preceded by four other A's (sequence AAAAA), but the same signal of type A was sorted into another bin, if it was preceded by four task irrelevant events of type B (sequence BBBBA), etc. The amplitude of P300 was systematically related to a priori probability of the two events, but beyond this it was also correlated with the alternation pattern. For example, a stimulus of type A evoked P300 with much larger amplitude when preceded by four stimuli of type B (pattern BBBBA) than when preceded by three of type B and one of type A (pattern BBBAA). In short, these studies have shown that the immediate context of an individual event in a longer series of events is of great importance for the generation of P300.

Introspective "Events". Another set of variables which seems to be correlated with endogenous ERP components can be derived from introspective reports of the subject. With introspective reports I mean all the information which can be gathered from the subject but which is not a response defined exactly in time. To illustrate the point, the subject's button press in a matching task whether two stimuli are same or different is not an introspective variable in this sense. On the other hand "confidence" operationalized with a rating scale is an introspective variable according to my definition, because the subject cannot simply react; rather he or she must "think about" his or her internal state, e.g., "how confident am I that this was a stimulus and not only noise?". The decision takes some time, and the introspective process necessary for this decision is only loosely related to the overt response. The very same holds for a guess about a future event; for example, "what do I think, which stimulus of all possible ones in this situation will be presented next?".

The fact that there exists a systematic relationship between the confidence variable and P300 amplitude was demonstrated by Squires and colleagues by means of a signal-detection paradigm (Squires, K., Hillyard & Lindsay, 1973b, Squires, K., Squires, N. & Hillyard, 1975a, b). After each trial, during which either a signal embedded in noise or just noise alone could have been presented, the subject was asked to report the confidence with which he or she had detected a signal or no signal. The scale provided for 8 levels of confidence. Thus, the ERPs were divided into 16 different event classes - 8 ratings times 2 states of the real world (signal presented vs. signal withheld). The results showed that events categorized with high confidence as either being signal present or signal absent events evoked prominent P300 waves; this was irrespective of whether or not a signal really was present or not. The amplitude of P300 waves to less confident events was almost negligible.

The paradigm just described is not the only way to relate subjective reports of confidence about a decision to P300 amplitude. The same confidence rating can be related as well in a "forward direction" to the amplitude of a P300 that is elicited by a feedback stimulus following the rating response. With this arrangement it has been shown that the P300 amplitude evoked by disconfirming feedback stimuli becomes larger and later the more confident the subject had been about his original decision (Campbell, Courchesne, Picton & Squires, 1979, Squires, Hillyard & Lindsay, 1973a).

Background variables. Finally, to complete the survey of variables which are directly observable and related to endogenous components, one has to mention "background variables". Under this heading I would subsume state variables such as "time of day" the measurements are taken (e.g., Kerkhof, this volume), or drug effects (e.g., Otto & Reiter, 1978), as well as subject variables such as age (e.g., Courchesne, this volume), personality (e.g., Plooij-van Gorsel, 1980), and psychopathology (e.g., Callaway, 1979, Roth, Ford, Pfefferbaum, Horvarth, Doyle & Kopell, 1979).

THEORIES, INTERVENING VARIABLES AND CONSTRUCTS

Up to this point I have presented a variety of variables which are defined by the immediate operations carried out by the experimenter. These operations can be used successfully to manipulate the characteristics of P300; however, taken directly, they do not provide a satisfactory explanation of the phenomenon itself. For example, relationships between P300 amplitude and a specific stimulus sequence or P300 amplitude and confidence ratings, are effects which only help to clarify antecedent conditions of the phenomenon; they do not specify the functional state of the brain or the cognitive process that is associated with this particular ERP component. It is just this "subroutine of the brain" which is of main interest to the ERP-researcher, not merely a more or less complete collection of antecedent conditions. For good reason, because only if an understanding of such a "subroutine of the brain" is attained by recording ERPs, then these ERPs can be used to monitor the otherwise hidden information processing activities of the human brain for applied or diagnostic purposes.

To make progress towards finding an exact definition of the functional state indicated by P300, one has to do something more than just manipulate or control input, context, or output variables. Therefore, the additional necessary step is to abstract from the directly observable variables and the particular experimental settings one common "denominator", an intervening variable or processing construct, which can be taken as the "real" trigger of P300. However, such an abstraction is only possible if one relys on another source of information besides the directly observable facts. One must add some assumptions about the information processing activities performed by the brain; in short, one must formulate, more or less explicitly, a cognitive theory. To illustrate the point, I will give some examples of cognitive theories or models with which the various antecedent conditions of P300 can be integrated.

An elegant approach, although still a relatively restricted one, was that of Squries and colleagues to explain the effects of a priori probability and stimulus sequence on P300 (Squires et al., 1976, 1977). To account for the variation of P300 amplitude in these studies they defined an internal or intervening variable to represent an abstraction called "expectancy". With this they meant that at a specified point in time the subject always holds an hypothesis about which of the several possible events in a situation will come next. They assumed that expectancy for a specific event can be expressed as a function of three entities. First, expectancy was seen as a function of the a priori probability of that event (an observable variable). Second, expectancy was seen as a function of an internally generated "alternation set" (a hypothetical construct). With this it was implied that a subject has a bias to expect a continuation of an alternating pattern with greater probability than a break of such a pattern. Third, expectency for a specific event was seen as a function of the rate of decay with which previously perceived events of the same type fade in memory (memory and decay are other hypothetical constructs). All three subfunctions of expectancy were added in a linear regression model to an overall expectancy score; the correlation with P300 amplitude was -.88. To explain the P300 phenomenon itself they said in their discussion (Squires et al., 1976, p. 1146): "The nervous system ... is a dynamic system that continously generates hypotheses about the environment. The P300 seems to be associated with the evaluation of such contextual hypotheses." This means that the specific cognitive process which is "responsible" for P300 generation is a comparison which relates an actually perceived event to the expectancy which has been computed from the previous context for this event for this moment in time. The higher the expectancy, i.e. the higher the subjective probability for the occurrence of this event the smaller will be P300 amplitude.

This model, which relates subjective probability to P300 amplitude, led Donchin to the idea that P300 might indicate in general a cognitive process of "context updating". He assumes that the "process manifested by P300 is invoked whenever data provided by a stimulus call for a revision of hypotheses, or models of the environment" (Donchin, 1979, p. 66). With this hypothesis more of the P300 amplitude effects can be explained than just those of experiments in which the probabilistic structure of the event series is manipulated explicitly. In particular the various effects observed in guessing paradigms (e.g., Chesney & Donchin, 1979, Ruchkin & Sutton, 1978a, Sutton, Braren, Zubin & John, 1965, Tueting, Sutton & Zubin, 1971) and feedback paradigms respectively (Campbell et al., 1979, Johnson & Donchin, 1978, Squires, Hillyard & Lindsay, 1973a) fit nicely into this more general model. These studies have shown that feedback stimuli giving information about the correctness or incorrectness of a prior decision (e.g., in a time estimation or a signal detection task), or the correctness or incorrectness of a previous guess, both evoke a big P300 whenever a subjectively very probable (confident) hypothesis is disconfirmed or a very improbable hypothesis is confirmed by the feedback information. No doubt, it makes sense to assume for both cases that something like context updating, like changing the schema about the external world takes place.

Ruchkin and Sutton (1978b) pointed out that another factor which they called "equivocation", might be important for the context updating process and the determination of P300 amplitude (s. also Sutton, 1979). The Squires' theory of expectancy and Donchin's generalization of it are primarily concerned with the problem of how well a particular bit of information picked up from the environment fits into the currently held model of the world. Whenever there is a mismatch between received and expected information the context updating routine will be invoked and P300 will appear in the EEG. However, such an updating of the internal model held by the CNS can only be performed adequately, if the information provided by a particular event has been received by the CNS unequivocally. In more introspective terms: to revise hypotheses about the external world only makes sense if there is no uncertainty about the meaning of the stimulus which has been perceived. Since an ambiguous stimulus does not help much to revise hypotheses, it can be discarded. Ruchkin's and Sutton's approach deals with this problem of how P300 amplitude might be related to the uncertainty about the meaning of a stimulus after it has been registered by the CNS. The greater this uncertainty, they say, the greater is the stimulus equivocation, and by reviewing the literature they present results from several studies which support the hypothesis that P300 amplitude is inversely related to equivocation. They summarize (Ruchkin & Sutton, 1978b, p. 177): "for a fixed level of prior uncertainty, P300 amplitude will be determined by equivocation - decreasing as equivation increases; for a fixed level of stimulus equivocation, P300 amplitude will be determined by prior uncertainty - increasing as uncertainty increases." (The term "prior uncertainty" is equivalent to "expectancy" as used by Squires et al. or Donchin.)

An experiment published by Johnson and Donchin (1978) seems to be a direct test of this hypothesis. Subjects had to perform a time estimation task and the discriminability of the two possible feedback stimuli, confirming or disconfirming the subject's estimate, was varied systematically. P300 amplitude was closely related to the discriminability of the two stimuli. The highest amplitude was recorded when the two stimuli could be discriminated easily, while the smallest amplitude was recorded when the two stimuli were difficult to distinguish. The amplitude of P300 contingent to the two types of feedback (confirming, disconfirming) did not differ substantially, obviously because their a priori probability was equated ($p(FB+) = p(FB-) = .5$). The results suggest that P300 is only elicited with substantial amplitude when the evoking stimulus does deliver useful information to the subject. If the stimuli do not provide real feedback, which can be used to revise the strategies for time estimation, then these stimuli might be "overlooked"; they are useless.

The results of odd-ball, feedback, and guessing studies together with the intervening variables of "expectancy" and "equivocation" and the processing construct of "context updating" converge to the hypothesis that P300 is triggered whenever information has been evaluated by the CNS as relevant for changing a currently held model of the external world. Thereby, the concept of equivocation plus the results given by the Johnson & Donchin study (see Ruchkin & Sutton, 1978b, for related results), reveal a very important point: the information provided by an external event must have been analysed completely with respect to its meaning and relevance before the functional state of the brain indicated by P300 is invoked. This aspect of timing, which says that the process manifesting in P300 must follow a complete analysis of the stimulus is supported by a completely different line of evidence. In several studies it has been shown that P300 latency, measured on a single trial basis, is related to the time which is necessary to evaluate the meaning of a stimulus (s. Duncan-Johnson & Kopell, 1980, Kutas, McCarthy & Donchin, 1977, McCarthy, this volume). These autors found that, compared to standard conditions, P300 peaks later when the related stimulus is more difficult to discriminate from others or when it has to be separated from a noisier background. In short, P300 latency is elongated when more must be done: more time-consuming processing to obtain the stimulus meaning.

Accepting the premise that P300 indicates a process which is invoked after the automatic analysis of stimulus meaning, one can try to clarify the issue a bit further and ask about the purpose of this functional state. To update the model of the external world is of course one obvious possibility; however, the concept of context updating is of limited heuristic value. It has emerged from the study of P300-effects in odd-ball and feedback paradigms, therefore well suited to account for these, but there are other effects for which an explanation in terms of context updating seems to be less suitable, e.g. changes of P300 amplitude in similarity rating (Rösler, 1978), discrimination-learning (Rösler, 1981), or memory scanning tasks (Adam & Collins, 1978, Gomer, Spicuzza & O'Donnell, 1976). I will present some further evidence from our laboratory (yet unpublished) which demonstrates that the concept of context updating is a one too specific to cover all effects observed with P300. Ironically, the experiment was set up as an explicit test of the context updating hypothesis, however, as it turned out, the results gave only partial support to it.

The basic idea of our paradigm is as follows. With a first cue or priming stimulus a specified attentional set is established, which is required to solve a task following two seconds later. Shortly after the first cue a second cue is

Table 1
Double priming paradigm to test the hypothesis that P300 is a manifestation of context updating. Sequence of events: Subject's task is to decide whether two letters of a quadratic letter-matrix are same (=) or different (≠). Cues 1 and 2 indicate which letter-pair has to be monitored, "/" means monitor diagonal "bottom left to top right", "\" means monitor diagonal "top left to bottom right". In the sequences (/ \) and (\ /) a change of the stimulus set, induced by the first cue, is demanded after the second cue. This means a shift or context updating operation has to be performed. This is not the case in the sequences (/ /) and (\ \). "*" has a different meaning for different subjects, in one group it denotes "no shift" in another "shift" (see Table 2).

Event: E0 Warning	E1 Cue 1	E2 Cue 2	E3 Task	E4 Reaction
PIP	/	/	A B A A	≠
PIP	/	*	A B A A	≠ OR =
PIP	/	\	A B A A	=
PIP	\	\	A B A A	=
PIP	\	*	A B A A	= OR ≠
PIP	\	/	A B A A	≠

presented, which tells the subject either to maintain or to alter the induced attentional set. Alternation of the previously induced set would imply that the "cognitive network" has to be restructured or updated, while maintainance would imply that no specific updating process is necessary. In the following, the two conditions will be called "shift-" and "no shift-condition" respectively.

The details of the experiment are given in tables 1 and 2. The subject's task was to decide whether two visually presented letters were same or different. The two letters which had to be compared were part of a quadratic matrix with four elements (cf. table 1, column "task"). Either the two letters of the main diagonal (top left to bottom right) or those of the secondary diagonal (top right to bottom left) had to be compared for match or mismatch. The subject indicated the response by lifting the index finger of either the right or the left hand. The first instruction which of the two diagonals had to be monitored was given two seconds before the matrix. The cue was a bar orientated as the two diagonals of the letter matrix. A left-tilted bar meant "monitor the two letters shown at top left and bottom right" and a right tilted bar meant "monitor the letters shown at top right and bottom left". The critical stimulus to test the hypothesis whether or not P300 is an index of "context updating" was presented in the middle of the interval between the first cue and the task matrix. This second cue could either confirm the perceptual set primed by the first cue or it could reverse it. If the second cue was the same as the first this meant "keep your stimulus set", if the second cue was opposite to the first this meant "change your primed stimulus set". To guarantee that the subject actually processes the first cue it is necessary to introduce a third quality for the second cue which gives no direct information about the direction of the diagonal as the left- or the right-tilted bar. For that an asterisk (*) was chosen.

For one group of subjects the asterisk had the meaning "keep the stimulus set induced by the first cue" (no shift), for another group it had the opposite meaning "alter the stimulus set" (shift). The probability structure of the design is given in table 2. The probabilities were chosen such that tests, critical for the context updating hypothesis, should not be biased by probability effects. We expected that the comparisons of event sequences 1 vs. 6, 2 vs. 5, and 3 vs. 4 respectively should always furnish a strong amplitude difference of P300 triggered by the second cue, with a large amplitude for the shift condition and a small and negligible amplitude for the no shift condition.

The Pz recordings of two representative subjects, one from each subgroup, are shown in figure 4. To begin with, consider the solid and the dash-dot lines. The solid lines mark the no shift conditions (sequence 1 in the left and sequence 4 on the right plot and the dash-dot lines mark the shift conditions (sequence 3 in the left and sequence 6 in the right plot). As

Table 2

Double priming paradigm; probability structure: The symbols "/" and "\" appear with equal probability (p=.50) as first cue. In column p(E) the probabilities are listed with which a specified event-sequence (S1=S2, S2=*, S2≠S1) does occur. p(E) is equal to the conditional probability p(S2|S1). In column p(O) the probabilities are listed with which either a "no shift" or a "shift.operation" is induced by the second cue.

Sequence of Events:	p(E2)	Operation Induced by E2:	p(O)
Subsample 1 (" * " means "no shift")			
1 S2=S1 (/ /) (\ \)	.25	no shift	> .50
2 S2=* (/ *) (\ *)	.25	no shift	
3 S2≠S1 (/ \) (\ /)	.50	shift	.50
Subsample 2 (" * " means "shift")			
4 S2=s1 (/ /) (\ \)	.50	no shift	.50
5 S2=* (/ *) (\ *)	.25	shift	> .50
6 S2≠S1 (/ \) (\ /)	.25	shift	

it can be seen clearly, in both subjects the shift condition provides a larger P300-complex after the second cue than the no shift condition. And this effect is independent from the *a priori* probability of the eliciting stimulus. Notice that for the subject on the left the probability of the stimulus which evokes the shift operation is twice as large as that of the stimulus which evokes the no shift operation. This effect between shift and no shift condition comes out as highly significant, whether tested with direct amplitude measures or with principal component scores. That is good support for the hypothesis that context updating, operationalized as in this experiment - as a change of previously induced stimulus set - has a substantial influence on the generation of P300. However, it is not the whole story. The clear difference between shift and no shift condition is only apparent, if the second cue states explicitly which stimulus set is valid. In the asterisk condition, when the subject is informed implicitly about the stimulus set which has to be chosen, the specific

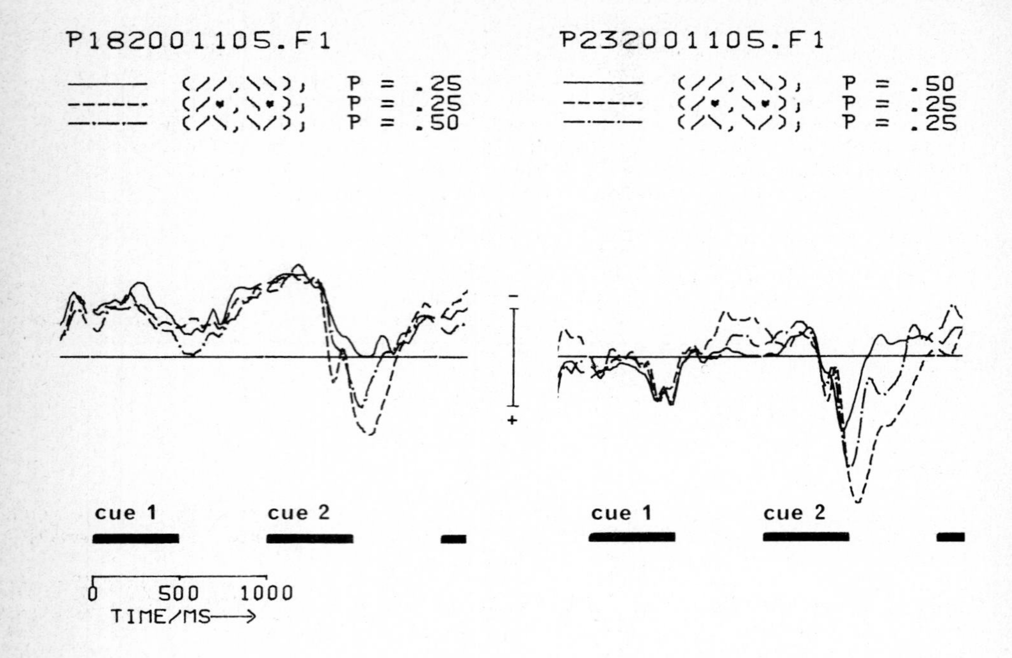

Figure 4
Double priming paradigm to test the context updating hypothesis. ERPs from Pz of two representative subjects. Left: Subject 18, member of subsample 1; Right: Subject 23, member of subsample 2. In each plot the solid line marks the no-shift condition and the dash-dot line marks the shift condition when the second cue explicitly told the direction of the stimulus set required to solve the letter matching task. These are the event sequences 1 and 3 in the left plot and the event sequences 4 and 6 in the right plot respectively (see table 2). The hatched line marks the asterisk condition when the second cue implicitly told which stimulus set is valid. For the subject on the left this condition meant "no-shift" (event sequence 2) for the subject on the right it meant "shift" (event sequence 5). Notice that in the left plot P300 after the second cue is larger in the shift condition even though the probability of the evoking stimulus is twice as large as in the no-shift condition. Length of bar between the two plots equals 10 Microvolts.

amplitude difference between shift and no shift condition is missing. Instead, another effect becomes evident. In both subgroups the asterisk evokes a much larger P300-complex than the two cues which state the stimulus set explicitly.

From that, one has to conclude that the cognitive operation of adapting or restructuring the brain of the new situation is not the only one which can evoke a substantial P300. The results suggest that in the condition with implicit presentation of information another cognitive operation comes into play. This operation might have something to do with memory activation or memory retrieval, because, when the asterisk is presented, the subject has to recall the orientation of the previously shown bar before he or she can decide definitely which diagonal has to be monitored. With the left or right tilted bar as second cue such a memory operation is not necessary. There the subject sees directly which stimulus set is required. As it seems, the memory retrieval operation is even more powerful to produce a large P300 than the context updating operation alone.

For a theory on P300, it would not be very parsimonious to add just another construct, so that one had to say P300 can be evoked either by a context updating operation or by a memory retrieval operation. A different and more general conceptual framework might be of help here in order to integrate and explain the various P300 effects. It is the heuristic distinction between *automatic* and *controlled* human information processing. The conceptual framework has been elaborated in detail on purely psychological grounds by Shiffrin and Schneider (1977, Schneider & Shiffrin, 1977). However, the same issue has been addressed by other authors too, Posner speaks of *automatic activation* of *pathways* and *conscious attention* (Posner, 1978, Posner & Snyder, 1975), and Neisser (1967) used the terminology of *preattentive processing* and *focal attention*. Disregarding the differences in semantics, the basic assumption by many is that there are two qualitatively different processing modes realized in the human brain which can be characterized as follows: Automatic processes are carried out without a person's intention and the person is not consciously aware of these processes. Further, they do not interfere with other automatic activities simultaneously performed by the CNS. Automatic in this sense are all processes which synthesize a percept from sensory input or which transform an action pattern into real movement. In contrast, controlled processes depend on the subject's intent, they can easily be monitored by introspection, and because they are dependent on a central processing device of limited capacity, they interfere with other cognitive activities of the same (controlled) type. Compared with automatic processes, controlled processes are much more time consuming. However, these disadvantages are compensated by the benefits deriving from the ease with which such processes may be set up, altered, and applied in novel situations for which automatic processing sequences are not available. Examples of controlled processes are rehearsal, search in memory, operations such as counting, decision making etc.. In the hierarchy of action systems of an organism, controlled processes are the top executing devices which are always invoked when the organism cannot cope with the environmental demands by routinized (automatic) action patterns (cf. Shallice, 1972, 1978).

If one analyses the phenomenology of P300-paradigms within this framework of automatic and controlled processing, it can be shown (cf. Rösler, 1980 (b), for details) that P300 is always evoked with considerable amplitude when the situation calls for controlled information processing and that P300 is absent or only picked up with negligible amplitude when the situation can be handled in a more automatic way with a routinized action pattern. Context updating or memory retrieval, seen in this framework, are only two examples of many which ask for the mode of controlled processing.

This hypothesis, which relates P300 to the mode of controlled processing, explains why in a particular situation a P300 is evoked at all, it does not account for the fact that the amplitude of P300 is sometimes larger sometimes smaller as e.g. in our experiment. To account for this, another theoretical assumption has to be made.

One possible solution is to assume that the amplitude of P300 reflects the extensiveness of control processes, which have to be performed by the CNS in a particular instant. Using Kahneman's terms (1973) one could say that P300 reflects the demands which are put on the central processor during a particular period of time. This idea that P300 amplitude could have something to do with capacity allocation or workload was introduced recently by Isreal, Chesney, Wickens and Donchin (1980a, b). They claimed that in a dual task situation P300 amplitude evoked by stimuli of a secondary task is inversely related to the workload demands of the primary task. Although their argumentation is in a certain sense inverse to the one presented here, this difference does not interfere with the general idea of relating P300 amplitude changes to differences in the extensiveness of cognitive control processes.

For the results of our double-priming experiment it would imply that maintaining a previously induced stimulus set puts only minimal demands on the central processor, while a greater amount of central capacity has to be allocated for the reversal of a fixed stimulus set. However, if it is necessary to retrieve information from memory before the stimulus set can be established, a qualitatively different situation is created. Then a maximum amount of capacity is allocated right away and the subsequent control processes for maintaining or shifting can be performed without further and diverse capacity "calls".

By summarizing the facts, one can try to give a more precise definition of the functional state of the brain which manifests itself in the P300-complex. Three theoretical deductions seem to be essential: (1) P300 appears whenever an event asks for the mode of controlled processing, (2) P300-latency is related to the termination of automatic stimulus analysis, and (3) P300-amplitude is related to the capacity demands which are put on a central processor. Therefore, P300 might indicate a process or a subroutine which invokes the limited capacity central processor. It might be a sign that the executive control is handed over to the top executing level of the organism.

Apart from the more specific interpretations worked out here to explain P300 effects the presentation should have made us aware of some more general points. It should have become evident that a satisfactory definition of the functional or cognitive state indicated by so called endogenous ERPs can only be attained if the researcher makes some assumptions about the information processing system, if he or she abstracts intervening variables, and if he or she introduces processing constructs. A purely behavioristic approach, although traditional in the field of evoked potentials, will fail to account for the effects found with endogenous event-related potentials. Since the behavioristic approach only operationally defines variables and relates them to variations in an endogenous ERP phenomenon, it can only help to clarify the antecedent conditions but it cannot by itself clarify the stage of processing which generates a given component. A fruitful integration of results will only be found after a grain of a theory is added. I dare say that for the study of endogenous ERPs, the psychological theory favoured by the researcher is as important as the experimental manipulations used to induce systematic variations in the ERP.

SOME PROBLEMS RELATED TO THE METHODOLOGY USED IN ERP RESEARCH

In section two of this paper, where I reviewed briefly the basics of ERP experiments, I made only passing reference to the fact that the triggering event is a prerequisite for applying one or the other signal extraction method to the raw data. In most ERP experiments the method used is that of averaging. This means that, depending on the signal-to-noise ratio, the EEG-activity of 20, 30 or more trials is combined into one averaged ERP. In the case of endogenous components with relative powerful amplitudes (P300, CNV) usually 20 to 30 trials are enough for a fair improvement of the signal-to-noise ratio.

The averaging method requires several well known formal assumptions: (1) random oscillation of the noise component around zero, (2) constancy of the latency, and (3) constancy of the amplitude of the signal. Beyond these another assumption has to be made when endogenous components are studied. By using the averaging method we have to imply that the specific series of information processing activities assumed to follow an input (stimulus) or preceding an output (response) is constant over the number of trials considered for one averaged event-related potential; the sample must be homogenous with regard to the psychological processes. Any systematic change in the information processing strategies over time might change the amplitude and/or latency of our peak under study and thus blurr the results.

Donald (this volume) gives some nice illustrations which show that characteristics of endogenous ERPs might even change within very short periods of time. Another convincing example has been presented by Courchesne (1978b). He showed that the scalp topography and amplitude of a "novelty P300" changed after the very first presentation of the critical event. If we accept an amplitude change of P300 as a sign of a change of the underlying psychological processes, then these data demonstrate that already during the first four repetitions of one type of event, marked changes can occur, and furthermore that an average over a number of repetitions of the same event might lead to fictitious results and improper conclusions.

However, graded changes in information processing such as habituation, learning etc. are not the only psychological "impairities" which violate some of the assumptions of the averaging method. The problem is even more severe when ERPs associated with qualitatively different "cognitive states" are not separated by the researcher and are instead mistakingly thrown into one averaging sample. To illustrate that the problem might be relevant in particular for the interpretation of graduated changes of amplitude, I will discuss a recent experiment of Isreal, Wickens, Chesney, and Donchin (1980b) in some detail. In this study subjects were asked to count infrequent tone bursts of a specified quality (secondary task while they monitored a visual display for occurrances of target events (primary task). The P300s to the secondary task tone bursts were smaller when both tasks were required than when only the secondary task was required. Moreover, when the perceptual load of the primary task was increased by requiring more events to be monitored for targets, the P300 amplitude elicited by the secondary task tones decreased further. The amplitude reduction with respect to the secondary-task-alone condition was about 30 % with low and about 60 % with high perceptual load.

As already mentioned, the authors of this study concluded that there is a functional relationship between the amount of perceptual effort invested into the primary task and the P300 amplitude. Although I favor the hypothesis myself that graduated changes of P300 amplitude might reflect graduated changes of the demands put on a central processor at a particular point in time, I will not conceal that a slightly different explanation of the results of the Isreal et al. study is also possible. For the sake of argument assume that the greater the perceptual load in the primary task the more tone bursts in the secondary task will not get access to the central processor. Assume that with greater load more tone bursts will be simply lost from the sensory register, because there is not enough free capacity to process them. If P300 is specifically related to the fact that an event gets access to the central processor, then the different load conditions will bring about different proportions of events which elicit a P300 and which elicit none at all respectively. This would result in the average P300 amplitude looking graduated even though in the single trial there are only two states of constant amplitude. The issue could be solved at least in part on a single trial basis, e.g. by examining whether the amplitudes of individual trials show a bimodal distribution in each separate condition.

Anyway, this theoretical deduction should evoke some skepticism against graduated amplitude changes observed in averaged ERPs (see also Rösler, 1980b). And it should make us aware that to a certain degree our <u>a priori</u> theory about information processing strategies will determine how we analyse our ERP trials and what we conclude from our averaged ERPs.

People who are adherents of single trial analysis, may say now, "okay, that's your problem, not ours, we are better off". I admit they are better off indeed, but they should be aware of the fact that they too make the same <u>a priori</u> assumptions. They too have to assume that throughout the total experiment the information processing activities are constant except for timing.

Take, for example, an experiment of Kutas et al. (1977), in which the latency of P300 was studied on a single trial basis and related to the correctness and incorrectness of the subject's response. The subject's task was to decide whether a particular word had been presented on a display or not (e.g., a male or a female first name). In most cases when an erroneous response had been made reaction time was extremely short while P300 latency was remarkably elongated, mostly longer than RT. With correct responses RT was usually longer than P300 latency. These facts led the authors to the conclusion that, in case of an error, the response was initiated prematurately before a complete stimulus evaluation had been carried out by the CNS. This, no doubt, is a convincing explanation for the dissociation of RT and peak latency in the two conditions. The explanation might be correct, nevertheless, if in such an experiment correct and incorrect trials involve qualitatively different routes of processing, then the method of Woodying would fail to picture reality.

The basis of this assertion is shown in figure 5. It depicts a model which assumes that in an error trial two P300s are initiated because the central processor is invoked twice - first to evaluate the stimulus with respect to a response and second to evaluate the outcome of the prematurately committed action. On the other hand, in a correct trial there is only one P300, because the second call for the central processor is missing.

Taking the model as valid, the temporal overlap of the two P300s and their relative strength of amplitude would determine whether or not they could be separated by a single trial peak detection method. If they showed too much of an overlap, the latency estimate of P300 in an error trial could represent the latency of the first peak, the latency of the second peak, or

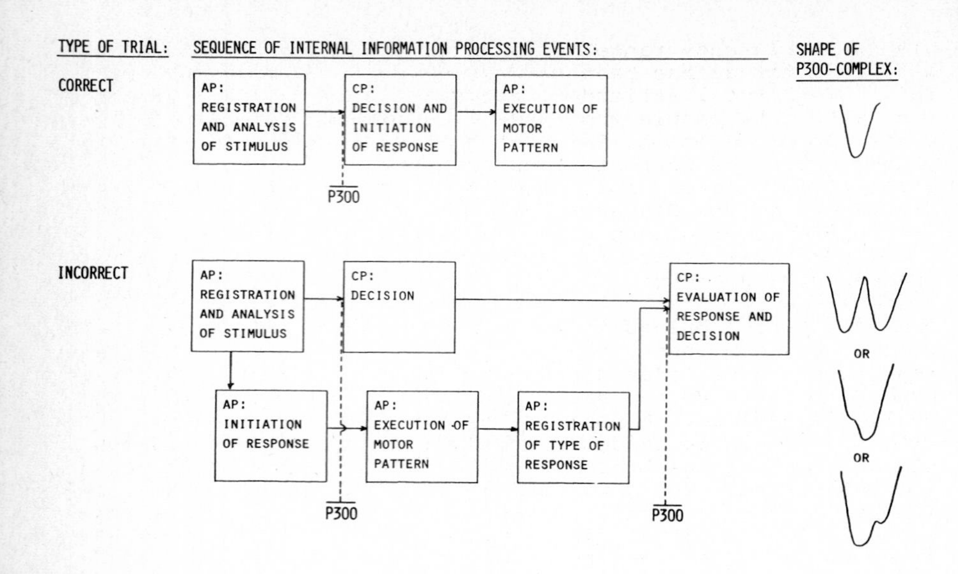

Figure 5
A model which assumes a different sequence of information processing events for correct and incorrect trials in a choice reaction time experiment. AP: automatic processing, CP: controlled processing (see text for further explanation).

an average of the latencies of both. Some possible situations are sketched in figure 5 too.

EXPLICIT AND IMPLICIT DEFINITIONS OF ERP-COMPONENTS

The last paragraph leads to the final point of this paper: What constitutes an ERP-component; what are the defining features of such an entity?

Explicit definitions. The traditional definition of a component has its origin in the study of sensory evoked potentials and it reads as follows: a component is a noticeable deflection in the averaged ERP trace with a specified polarity and latency, and the latency range within which the component can vary is relatively narrow. This polarity-latency definition was also used when endogenous components began to be studied. One of the first endogenous components discovered was P300 and it was often explicitly defined as the most prominent peak within the latency range of 250 and 350 msec. But now, its latency interval has been broadened up to 800 msec and more. This more liberal attitude towards the latency range is especially held by people who study P300 latency on a single trial basis (Duncan-Johnson,1981, Kutas et al., 1977, McCarthy, this volume). In the current literature the definition of P300 often reads: after application of a low pass-filter to the single epoch (see Glaser & Ruchkin, 1976, pp 79) define as P300 the most prominent positive peak in the Pz trace whose amplitude maximum does not occur earlier than 300 msec after the triggering event.

That definition brings in another characteristic which is used to define a component, that is, its topography. In this matter researchers have taken different positions. Some take a relatively strict view and say that P300 must have a parietal amplitude maximum. A positive peak with a similar latency but with a frontal amplitude maximum indicates not the same functional state of the brain as the "classic" parietal P300 but something else. The consequence of that position is that topographical changes cannot be used as a dependent variable and a whole family of P300s, distinguished by their topography, must be accepted. The other position is more liberal, and, as I think, scientifically more fruitful. For this, the topographical range within which P300 may have its amplitude maximum is enlarged (the total fronto-parietal area) and it is assumed that locally different peak maxima can tell us something about the localization of the underlying process. For instance, Courchesne (1978b) interpreted P300 as being involved in categorization processes and that topographical differences reflect different aspects of these processes: A more frontal P300 occurs whenever new concepts are required for stimulus categorization and a more parietal P300 occurs whenever already existant concepts are appropriate for stimulus categorization (cf. also Rösler, 1977, 1980b).

In recent years another definitional characteristic has been added: The experimental variables (see Donchin et al., 1978). For example, this characteristic along with polarity and latency would define P300 as follows: "P300 is the peak that is positive, has a maximum amplitude later than 300 msec after the triggering event and which follows rare and task relevant stimuli". Stated in this way, the definition looks a bit silly: First, the circularity of the definition is obvious (what process is indexed by P300? Well, P300 is a manifestation of task relevance. Why, how do you know? Well, because the wave that is elicited by task relevant stimuli is called P300!). Second, if that definition is taken literally, we would have to accept as many components as there are paradigms, possibly an infinite number.

Moreover, the latter point is obviously in contradiction to the basic logic of research done with endogenous ERPs. The premise, as stated above in section 2, is that an ERP component defined by its phenomenology (e.g. polarity, latency, topography) is a manifestation of one particular functional state of the brain. When two different experimental paradigms (e.g. an odd-ball task and a memory scanning task) both produce an ERP peak which is phenomenologically similar then, it is assumed that a common psychological and/or physiological process is involved in the two tasks. The problem to be solved is to fixate the quality and/or significance of this common state. This basic logic is turned upside down if specific experimental operations are accepted as defining properties of an ERP component.

Implicit definitions. Unfortunately, this error in logic occurs when researchers who use principal components analysis implicitly define ERP components in terms of specific experimental operations.

A specific principal component reflects a source of covariability in the data. The substantial loadings on a component specify the variables which show covariation. In the case of ERPs, these are the time points at which the measured amplitudes covary. This covariation depends on the sources of variance considered in the matrix submitted to the PCA algorithm. With ERPs usually the following sources of variation are assembled in one matrix of raw data: different experimental conditions, different electrode sites, and different subjects. Therefore, the loading pattern depends on the autocorrelative nature of the amplitude of successive time points, but also on the covariation that is due to the particular sample of variables considered in the study. The loading pattern depends heavily on which source of variation is dominant at a specific time point. As we have discussed elsewhere (Rösler & Manzey, 1981) the most important factors for the determination of principal components in ERP studies are most likely the experimental manipulations and electrode locations. Thus, ERP components defined by principal components analysis are implicitly defined by the particular pattern of manipulations considered in each particular experiment (see also Wastell, this volume).

One may object that this is not a real problem as long as we can find in different studies comparable principal components, i.e., components which show a similar or almost identical loading pattern. One could assume that components coming from different studies but showing the same loading pattern can be taken as equivalent and as entities which indicate the same physiological or psychological process. However, this assumption is not valid. Principal components from different studies which show identical loading patterns may not necessarily be related to identical variations in the original waveforms. Imagine that in several studies a component is found which has high loadings within the time region of 300 and 600 msec. This reveals that the amplitudes of the time points in the specified region covary, possibly because in each experiment they are affected by the experimental manipulations in similar manner. However, in one case the covariation between 300 and 600 msec could be due to amplitude variations of one single peak, in another the same covariation could be due to amplitude variations of two peaks (e.g., a P3 and a P4, see Stuss and Picton, 1978), and in still another study there might be no peak at all but just a rising or falling flank of a peak which causes the covariation of successive time points. In short, in a particular study, a component defined by PCA, i.e., by a pattern of substantial loadings of time points, may not necessarily have a counterpart in the ERP which can be identified as one single peak in the traditional sense, and between studies, similar principal components may not necessarily reflect the same ERP peak structure.

It is questionable whether a definition of an ERP component which only considers the statistical nature of successive time points and which neglects the particular waveshape, is really desirable (cf. Donchin & Heffley, 1979, vs. Wastell, 1979, Donchin & Heffley, 1978, Rösler & Manzey, 1981, Wastell, this volume). The decision for one or the other approach will depend on personal preferences of the researcher. Whether a researcher is willing to accept directly observable peaks as more meaningful physiological entities than statistically defined components or not will, for the time being, depend upon personal preferences. Since little is known yet about the physiological generators of endogenous components the choice of one over the other can hardly be decided objectively at the present.

CONCLUDING REMARKS

The intent of this paper was to show how intimately theoretical assumptions and experimental manipulations are interwoven when research is done with endogenous ERP phenomena. Although the points were elaborated for the most part by referring to P300 studies, I feel that what has been said is also valid for other ERP components accepted as endogenous in origin, e.g. N100, N200, slow wave, CNV, RP (for an overview see Hillyard, Picton & Regan , 1978, Donchin, Ritter & McCallum, 1978) or N400 (Kutas & Hillyard, 1980c). Studies dealing with these components are all similar with insofar as cognitive states of the brain are infered from ERP phenomena.

Whenever such an association between physiolgical phenomena and psychological constructs is attempted, the theoretical assumptions made by the researcher - either explicitly or implicitly - are of outstanding importance. This is so for two reasons: Firstly, since the very same ERP component can be elicited and manipulated in its shape by phenomenologically very different experimental operations, the functional state of the brain related to this component has to be abstracted from directly observable entities (variables, instructions etc.). Such an abstraction is only possible if assumptions about the information processing activities of the brain are made and if intervening variables and/or hypothetical constructs are defined. Secondly, the theoretical assumptions made have strong implications for the methodology used to detect the ERPs in the EEG. For example, when methods such as averaging or peak detection by "Woodying" are applied, it must be taken for granted that all EEG epochs considered for one sample are homogeneous with respect to the cognitive state. This implies that the researcher has to make very precise assumptions about the information processing activities performed by the brain in a particular trial, otherwise epochs differing in their "cognitive quality" will be lumped together. This can, at least, blur effects or in more severe cases it can even lead to fictitious results and invalid conclusions about psychophysiological relationships.

Before concluding this paper I want to point out that the more critical remarks presented here were not meant in any way to discredit particular experiments, particular methods, or even the approach in general. Particular experiments were chosen under didactic considerations just to disclose some general problems inherent in this kind of research. By pointing to these problems, my aim was to show where and why already excellent experimental and theoretical approaches should be improved so that some more pieces of the mind-body puzzle can be fitted together successfully.

Notes

This work was supported in part by grant Ro 529/1 from Deutsche Forschungsgemeinschaft (German National Research Foundation). I wish to thank Eric Courchesne, Tony Gaillard, Walter Ritter, and Tom Roth for their helpful comments on an earlier draft of this paper. Address requests for reprints to: Dr. Frank Rösler, Institut für Psychologie, Neue Universität, Geb. N30, D-2300 Kiel, West Germany.

[1] Although the question whether intervening variables and hypothetical constructs should be differentiated might be a rather academic one to some people, I think, the distinction is of heuristic value, especially in the field of psychologically orientated ERP research. According to the original definition of McCorquodale & Meehl (1948) intervening variables refer to theoretical words which merely abstract empirical relationships from observable facts. Hypothetical constructs, on the other hand, are theoretical words which involve the supposition of entities or processes not among the observed facts. In this sense entities as "processing time" or "subjective probability" would be labeled intervening variable, while entities as a "mismatch detector", a "memory search process", or a "context updating operation" would be labeled hypothetical construct.

[2] Notice that if the mediating processes are defined psychologically as "cognitive" (template matching, context updating etc.), then this epiphenomenon assumption becomes an exact reversal of classic materialistic epiphenomenalism. In materialistic epiphenomenalism, cognitive events (thoughts, percepts

etc.) are seen as epiphenomena (by products) of physiologically definable states of the "brain machinery". In psychologically motivated ERP research it works the other way round: Physiologically definable states of the brain (ERP components) are accepted as epiphenomena of cognitive events.

[3] Behavioristic analysis is understood in the sense of classic behaviorism, where the organism is treated as a "black box" and only input or output variables, which are directly observable, are considered for an explanation of behavior changes. Behavior is understood in the broad sense of any observable entity of the organism that is measurable in terms of quantifiable effects. Thus, an ERP is accepted as behavior.

[4] According to the data and theory presented by cognitive psychology the analysis of stimulus meaning and relevance is most probably an automatic process. Only if the organism is unfamiliar with a stimulus or a stimulus pattern, then a controlled synthesis of the meaning has to be performed (see e.g. Anderson, 1980, Posner, 1978).

Tutorials in ERP Research: Endogenous Components
A.W.K. Gaillard and W. Ritter (eds.)

3

NEURAL SELECTIVITY IN AUDITORY ATTENTION: SKETCH OF A THEORY

Merlin W. Donald

Department of Psychology
Queen's University
Kingston, Ontario,
Canada

Introduction

This paper represents an attempt to create a model of selective attention which accounts for recent evidence derived from human electrophysiological experiments. The definition of attention adopted here follows that of Hebb (1948) and Broadbent (1958) in emphasizing selective processing of the input as the critical variable. Attention in this definition is concerned with selective perception and not with general alertness, physiological activation, motor readiness, or response preference. It is concerned with how the organism can apparently exclude certain stimuli from awareness and memory, while focussing on other equally salient aspects of the environment. Stimuli singled out by attention are analysed, rehearsed, integrated into the current perceptual interpretation of the environment, and stored in memory in a selective manner. Rejected or unattended channels of input receive more limited processing, and in a sense the problem of selective attention can be reduced to one of comparing the processing of attended and rejected inputs. How are inputs excluded, at what level, and what does attention add?

The advantage of electrophysiological recording in this context is that it allows the simultaneous examination of neural processing both inside and outside the focus of attention. Voluntary behavior is closely tied to attention, and behavioral access to the rejected channel is very indirect and limited to measures of delayed recall or recognition, which inevitably confound problems of memory retrieval with problems of initial input processing. Measures of autonomic and EEG arousal have been used to probe the rejected channel, but such measures involve very indirect, delayed, and unreliable responses. The evoked potential technique represents a method of probing the on-line neural response to inputs. Although this technique also has great limitations, it appears to be much closer to the actual processing of the input, given the anatomical distribution and relatively short latency of the components.

This paper will deal primarily with human evoked potential data, and within that literature it will focus largely on high-speed multichannel experiments. The reason for this is that the high-speed multichannel design, pioneered by Hillyard and his associates, constitutes the best available means of controlling nonspecific effects which tend to be confounded with attentional selection: resting dc levels, ongoing EEG rhythms, and time-locked events due to the subject correctly anticipating the time of stimulation. As observed by Näätänen (1970), Karlin, (1970), Wilkinson and Lee (1972) and others, at rates of stimulation slower than 1Hz and with regular interstimulus intervals, it becomes possible for the subject to anticipate the time of stimulation, and it is difficult to ensure that attention is restricted to a specified class of stimuli. Single-channel experiments by their very nature do not require the exclusion of alternate input channels and do not directly address the problem of selectivity of response. The optimal design involves at least two equally salient channels of stimulation presented at irregular intervals at speeds which make it difficult for the observer to monitor all input channels at the same time. Fortunately, there are now a considerable number of experiments in print utilizing this type of design, forming a literature sufficiently coherent to allow comparisons across studies.

The greater part of this literature consists of studies of auditory selection. There is evidence for selective tuning of somatosensory evoked potentials (Desmedt and Robertson, 1978) but this field is still very limited. Visual attention has generally been approached with single-channel designs, with relatively slow rates of presentation, which are thus vulnerable to nonspecific effects. Possible exceptions to this objection are studies by Eason et al (1969), Van Voorhis and Hillyard (1977) and Harter and Previc (1978). In the study of multichannel auditory selection the stimulus 'channels' are usually defined spatially. Dichotic designs, in which each ear receives a different sequence of stimuli, are most common, although there have been studies in which stimuli were delivered in up to four or five different locations in auditory space (Schwent and Hillyard, 1975b; Hink et al, 1978). Pitch is another widely-used attribute for defining channels. The best results are obtained by combining pitch and spatial location, that is, by delivering streams of stimuli with different fundamental pitches in each spatial locus (Schwent et al, 1976). Rates of stimulation have varied from 4 Hz to .5 Hz per channel, but evidence collected by Schwent et al (1976a) indicates that selectivity of response becomes less evident, at least in the 100 msec latency range, at rates slower than 1 Hz, and the majority of studies have employed stimulation rates of about 3 Hz per channel. In a two-channel study, this translates into an overall rate of input of 6 per sec, approaching the range in which it may be concluded, from objective behavioral evidence, that subjects find it easier to listen to one ear than both at once (Harvey and Treisman, 1972).

The experimental literature prior to 1979 has been reviewed thoroughly by Näätänen and Michie (1979), as well as Hillyard and Picton (1979), and by Hillyard (1981). I will attempt to take into account the data covered in those reviews, but in addition I will try to account for more recently acquired data, some of which has been collected in my laboratory, concerning several additional aspects of attentional selection. One of these aspects is the time course of selection; how long does it take the brain to "tune" its response to the attended channel, and how does it reject the unattended channel or channels? Can it pre-select, that is, bias its response prior to receiving any stimulus, or must the stimulus be present for some time before it can be selectively processed? A preliminary report (Donald and Young, 1980) has briefly described the approach taken by this author, and a more recent report deals with the time-course analysis of attentional selection in more detail (Donald and Young, 1982).

Another aspect is the processing of stimulus probability. The probabilistic relations between environmental stimuli are obviously processed in the focus of attention, but how are they treated in the rejected channel? Probability information requires some storage of previous stimuli, and thus the sensitivity of a sensory channel to changes in stimulus probability gives some indication of recent memory storage in that channel. What kind of recent storage occurs in the rejected channel when attention is directed elsewhere? A recent experiment on this question was reported in Donald and Little (1981).

Finally, there will be some discussion of the ontogeny of auditory selective attention. Does auditory selection develop all at once, or do different aspects of neural responses show selectivity at different ages? Does selection involve a net suppressive or net excitatory effect? A dissertation by Brooker (1980) examined this problem in 6-to-13 year old children.

This discussion will focus primarily on two components which will be called, for historical reasons, N_1 and P_3. N_1 is defined as the peak negativity between 70 and 130 msec, and P_3 as the peak positivity between 250 and 500 msec. This provides continuity with the earlier literature, although it leaves open some of the evidence assembled by Näätänen, Gaillard & Mantysalo (1978) that attentional selection is evident in other latency ranges as well. The question of multiple N_1 and P_3 peaks will be discussed later in the paper.

(a) The time course of N_1 and P_3 tuning

Evoked potentials, and single neuronal units, can show several kinds of systematic trend over time, the principal tendencies being either gradual decrement or increment, or cyclical variation. They may also, of course, remain stable for considerable periods of time. Repetition of a stimulus at a rate slower than the recovery cycle of the evoked potential usually results in a fairly stable response in primary sensory pathways, whereas in the reticular formation and hippocampus the neural response is most likely to show either a rapid exponential decrement in amplitude, called habituation, or an initial increment, called sensitization, followed by a decrement (Groves, de Marco and Thompson, 1969).

Assuming that N_1 and P_3 tuning are superimposed upon these fundamental neuronal properties, there would be at least three quite different possible outcomes to an experiment which described the time course of the emergence of N_1 and P_3 tuning: (1) stability over time with instantaneous onset; (2) increment in the attended channel; (3) decrement in the rejected channel. These three possibilities are illustrated schematically in Figure 1. It is evident that an average across all of the successive stimuli in any of these series would in each case produce a larger EP amplitude in the attended channel.

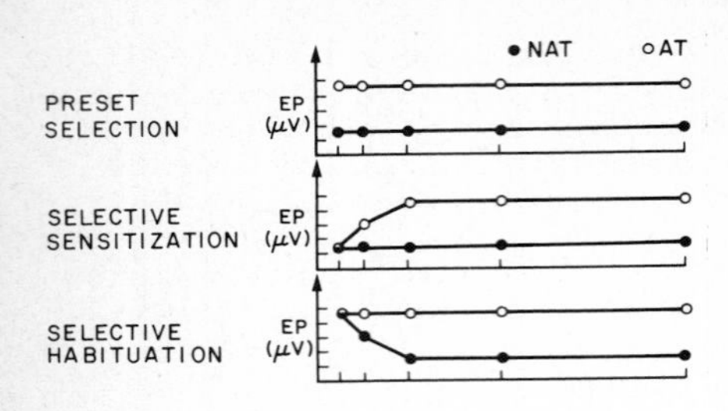

Figure 1. Three hypotheses about the temporal emergence of the superiority of the attended channel. The abscissa indicates time.
NAT = unattended channel
AT = attended channel

An experimental paradigm was devised with a view to mapping the emergence of N_1 and P_3 tuning in time (Donald and Young, 1982). The paradigm, described in an earlier report (Donald and Young, 1980), was somewhat similar to the high-speed two-channel selective listening task originally utilized by Hillyard et al (1973). Two sequences of pure tone stimuli were presented, one to the right ear, one to the left. The average rate of presentation was approximately 3 stimuli per sec per ear. The basic pitch was different for each ear (800 vs 1500 Hz), and rare stimuli (10% probable) of a slightly different pitch were inserted at random in each sequence. The stimulus sequence was broken up into brief trials of 10-11 sec duration. Each trial was followed by a 6 sec intertrial rest period, during which the subject wrote down the number of rare tones, or targets, that he detected in the assigned ear channel. The subject knew in advance which ear to monitor, and the assigned channel was never changed during an experimental run.

At the outset, it was not known what time parameters would be appropriate to map the emergence of N_1 and P_3 tuning: they could have emerged instantly on the first stimulus, or in a matter of seconds, or tens of seconds, or possibly over a longer period. It was also not known whether N_1 and P_3 tuning would have to be re-established on each brief trial, or whether they would "carry over" from trial to trial. To test the widest possible range of time scales, two types of analysis were attempted, a within-trial analysis, and a between-trial analysis. The former revealed the time course of N_1 and P_3 amplitude change within the average 10 sec trial, the latter revealed longer-term changes from trial to trial, over a performance period of about fifteen minutes.

The within-trial analysis indicated that N_1 amplitude was consistently larger in the attended channel throughout the average 10-second trial from the first stimulus to the last; in other words, N_1 tuning was carried through the intertrial interval and did not have to be re-established for each individual trial. The between-trial analysis, which examined amplitude trends over a longer period of some fifteen minutes, (Figure 2, left side) revealed a significant trend towards amplitude decrement in both the attended and rejected channels over a period of minutes, with the rejected channel reaching asymptote earlier than the attended channel. This result allowed certain of the initial possible outcomes to be discarded: N_1 amplitude was not stable over time, and it did not increment in absolute terms from the initial value in either channel. N_1 tuning seemed to reflect a different rate of habituation in the attended and rejected channels, but presumably some factor other than habituation was required to explain the complex time course shown in the attended channel.

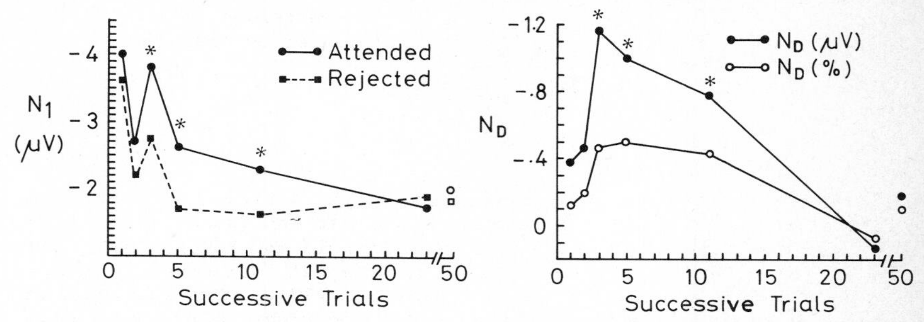

Figure 2. Left, temporal pattern of N_1 change over a typical performance run of 50 brief 10-second trials. Right, absolute (μ) and percentage (%) difference between N_1 amplitudes of attended and rejected channels (N_d) over successive trials. Asterisks indicate N_d significantly different from zero. From Donald and Young, 1982.

The channel separation, or amplitude difference, between the attended and rejected channels (N_D) is illustrated in Figure 2, right side. The amplitude scale has been increased to clarify the time course of N_D, and the same data have also been plotted as percentages of the N_1 component in the rejected channel, in the manner formerly used by Hillyard et al (1978). The resulting function has a simple form, N_D being maximal and significantly different from zero, from trials 3 through 12. The averaged N_D amplitudes for trials 1 and 2 were not significantly different from zero, although moving in the right direction. It was not until trial 3 (after about 35-45 seconds of stimulation) that the attended channel achieved significant amplitude superiority over the rejected channel.

An examination of the grand-averages for trials 1-3 for all subjects (Figure 3) shows that the N_1 tuning effect coincided in latency with the N_1 component. No significant difference occurred in the 200 msec latency range, but this may have been due to the time-constant (.1 sec) of our amplifiers. Thus N_1 tuning was not "preset" prior to the first trial; it took, on the average, about 40 seconds to emerge and was superimposed upon a general tendency towards reduced N_1 amplitude over time. Judging from the rapid rate of decline from trial 1 to trial 2 in this experiment, both auditory channels showed approximately the same initial rate of decrement. The emergence of N_1 tuning on trial 3 was not simply the result of a more rapid decrement in the rejected channel, but rather the result of a large increment in the amplitude of the N_1 component in the attended channel. An examination of the data of individual subjects indicated considerable differences in the relative importance of incremental and decremental factors.

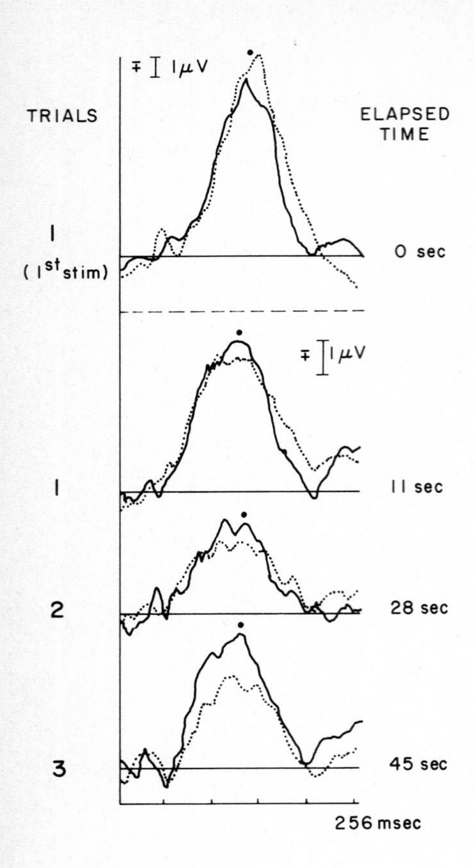

Figure 3. The emergence of N_1 superiority in the attended channel during the first three trials. Each trace is a grand average of data from 8 subjects. Upper trace: response to first stimulus of trial 1, N = 32. Lower three traces: average EP during trials 1, 2 and 3, N = 512. Solid lines = attended channel; dotted lines = rejected channel. Dots indicate where N_1 peaks were measured. From Donald and Young, 1982.

The simplest explanation of the results shown in Figures 2 and 3 would be the hypothetical events shown in Figure 4: a rapid negative exponential decline represents the time course of EP change without attention; and a different function represents what is added in the attended channel. Since the latter was not simply derived from subtracting two negative exponentials, it is unlikely that it represents simply a delay in the habituating neural elements in the system. Rather it appears also to involve a sensitizing, or incremental event, with a time course of its own. Perhaps N_D could be extended further in time with a different experimental paradigm, but in this study it disappeared after about 6-7 minutes, (about 24 trials), for the standard stimuli. For the rare target stimuli, as far as could be determined, N_D did not show the same decline in later trials.

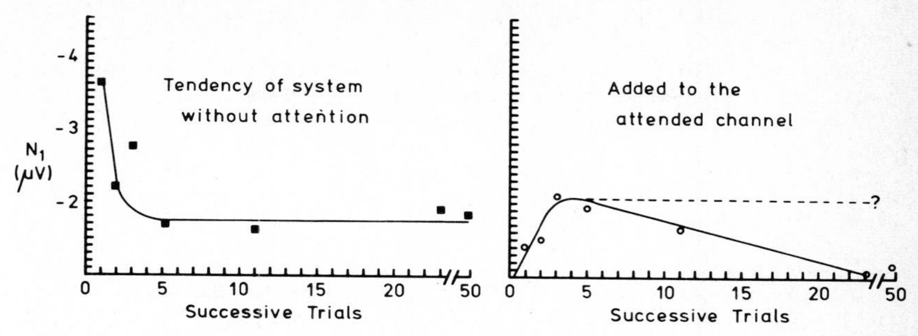

Figure 4. Hypothetical processes underlying the emergence of N_1 tuning.

The time course of the P_3 component could not be examined in much detail, since the rare stimuli which elicit P_3 do not occur frequently enough to enable averaging over short time periods. Nevertheless, an estimate was obtained by averaging over longer time periods (the first and last 5 minutes of each run). An estimate of the P_3 to the very first rare stimulus in each channel in each run was attempted by taking single-trial samples from each subject-run. This procedure yielded a clear P_3 component, presumably because the P_3 component was unusually large, at least in the attended channel, on the first stimulus. The resulting estimate of the time course of P_3 is shown in Figure 5: P_3 amplitude is stable at a low level (not zero voltage, since it was easily identifiable in most subjects) in the rejected channel, from the very first stimulus. In contrast to N_1, P_3 was much larger in the attended channel from the start, and after an initial decrement (to be expected on the basis of the results of Ritter et al, 1968) remained at a higher level than the rejected channel P_3 throughout the run. It was concluded that, at least with this kind of paradigm, N_1 tuning is a gradual selection process, coinciding with both rapid habituation of N_1 in the rejected channel, and an incremental event in the attended channel, whose combined effect appears to be that N_1 becomes larger in the attended channel in about 30-40 seconds. In contrast, P_3 tuning is very rapid, preset by an internal switching mechanism prior to the occurrence of the first rare stimulus in either channel.

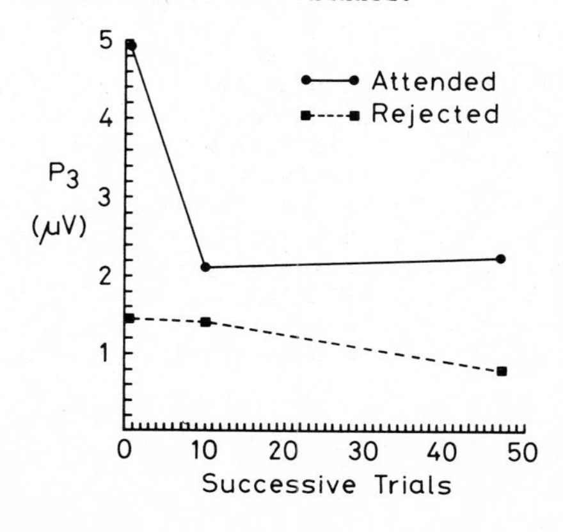

Figure 5. Group means for P_3 amplitude over successive trials. The first data point for each channel is the mean response to the first rare stimulus, comprising single-trial P_3 samples taken from each subject-run, N = 32. The subsequent two data points are means of conventional within-subject averages (N=2048) for the first and last half of each run. From Donald and Young, 1982.

(b) Probability and Attention

Organisms must maintain a record of the probabilistic relationships between environmental events; this serves as the basis for most types of perceptual learning, from elementary perceptual constancies to the modelling of three-dimensional space. On theoretical grounds, it is necessary for the system to continuously "match" its inputs to the ongoing perceptual record, and thus "mismatch" signals are to be expected in the nervous system at many levels (Mackay, 1964; Grossberg, 1978). In one sense all transient components of evoked potentials are mismatch signals, being primarily sensitive to stimulus change (Clynes, 1964). One component, P_3, has a special relation to stimulus probability, and although P_3 is also determined by other factors, the notion that P_3 might be, at least in part, a mismatch signal, has been in the literature for some time (cf. review by Pritchard, 1981).

One issue of interest is to determine whether, and how, the probability record is modified by attention. Presumably some record of events is maintained even peripherally to the attentional focus, and a substantial experimental literature attests to that fact (Moray, 1969). The question is, does attention simply damp or reduce the incoming signal in amplitude, leaving the ongoing probability record intact, or does it also switch out the latter in the rejected channel? And if such a switch exists, at what level does it operate?

This question was put in a recent experiment (Donald and Little, 1981) which utilized the same selective listening paradigm used in the analysis of the time course of N_1 and P_3 tuning, except that the probability of a rare tone was varied from one in sixteen (.0625), through one in eight (.125), to one in four (.25). It was noted in our previous work that P_3 was not completely absent from the rejected channel, and the question was asked, would the P_3 in the rejected channel increase in amplitude if the rare stimuli were made less probable? There were two possible outcomes, illustrated in Figure 6.

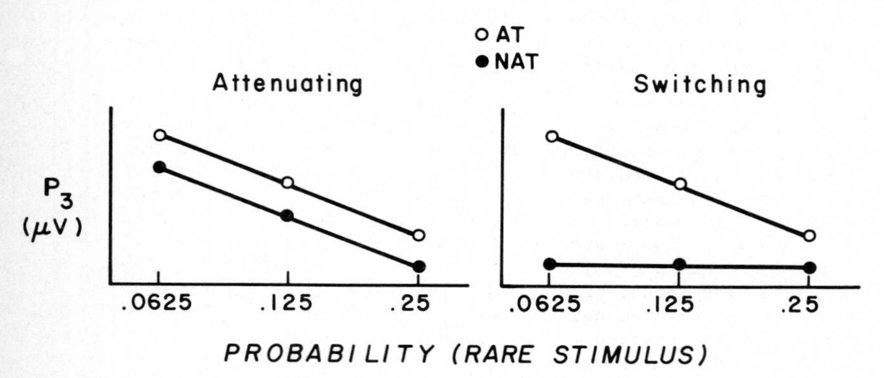

Figure 6. Two possible relationships of P_3 amplitude to probability, either of which would predict P_3 superiority in the attended channel.

If the rejected channel was simply attenuated, it should remain sensitive to stimulus probability, albeit at a reduced overall amplitude, and the regression against probability ought to be similar, or at least significant, in both channels as shown on the right side of the figure. If P_3 in the rejected channel was merely a residual due to unsystematic failures of attention, it ought not to change with stimulus probability. Although N_1 is less sensitive to probability than P_3, it was nevertheless of some interest to examine the same relationships with N_1 as the dependent variable.

The outcome of the experiment was that P_3 showed no significant change in amplitude in the rejected channel as rare stimuli were altered in probability, whereas the attended channel P_3 showed the familiar linear relationship to probability (Figure 7). Thus the residual P_3 in the unattended channel appears to be merely a result of unsystematic lapses of attention, and the rejected channel had effectively been "switched out" of the matching process at the P_3 level. Once again, a switch-like mechanism was implied for P_3.

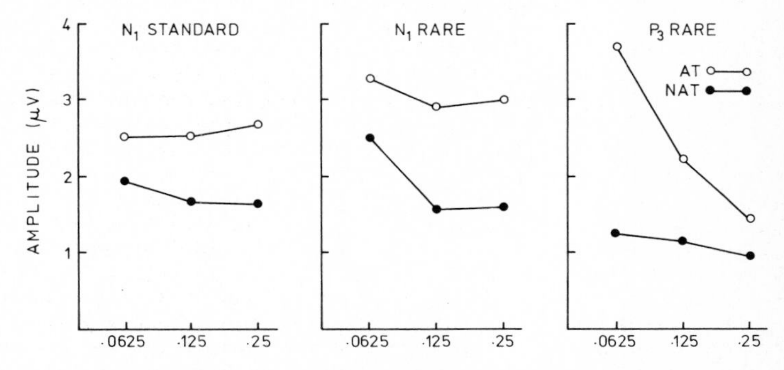

Figure 7. Relation of EP amplitude to probability level shown for the attended channel (AT) and unattended (NAT) channels. The left figure refers to standard tone N_1 amplitude, the middle and right to rare tone N_1 and P_3 amplitude. The markers on the ordinate are spaced at μV intervals. From Donald and Little, 1981.

With regard to N_1 there was a rather surprising outcome (Figure 7). The rare tone N_1 did change significantly in amplitude as a function of probability, albeit in a nonlinear manner. The lowest-frequency condition (.0625) produced larger N_1 peaks in both the attended and rejected channels, with the rejected channel showing the larger increase. Thus N_1 did prove sensitive to the relative probability of environmental inputs, and this sensitivity was not switched out in the rejected channel. This calls for an attenuator, or graded-bias model of N_1 tuning, rather than a switch model. In effect, the rejected channel was inhibited relative to the attended channel, but continued to be sensitive to the relative frequency of events in the environment, at least at the level of processing reflected by N_1. The main conclusion is that the rejected channel stores information about immediately-previous stimulation, but only up to the level of processing which produces N_1; the circuitry producing P_3 contains no such information, at least not in the paradigm used in this study.

One other result of this study must be noted here: P_3 tuning dropped out in the .25 condition; P_3 amplitude in the attended channel was not significantly different from the rejected channel (Figure 7, far right). Performance was maintained at a high level, and N_1 tuning was actually increased in this probability condition (.25). In other words, while the overall rate of stimulation (standards plus rares) was constant and the subject was performing efficiently, the N_1 response was highly selective, but P_3 was virtually suppressed in the attended channel. This shows that attention, although necessary, is not sufficient to elicit P_3. The physical patterning of the stimulus stream in the .25 condition somehow caused the suppression of P_3, and by extension P_3 amplitude was determined by stimulus events, not exclusively by endogenous factors. The physical feature most probably related to the suppression of P_3 in the .25 condition might be the high rate of target presentation per unit of time (an average of about 1 per sec) in that condition. This point will be expanded later.

(c) The ontogeny of attentional selection

In a recent doctoral thesis Brooker (1980) examined the developmental course of N_1 and P_3 tuning in normal boys aged 6 to 13. Since the recording techniques and stimulus parameters employed were identical to the .0625 probability condition in the Donald and Little (1981) study, it was possible to compare the three younger age groups (with mean ages of 8, 10 and 12 years, respectively) with the college students used in the latter study, whose mean age was 23. There were 12 subjects in each age group.

The principal result of the developmental analysis was that P_3 tuning developed earlier than N_1 tuning. Figure 8 shows the difference in N_1 and P_3 amplitude between the attended and rejected channels for the four age groups. Analysis of variance indicated a significant relationship of both variables with age, and asterisks indicate where the differences between channels exceeded zero. P_3 was

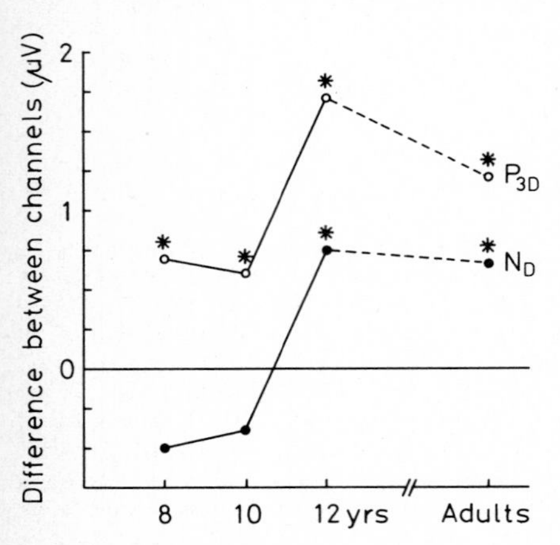

Figure 8. Developmental emergence of amplitude superiority of N_1 and P_3 in attended channel. N_D = standard tone N_1 difference in amplitude between channels. P_{3D} = amplitude difference in P_3 between the attended and rejected channels. Asterisks indicate difference is significantly different from zero. N_D is reversed in polarity, so that both plots show the tuning effect in the same direction.

larger in the attended channel at all ages, whereas N_1 showed no selectivity until age 12. Both N_1 and P_3 tuning improved greatly between the 10 and 12 year-olds, and in this sense the two types of selection improved at the same time, but P_3 selectivity was already present in the younger children, whereas N_1 tuning was not. It should also be pointed out that N_1 latency in the two youngest age groups was significantly longer than in the 12-year-olds and adults. The peak occurred between 150 and 200 msec in most subjects. The N_1 tuning effect was always measured at the peak latency of N_1. Conceivably the results might have been different with a different task, but with this task N_1 selection emerged later, at least 4 years later, than P_3 selection.

Another trend to emerge in this study was the tendency for the rejected channel evoked potential to decrease in amplitude with age; this was true of both N_1 and P_3 which showed statistically significant amplitude decreases in older children. Age trends did not appear in the attended channel (Figure 9) for any of the three variables. This suggests that the net ontogenetic trend is towards suppression of the rejected channel, rather than augmentation of the attended channel.

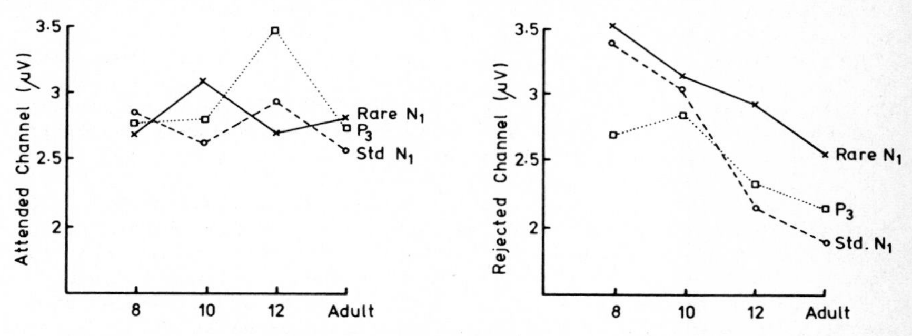

Figure 9. Age trends of N_1 and P_3 amplitude. Rejected channel shows significant linear decline, attended channel shows no significant trend over time. Thus, the emergence of N_1 and P_3 tuning reflects a net decline in amplitude in the rejected or unattended channel.

(d) Summary

The foregoing results, when added to the experiments reviewed by Hillyard and Picton (1979), yield the following list of features attributable to N_1 and P_3 tuning (Table 1). The first four items in each column of Table 1 represent those items from Hillyard and Picton's review which are not in conflict with our results. The early latency of onset of N_1 is a controversial item (cf. Näätänen and Michie, 1979) but with the high-speed stimulation employed in our studies the onset of N_1 tuning is in fact in the 50-100 msec latency range. The subsequent items were derived largely from the three studies summarized above. The conviction that N_1 and P_3 tuning are different aspects of attention is reinforced by a glance at the table: they are different on virtually all aspects reviewed.

Table 1. Features of N_1 and P_3 selection: Summary of current information.

N_1 Selection	P_3 Selection
- Simple physical cue (space, pitch, duration)	- Simple or complex cue
- Affects all stimuli in channel	- Only rare stimuli
- Early latency of onset (50-100msec)	- Late (200 msec +)
- Best with low intensity stimuli	- Independent of intensity
- Effective at all levels of target probability	- Low probability targets only
- Best when targets are frequent	- Not present with frequent targets
- Rejected channel stores information about relative frequency of previous stimuli	- Rejected channel does not store such information at this level of processing
- Gradual onset of selective tuning	- Very rapid onset; possibly preset
- Ontogenesis involves a net reduction of N_1 amplitude in the rejected channel	- Same as N_1 - net suppression of the rejected channel
- Emerges between 10 and 11 years	- Emerges prior to 6 years

This reinforces the notion that attentional selection results in two quite distinct effects on the neural processing of input. This result remains in fundamental disagreement with "upper-end" theories of attention which attribute attention to a single limited-capacity central processor (e.g. Schneider and Shiffrin, 1977; Shiffrin and Schneider, 1977). This aspect of Hillyard and Picton's (1979) theoretical position is still supported by the available data.

With regard to the mechanisms which underlie N_1 and P_3 tuning, a separate theoretical treatment of each would seem necessary, since they have such different properties. In particular the demonstration that they appear at different ages in the ontogenesis of tuning to the same task, and that they follow distinctive time courses in performance, emphasizes their independence from one another.

The N_1 Tuning Effect

(a) The definition of N_1 Tuning

Definition of the component structure of N_1 tuning has become more difficult since Näätänen and Michie's review of the field in 1979. The original definition of the N_1 tuning effect (Hillyard et al., 1973) simply identified N_1 as the largest peak in the 70-130 msec latency range, and the great majority of studies in this area have employed this criterion. In this latency range, N_1 typically peaks at the vertex or slightly anterior to the vertex (Goff et al., 1969; Picton et al., 1974). Naatanen and Michie (1979) have demonstrated that under some conditions the increased negativity resulting from selective attention does not coincide in latency or distribution with N_1, and that the N_1 tuning effect is sometimes caused by an endogenous negative wave which overlaps and summates with N_1, producing an apparent change in N_1 amplitude. Their concept is similar in principle to the "parallel late waves" known to summate in the late somatosensory evoked potential (Donald, 1972; 1976).

In his rejoinder to this claim, Hillyard (1981) cited new evidence (Hansen and Hillyard, 1980) on this point. Hansen and Hillyard confirmed that, with dc recording, and longer averaging epochs, the N_1 difference between channels did extend in time well beyond the latency of N_1. Like Näätänen and Michie (1979) they were able to differentiate at least two sub-components in the difference record, one which coincided in onset latency and distribution with the auditory N_1, and another which was longer in latency and distributed more frontally. The former was largest when the physical difference between channels was large, and the discrimination was easiest (as in these experiments), whereas the later, more frontal, component dominated when the physical difference between channels was made smaller, thus making the discrimination more difficult. Näätänen et al. (in press) have verified that, with short interstimulus intervals (250 msec) the distribution of the N_1 tuning effect coincides with that of the N_1 component itself, as recorded in a passive subject. The duration of the effect, however, extends beyond the latency of the N_1 component as elicited in a passive subject.

The existence of a longer-latency "processing negativity" in these experiments, although it provides another phenomenon to explore, does not appear to undermine completely the traditional N_1 tuning effect. The great majority of studies in the field, including this author's, employed large differences between channels (usually a combined pitch and spatial cue) to maximize the effect. Moreover, short interstimulus intervals were the rule, especially after Schwent et al., (1976a) demonstrated that the N_1 effect as traditionally defined, disappeared at longer interstimulus intervals.

In the author's laboratory, consistently utilizing stimulus parameters very close to the first Hillyard et al (1973) experiment, we have found that the peak difference between channels has corresponded in latency to the N_1 peak. However, the time constant used in our past experiments (.1 sec) would pass a 2Hz sine wave undistorted, but would attenuate very slow waves, particularly below 1Hz. Given the relatively fast rise-time of N_D, it is likely that the onset latency of N_1 tuning was accurately reflected in our results. However, the duration of the effect may have been cut back by our time constant, and our failure to find a significant N_D in longer latency ranges should be qualified by this difference in methodology. The existence of prolonged negative slow waves has been confirmed in a few of our subjects, which suggests that our recording parameters were not entirely preventing such an observation. It is possible, however, that with dc amplification a reliable longer-latency N_1 effect might have been revealed, consistent with that reported by Hansen and Hillyard (1980).

On this evidence it would appear that the N_1 tuning effect coincides in onset latency and distribution with the auditory N_1

component, but only when physical differences between channels are maximized, and when high-speed (i.e. better than 2-3Hz) stimulation rates are used. This conclusion would remain in substantial agreement with Hillyard and Picton's (1979) review, which was written before much of these data were available. However, the peak latency and duration of N_1 tuning appear to be different from the auditory N_1 component. The longer-latency, frontally-distributed slow wave, whose peak latency appears highly variable, will require further research. Meanwhile, it does not appear likely that the frontal slow wave can entirely account for the substantial literature on N_1 selectivity. It will not be possible to determine conclusively whether N_1 tuning sometimes represents a direct effect on the neuronal generators of the auditory N_1 component, or the summation of two negative components in the 100-200 msec latency range, without additional topographic studies using the high-speed selective listening paradigm. In either case N_1 tuning involves a substantial amount of endogenous control over the N_1 amplitude of negative peaks in the 100 msec range.

Another factor in defining N_1 concerns the multiple peaks which can appear between 70 and 130 msec, even in the passively-evoked auditory evoked potentials. McCallum and Curry (1980) recently addressed this problem and confirmed the existence of several "N_1 components". In their nomenclature, it appears that the effects observed in the literature on N_1 tuning are restricted to the "N_{1B}" component, which peaks at the vertex at a latency of about 110 msec. The other two components they observed peaked at temporal electrodes at 65 and 150 msec, respectively, and were generally absent or very small at the vertex electrode. The possibility that N_{1A} and N_{1C} are myogenic cannot be ruled out, given their distribution. Curry and McCallum (this volume) have carried this investigation further with Principal Components Analysis, confirming the existence of several negative auditory components in the 100 msec latency range. However, there is no reason to attribute the traditional N_1 tuning effect to either N_{1A} or N_{1C}, whose latency and distribution differ markedly from those of the N_1 tuning effect.

(b) Conditions for eliciting N_1 tuning

Reviews by Hillyard et al (1978) and Hillyard and Picton (1979) on the N_1 tuning effect concluded that the induction of a "stimulus set" in the subject results in the selective modification of N_1. Stimulus set was defined in Broadbent and Gregory's (1964) terms, referring to selection on the basis of source of stimulation, the latter defined by means of a prior physical cue such as location or pitch. The fact that N_1 superiority in the attended channel is maximized at rapid rates of stimulation, that is, roughly at rates faster than 3 per second per channel, is consistent with this interpretation (Schwent et al, 1976a). Treisman (1970) and Harvey and Treisman (1973) demonstrated that at such rates, with dichotic listening tasks, subjects found it easier to monitor one channel than both; performance deteriorated when both channels were monitored. This result is consistent with studies reported by Hink et al (1977), and Parasuraman (1978), who found that both N_1 tuning and performance were degraded when attention was split between channels.

Further support for this position came from evidence that N_1 tuning was increased when the physical differences between channel were greatest. When the spatial separation of channels was reduced or eliminated, N_1 tuning was accordingly reduced or eliminated; the same was true of pitch separation (Hansen and Hillyard, 1980; Schwent and Hillyard, 1977). This result is compatible with the stimulus set hypothesis, and contrary to what would be predicted if the psychological correlates of N_1 tuning were, for instance, information load or task difficulty. If N_1 tuning reflected task demands in direct proportion, it should have been diminished when stimuli were made more easily distinguishable. Instead, it increased.

Results reported by Donald and Little (1981) further support the stimulus set hypothesis. In their experiment, interstimulus interval was held constant, while the probability of a rare (target) stimulus was varied. The N_1 difference between channels increased significantly as targets were made more likely, that is, closer together in time (Figure 10). Thus more frequent presentation of targets had substantially the same effect on N_1 tuning as shortening the overall interstimulus interval; presumably the temporal proximity of stimuli facilitated attention to a single source.

Another aspect of the results shown in Figure 10 might support the stimulus set hypothesis. Increases in target probability improved N_1 tuning, not only for rare target stimuli, but also for standard stimuli. In fact, the difference in standard tone N_1 amplitude between channels showed a significant linear relation to target probability. Thus the improved N_1 tuning effect was not restricted to targets, but generalized to all stimuli in the channel, which is consistent with the notion of a stimulus set, generalized to all inputs sharing a similar spatial location.

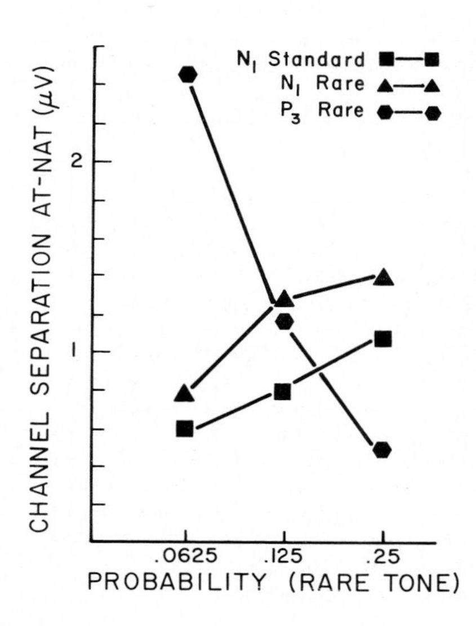

Figure 10. Channel separation, the amplitude difference between the attended (AT) and unattended (NAT) channels, as a function of the probability of a rare stimulus. Interstimulus intervals varied randomly, with the average rate of stimulation set at about 3 Hz for each channel of input. All points represent significant tuning effects except the .25 condition for the P_3 to rare tones. From Donald and Little, 1981.

The term stimulus set has a fairly precise meaning in that it describes the conditions which elicit N_1 tuning. It has no specific implications regarding the mechanism of N_1 tuning in the nervous system, or on its functional role in cognitive processing. These require separate consideration.

(c) Perceptual and cognitive correlates of N_1 tuning

An additional psychological question concerns whether N_1 tuning bears any direct relationship to perception or cognition. For instance, is it directly tied to a reduction in the effective intensity of the rejected channel? Is it a function of the relative salience, or perceptual clarity, of the attended channel? Is it related to the detection, recognition and recall of the stimulus material? Or does it reflect a physiological system whose role in perception is indirect?

There is less evidence on these points than on the issue of stimulus set, but there are a few studies which might be considered relevant. First, there is the audiometric literature on the relation of N_1 to stimulus intensity, which has been reviewed by Davis (1976) and by Picton et al (1977). The latter paper summarized the results of five studies on the relationship of the vertex N_1 - P_2 complex to stimulus intensity. From these data it can be estimated that at 55 db nHl a reduction of 40% (the average N_1 tuning effect) in the vertex negative wave translates into an effective 30 db reduction in stimulus intensity. That is, to produce a similar reduction in N_1 - P_2 amplitude, stimulus intensity would have to drop by 30 db.

This might be extrapolated to the dichotic listening condition, to suggest that the rejected channel is typically down 30 db relative to the attended channel, but there are severe difficulties in doing this. First, there are no similar parametric data available for N_1 alone, measured base-to-peak, as it was quantified in the audiometric experiments. Second, there are no similar data at short interstimulus intervals; the parametric data summarized in Picton et al (1977) were all obtained with intervals greater than one second so that the amplitude of the vertex response was much greater than in the high-speed dichotic listening experiments. The relation of amplitude to intensity may not be linear across drastically different N_1 amplitude levels. Finally, there is still some question, reviewed above, as to the generator underlying N_1 tuning, which might not be identical to that of the auditory N_1 component.

Regarding perceptual clarity, the only evidence available comes from the subjective reports of individual subjects. Subjects in the author's laboratory report that stimuli in the rejected channel are less clear than in the attended channel, although this impression could be confounded with a reduction in subjective intensity. There are no studies which have attempted an objective approach to this issue, yet improving both the intensity and clarity of perception would appear to be prime functional reasons for the existence of N_1 tuning and stimulus set.

The confidence of a subject in the detection of weak signals is correlated with N_1 amplitude and latency in a single-channel paradigm (Squires et al, 1973; Parasuraman and Beatty, 1980), suggesting that N_1 reflects the strength of the transmitted signal. In a multichannel selective listening paradigm, the superiority of the attended channel (i.e. the amplitude of N_D) is greater on those trials showing the greatest accuracy of signal detection (Schwent et al, 1976b). A divided set dilutes both N_1 tuning and performance (Hink et al, 1977, Parasuraman, 1978). In combination these results suggest that N_1 tuning reflects a change in the relative strength of the transmitted stimulus.

In several experiments carried out in this author's laboratory (Donald and Little, 1981; Donald and Young, 1982; Brooker, 1980; Broekhoven et al, in press) the method of maximum likelihood estimation has been applied to the target detection accuracy of subjects performing selective listening tasks. The advantage of the technique is that it does not require an on-line motor response by the subject, with consequent risks of time-locked readiness potentials, motor positivities, or electromyographic artifacts. Instead of responding to every stimulus, the subject simply counts targets and periodically writes down a number, his estimate of the number of targets presented in the previous trial. Any single such estimate is ambiguous, since it might be a guess. But after a sufficient number of trials (about 20-30 trials minimum, depending upon the variance), the subject's written responses can be used to estimate two parameters, π, which is the hit rate, and μ which is an estimate of guessing or inventions, including both false alarms and mis-counts. The significance of the fit of π and μ to each subject's data can also be estimated (Broekhoven et al, in press).

Figure 11 illustrates how the matrix of maximum likelihood functions can be used to derive specific π values for an individual subject. If the behavior of the subject follows a consistent strategy, the matrix will be characterized by a single negative peak, from which π can be extrapolated. A 95% confidence interval can be calculated around the peak value, to estimate the significance of the

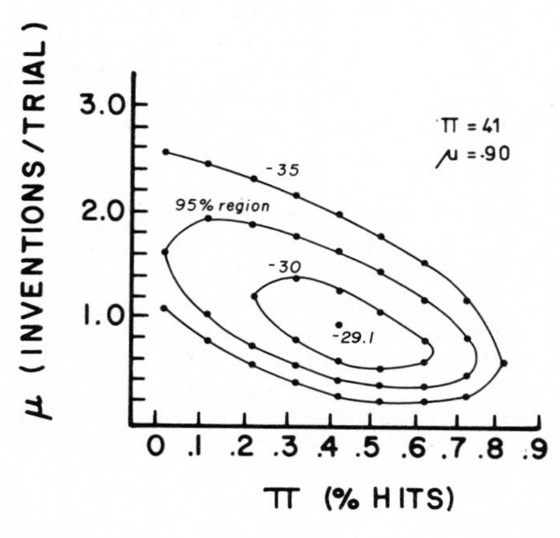

Figure 11. Matrix of maximum likelihood values for one subject, for one run, showing the confidence intervals surrounding the peak of the distribution (29.1). Extrapolation from the peak to the abscissa yields Π for this run; from the peak to the ordinate, the estimate of μ.

fit to the subject's data. The figure shows an isocontour map of maximum likelihood values; the lowest value in the matrix was 29.1, occurring at a point which represents a π value of about 41% on the abscissa, and a μ value of .90 on the ordinate. The matrix had a single negative peak, with the likelihood values increasing in concentric elipses around this value. In fact π and μ values can be estimated directly, without the use of the matrices, by means of an iterative program. Significant fits of π and μ have been made for all adult subjects (N = 26) whose performance has thus far been evaluated. Children (N = 36) are more difficult to model; but there is a systematic linear improvement in behavior, as shown by both π and μ values, as a function of age (Brooker, 1980), a result which supports the construct validity of the method. The correlation of π with conventional measures of total error is generally high, ranging above an r of .80.

The relationship of π and μ values of N_1 tuning has been examined in a few studies. The overall inter-individual correlations between π and N_1 tuning were found to be positive and significant in one study of adult subjects (Donald and Young, 1982). Correlations between μ and individual N_1 tuning parameters did not reach significance, possibly because guessing behaviour is more variable than target detection rate in this type of task. In a population of juveniles (Brooker, 1980) N_1 tuning did not correlate significantly with π or μ, but both measures of performance and N_1 tuning followed similar, but phase-lagged, developmental trends. The meaning of inter-individual correlations is not clear in this context, since individual differences in evoked response amplitude could reflect to an unknown degree many non-functional variables such as attenuation of the signal by the scalp and cranium, orientation of the generator, recovery functions, etc. Moreover, individual differences in performance accuracy were probably minimized by the use of large physical differences in stimulation between channels, which increased N_1 tuning. Within-subject relationships between behavioral and electrophysiological variables are probably a more valid test of their functional meaning.

When within-subject comparisons are made between the time course of N_1 and the time course of π, it is possible to dissociate behavioral measures from some aspects of N_1 tuning (Figure 12). Signal detection, or π improved during early trials in our studies, staying at an asymptotic level for the rest of the typical run. In

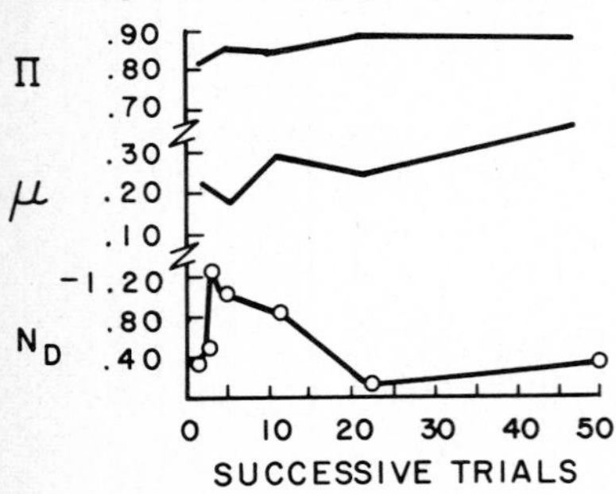

Figure 12. Comparison of the time course of N_D for standard tones and the two behavioral measures, Π (% hits) and μ (inventions or guesses), over a run of fifty 10-sec trials. Generalization of N_D to standard stimuli was a transient effect disappearing after about 20 trials.

contrast, N_1 tuning of standard stimuli peaked during the first few minutes, and then gradually disappeared during later trials, while detection accuracy remained constant. Hillyard (1981) suggested that the results of Hink et al (1978) indicate that the N_1 tuning effect for standard stimuli is robust over long periods of time. However, in that study no fine-grained time course analysis was done within any given run or trial. Had Hink et al performed a time course analysis on each run, they should have found both the onset and decay of N_1 selection over time, although the specific times of onset and decay might have differed from those found with our task parameters.

One interpretation of the temporal dissociation of N_1 tuning and task performance might be that the neural focus of attention gradually narrowed to the rare target tones with the passage of time. This would be consistent with the fact that while N_1 tuning to standard stimuli declined, N_1 tuning to rare tones showed no significant decay over time. Since estimates of the subjects' detection behavior were based solely on their detection of rare tones, N_1 tuning was not really dissociated from their detection behavior, but became more narrowly focussed upon it. This finding does illustrate that grand averages across long periods of time can leave out important information: the N_1 tuning effect was not static, but rather became more focussed over time. The behavioral dissociation was restricted to standard tone N_1 tuning, which was apparently not essential to the detection of targets in the attended channel.

In summary, the best current hypothesis would appear to be that N_1 tuning represents a change in the relative strength of the transmitted signal in the higher auditory system. The principal evidence for this ultimately rests with single-channel studies of the detection of weak signals, which leaves the chain of reasoning open to certain objections. It may be objected that the component structure of N_1 change in a single-channel paradigm may differ from that of N_1 tuning; specifically, the processing negativity has not been defined in a single-channel context. There is no evidence in print on this question, but it is possible that the N_1 observed with weak stimuli in single-channel studies is composed in part or even entirely, of the processing negativity. Conversely, it is possible that the "auditory N_1" is the common element underlying the correlation of N_1 with signal detection in various task paradigms.

It would seem reasonable to associate the relative strength of the transmitted signal with psychological constructs such as perceptual clarity and subjective intensity, but no direct evidence exists on this point either. If N_D reflects relative signal strength, it might be expected that its correlation with behavior would break down above a certain critical level, particularly in simple tasks. This appears to be the case; N_1 tuning is not correlated with reaction time (Hansen and Hillyard, 1980) or with target identification (Donald and Little, 1981) in tasks which used stimuli well above threshold, and contrasted stimulus conditions which modified N_1 tuning without corresponding effects on behavior. Correlations of N_1 tuning with behavior should only be expected when signal strength is critical to performance.

(d) Functional properties of the mechanism of N_1 tuning

Thus far, the anatomical mechanism of N_1 tuning has resisted direct investigation for several reasons. The absence of infrahuman data collected under similar conditions, and consequently the impossibility of controlled single unit recordings and lesion studies, combined with the uncertainty over the number of physiological generators underlying N_1 have left the anatomical mechanism obscure. However, there is another way to study this mechanism, and that is to define its functional properties. This can be done utilizing the same kind of logic used in experimental studies of cognition. Just as it is possible to describe many of the functional properties of human information processing by studying the factors influencing behavioral outputs, like reaction or free recall, it is possible to study evoked potentials as system outputs which reflect the functional properties of underlying anatomical mechanisms. Some of the functional properties of N_1 tuning can be inferred from the existing literature, although in its present form, the list should be regarded as tentative.

1. N_1 is a relatively slow-moving variable. With the data now available on the time course of N_1 tuning (Donald and Young, 1980; 1982) it appears that the superiority of N_1 in the attended channel emerges gradually. With a dichotic paradigm similar to the majority of studies in the literature, the average subject takes 30 to 40 seconds to establish N_1 tuning. This time course varies and our most recent data suggests it may be shorter under some circumstances. However we have not yet seen evidence suggesting that N_1 tuning is instantaneous or preset. Moreover our subjects knew in advance the properties of the assigned stimulus channel - that is, its location and pitch. It would presumably take longer to achieve N_1 tuning if subjects did not have advanced knowledge of the assigned channel. Further experiments on the variables influencing the time course of N_1 tuning will be needed to clarify these issues. The 30-40 second emergence of N_1 tuning was followed in our study by a period of some stability in which the N_1 response remained steadily higher in the attended channel. During this period, as the within-trial analysis showed, the amplitude difference between channels "carried through" the 6-second silent intervals between trials, supporting the notion of a slow-moving mechanism which did not have to re-tune on every trial. Thus both onset and decay of N_1 tuning are relatively slow-moving effects with N_D emerging gradually over a period of seconds, and remaining steady, once established, through periods of no stimulation for up to 6 seconds.

2. N_1 tuning is a quantum effect; that is it represents a constant increment in the attended channel rather than a percentage of N_1 in the unattended channel. N_D is not proportional to the absolute amplitude of N_1. In Donald and Young's (1982) results, N_1 amplitude habituated very rapidly during the first few trials in the rejected channel, then remained stable at an asymptotic level throughout the average session. N_D, that is the amplitude difference between channels, followed a different time course. Thus the moment-to-moment amplitude of N_1 was not predictive of the magnitude of N_D. Other

evidence from the same study attests to this point. A within-trial analysis of the time-course of N_1 tuning demonstrated that the difference in amplitude between the attended and rejected channels was the same for the response to the first stimulus as for subsequent stimuli in a trial, despite a threefold difference in N_1 amplitude due to rate effects. Thus a constant microvolt value was added to N_1 during any 10-second trial, regardless of its base amplitude. This is a quantum increment in voltage, rather than an amplifier effect.

At first glance this finding appears to support Näätänen and Michie's (1979) claim that N_1 tuning is the product of the summation of two events, the N_1 component and an endogenous negative wave. If the latter was fairly constant in amplitude, the variations in N_1 itself would not be expected to affect the size of the N_1 tuning effect. On the other hand, it is possible to design a filter or attenuator which adds or subtracts a constant from the total output of a system, rather than adding or subtracting a proportion of output. In neuronal circuitry this could take the form of constant amounts of inhibition or excitation, whose net effect on amplitude would not change if total system output suddenly increased. Thus the quantum nature of N_1 is compatible with either direct N_1 modification, or the notion of an endogenous processing negativity.

3. N_1 selectivity does not exclude the storage of information about unattended stimuli at this level. Thus N_1 tuning involves a bias, not a switch. The property of a switch is that it can introduce a gap in transmission in one channel, while directing information elsewhere. A bias simply alters the relative intensity of a signal, but leaves the information relatively constant across a wide range of magnitudes. N_1 is the output of a system which increases its response if it receives a low-probability stimulus, implying some form of memory response of previous stimulation. A switchlike mechanism would eliminate the registration of this information in the rejected channel, whereas an attenuator would reduce the system output of the rejected channel, leaving the probability information intact. The latter was the outcome of the Donald and Little (1981) experiment: N_1 amplitude was reduced in the unattended channel, but probability information was intact, supporting the attenuation hypothesis.

Unfortunately for the cause of simplicity, if N_1 tuning was due to an endogenous processing negativity which was generated exclusively in the attended channel, roughly the same outcome would be predicted. N_1 would continue doing whatever it was doing in response to stimulus probability, while the processing negativity would summate with it in the attended channel to produce the N_1 tuning effect. This hypothesis does not fit the outcome of the experiment perfectly: the plots of N_1 amplitude against probability in the two channels are not parallel. But the Donald and Little (1981) results cannot in themselves rule out this interpretation. In either case, information about previous stimulation is being stored in some form in the unattended channel so as to feed into the system generating N_1.

4. The N_1 difference between channels depends partly upon a decrement in N_1 in the rejected channel. The evidence for this comes from two sources. First, in the time course analysis (Donald and Young, 1980;

1982) N_1 amplitude never exceeded, or even equalled, its initial value in either the attended or rejected channels. The rapid decrement in the rejected channel appeared to be a necessary preliminary step, without which the attended channel would not have attained superiority.

This in itself might not be conclusive, but the second source of evidence confirms the importance of the amplitude decrement in the rejected channel: the developmental data reported in Brooker (1980) showed that the emergence of N_1 channel separation in children involved a net reduction of N_1 in the rejected channel (Figure 8).

This evidence confronts Näätänen and Michie's theory with a serious difficulty. If N_1 tuning resulted from the addition of a parallel endogenous negative wave to N_1 in the attended channel it would be predicted that the ontogeny of N_1 channel separation should involve a net increase in N_1 in the attended channel. On present evidence, and under the particular stimulus conditions specified in the Introduction, this does not occur. There was no age-related change in N_1 amplitude in the attended channel. One could postulate, of course, that the latter resulted from the summation of an age-related decrement in N_1 which was cancelled perfectly by the growth of the processing negativity, but this appears to be an unnecessarily complex and unlikely proposition. It cannot be argued that the processing negativity was generalized to both channels in younger children because of a complete failure of selection on all levels, since P_3 showed a selective response to the same stimuli in that same population. Thus the simplest explanation of our developmental results is to attribute N_1 tuning, at least in part, to a genuine decrement in N_1 in the unattended channel.

5. An independent incremental process in the attended channel must also be postulated. The emergence of N_1 channel separation in the time-course analysis reported by Donald and Young (1982) is most easily explained in terms of an independent process which reverses the rapid initial decrement in the attended channel (Figure 3). The incremental process might represent either a sensitization of the auditory N_1 component, or the superimposition of a processing negativity in the attended channel, or perhaps both of these, under some circumstances.

The balance between decremental and incremental processes would determine the time course of N_1 amplitude in a specific input channel. The hypothesis that the balance between these two tendencies determines N_1 amplitude is also compatible with the results of Donald and Little (1981), shown in Figure 6. In that experiment, the N_1 component to rare stimuli in the rejected channel grew much larger in amplitude with very rare (one in sixteen) stimuli, whereas this trend was considerably smaller in the attended channel. This is possibly because of a "ceiling effect", which limited the N_1 incremental process in the attended channel. One way to interpret this might be that N_1 distribution following a rare stimulus would depend on the depth of previous habituation, which was greater in the unattended channel. Otherwise why would the attended channel appear less sensitive to stimulus rarity than the rejected channel?

6. The neural focus of attention is dynamic, rather than static. The result of averaging across all the stimuli of an experimental run is to convey the impression that selective attention results in a static alteration in the evoked potential - for instance, condition A might alter N_1, condition B might alter P_3, etc. However our time-course analysis (Figures 2, 3, and 4) revealed that the evoked potentials elicited by stimuli in a given dichotic task undergo a series of changes over time. The neural focus of attention, defined as the site of alterations in the evoked potential produced by attention, systematically evolved during performance of the task, as shown in Table 2.

Table 2. Development of the neural focus

Time	Tuned component	Neural focus
0-30 sec	P_3	rares only
.5-7 min	N_1 P_3	generalized rares only
7 min +	N_1 P_3	rares only

The very first standard stimulus was not processed selectively. As far as could be determined with these techniques, the first rare stimulus showed a selective P_3 response, but no N_1 tuning; thus the neural focus of attention was restricted initially to long-latency processing of targets. Subsequently both standard and rare N_1 components were tuned - whether at the same time we cannot say - at this stage, the neural focus of attention encompassed all stimuli with the same location in auditory space as the designated target tones. In later trials, the tuning of standard stimuli dropped out, and the focus returned to the targets above, this time including both the N_1 and P_3 components.

(e) Conclusion

The initial model of N_1 tuning postulates a quantal, slow-moving, bias which alters the balance between decrementing and incrementing neuronal elements in the auditory system. The biasing mechanism is guided by the subject's ability to select channels in terms of their primary auditory features - space and pitch. Attention to source, or stimulus set, is a peculiar kind of attention precisely because it does not treat all stimulus features as equals, but favors simple physical cues (Broadbent, 1970). This strongly suggests that the elementary channel-structure of audition is maintained at a higher level in the system, and in turn, that this system maps back onto the input channels themselves. In fact, there is now anatomical evidence for several levels of highly specific cortico-thalamic feedback loops in the auditory system (Ravizza and Belmore, 1978). Conceivably "top-down" control over the auditory system could be exerted by inputs from association areas to these high-level feedback loops, which have very specific connections with the thalamus and with lower structures in the auditory system.

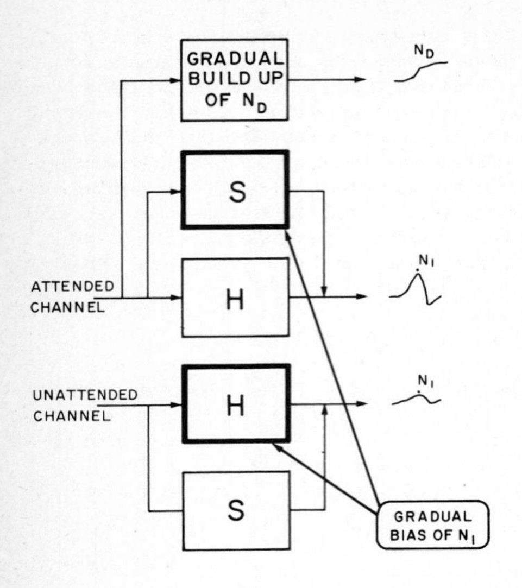

Figure 13. Schematic of the model of N_1 tuning. Tuning Effects on the auditory N_1 might involve altering the balance between habituating (H) and sensitizing (S) neurons, in the circuitry generating N_1. The "processing negativity" is generated in parallel, but only by stimuli in the attended channel of input, and is characterized by a longer latency and different distribution, from those of the auditory N_1.

In Figure 13 a schematic diagram of N_1 tuning is shown. The existence of decrementing and incrementing elements in the system is shown, as well as the existence of an independent process which can bias these elements according to the pitch and location of the input. Habituating and sensitizing neuronal units are the most obvious candidates on the independent incrementing and decrementing elements; they also fit in with Megela and Teyler's (1979) evidence that N_1 habituation and dishabituation accord with Groves and Thompson's (1970) two-process model of habituation. The evidence from the time-course analysis suggests that it may be at the level of these units that attention is acting on N_1. On evolutionary grounds it would appear more parsimonious to build attentional selection on some fundamental property of the system, such as habituation and sensitization, rather than to develop some completely new principle to achieve selection. However, the incrementing process in particular may involve some different principle, and act on a different level, from N_1 habituation.

This possibility is shown in Figure 13 as a gradual building of N_D in the attended channel, as suggested by Näätänen and Michie (1979). If this process were exclusively responsible for N_1 tuning because of temporal overlap, the habituation and dishabituation of the auditory N_1 component would occur in parallel, unaffected by attention. A more likely view might be that both kinds of explanation are partly true. At high rates of stimulation N_1 tuning may reflect a true change in the amplitude of the auditory N_1 component, and a processing negativity may occur at longer latencies as well. At slower rates (less than 2 stimuli/sec) the processing negativity may predominate in the difference record.

Recent claims of attentional influences on the auditory input in earlier latency ranges (Lukas, 1980; Curry and McCallum, this volume) have suggested that selection may alter auditory evoked responses within the first 20 msec of auditory processing. These findings conflict with the negative results of Picton and Hillyard (1974) and Woods and Hillyard (1979). The reasons for these conflicting outcomes are not yet clear. The experiment by Lukas, reporting an effect of attention on waves I and V of the BAER, compared the BAER across two successive attention conditions, rather than making simultaneous comparisons of the attended and rejected channels. Thus, his results might have been vulnerable to various nonspecific effects, such as myogenic artifact, changes in effective stimulus intensity due to alterations in neck position, etc. or changes in arousal. A between-modality paradigm was used, introducing another possible source of the difference in results. Most importantly, perhaps, Lukas used a faster rate of stimulation (10 Hz) and sine-wave, rather than click stimuli. Curry and McCallum's results also await replication; their experiments were run at a much slower rate, using free-field stimulation, and differences in procedure hinder comparisons with the present literature.

If earlier selective influences exist, they may or may not have any direct relation to N_1 tuning. To demonstrate that earlier auditory gating "explained away" N_1 changes, the earlier effects would have to share all of the properties of N_1 tuning documented above. In the absence of such evidence, and considering the present uncertainty about the existence of lower-level auditory selection, it would be premature to incorporate such a mechanism into the model of N_1 selection at this time. However, the possibility should be acknowledged that selective attention might alter processing at earlier stages under some conditions; the question is not as definitively settled as it appeared to be several years ago.

The P_3 Tuning Effect

(a) Definition of P_3 Tuning

The definition of P_3 has become as complex and difficult as the definition of N_1. The original P_3 component described by Sutton et al (1965) has been succeeded by a variety of late positive components, some of which have been verified with the use of principal components analysis (çf. Pritchard, 1981). It has not been convincingly established that all the late positive components possess the same properties. In particular, it is possible that the very late (450-900 msec latency) positive reponses elicited with certain on-line decision tasks (e.g. Kutas et al, 1977) and with highly novel stimuli (e.g. Squires et al, 1975; Courchesne et al, 1975) are outputs of mechanisms which are different from the system generating the P_{3B} component, whose average latency is shorter and less variable (Tueting, 1978; Donald, 1979). This initial discussion will not encompass the whole of the literature on late positive components, but will concentrate mostly upon the P_{3B} component elicited with the oddball paradigm.

In the auditory selective attention literature, P_3 has been elicited in most studies with the oddball paradigm, and on-line motor responses have generally been avoided. Highly novel stimuli have not been used. The observed P_3 component has consistently fallen within the 250-400 msec latency range, and its distribution has been centro-parietal, resembling that of the "classic" P_{3B} component (Squires et al, 1975). This discussion will assume that P_3 recorded under these circumstances in a multichannel listening paradigm is equivalent to P_{3B} recording in a single-channel paradigm.

Unlike N_1 selection, P_3 selectivity has two aspects. In a multichannel selection task, P_3 amplitude is much larger in the attended channel than in the rejected channel or channels, and in this it is similar to the N_1 tuning effect. However, unlike N_1, the P_3 component does not accompany every stimulus in the attended channel, being confined in most experiments to relatively rare stimuli. Thus it is doubly selective, being influenced by both the instructional set of the subject, and the rarity of the stimulus. Our initial model of P_3 selection must account for both of these features. In addition it should take into account current theories of P_3 function, but this part of the discussion will be postponed to a later section.

(b) Psychological conditions for P_3 selection: stimulus set, not response set?

Hillyard et al (1978) and Hillyard and Picton (1979) have attempted to account for the conditions leading to P_3 selection with a single psychological concept, the notion of response set (Broadbent and Gregory, 1964; Broadbent 1970, 1971). This hypothesis postulates a hierarchy of selection, involving two stages:

> "Stimuli are first selected on the basis of their simple cue (channel) characteristics, with targets and nontargets treated equivalently (as indexed by N_1). Only after this initial selection (stimulus set) are stimuli selected for their target properties.... as indexed by P_3". (Hillyard et al, 1978, p. 292).

In this theory, target stimuli are subjected to two stages of selection, with stimulus set altering the N_1 component, and the response set of the subject determining whether or not a cognitive response, and thus a P_3 component, will occur. The targets elicit additional cognitive processing, following the instructional set, and this additional processing is accompanied by a P_3 component. Hillyard (1981) has tempered his position somewhat, but still insists on a special "target selection stage" following initial selection.

The response set hypothesis is based on the assumption that it is the subject's cognitive response to the stimulus, not its rarity, that determines P_3 amplitude. It is also considered to be part of a hierarchy of selection which can be best described as a "bottom-up" model: first the source (channel) of stimulation is isolated (stimulus set) and then the target events within that channel are selected for special processing (response set).

Neither of these assumptions fits well with the results of studies recently carried out in our laboratory. The results of Donald and Little (1981) suggest that, contrary to Hillyard et al's thesis, it is the relative frequency of the target stimulus, not the cognitive response of the subject, which sometimes determines P_3 amplitude, at least in the context of a high-speed selective listening task. As target probability was increased per unit time, P_3 disappeared, even though the subject's response set did not change, and performance of the task was maintained at a high level. Data on the time-course of P_3 tuning (Donald and Young, 1982) and on the age of onset of P_3 selection (Brooker, 1980) both suggest a "top-down" hierarchy in which the later components of the neuronal response are tuned independently of, and prior to, the earlier components. In fact, it appears as though children perform these particular auditory selection tasks with P_3-type selection alone for at least 4 or 5 years before N_1 selection appears. These results suggest that the relationship between N_1 and P_3 tuning is probably much more fluid and indirect than a model such as Hillyard's would indicate.

With neither of these assumptions holding up to analysis, the logic of the response set hypothesis needs re-examination. Strictly speaking, the two-channel pure-tone discrimination paradigm which is still widely used in this field is a stimulus set paradigm, nothing more. The two or more input channels are initially discriminable on the basis of their simple physical properties - usually pitch and/or location. Differences between channels, whether in N_1 or P_3, could thus be attributed to stimulus set. Even within each channel, targ s can be singled out on the basis of predictable physical properties. In order to insist on a response set interpretation, the selection between channels should be forced to a higher level, eliminating the possibility of a stimulus set strategy. One way to do this is to insist on conditional processing of stimuli, for instance in the instruction "respond to the high tones on the left and the low tones on the right". In such a task a stimulus set cannot be adopted, and selection has to be delayed until after a full analysis of incoming features (Broadbent, 1970). However none of the dichotic selection paradigms employed thus far has forced a response set strategy, and there is no compelling reason to invoke response set as the psychological process underlying P_3 elicitation in the attended channel.

Hillyard (1981) has recognized that the paradigms used in this field of research were not forcing a response set strategy on his subjects. However, he has stated his preference for this type of interpretation because it appears compatible with single-channel experiments which have elicited P_3 components in contexts requiring a response set, such as discriminations based on word meaning or phonemic features (e.g. Kutas et al, 1977; Courchesne et al, 1977). While selection of targets may have been attributable to a "response set" in some of these studies, it does not necessarily follow either that P_3 is always contingent on that level of processing, or that subjects were adopting a response set strategy in auditory selection studies utilizing simpler discriminations.

The link between response set and within-channel target selection is very tenuous in a simple pitch or place discrimination. The definition of a target is somewhat ambiguous here: conceivably every stimulus in the series presented in the attended channel, not only the rare ones, could be treated as a target, and fully analysed, (Picton et al, 1978). Conversely, every rare stimulus might be singled out on the basis of its pitch and location, with the subject adopting a much more specific stimulus set than previously supposed (Donald and Young, 1980) and just "listening to" targets. In other words, the subject might have a response set for every stimulus in the attended channel, or a stimulus set attuned only to targets. The behavioral data give us no compelling basis on which to choose between these extremes, or to choose Hillyard's compromise position, namely that a stimulus set applies to all stimuli in the selected channel, and a response set only to the rare ones. The electrophysiological data on N_1 suggest that the subject's stimulus set is generalized to all stimuli in the attended channel, at least some of the time. However, the data do not directly support the response set hypothesis for P_3 tuning. In summary, there is no evidence to contradict the notion that stimulus set is responsible for both tuning effects, while there is evidence against the response set hypothesis for P_3.

The most conservative interpretation of the available data on auditory selection would be to attribute differences between channels to stimulus set alone for both the N_1 and P_3 components. This will be the position adopted here. It is stimulus set which is responsible for the virtual suppression of P_3 in the rejected channel, and its presence in the attended channel. Thus, stimulus set controls access to the P_3 system. The factors determining which stimuli within the attended channel elicit P_3, require further discussion.

(c) Echoic Memory, Feedforward, and P_3

Thus far our interpretation of the dichotic listening data is that stimulus set, that is, attention to the selected source of input, serves to produce a slow-moving bias of the N_1 component from that source, while also controlling access to the neuronal systems which generate P_3. Our time-course analysis suggests that it is access to the latter that is first evident upon directing attention to the source, and that N_1 tuning develops later. In this model, all stimuli in the selected channel address the system generating P_3. What factors determine which of those stimuli produce a P_3 response?

One of those factors is an event of low probability in the stimulus stream, whether a rare stimulus, or a rare omitted stimulus. Such an event is a physical property of the stimulus stream, taken as a pattern of input over time. In this sense some of the P_3 amplitude is exogenously determined. Even an omitted stimulus is a physical event, when seen in its temporal context, just as a gap in a visual pattern is part of the physical stimulus, when seen in its spatial context. Thus the two classic paradigms eliciting P_{3B} both involve relatively rare events in the stimulus stream.

To explain how stimulus rarity could affect P_3 amplitude, it is necessary to place P_3 at the output of a memory-based system, one which maintains a record of recent stimulation. One possible version of such a memory record (we will call it the echoic trace, following Neisser (1966)) would register a template of the most recent stimulus features; this template would feed back onto a comparator which was also receiving each new input. A mismatch would be registered whenever a low-probability stimulus occurred, since an echoic match to that stimulus would be unlikely to be in storage. One of the products of the mismatch would be P_3.

The echoic mismatch model can account for a good deal of the available data on P_3. It can explain the relation of P_3 to both the oddball paradigm (so long as the oddball stimulus differs on some elementary physical dimension) and the omitted stimulus paradigm originally discovered by Sutton et al (1965; 1967). It is compatible with the fact that P_3 is very large following the first stimulus in a new series (Ritter et al, 1968; Vaughan and Ritter, 1970), since, prior to the first stimulus, there would be no template or trace of previous stimulation in the memory record, and any attended stimulus gaining access to the system would elicit a mismatch response. The model also fits in with the fact that the overall rate of stimulation is inversely related to P_3 amplitude, holding the a priori probability of the stimulus constant (Fitzgerald and Picton, 1981). At higher rates of stimulation more targets would be entered into the ongoing memory record, increasing the probability that targets would occur before the echoic trace decayed, thus reducing the likelihood of a mismatch, and the P_3 component would be reduced accordingly.

The echoic mismatch model is also compatible with the results of Donald and Little (1981), who found that, as target probability increased, P_3 declined in amplitude, in spite of good performance of the task, and continued N_1 tuning. In that study the a priori probability of a rare stimulus was .25 or less, well within the probability range used to elicit P_3 in many previous studies, and the rate of stimulation was invariant, across the three probability conditions. However, the net temporal density of a rare stimulus was increased as a function of increasing its a priori probability. This would result in more targets per unit time. In the 10-second trials employed in that study, there were 32 stimuli per channel. Thus in the highest-probability (.25) condition, targets were likely to occur about once per second, while in the lowest, there was an average delay of almost 5 seconds between targets. If the duration of the echoic trace were of the order of 5 or 10 seconds, the template of the target would have been "refreshed" 5 or 10 times in the condition where P_3 was virtually absent; thus the mismatch with echoic memory would be minimized. In contrast, it would have received only one target stimulus in the other lower probability conditions, and on the average, the mismatch should be considerably greater.

The demonstration by Squires et al (1977) that the actual pattern of stimulation in the few seconds prior to a given target influences P_3 amplitude, holding a priori probability constant, also fits in with this model. Given that an echoic mismatch involves a comparison of the input with the precise pattern of recent stimulation, rather than with a long-term grand average or cumulative estimate of probability, those trials immediately preceded by the least number of targets would be expected to produce the largest P_3, and those preceded by the most would produce a smaller P_3. This was the outcome of their study. Squires et al (1977) interpreted their results in terms of the constantly-revised subjective probability, or expectancy, of the subject; but the echoic mismatch model fits their data just as well.

The echoic mismatch model, combined with the control of access to the echoic storage system by stimulus set, can account for the large amount of evidence that stimulus relevancy has an independent effect on P_3 amplitude (see Donchin et al, 1978, for a review). In most cases, the effect of the relevancy instruction could simply be attributed to attention to source or stimulus set; thus the 'ignore' condition would be equivalent to the rejected channel in the multichannel paradigms, with access to the echoic trace system reduced by directing attention to another source of stimulation. Neisser (1966) raised the possibility that access to echoic storage was under attentional control, and although there is no perfect isomorphism between the evoked potential literature and data obtained in a purely behavioral level, the same sort of model appears to be required here.

Sutton (1977) has pointed out that the effect of stimulus rarity on P_3 cannot be accounted for solely in terms of the subjective uncertainty of the observer. Experiments by Tueting et al (1970) and Friedman et al (1973) demonstrated that the effect of stimulus rarity persists even when the subject has accurate prior knowledge of the stimulus - in other words even when there is no subjective uncertainty. This result substantiates the echoic memory model. However, the effect of prior knowledge was very large in both studies, reducing P_3 amplitude by 60-70% across all levels of stimulus rarity. How can this be reconciled with an echoic memory model? Sutton has interpreted this result to mean that uncertainty has a potentiating effect on P_3, independently of the effect of stimulus rarity. While this is difficult to question, the interpretation can be turned around; perhaps prior knowledge effectively suppresses P_3 under some circumstances. In this view, an input from long or short-term memory, in the form of feedforward to the comparator, is required to mediate the effect of prior knowledge on P_3.

A distinction should be made here between the two sources of mismatch information. Feedback of the record of immediately previous stimulation onto a new input, such as proposed in the echoic mismatch model, is a function of the most recent features extracted from the physical stimulus stream, and thus is heavily determined by exogenous factors. Feedforward, on the other hand constitutes a completely endogenous readout which continuously "predicts" the sensory consequences of action, as well as the spatiotemporal patterning of passively received input. Feedforward from memory can be exquisitely detailed; for example, in listening to, or playing, a familiar piece of music the slightest variant in pitch or timing can be instantly picked up. Here, too, a mismatch occurs, but the source of the mismatch is different, and its temporal dimension is less constrained than in echoic memory.

There is evidence that P_3 is affected by learning the temporal pattern of stimulation. Donchin et al (1973) presented subjects with different sequences of stimuli possessing the same a priori probabilities, but varying in the degree of regularity. Learnable sequences, in which the stimuli alternated in some predictable order, produced smaller P_3 responses than irregular sequences which contained some element of unpredictability or unfamiliarity. These data also suggest that feedforward from memory can play a role in determining P_3 amplitude. If memory correctly predicts the input, the P_3 response is reduced; if it cannot, a mismatch is registered. The idea that both the current record of recent stimulation and memory readout must converge in error-detection is an integral part of current theories of sensorimotor learning (e.g. Schmidt, 1975).

If feedforward from memory is active in producing P_3, why did the Donald and Little (1981) results not reflect a mismatch with long-term memory readout, even when the temporal density of targets was increased? This illustrates one fundamental difference between mismatches in echoic memory, as opposed to those from long or short-term memory, at least with regard to P_3. The echoic trace, as proposed, has no structure other than the actual order of stimulation; thus a mismatch is determined by <u>the absence of a recent record</u> of the stimulus. In contrast, feedforward has an elaborate structure, and contains no literal record of prior stimulation; a mismatch consists of <u>a violation of predicted structure</u>. In the absence of structure, no prediction is made, and memory plays no role in the generation of P_3. Since, in Donald and Little's study, and in the majority of experiments in the selective listening field, the stimuli had no temporal structure, no precise prediction could be made for any given stimulus, and accurate feedforward from memory was nil. All of the amplitude variance in the attended channel P_3 was contributed by the combination of echoic mismatch features embedded in the stimulus sequence, and the stimulus set of the subject.

Donchin et al (1975, 1978) proposed a rather different view of P_3, one which emphasizes a continuous active on-line estimation of stimulus probability. These estimations generate subjective expectancies on the basis of recent stimulation. When expectancies are violated, a mismatch occurs, generating a P_3. Mismatches indicate that the model be updated, and the revised expectancy is then used to predict the next stimulus. The model of the environment thus constructed is continuously changing, and evidently involves a great deal of computation, which generates "strongly held" hypotheses about the environment. Unexpected stimuli generate "disconfirmations", and then result in "model updates" (Donchin, 1975); the latter events are accompanied by a P_3 response, which signals the disconfirmation of the hypothesis.

One difficulty with this model is the enormous computational load it would represent to a human being, as well as the inherent futility in producing such computations endlessly, long after the observer knows the stimulus sequence has no predictable structure. In contrast to the computational load imposed by such a model, the proposal put forth in this paper involves a simple controlled-access echoic mismatch system, which involves no processing space in short-term memory. The comparator in the system also receives feedforward from memory, which involves no on-line computation. The observer's conscious processing-capacity is thus usually left free to perform the task. There are particular situations where the observer may be forced to consciously estimate the likelihood of the next stimulus, and in these cases processing space may be dedicated to predicting stimuli on a moment-to-moment basis. Tueting and Sutton (1973) and Tueting et al (1971) utilized paradigms which demanded such estimates as part of the task, and under such conditions the surprise of the observer, (a violation of his prediction) affects P_3 amplitude. But such on-line computation of probabilities would appear unusual in sensorimotor behavior, and it is not even clear that prediction under such conditions involves computation of probabilities at all. The purpose it would normally serve is not obvious, particularly in situations where the sequence of stimulation lacks a coherent structure. Tueting and her colleagues used slow rates of stimulation, so that every target stimulus was rare enough to produce an echoic mismatch. Possibly it was memory feedforward that reduced the mismatch in cases of correct predictions.

The amplitude of P_3 in the attended channel is largely the product of controlled-access echoic storage system in this model. Where stored structural information about the stimulus sequence is possible, there can be some involvement of memory in determining a mismatch. In the high-speed selective listening paradigms typically used, the involvement of any on-line computation of probability is thought to be minimal. The "tuning" of the P_3 mechanism so that the rejected channel becomes relatively insensitive to echoic mismatches is attributed to stimulus set, which controls access to the comparator.

(d) P_3 and Stimulus Evaluation

It has been proposed that P_3 indexes the completion of stimulus evaluation (Kutas, McCarthy and Donchin, 1977). The evidence for this is extensive, and some of the most compelling data concern the variables which affect P_3 latency. Kutas et al (1977) and McCarthy and Donchin (1980) showed that prolonging the period of stimulus evaluation delays P_3, whereas prolonging the period of response selection does not affect P_3 latency. Gomer et al (1976), Ford et al (1979) and Adams and Collins (1978) showed that P_3 latency is prolonged by increasing the number of items to be scanned in the Sternberg paradigm. Parasuraman and Beatty (1980) showed that P_3 amplitude and latency are a function not only of the detection, but also of the correct identification of low-level auditory signals embedded in noise. The implication of all of these studies, and of several other studies reviewed by Pritchard (1981) would appear to be that P_3 is not generated until the completion of stimulus evaluation.

By way of contrast, if it indexed the <u>initiation</u> of stimulus evaluation, the latency of P_3 in particular would not vary systematically with so many cognitive paradigms designed specifically to manipulate stimulus evaluation time.

How can this evidence be reconciled with a mismatch-to-memory model of the P_3 generator? There are some similarities between the two positions. The detection of mismatches, whether with the echoic trace or with feedforward from memory, must necessarily imply some stimulus evaluation to the level involved in detecting the mismatch; and it would imply a corresponding level of memory storage. To this extent the mismatch model is compatible with the notion that P_3 occurs after stimulus evaluation.

However, there are differences in the predictions derived from the two approaches. First, there are circumstances in which stimulus evaluation is carried out and P_3 is not generated; the results of Donald and Little (1981) and of Fitzgerald and Picton (1981) are examples of this. The echoic mismatch model has no difficulty dealing with these experimental results, by postulating a rapid decay of the echoic trace, but the stimulus evaluation model cannot account for such an outcome.

Second, there are circumstances where P_3 is apparently produced by a match, rather than a mismatch. An example of such a case is the study by Parasuraman and Beatty (1980) where the largest P_3 components were produced on trials where subjects correctly identified a low-level stimulus. As their certainty about the correctness of their response declined so did P_3. The stimulus evaluation model is compatible with this result, but the echoic memory model would appear to run into some difficulty here. The echoic memory model, however, is in fact compatible with Parasuraman and Beatty's (1980) outcome. The certainty of the subject regarding the occurrence of a target stimulus might depend upon how vividly the stimulus was registered in echoic memory. At or near threshold, random variations in signal-to-noise ratios might determine entry into echoic memory. Such variations would determine P_3 amplitude indirectly, and would correlate with perceptual identification as well. Given the low rate of target occurrence in their study (about 1 target every 5 sec), as well as the low level of stimulation, the existence of an active echoic record of previous stimulation would be unlikely, and a mismatch would occur. The level of uncertainty was high - that is, precise feedforward wasn't possible. Finally, the search for a "match" to a predicted target in their study could more easily be seen as a stimulus set, producing access to the P_3 system. In short, the target stimulus must "match" the stimulus set, and produce a mismatch at the comparator; their stimuli met both criteria. The same argument applies to data reported by Thatcher (1977) and John (1977) on increases in P_3 to stimuli which fit an internal representional system.

The theory that P_3 is the output of a comparator in a controlled-access memory system is not incompatible with the idea that P_3 latency sometimes covaries with stimulus evaluation time; however the former theory is more specific about the properties of the system, and thus should prove testable in detail.

(e) Properties of the P_3 tuning mechanism

The functional properties of the P_3 selection process can be summarized as follows:

1. P_3 tuning is a fast-moving process. Whereas N_1 tuning takes time to develop and, on present evidence, is dependent upon the repetition of the stimulus, P_3 is present from the first target stimulus in the attended channel. More importantly, P_3 is not present on presentation of the first rare stimulus in the unattended or rejected channel. Note that the rare stimulus was equiprobable in either channel in the Donald and Young (1980; 1982) studies. The rare stimuli were never the very first stimuli in a series, since it was thought this would confound the P_3 component, and might elicit the "novelty" P_3 observed by Courchesne et al (1975). In retrospect this may have been an unfortunate decision since it is not clear that the rejected channel was completely excluded prior to the first standard stimulus. However, on the average, the first rare stimulus of a series occurred within 2 seconds of the onset of stimulation, and by this time P_3 had been virtually gated out in the rejected channel, while it was already very prominent in the attended channel. In contrast, N_1 was not selectively tuned until 30-40 seconds later, on average.

2. P_3 tuning reflects the action of an attentional switch capable of excluding unattended parts of the stimulus field from echoic storage. This would account for the insensitivity of P_3 to stimulus rarity in the rejected channel, as well as the very rapid exclusion of P_3 in the rejected channel at the beginning of a stimulus series. The switch is controlled by the stimulus set of the subject.

The use of the term "switch" is here intended only to convey the functional properties associated with a fast-moving device for directing information into specific channels. The neuronal device by which this is achieved could take a variety of forms, of which the most likely is probably central summation of the input with its internal representation, a device favored by Hebb (1948). Once such summation had occurred, the production of P_3 would depend upon the rarity per unit time of the stimulus, when matched to the active stimulus trace in recent echoic memory; or upon a violation of structure, where such structure was predicted by feedforward signals from memory.

3. The P_3 difference between channels depends upon a reduction in the amplitude of P_3 in the rejected channel. The developmental data (Brooker, 1980) demonstrated that, as children grew older, P_3 declined in the rejected channel, rather than showing a net increase in the attended channel. This would suggest that the comparator in the memory system proposed here matured earlier than the switching mechanism controlling access to it. Without controlled access, the comparator responded in a relatively less specific way in younger children. As the representation of stimuli in space-time became more specific, so did the summation of that representation with incoming sensory signals, and access to the P_3 comparator system. This would be a more economical solution than to propose active inhibition of all possible irrelevant sensory signals.

4. Once access to the P_3 system has been gained, P_3 amplitude in the selected channel is primarily a function of a mismatch to memory. This feature of the system accounts for most of the data on high-speed multichannel selection, as well as the single-channel work on the classic P_{3B} component. Feedforward from memory to the same comparator which receives input from echoic memory must be postulated.

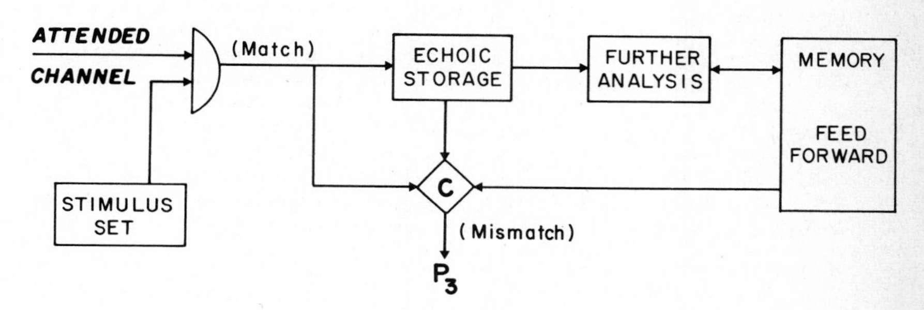

Figure 14. Schematic of the model of P_3 tuning. Two mechanisms are required: (1) Control of access to memory by the subject's stimulus set, and (2) a comparator system (C) which compares the input with the contents of a rapidly-decaying echoic storage system, and with the feedforward signal, which may originate in various levels of memory. To generate a P_3, a stimulus must gain access to the system, i.e. match the stimulus set; however, gaining access will not in itself generate a P_3 unless a mismatch is registered at the comparator, against either the recent echoic record, or against the feedforward signal.

(f) Conclusion

The main features of the P_3 selection model are summarized in Figure 14. The two primary elements are a comparator which compares the input to the echoic record of the previous 5-10 seconds (approximately) of input, and which also compares the input to feedforward from memory; and a rapid switch-like mechanism which controls access to the comparator. The independence of these two elements is an important feature: it is assumed that all stimuli selected by the subject's stimulus set address the comparator system, but of these, only mismatches to memory elicit P_3.

This model is an attempt to synthesize most of the available information on P_3 and P_3 tuning. It has taken into account both the omitted stimulus paradigm, the oddball paradigm, and experimental designs which manipulate subjective uncertainty independently of stimulus probability. It can account for the sensitivity of P_3 to target density per unit time, independently of overall stimulation rate. It can also account for the ability of the nervous system to restrict access to the P_3 system by attentional selection when several competing sensory channels are active. Finally, it can account for the serial effects of stimulus sequences, including the effect of learning a repetitive sequential pattern, on P_3 amplitude.

The model also allows for an account of an apparent anomaly, that P_3 can be elicited by rare nonsignal stimuli in the attended channel (Roth et al, 1976; Courchesne et al, 1975, 1977). These would also constitute an echoic mismatch, and so long as the rare nonsignals gained access to the memory comparator they would elicit P_3. Presumably fine-tuning of the access mechanism (i.e. a more specific stimulus set) could block P_3 components to rare nonsignals, although this has not yet been tested experimentally. Further specification of the parameters which can be used to control access to the P_3 component system should constitute an important direction for future research.

Combined Model

The distinctive properties of N_1 and P_3 tuning lead to different models of the systems underlying them, but both tuning effects should also be examined in the context of the supraordinate systems controlling attention. What are the properties of the systems which control access to the P_3 system, and which produce the bias observed in N_1 tuning?

The systems controlling both the N_1 tuning effect and access to the P_3 system are intelligent in the sense that they utilize a knowledge of the relevant features of stimulation and their location in space and time. They are also programmable by means of language, which is evident in the many experiments with human subjects in which instructions are delivered verbally. This does not imply that the N_1 effect or P_3 are particularly high-level responses in the nervous system; it only implies that they are influenced by complex respresentational systems.

An important factor in defining the nature of N_1 and P_3 tuning might be the familiarity of the observer with the experimental environment. The typical dichotic listening experiment allows a period of familiarization prior to recording, in which the subject learns the location, pitch and variability of the stimuli to be employed. Once familiar with the stimulus environment, the subject is asked to focus on one input channel, which is defined by specifying one or several of the stimulus dimensions he has registered. At some later stage of the experiment he is asked to move his focus to another channel: in other words, to reconfigure his set according to a new list of features. The verbal instructions must have a referent to be executed. Therefore this type of control requires prior knowledge of the environment, and the ability to place or fix each new input into an established reference system.

A reference system in which objects are represented in their current spatial and temporal configuration is sometimes called a cognitive map. O'Keefe and Nadel (1977) have assembled persuasive evidence that such a mapping system must exist, that it integrates information from all major sensory modalities, and that it is closely tied to locomotion in three dimensional space. Even without supporting the anatomical hypothesis in their proposal, that places the hippocampus at the center of the cognitive mapping system, at least in the cat, a concept similar to theirs would have to be invented to contain economically what an attentional guidance system must know about the environment. A cognitive map would provide the equivalent of a set of coordinates by means of which inputs to the sensory systems could be placed into a stable spatial and temporal context. One consequence of this would be that a prior set towards a particular input channel would become possible, with a sufficiently high degree of specificity to permit the kind of within-modality tuning observed in both the N_1 and P_3 effects. The cognitive map, tied closely to the actual environment, would also serve to mediate the effects of more remote representational systems, such as language, on the processing of input.

The flow of attentional control in such a system is illustrated in Figure 15. The chain of command is top-down: the verbal instructions serve to establish a set which preselects a target item contained within the current cognitive map. Control over the processing of new data in the sensory field would originate in the mapping system, rather than at a higher level.

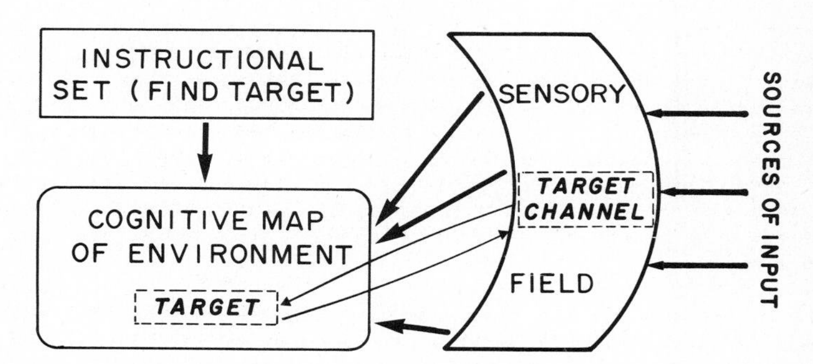

Figure 15. Schematic utilizing the concept of a cognitive map to mediate the effect of an instructional set on the processing of input. Once a set is established, the amount of information to be processed on a single trial is greatly reduced, and the involvement of higher levels of processing is minimized.

One of the properties of such a system is that, once a set is established, the amount of processing required to identify a target stimulus is greatly reduced. In effect, most of the properties of the environment are known and classified in advance, and the identification of a target may involve no more than a few confirmatory bits of information. The concept of stimulus evaluation is thus simplified. There would be no need to perform an exhaustive analysis of a familiar stimulus, and the information load imposed by most repetitive listening tasks would be minimal.

Although the cognitive map, or a facsimile thereof, might provide a coordinate system for attentional selection, it could not constitute the source of selection. The cognitive map must contain potentially relevant, but momentarily unattended, features of the environment. It is the internal space within which the attentional focus can move, but selection would be controlled from without, from higher representational systems.

Referring back to our discussion of N_1 and P_3 tuning in selective listening, we specified two kinds of selection on the basis of presently available data: (1) an instantaneous control of access to echoic and short-term memory, and (2) a gradual change in the relative strength of signals transmitted in the selected channel. The first type of selection is manifest in the tuning of the P_3 system, and the second in the tuning of N_1, regardless of whether the latter truly represents a change in the auditory N_1 component.

If the cognitive map is cast in the role of mediator of both tuning effects, the process of tuning could be described in the following way. Items in all incoming channels are analysed with reference to the cognitive map. Any item which does not fit with the current version would be disruptive and require a complete system update, in the manner of a program interrupt signal. During a typical experiment, there would be no such disruption, except through accident, since the subject is familiar with the environment and stimulus variability is highly constrained. Certain items, selected with reference to the map, gain access to echoic memory and higher levels of processing. Gradually, processing of all stimuli in the selected channel becomes tuned at a lower level, and this tuning occurs under the guidance of the reference system provided by the cognitive map. A corollary of this model is that the system could only reject what it could model, i.e. stimuli already included in the cognitive map.

The notion that the cognitive map might serve as a gate to memory storage is compatible with Olton's (1980) view of the hippocampus as an essential structure in the animal's working memory, and with extensive evidence on the importance of the hippocampus in human memory storage. The coincidence that the same anatomical structures are implicated in both cognitive mapping and memory storage strongly suggests that the two functions are closely interconnected. When it is added that P_3 appears to peak in amplitude in the region of the hippocampus, that it is absent in H.M., the hippocampectomized case extensively documented by Milner (Squires et al, this volume), and that on other grounds P_3 appears to have a special relationship to memory (Donald, 1980), it appears feasible that P_3 tuning may reflect the control of access to memory by a system in which the hippocampus plays a major role.

The tuning of lower structures, presumably reflected in the N_1 tuning effect, would require that the cognitive mapping system possess reciprocal connections with the sensory systems. For instance, if the subject can increase transmitted signal strength at a specific location in auditory space, then the cognitive map must project back onto the auditory system in a spatially-specific manner. The same would have to hold for any stimulus dimension which could serve as a stimulus set. Although no direct evidence is available on this question, it is worth reviewing the known anatomical pathways by which neocortical regions might influence the activity of the auditory system. These are examined in Figure 16. The direct pathway from the ear is illustrated separately from the indirect pathways, whose inputs do not come directly from the ear. No structure lower than the colliculi is shown, due to the absence of documented paths from cerebral cortex to either the cochlear nucleus or the superior olivary complex. In the direct path in the cat auditory system, neurons in primary cortex (A1) receive inputs specific to layers III and IV, which are tonotopically organized; reciprocal projections exit from layer VI and return to the same neurons in the ventral medial geniculate nucleus (MGNv), with the same tonotopic organization. Similar reciprocal paths exist in the indirect path, between AII and the dorsal and medial MGN. In addition, several cortical areas project to the inferior and superior colliculi and the MGN, and share reciprocal projections with the hippocampal formation.

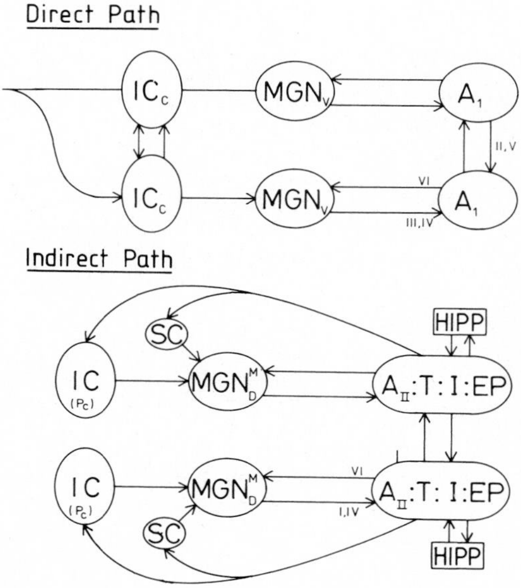

Figure 16. Some corticofugal paths which might influence processing in the auditory system. There are specific tonotopic projections from auditory cortex (AI and AII) to the medial geniculate (MGN). The inferior (IC) and superior (SC) colliculi receive projections from temporal (T) insular (I) and post. ectosylvian (EP) cortex, as well as from AII. HIPP: hippocampus. Adapted from Ravizza & Belmore, 1978.

Although the function of these corticofugal projections is not understood, their existence points to the possibility of higher cortical control over activity in the auditory system. This picture leaves out other possible sources of attentional influence, for instance through the midline thalamic nuclei, which are not cortical in origin but which might also exert an influence over the processing of inputs (Skinner and Yingling, 1978). The existence of these pathways suggests that ample opportunity exists for cortical association areas and secondary regions to modify auditory processing at collicular and post-collicular levels. It remains to be seen whether pitch- and location-specific effects could be mediated through these paths.

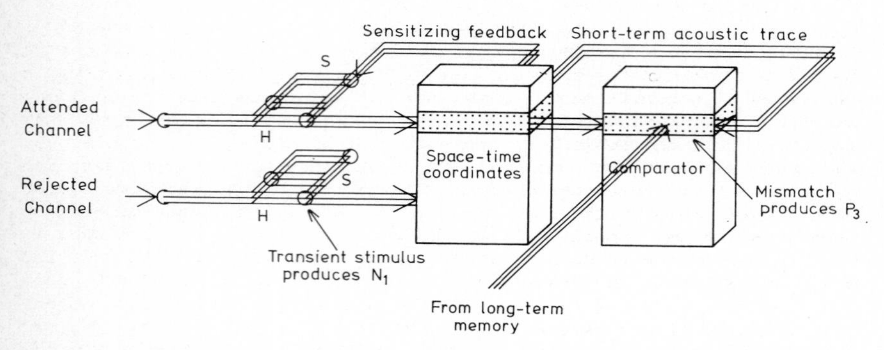

Figure 17. Summary diagram indicating how the space-time coordinates set up in the cognitive map might both control access to short-term acoustic storage, and provide sensitizing feedback to gradually improve signal transmission in the attended channel.

Figure 17 attempts to illustrate how a coordinate system set up for a particular experimental situation by the process of cognitive mapping might control access to short-term acoustic storage, including echoic memory and more elaborate forms of auditory short-term memory. The same system must also exert control over N_1 tuning, in the form of sensitizing feedback which is allocated on the basis of preselected stimulus properties. The feedback thus provided might directly alter the relations between habituating (H) and sensitizing (S) neurons in the auditory system, or might produce an independent output (N_D).

The reasons for placing both types of tuning under the control of the same system derive from our discussion in previous sections of this paper, of the conditions eliciting N_1 and P_3 tuning. Both types of tuning appear to be related to adopting a stimulus set, and under appropriate conditions, a single verbal instruction will lead to both types of tuning. This suggests that a single supraordinate system is in control, although on a lower level N_1 and P_3 tuning are mediated by different mechanisms.

If a single mechanism is in control why are the two types of tuning independently variable? The existence of N_1 tuning doesn't always imply P_3 tuning, and vice-versa. On the present evidence, the reason for this lies largely in the rate-dependence of N_1 tuning, and in its different time course. N_1 tuning appears to increase directly with the rate of target presentation, i.e., as the selected input approximates continuous stimulation. This may also be true of the access mechanism controlling P_3 tuning, but P_3 cannot be recorded in high-rate paradigms because of other features of the P_3 mechanism, particularly the tendency for an echoic match to squelch the P_3 response. The different time course of N_1 tuning might, in a similar manner, reflect the properties of the N_1 tuning system itself, rather than those of the supraordinate system proposed in this section.

In summary, the picture of auditory attention that is gradually emerging from this research is one of a fluid, complex process that varies in its neural manifestations with the stimulation conditions and processing demands of a particular experiment, as well as with the amount of time the subject has been listening to the selected channel. No model of attention with these properties has emerged from the purely cognitive research on this question, and it does not appear possible at this time to simply adopt a pre-existing theoretical framework from cognitive psychology, although ultimately any comprehensive theory of attention should take into account both the neurophysiological and behavioral data which become available.

The type of supraordinate system proposed to control N_1 and P_3 tuning has some points of agreement with the "top-down" model of attentional control recently proposed by Treisman (1980), but it is not yet known whether the control of N_1 and P_3 tuning is as flexible as her model would suggest it should be. The details of the N_1 and P_3 systems themselves have no parallel in cognitive psychology, and cannot be tested without recourse to electrophysiological recording. The behavioral evidence, and the models based on it, from Broadbent's early model (1958) to more recent attempts such as those of Schneider and Shiffrin (1977), appear to deal largely with the properties of the upper end of the system, without providing specific predictions about how selection acts on the neural processing of inputs. Electrophysiological evidence will allow us to describe the effects of attentional control in more detail than previously possible, and hopefully to unify the two levels of inquiry.

Acknowledgement

The author gratefully acknowledges support received from the National Science and Engineering Research Council of Canada, and from the Department of Psychology, University College, London; the helpful comments of the editors, and of Drs. R. Näätänen, D.E. Broadbent, and M. Rugg; and the help of K. Jackson and P. Gage, in the preparation of this paper. Address requests for reprints to Dr. Merlin W. Donald, Department of Psychology, Queen's University, Kingston, Ontario, Canada, K7L 3N6.

Tutorials in ERP Research: Endogenous Components
A.W.K. Gaillard and W. Ritter (eds.)

4

THE PRINCIPAL COMPONENTS OF AUDITORY TARGET DETECTION

S.H.Curry, R. Cooper, W.C.McCallum, P.V.Pocock,
D. Papakostopoulos, S. Skidmore and P. Newton

Burden Neurological Institute
Bristol, England

In recent years the application of principal component analysis (PCA) to brain event-related potential (ERP) data has come into vogue. Much recent work has employed PCA as a primary analytic tool (Donchin et al., 1975; McCarthy and Donchin, 1976; K.C.Squires et al., 1977; Ruchkin et al., 1980a,b; Sanquist et al., 1980). There seem to be several benefits to be obtained by the application of PCA and related techniques to ERP data. The first is objectively to reduce the usual vast quantity of data to a more reasonable size by reducing the dimensionality in such a way that the salient features of the data are retained. A typical ERP study can generate several thousand waveforms which are each composed of from 64 to 1024 points. PCA attempts to describe the underlying structure of this sort of large data set in terms of relatively few "basic waveforms". These 'basic waveforms' are also variously called the 'principal components', the 'factor loadings' or the 'factors'. The 'basic waveforms' are computationally determined from the cross-products, covariance or correlation matrix of the data points in such a way that the first 'basic waveform' accounts for the most variance and all subsequent waveforms account for the largest amount of residual variance in the data. Therefore the 'basic waveforms' are orthogonal and represent independent dimensions of the data. This ability of PCA to map independent sources of variance has allowed experimenters to separate temporally and spatially overlapping brain processes such as the CNV, P300 and various slow waves (McCallum and Curry, 1981; Ruchkin et al., 1980a,b). After extraction it is the usual procedure to 'rotate' the 'basic waveforms' to facilitate interpretation. This is most commonly done using the 'varimax' procedure which attempts to simplify the internal structure of the 'basic waveforms' at the same time retaining the lack of correlation between 'waveforms'. After extracting and rotating the 'basic waves' each of the original data waveforms can be quantified in terms of its similarity to each of the rotated 'basic waves'. These quantities, variously called weighting coefficients or factor scores, can replace the original much larger set of variables (time points). The weighting coefficients are an unbiased metric that can be used in additional analyses to examine the known sources of variance (electrode, condition, stimuli, etc.) in the original data set.

As with any other technique PCA has its limitations and problems. The most obvious limitation is that the latency of any of the data features should be relatively constant. It is not quite clear how much latency jitter can be tolerated before misleading factors are produced. The point at which latency shifting could introduce a fallacious component is no doubt related to the frequency of the particular component and the scalp latency distribution. With the sort of within-condition latency variability that we have seen across the scalp PCA seems only to produce a slightly broader factor. Another of the problems with PCA is that it is often confounding sources of variance. If the PCA was performed across subjects, electrodes, conditions, and types of stimuli it is important to examine the weighting coefficients for each of these dimensions to determine what the 'basic waveforms' are representing. A further problem with PCA is that the resulting set of 'basic waveforms' are likely to differ notably depending upon the matrix (cross-products, covariance or correlation) that was originally factored. These differences have not been systematically documented with the same data sets. Furthermore although the principal components are themselves unique, the varieties of rotation are infinite. Therefore there are theoretically an unlimited number of sets of 'basic waves' to describe any one data base. It is not documented what differences in the 'basic waveforms' one should expect with various rotations.

To examine critically both the usefulness and limitations of PCA in the analysis and interpretation of ERP data we have recently applied various principal component analyses to several large data sets. In this chapter we will present summary results from two different auditory 'target detection' tasks which were both similarly analysed with PCA. We will describe the two studies and their numerous PCA results separately and at the end attempt to compare and contrast these.[1]

EXPERIMENT I

The first experiment was designed to examine auditory 'target detection' in a free-field localization paradigm. The subjects were fixed in the centre of a sound-dampened room and presented with a pseudo-random sequence of 30 msec 86 db SPl 'white noise' bursts from speakers placed at the front,back, left or right. For each run the sequence consisted of a different string of 32 events which was continuously repeated until the required number of trials were collected (40 correctly responded targets). The randomness of the sequence was compromised by the desire to have successive rare stimuli separated by as variable an interval as possible (0 to 9 intervening events) using a relatively short repeatable sequence. The ISI was a constant 1.2 seconds. In any one condition the subject's task was to detect the occurrence of the sound from the designated target location. Detection was indicated by an 'as rapid as possible' right-hand button press. Each subject served in 8 conditions - twice for each location designated as target. In half the conditions the subjects faced one way and in the other half they were rotated 180 degrees. This rotation was to counterbalance the possible influence of a rectangular room on sound localization cues. The probability structure was .1 for 'targets', .1 for sounds from the 180 degree placed sound source ('rare non-targets') and .4 for each of the orthogonal speakers ('frequent non-targets'). This structure ensured that differences due to rarity could be separated from those due to target selection.

24 normal males between the ages of 17 and 30 served as paid subjects. ERP data was collected from Ag/AgCl electrodes placed at FPz, F7, Fz, F8, T3, Cz, T4, T5, Pz, T6 and Oz each referred to a balanced non-cephalic electrode pair (Stephenson and Gibbs, 1951). The recording characteristics were 50 microvolt/cm gain, 1.2 second time constant and upper frequency cut-off at 2.2 KHz. The output of the recorder was interfaced to two PDP-12 computers to sample simultaneously the data at two different rates to allow adequate resolution for both the early (up to 100 msec) and late (up

to 700 msec) components. Computer 1 sampled all 12 channels at a rate of .5 msec/point for an epoch length of 128 msec. This data set will be referred to as the 'fast' set due to its fast sampling rate. The second computer sampled in parallel all channels for 768 msec at 3 msec/point. This data will be referred to as the 'slow' set. On both time bases separate averages were computed on-line for each of the stimuli in each condition, producing 32 12-channel averages for each subject on each time base. Trials on which there was significant eye movement or behavioural error were automatically excluded from the average. Reaction times were recorded for all 'target' trials.

PERFORMANCE DATA

Table I

PERFORMANCE DATA

	MISSED TARGETS	FALSE ALARMS	RT	RTSD
FRONT	19.0	8.0	439	102
BACK	22.5	8.4	459	100
LEFT	1.3	.3	362	78
RIGHT	1.4	.2	360	77
F-VALUE	13.4	17.8	25.3	16.2
PROB	≤.001	≤.001	≤.001	≤.001

Table I presents the mean performance measures for 'target' stimuli from each location. The values in the columns labelled 'missed targets' and 'false alarms' are the average number of such errors in each run. As can be seen the subjects had more difficulty detecting 'targets' from either the front or back than from left or right. There were no significant differences between either front and back or left and right performance measures. Quite clearly 'target' detection and hence sound localization were much more difficult for sounds in the anterior-posterior dimension.

'FAST'DATA RESULTS

Figure 1 illustrates the 'fast' ERP data averaged across the 24 subjects for all 'targets', 'rare non-targets' and all 'frequent non-targets'. These averages are constructed over all conditions and hence all speaker locations. On the midline the major peaks are clearly visible and can be identified as N19, P26, N33, N47, P55, N80 and the ascending limb of the N107. Both the N80 and the N107 are parts of the so-called N1 or N100 response. From the figure it is obvious that the later part of the N1 complex shows marked amplitude differences dependent upon the class of stimuli - whether they were 'targets', 'rare non-targets' or 'frequent non-targets'. Amplitude is smallest to 'frequent non-targets', intermediate to 'rare non-targets' and largest to 'targets'. This is the much reported N1, or early negative enhancement with selective attention (Hillyard et al., 1973; see review by Näätänen and Michie, 1979). It should be stressed that it is not only the ERPs to 'target' stimuli that

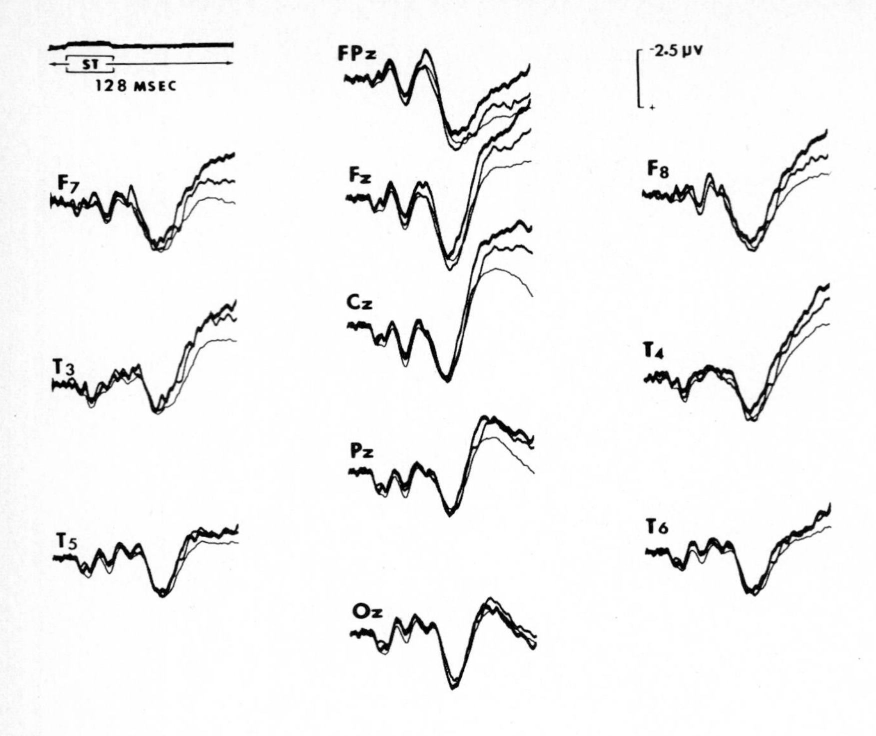

Figure 1
Grand-averaged ERP data for 'targets' (thick), 'rares' (intermediate) and 'frequents' (thin). Negativity in this and all other illustrations is up.

display the N1 amplitude augmentation; the 'rare non-targets' produce a similar but smaller negative enhancement. It should be noted that at electrode locations such as Oz, T5 or T6 that clearly display only the earlier negativity - the N80, the amplitude enhancement with 'target' selection is minimal. This is another demonstration of the existence of multiple N1 peaks having differing distributions and sensitivities to task demand (McCallum and Curry, 1980). An examination of the frontal traces reveals that the effect of directed attention is influencing components earlier than the N1 complex. P55 is clearly smallest for 'targets'. N33 and N47 are both largest for 'targets'. The P26 peak is of smallest amplitude in the 'target' ERPs. In fact, although it is difficult to see, N19 is also slightly largest for 'targets' over the frontal location. These small but consistent amplitude differences are illustrated more clearly in figure 2. It is interesting to note that the amplitude differences due to stimulus class are all maximal over the frontal and pre-frontal areas for at least the first 60 msec.

The central point here is that the cerebral processes subserving stimulus selection seem to manifest themselves, at least in this situation, much earlier than the usually quoted 60 to 70 msec. A caution must of course be inserted about the origins of some of these early components. Several

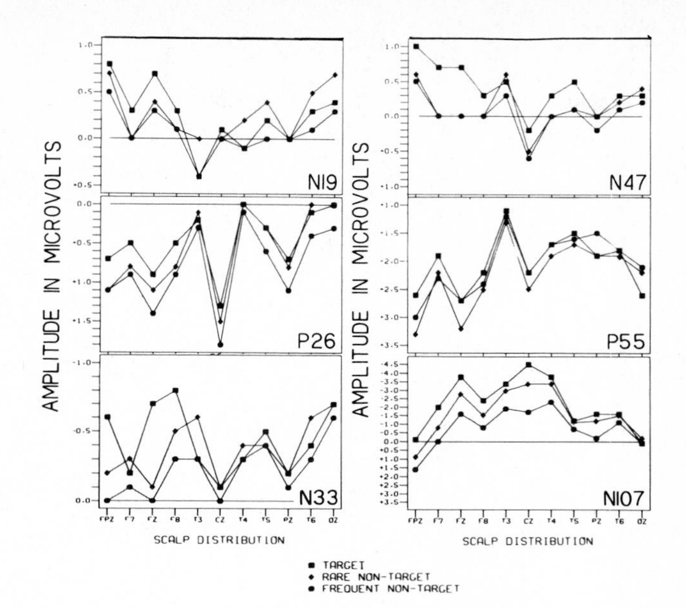

Figure 2
Average amplitudes of indicated peaks as a function of electrode and stimulus class.

have been previously reported as myogenic (Goff et al., 1977; Bickford et al., 1964). Even if this were the case - which seems doubtful for all but the N33 - one is still left with the need to explain the selective enhancement of myogenic responses as early as 19 msec post-stimulus.

To determine the nature and onset of this early attention effect difference waveforms were constructed by subtracting on a point-by-point basis one waveform from another. The resulting waveform has the common or obligatory components of the ERP removed. Difference waveforms were constructed for 'targets' and 'frequents', 'rares' and 'frequents' and 'target' and 'rares'. The difference waveforms were calculated separately for the ERP data from the front or back (F/B) and from the left or right (L/R). Figure 3 presents these difference waveforms for the midline electrodes. In the left column are the waveforms resulting from the subtraction of the 'frequent' from the 'target' data. The solid lines illustrate the class differences when the sounds were from either the left or right. The dashed lines illustrate the similar difference waveforms for sounds from either the front or back. The first point is that the class differences appear to be due to a sustained shift as much as to changes in the series of peaks and troughs. This shift has an onset as early as 10 msec in some of the channels. Secondly the differences between 'target' and 'frequent' ERPs are larger for L/R stimuli. This is most marked over the 50 to 90 msec period. In the middle column the difference waveforms between the ERPs to 'rare' and 'frequent' non-targets are displayed. For L/R stimuli the effect of rarity seems to manifest itself as a sustained negativity over the 50-110 msec period. In contrast the difference waveforms for the F/B stimuli clearly show both an early and late shift similar to that seen for 'target'. In this instance

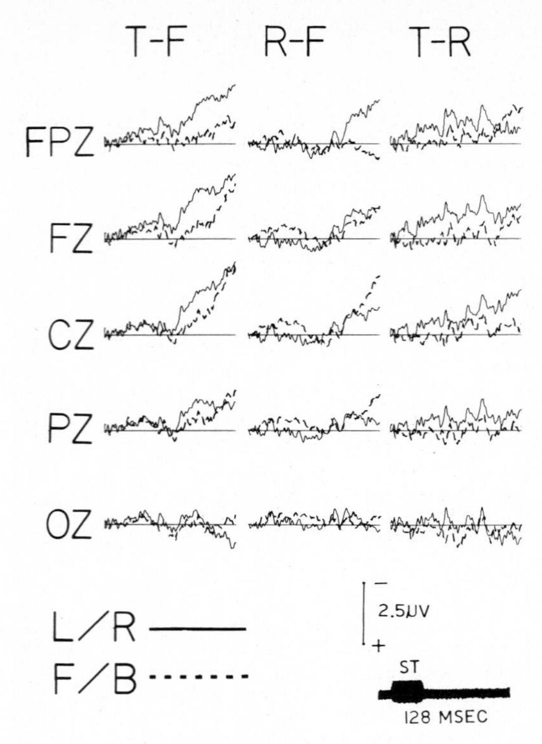

Figure 3
Midline difference waveforms obtained by subtracting the ERPs to one class of stimuli from those to another (see text).

the later shift is not readily distinguishable from an N1 amplitude increment. The right column, presenting the difference waves obtained between 'target' and 'rare' ERPs, merely reinforces what was previously observed. For L/R stimuli the target effect consists of a sustained shift with a very early onset. In contrast there are really no consistent differences between the ERPs to 'rare' and 'target' stimuli from the F/B orientations.

These results are interesting in 3 ways. First of all they suggest that the onset time for attention related ERP effects may be around 10 msec. Secondly the different difference waveforms for F/B and L/R ERPs indicate a strong interaction between stimulus class and plane of localization. The lack of differentiation between the F/B 'targets' and 'rares' ERPs suggests that they were indistinguishable over that period of time. The performance data would suggest that this probably was the case. Thirdly, the sustained negative shift observed in this study is at least superficially similar to both the 'processing negativity' of Näätänen (Näätänen et al., 1978; Näätänen and Michie, 1979) and the Nd shift of Hillyard (1981). However both the 'processing negativity' and the Nd are reported to have a much later onset than 10 msec.

FAST PCA RESULTS

To determine the underlying structure of the 'fast' data and to attempt to confirm the supposition that some of the observed amplitude differences were due to the superimposition of one or more sustained shifts, this data set was subjected to PCA. Prior to PCA each of the original 256 point waveforms were reduced to 85 points by 3-point averaging. The PCA was performed on the covariance matrix of the total data set of 8448 waveforms (24 subjects x 8 conditions x 4 locations x 11 electrodes). The PCA produced 8 interpretable 'basic waveforms' or 'factors' which accounted for 84% of the total variance. These 8 obtained basic waves were then rotated using the varimax procedure. Factor scores (weighting coefficients) were calculated for each waveform and stored for later analyses.

Figure 4 presents the 8 rotated basic waveforms produced by the PCA and illustrates the variation of the factor scores by electrode and by class of stimuli. The factors are presented in the order in which they were extracted (1-8). The "polarity" of the factors has been altered as necessary to keep with the convention of 'negative up'. The label to the left of each trace indicates the data feature(s) most closely corresponding to the basic waveform. The figure below the trace indicates the proportion of the total variance accounted for by that basic waveform. The first basic wave primarily corresponds to the N107 peak in the data. It should be noted that there is a consistent low level of loading from about 10 msec. This suggests that part of the sustained shift observed in the data is inseparable from the later N1 peak. The distribution of the weighting coefficients is primarily coronal extending anteriorly to Fz. This corresponds well with the amplitude distribution illustrated in figure 2. The rightmost column presents the mean factor score values for each class of stimulus ('target', 'rare' and 'frequent') pooled across all subjects, electrodes and locations. For the first basic wave the scores are significantly ($p \leq .0005$) more negative (equivalent to larger amplitude for a negative factor) for the 'targets' and 'rares' than for the 'frequent'. Factor 2 - the second basic waveform appears to represent the P55 peak. The weighting coefficient and amplitude distributions are similar in shape and both display a clear Fz maximum. Factor 2 was not significantly related to stimulus class. The third basic waveform quite clearly corresponds to the N80 data peak. The distribution of this factor is central-posterior midline. This corresponds well with what was observed in the data. It is supportive of our contention as to the multiplicity of N1 peaks that the analysis pulled out two independent factors peaking over the usual N1 bracket of time. Factor 3 was significantly ($p \leq .001$) more prominent for 'targets' than for 'non-targets'. The fourth factor corresponds to the N47 peak. The distributions of both the weighting coefficients and amplitude measures are fronto-temporal. This 'basic wave' is significantly ($p \leq .005$) less prominent for 'rares' than for the other 2 classes. Basic wave 5 can be equated with the N33 data peak. The flat distributions of both the amplitude and factor scores suggest that this component may be myogenic. Furthermore the weighting coefficients are not at all related to stimulus class. Factor 6 corresponds primarily with the low amplitude N19 data feature. The frontal distributions of both the amplitudes and weighting coefficients are similar. In addition to the N19 peak this 6th basic waveform loads on all time points across the epoch, confirming the presence of a small sustained negative shift. This 'basic waveform' is significantly ($p \leq .0002$) more prominent for rare stimuli ('targets' and 'non-targets') than for frequents. P26 came out as the main feature of basic wave 7. Both

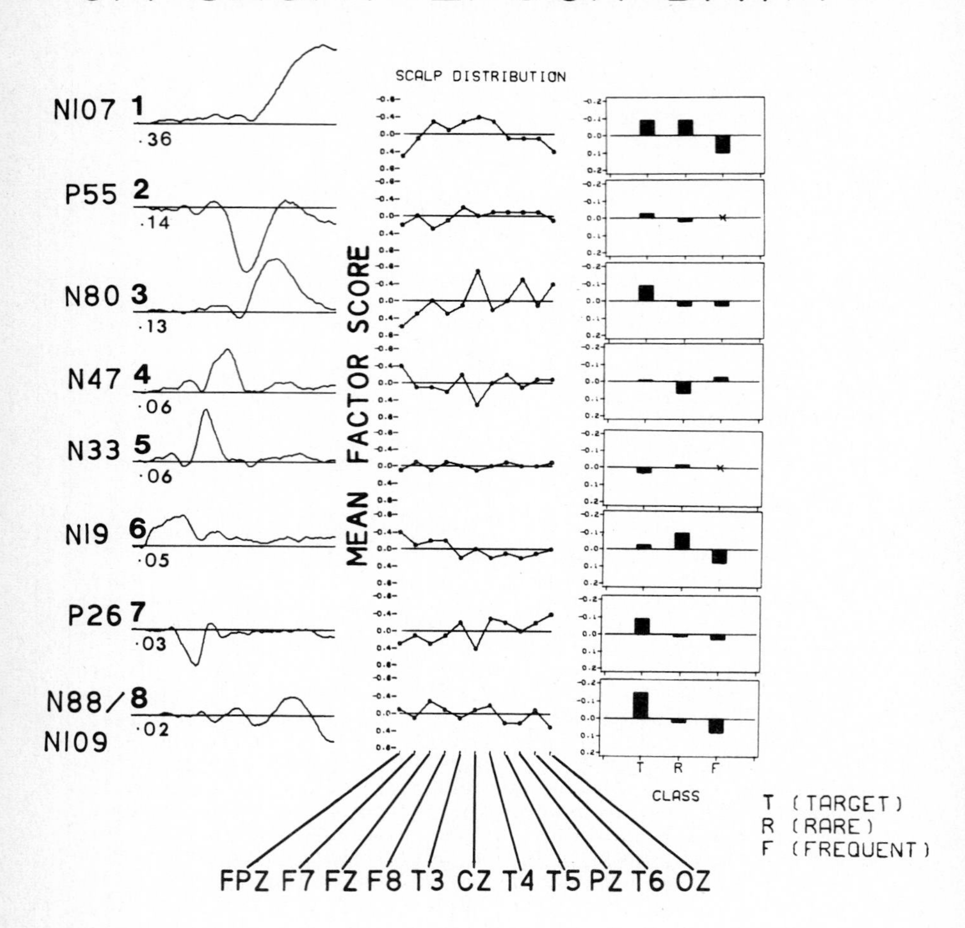

Figure 4
Results of the PCA of the 128 msec data. In the left column are the 8 rotated basic waves. The scalp distributions and task sensitivities are illustrated in the next two columns.

the amplitude and weighting coefficient distributions have a clear fronto-central distribution with Cz maximum. The P26 factor is significantly ($p \leq .003$) less prominent for 'targets' than for 'non-targets'. The last basic waveform is more complex than the preceding ones. It is basically an inversely related early and late N1 factor. This is primarily a distributional feature of the N1 complex. At the posterior midline electrodes the N1 peaks early and then falls off rapidly. This factor is also related to class being significantly ($p \leq .0004$) more prominent to

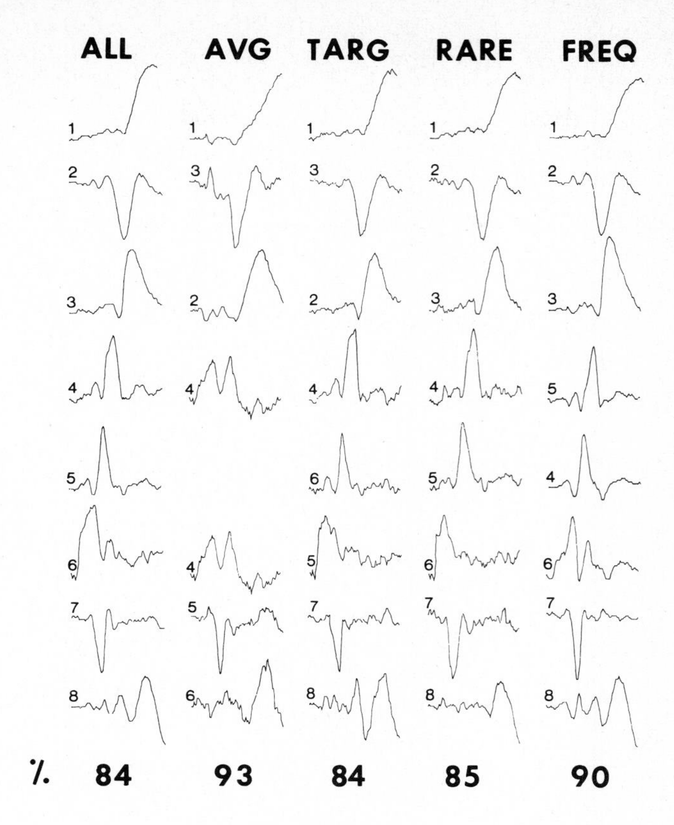

Figure 5
Comparison of rotated basic waveforms obtained from PCAs of the full data set (ALL), of each of the class subsets (TARG, RARE and FREQ) and of the across-subject averaged data (AVG).

'targets' than 'non-targets'.

To examine the stability of the obtained factor structure similar PCAs were also performed on the covariance matrices of the across-subject averaged data and on the class subsets of the full data set. Figure 5 presents the varimax-rotated results of these analyses. From left to right the columns depict the basic waveforms from the full data set, from the across-subject averaged data and from the class subsets - 'targets', 'rares' and 'frequents' - of the total data set. The primary point is the similarity of the extracted basic waves in each of the 5 analyses. This comparability highlights the strong internal consistency of the data and the ability of PCA to extract consistently the salient features. It is interesting to note that the PCA of the averaged data yielded only 6 basic waves accounting for 93% of the variance. This is a substantial improvement over the across-subject analysis. The six factors contain all of the features seen in the PCA of the full set with the exception of N33 - basic wave 5 of the first set. This suggests that the N33 was relatively more variable across subjects than it was across electrodes and conditions. This high inter-subject variability is interpreted as additional support for the myogenic origins of this peak. Factor 4 of the averaged set corresponds to both the N19 and N47 peaks and as such is a composite of the original basic waves 4 and 6. The similarity of the factor structures of the across-subject and averaged data sets is interesting in that it suggests that the sizable between subject variance is treated essentially as 'noise'. By extracting basic waves representing the consistent patterns of covariance, PCA is able to work through this 'noise' to extract the 'signal' which in this analogy is the underlying data structure.

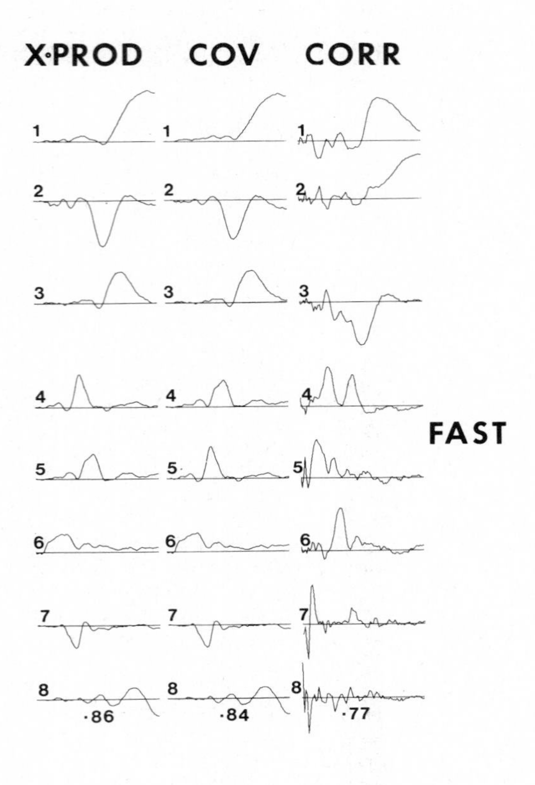

Figure 6
Comparison of rotated basic waveforms obtained from PCA of the cross products, covariance and correlation matrices.

Figure 6 presents a comparison of the varimax-rotated factors obtained from separate PCAs performed on the cross-products, covariance and correlation matrices of the full data set. The PCAs on the cross-products and the covariance matrix produced an essentially identical set of basic waveforms. Each of these factor sets accounted for about 85% of the total data variance. In contrast the factoring of the correlation matrix accounted for only 77% of the variance and produced much more complex basic waves that could not be readily identified with any one particular data feature.

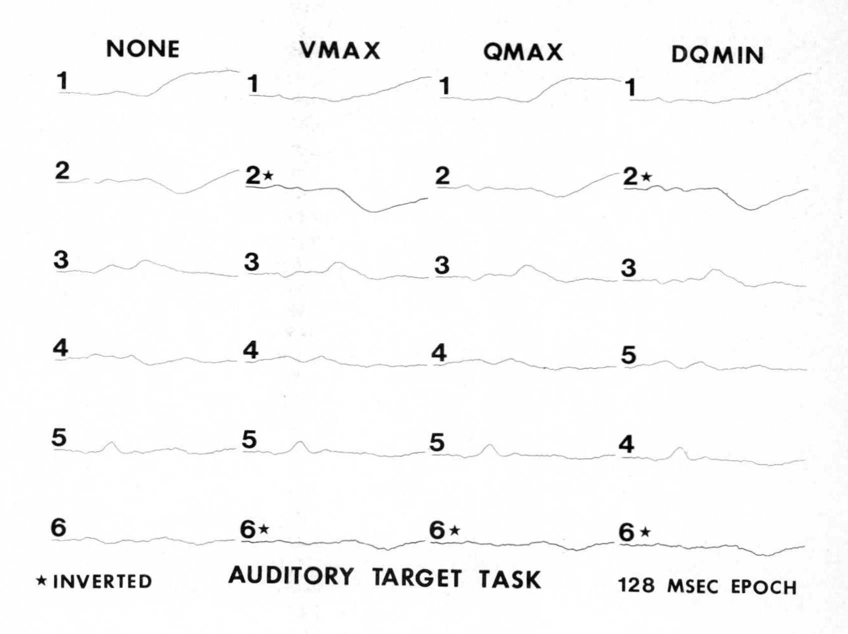

Figure 7
Comparison of various rotations performed on the same original basic waves (VMAX=varimax, QMAX=quartimax, DQMIN=direct quartimin, an oblique rotation).

Figure 7 demonstrates the relative invariance of the factor structure to objective rotation. From left to right the columns present the unrotated, varimax-rotated, quartimax-rotated and direct quartimin-rotated factor loadings extracted from analyses of the averaged data set. The last rotation is an oblique rotation which allows the basic waveforms to become correlated. The striking similarity of the unrotated and rotated factors is obvious from the figure. Thus in this data set various objective rotations of the factor set did not substantially alter the relationships between the data features and the basic waves. Nonetheless there are another infinite number of rotations (primarily subjective) that would have substantially changed the basic waves.

'SLOW' DATA RESULTS

Figure 8 begins the presentation of results for the ERPs collected on the longer time base of 768 msec. This figure illustrates the grand-averaged 'slow' data for all 'targets', all 'rare non-targets' and all 'frequent non-targets'. These averages are constructed over all sound locations.

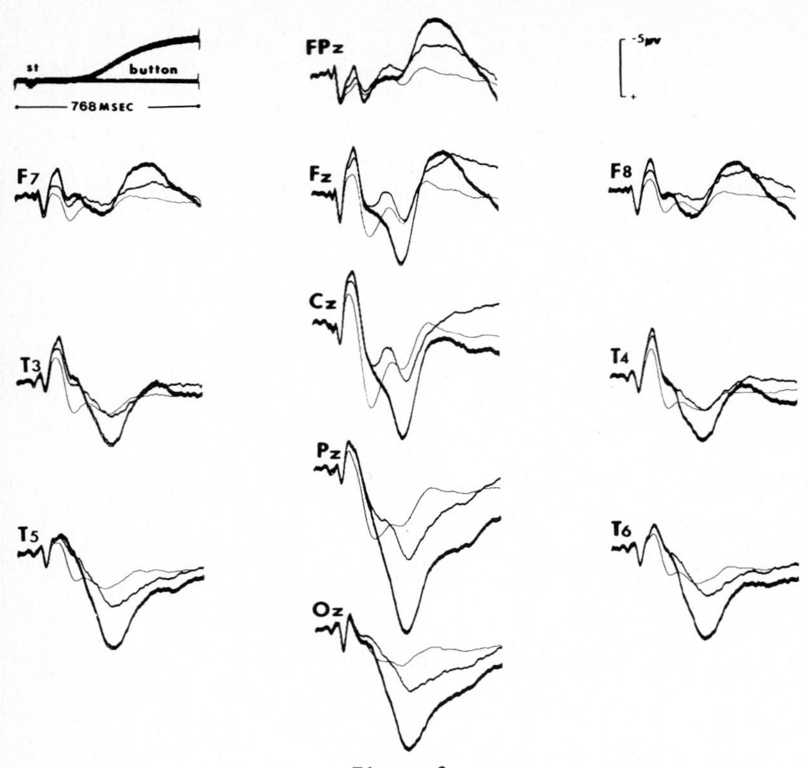

Figure 8
Grand averaged ERP data for 'targets' (thick), 'rares' (intermediate) and 'frequents' (thin).

It is immediately apparent that these ERPs are markedly influenced by the class of stimuli. The peaks which can each be identified in some of the waveforms are P55, N80, N107, P177, N232, P335 and some late slow waves. As seen before with the short epoch data the N1 complex shows a clear amplitude differences between classes of stimuli. 'Targets' are largest, 'rare non-targets' are intermediate and 'frequent non-targets' lowest in amplitude. P177 amplitude bears the converse relationship to class. At midline electrodes the N232 peak is largest to 'rares', intermediate to 'frequents' and nearly vanished to 'targets'. The P335 is clearly largest to targets, intermediate to 'rare non-targets' and smallest to 'frequent nontargets'. Following the P335 at pre-frontal and frontal locations there is a prominent negative slow wave, peaking at about 420 msec, which is largest amplitude for 'targets'. This frontal negative slow wave seems to be mirrored by simultaneous posterior positivity. At posterior electrodes there seems to be an even later slow wave extending through the end of the epoch. Figure 9 presents the amplitudes of the P177, N232, P335 and N420 peaks for each stimulus class at each electrode.

Figure 10 illustrates the class by plane of localization interaction for the 'slow' data. In this figure the combined F/B and L/R ERPs are over-

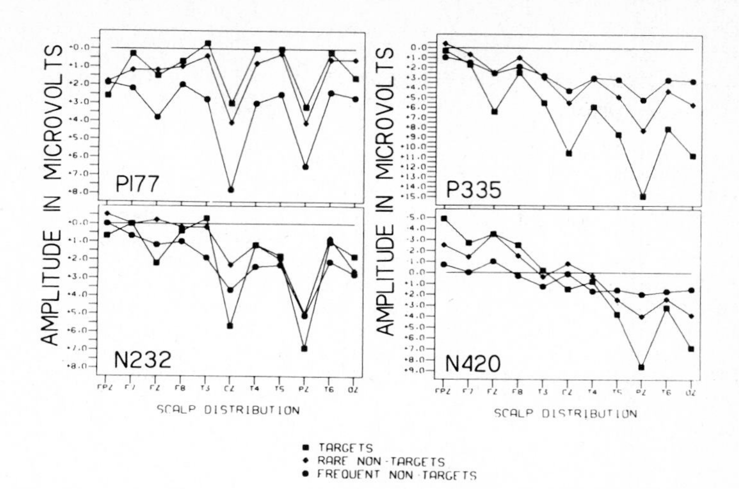

Figure 9
Average amplitudes of indicated peaks as a function of electrode and stimulus class.

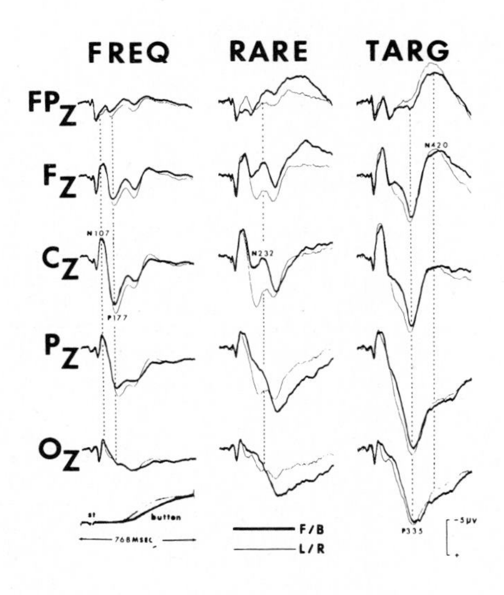

Figure 10
Midline grand-averaged ERP data illustrating the class by plane of location interaction. F/B ERPs are illustrated with the thicker trace.

plotted for each of the 3 stimulus classes. For the 'frequent non-target' ERPs (left column) the differences between F/B and L/R are restricted to small but consistent amplitude differences over the 150 to 300 msec part of the epoch. F/B traces display larger negativity and less positivity over that period. This appears to be due to a change of overall level over that interval. These F/B-L/R differences are magnified in the ERPs to 'rare non-targets' presented in the centre column. The F/B ERPs produced a much attenuated P177 and a much increased N232 peak. Also in the F/B traces the P335 was smaller frontally but larger posteriorly. Following the P335 the bipolar slow wave was much augmented to F/B 'rare non-targets'. It appears that many of these differences are due to the superimposition of a frontal-negative slow shift over the bulk of the epoch (see below). This sort of sustained shift is perhaps comparable to Näätänen's 'mis-match negativity' (Näätänen and Michie, 1979). The differences between F/B and L/R ERPs are much reduced to 'target' stimuli. The two primary features of the 'target' ERPs - the P335 and slow wave - are both larger to L/R targets. It is interesting to note that the P335 latency is identical for the F/B and L/R 'target' ERPs. This is despite the 80 msec difference in reaction time.

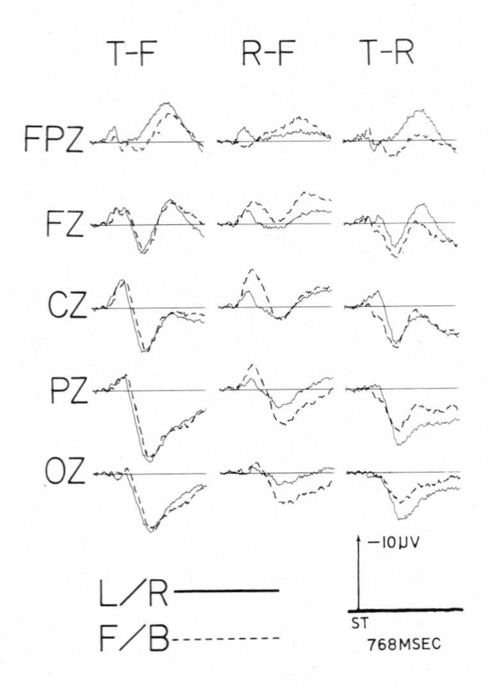

Figure 11
Midline difference waveforms obtained by subtracting the ERPs to one class of stimuli from those to another (see text).

Figure 11 presents difference waveforms constructed to examine the supposition that many of the differences between the ERPs to F/B and L/R stimuli as a function of class are due to a sustained negative shift. These difference waves are constructed by subtracting one waveform from another on a point-by-point basis. The left column illustrates the results of subtracting the ERPs to 'frequent non-targets' from the ERPs to 'targets' for both the F/B and L/R data. The L/R difference waves are depicted with the solid trace. The primary features of these difference waves are the P335 and slow wave components that appear following correctly detected target stimuli (either F/B or L/R). Of more interest though is the earlier negative difference that corresponds to the N1 complex. The onset of this earlier negativity is delayed for F/B targets and this delay is apparent in the difference wave continuing until the peak of the P335. The centre column presents the difference waves obtained by subtracting the ERPs to 'rare non-targets' from the ERPs to 'frequent non-targets'. For the L/R ERPs the only differences between 'rare' and 'frequent non-targets' are slightly augmented N107 and P335 components. For F/B stimuli the differences are much larger over the whole epoch. The 'rare non-targets' produce a large sustained negative shift over the 60 to 250 msec period. This is interrupted by the appearance of a larger P335 component and followed at frontal locations by a negative slow wave. The rightmost column presents the differences between the ERPs to 'targets' and 'rares'. As expected there is a larger difference between 'targets' and 'rare non-targets' for L/R than for F/B stimuli. These difference waveforms suggest the presence of at least one sustained negative shift overlapping the sequence of peaks and troughs. This shift is apparent in the ERPs to both 'targets' and 'rares' from the F/B orientations but only to the L/R 'targets' The onset of this shift is earlier for the L/R 'targets'. Furthermore there are much larger differences between 'targets' and 'rares' for L/R than for F/B stimuli.

'SLOW' PCA RESULTS

The 'slow' data set was subjected to a PCA identical to that described before for the 'fast' data set. The original waveforms were reduced to 85 points by 3-point averaging giving a 9 msec/point resolution. The PCA on the 8448 waveforms extracted 8 basic waves accounting for 89% of the total variance.

The results of the PCA are presented in Figure 12. In the left column are the 8 rotated basic waves with the corresponding data feature(s) indicated to the left. The value underneath each trace indicates the proportion of the total variance accounted for by that factor. The loading plots have been 'inverted' as necessary to maintain the convention of negative up. In the centre column the factor score distribution is indicated for each basic wave. The right column presents the mean weighting coefficient for stimulus class (target, 'rares' and 'frequents') pooled over subjects, electrodes and runs. The first basic wave is a broad negative going slow wave with a peak latency of 420 msec. The distribution of the slow wave factor is anterior negative and posterior positive. Overall this factor had significantly ($p \leq .0005$) more negative weighting coefficients for both the 'targets' and 'frequents' than for the 'rares'. The basic wave clearly corresponds to the P335 peak. The distribution is identical to that of the classic parietally maximal P300. The latency of the factor is at 288 msec - earlier than the peak scalp latency confirming the presence of overlapping activities in this part of the epoch. As expected basic wave 2 is significantly ($p \leq .0001$) more prevalent to 'targets' than to both 'non-

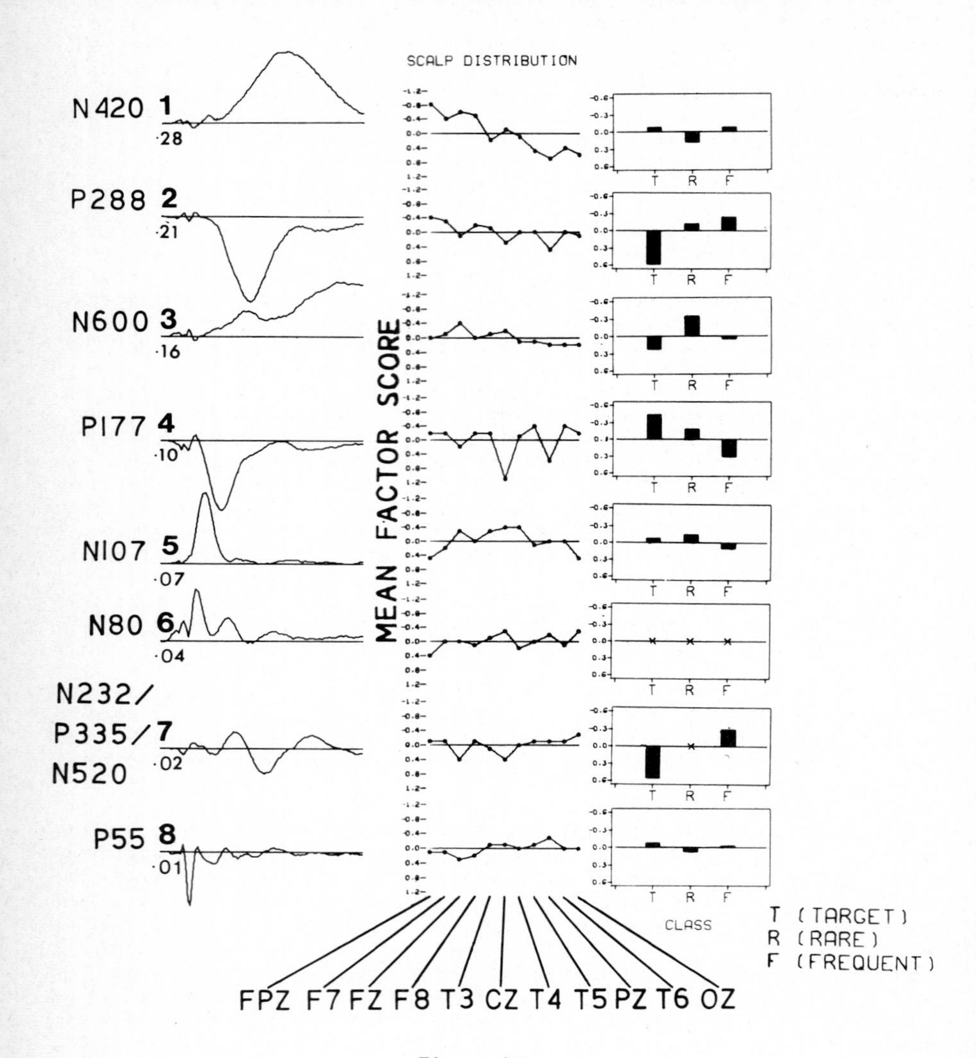

Figure 12
Results of the PCA of the 768 msec data. In the left column are the 8 rotated basic waves. The scalp distributions and task sensitivities of the weighting coefficients are illustrated in the other 2 columns.

targets'. Furthermore 'rare non-targets' displayed more of this factor than did 'frequent non-targets'. Factor 3 is a long protracted negative slow wave with peaks at 264 and 620 msec. The distribution of basic wave

3 is midline, maximal at Fz. The amplitude distributions of the two most likely component data features - the N232 and a very late slow wave - are not really similar. This is an example of PCA revealing significant features in the data not readily apparent to visual inspection. This third factor is significantly ($p \leq .0001$) more prominent in the 'rare non-target' data than in the data from the other two stimulus classes. Furthermore the weighting coefficients for the 'frequents' were significantly ($p \leq .0004$) more negative than for the 'targets'. Closer analysis of the scores for this factor suggest that this basic wave corresponds most closely to the apparent sustained negative shifts seen in the difference waveforms in Figure 11. Basic wave 4 corresponds well with the P177 peak in shape, latency and distribution. This factor is significantly ($p \leq .0001$) more prominent for the 'frequents' than for either the 'rares' or the 'targets'. Furthermore the 'rares' had significantly ($p \leq .0003$) more positive (larger) scores than did the 'targets'. Factor 5 can equally be directly related to the N107 peak. The factor score and amplitude distributions are nearly identical. This factor is significantly ($p \leq .0009$) more prominent for both the 'targets' and the 'rares' than it is for the 'frequents'. The next basic wave (6) primarily represents the earlier N1 component peaking between 80 and 90 msec. This component is distributed maximally along the central-posterior midline and is unrelated to stimulus class. The seventh factor is triphasic having peak loadings at 225, 342 and 522 msec. In terms of the data waveforms this factor corresponds to a short latency N232 coupled with both the P335 return to baseline and the slow wave. This is primarily a 'target' basic wave with a distribution that peaks at Cz and Fz. The factor scores are significantly ($p \leq .0001$) more positive for 'targets' than for either the 'rares' or 'frequents'. Furthermore the mean factor scores for the 'rares' and the 'frequents' are also significantly ($p \leq .0001$) different. Basic wave 8 represents the P55 peak. The weighting coefficient distribution is similar to both the amplitude distribution and the factor distribution of the corresponding factor 2 of the 'fast' data set (Fig.4).

Figure 13 compares the basic waveforms obtained on the full 'slow' data set with those obtained from PCAs performed on the grand-averaged data (removing subject variance) and on class subsets of the total data set. The original factors are presented in the leftmost column. The column labelled 'AVG' illustrates the results of the PCA of the across-subject averaged data set. This PCA extracted 5 basic waves accounting for 97% of the variance in the data. The 'AVG' factors are remarkably similar to the first few original factors with the exception of factor 4. This basic wave links the N107 with a late P335 peak. As before each of the subset PCAs produced a factor structure very similar to the ones from the full data set. This seems to again confirm the robustness of the data and the strength of the method.

Figure 14 presents the varimax-rotated basic waves obtained from PCAs performed on the cross-products, covariance and correlation matrices of the full 'slow' data set. As before the factor structure of the cross-products and covariance matrices were essentially similar. The only real difference is that the seventh and last factor from the cross-products PCA is a combination of the covariance factors 7 and 8. The basic waves obtained from the PCA of the correlation matrix were generally more difficult to relate to the data features. They also accounted for less of the variance in the data.

Figure 15 presents the results of various rotations upon the PCA of the 'slow' averaged data set. As before the rotations all produced an

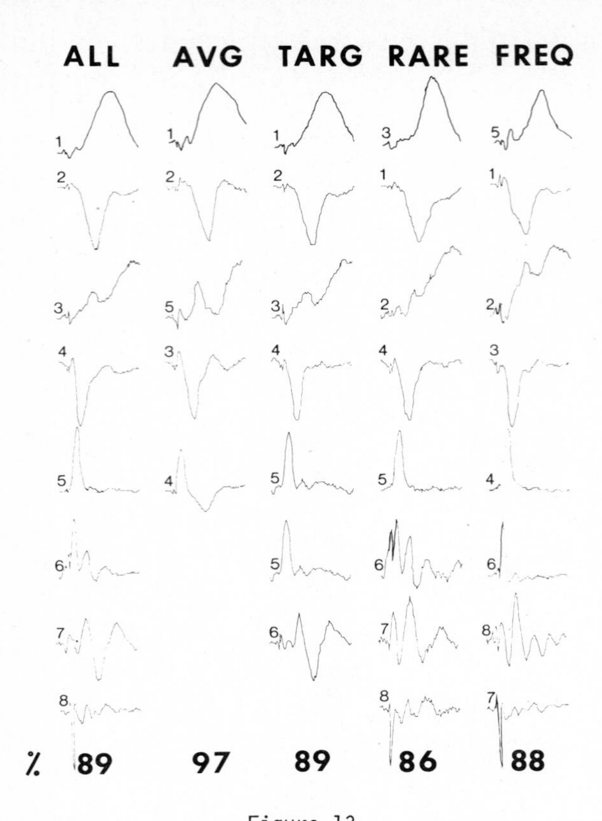

Figure 13
Comparison of rotated basic waves obtained from PCAs of the full data set (ALL), of each of the class subsets (TARG, RARE and FREQ) and of the across-subject averaged data (AVG).

essentially identical set of basic waveforms. The differences between the rotations are restricted to the ordering of the factors.

EXPERIMENT II

The second 'target' detection experiment based stimulus selection on ear and tone criteria. For this study the ISI was a fixed 1.8 seconds as compared with the 1.2 seconds of the previous study. The stimuli were 30 msec 90 db tone pips of either high (1600 Hz) or low (800 Hz) frequency presented in either the left or right ear of a stereo headphone set. In any one condition the subject's task was to press a button to the occurrence of the designated target, which was an ear/pitch combination such as left/low. The probability structure was identical to that in the previous study. The probabilities were .1 for 'targets', .1 for the equally rare 'same pitch non-targets' (SPNT) and .4 for both the opposite pitch 'same ear frequents' (SEFQ) and the opposite pitch 'opposite ear frequents' (OEFQ). Each subject served in 4 conditions; once with each ear/tone combination

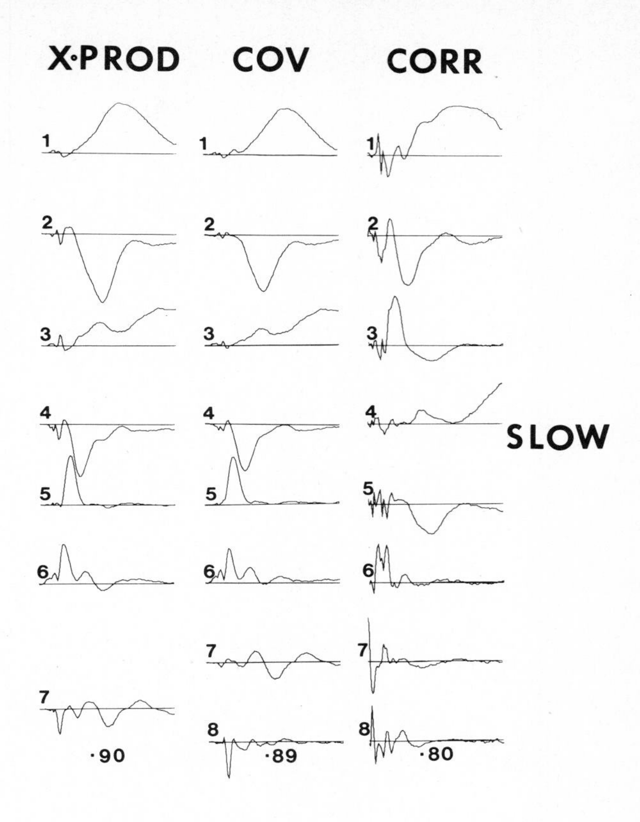

Figure 14
Comparison of the rotated basic waveforms obtained from PCA of the cross products, covariance and correlating matrix.

serving as 'target'. Stimuli were delivered according to the same pseudo-random schedule as in the first experiment.

56 neurologically normal subjects between the ages of 16 and 60 served as paid volunteers. There were an equal number of males and females. ERPs were recorded from electrodes placed at FP1, FP2, F7, F8, Fz, T5, T3, C3, Cz, C4, T4, T6, O1, O2 and Pz. Each of these electrodes was referenced to a balanced non-cephalic pair.

Brain electrical activity was sampled at a rate of 3 msec/point for an overall epoch length of 768 msec. This is an identical sampling rate to that used for the 'slow' data from the first study. Stimuli were presented 50 msec into the epoch. Data were collected as single-trials and off-line averaging employed to form averaged ERPs to each stimulus in each condition. An automatic artifact rejection procedure was utilized at this stage to remove trials contaminated with ocular or other artifactual activity. Subsequently across condition averages for 'targets', 'SPNTs', 'SEFQs' and 'OEFQs' were formed for each subject. It is these averages which were subjected to PCA.

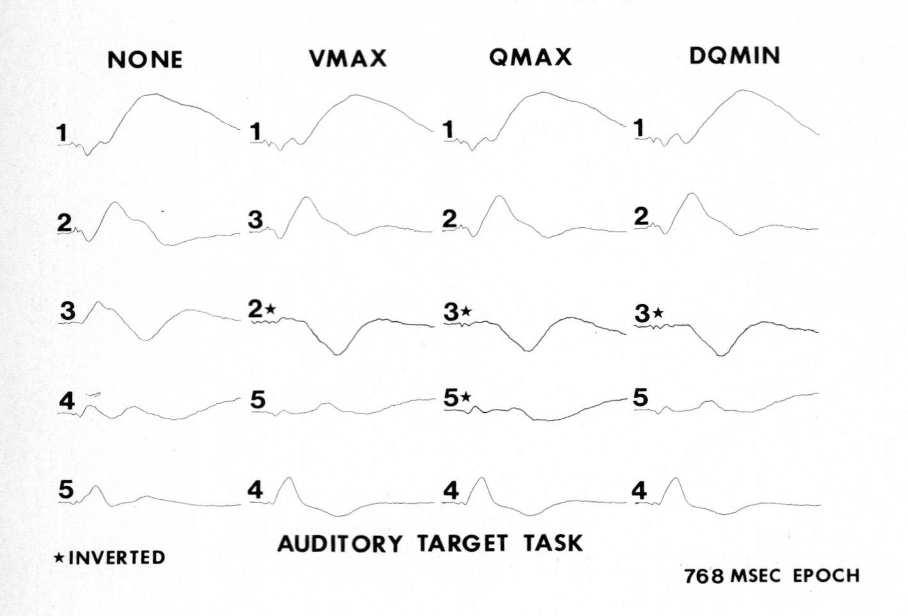

Figure 15
Comparison of various rotations performed on the same original basic waves (VMAX=varimax, QMAX=quartimax and DQMIN=direct quartimin, an oblique rotation).

PERFORMANCE DATA

The mean reaction time was 396 msec. This is 30 msec longer than the mean RT for the left and right location 'targets' of the previous study. The mean number of missed 'targets' was 1.3 per condition (32 target trials per condition). There were on average .7 false alarms.

DATA RESULTS

Figure 16 presents the superimposed grand-averaged data for all 'targets', 'SPNTs' and combined 'SEFQs' and 'OEFQs' (ONT; other non-targets). The principal features of this data are the N1 complex, the P180, the N240, the P315 and the frontal negative slow wave. In this data set the N1 complex can be seen to be composed of three spatially distinct peaks at 70, 96 and 130 msec. Both the N96 and N130 amplitudes are augmented to 'target' stimuli. However this increased amplitude applies equally to the 'SPNTs'. This finding suggests that in this situation the effect is due primarily to the low probability of these stimuli. The P180 is also reduced in amplitude

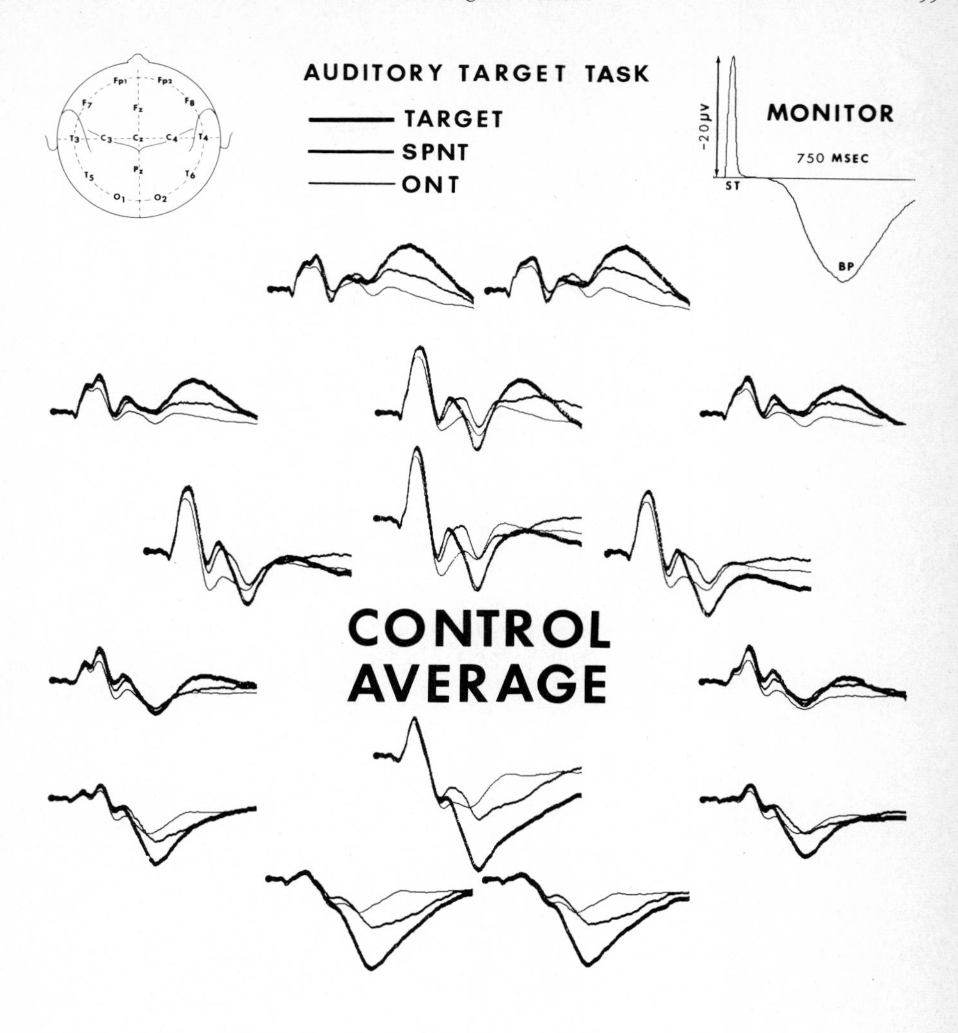

Figure 16
Grand-averaged ERP data for 'targets' (thick), 'rares' (SPNT;intermediate) and 'frequents' (ONT;thin).

for all 'rare' stimuli relative to the'frequents'. The N240 is markedly affected by class of stimuli. It is smallest to 'frequent non-targets' at most scalp locations. It is largest generally in 'target' ERPs and peaks at latencies as short as 213 msec. This shortening of latency is

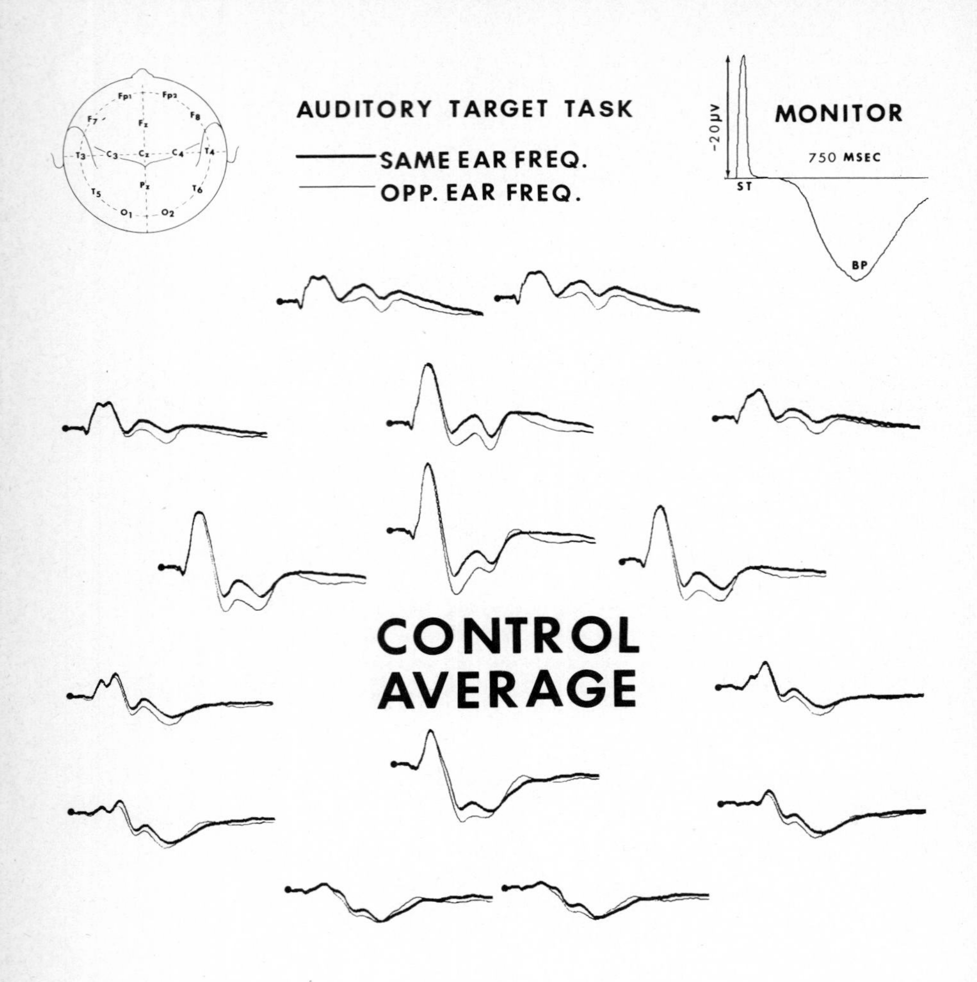

Figure 17
Comparison of grand-averaged ERP data for frequent non-targets in the attended ear (SEFQ;thick trace) with frequent non-targets in the opposite ear (OEFQ;thin trace).

presumably due to the overlap of the following P315. The P315 itself is clearly largest to 'targets', intermediate to 'SPNTs' and smallest to 'frequent non-targets'. Overlapping the P315 there is a pronounced negative slow wave over anterior locations peaking at about 490 msec. This slow wave is clearly largest to 'targets', much smaller to 'SPNTs' and just barely visible in the ERPs to 'frequent non-targets'.

Figure 17 illustrates the relative lack of an N1 complex enhancement effect for non-target material in the same channel (ear) as the 'targets'. This figure overplots the grand-averaged 'SEFQ' and 'OEFQ' data. It is clear from the figure that the vertex maximal N96 peak is very similar in the two averages. However at posterior temporal locations the later N130 peak is slightly larger to attended channel 'non-targets'. This is further evidence to support the multiplicity of negativities present over the usually defined N1 bracket of time. Despite the stability of the N96 peak there are differences in the P180, N240 and P315 parts of the waveform. Attended channel 'non-targets' produced a smaller amplitude P180 and a larger N240 at all electrode locations. These differences are very similar to the differences between F/B and L/R 'rare non-targets' observed in the first study.

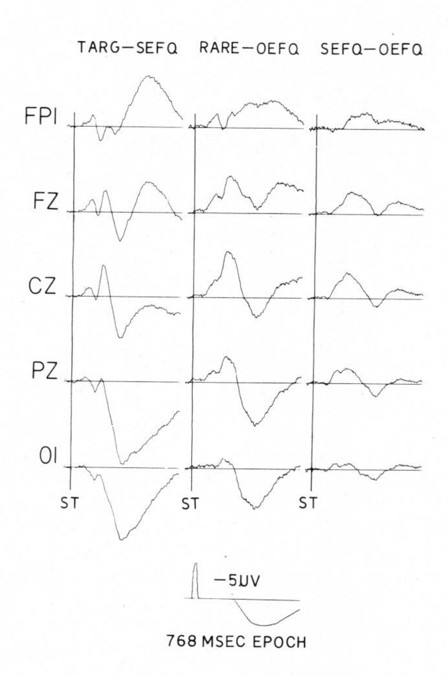

Figure 18
Midline difference waveforms obtained by subtracting the ERPs to one stimulus from those of another (see text).

To examine if these differences are also due to an apparent sustained negative shift, difference waveforms were constructed for several of the class comparisons. Figure 18 depicts the difference waveforms for the medial electrodes. In the left column are the difference waves obtained by subtracting the ERPs to the SEFQ tones from the ERPs to the 'target' tones. The primary features of these difference waves are the double negativities over the N100 and N240 periods of time and the following P315 and N490 (slow wave) components. The two negativities appear similar to the processing and mismatch negativities of Näätänen (Näätänen and Michie, 1979). The centre column presents the differences between the ERPs to 'rare non-target' and OEFQ stimuli. These difference waves are similar to those in the first column. The two negativities are somewhat merged with the second one increased in amplitude. The parts of the difference waveform corresponding to the P315 and slow wave components are reduced in amplitude. The rightmost column depicts the difference waveforms obtained by subtracting the OEFQ from the SEFQ ERPs. In this column the two negativities are now completely merged into one broad wave. As expected, there are only relatively trivial differences between the SEFQ and OEFQ waveforms over the latter part of the epoch.

PCA RESULTS

To examine the underlying structure of this data PCA was applied to the covariance matrix calculated over 3360 85-point waveforms. This number of waveforms comes from 56 subjects x 4 classes x 15 electrodes. The extracted factors were subjected to varimax rotation as previously described.

This analysis extracted 7 basic waves accounting for 94% of the variance in the data. Figure 19 presents the obtained basic waveforms, the weighting coefficient distributions and the amplitude distributions of the corresponding data features. Both the factor score and amplitude distributions are displayed as density head plots. Greater density corresponds with larger amplitude. Positive and negative values are plotted separately. The amplitude values used in this figure are mean deviations to make them more comparable with the factor scores. Basic wave 1, which alone accounts for 39% of the variance, is a combined P315 and slow wave factor. The peak loading is at 370 msec. The distribution of the factor is bipolar - being negative anteriorly and positive posteriorly. The amplitude distributions of both the P315 and slow wave are identical in shape. The second factor is a long slow negative wave starting at about 150 msec and peaking at about 600 msec. Both the factor score and amplitude distributions are clearly frontal. Basic wave 3 is quite clearly the P180 peak. Factor 4 is equally clearly the N96 peak. Both factors 3 and 4 have distributions identical with the corresponding amplitude distributions. Basic wave 5 corresponds to a long latency broad N240 - the mismatch negativity - with a frontal prominence. The sixth basic waveform is a combination of the later N1 component and a late slow wave peaking at 510 msec. The last factor is another complex waveform similar to basic wave 7 of the 'slow' data from the previous study. This factor is believed to correspond to an early N240 coupled with the P315 seen primarily at mid-frontal locations in the ERPs to 'target' stimuli. Across conditions the factor distribution is highly asymmetric over the posterior scalp. There is no corresponding amplitude distribution presented as there is no directly related data feature.

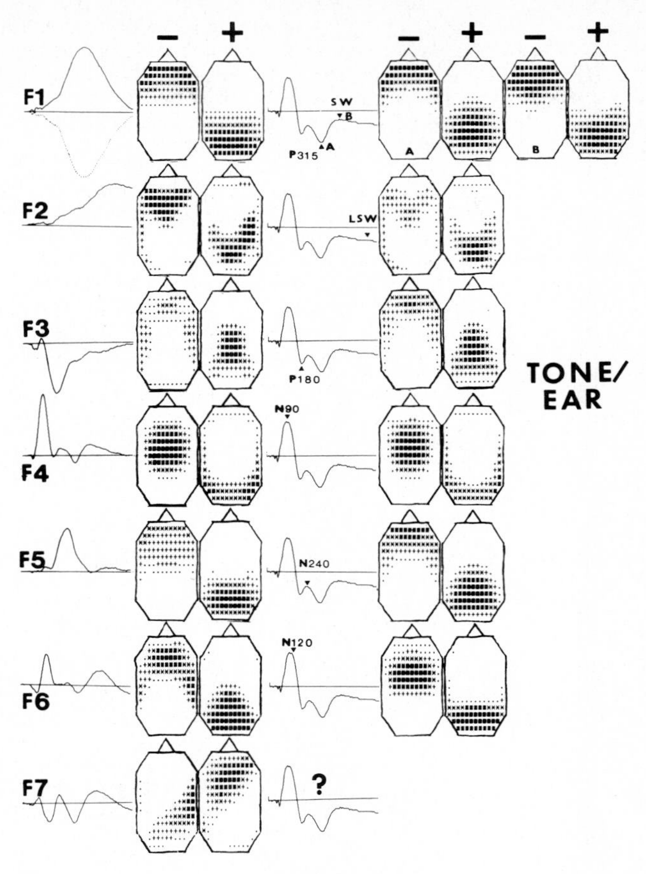

Figure 19
Illustration of the 7 rotated basic waves and a comparison of the scalp distribution of the weighting coefficients with that of the corresponding amplitude measures (see text for details).

The left half of figure 20 illustrates the weighting coefficient distributions as a function of stimulus class. The right half of the figure presents a summary of the class main effects obtained by collapsing across the 15 electrode positions. Basic wave 1, the combined P315 and slow wave, displays both large frontal negative and larger posterior positive factor scores for 'targets' and 'rares'. Over all electrodes the weighting coefficients for 'targets' are significantly ($p \leq .0001$) more positive than for the other three classes. Additionally the 'rare non-targets' are significantly different from the 'frequent non-targets'. Factor 2, the sustained negative slow wave, is largest at anterior locations to both types of rare stimuli. Overall electrodes the 'rare non-targets' displayed significantly more negative weighting coefficients than did the 'targets' ($p \leq .009$) or 'frequents' ($p \leq .0001$). The third basic wave corresponds to the P180 peak and shows a similar distribution

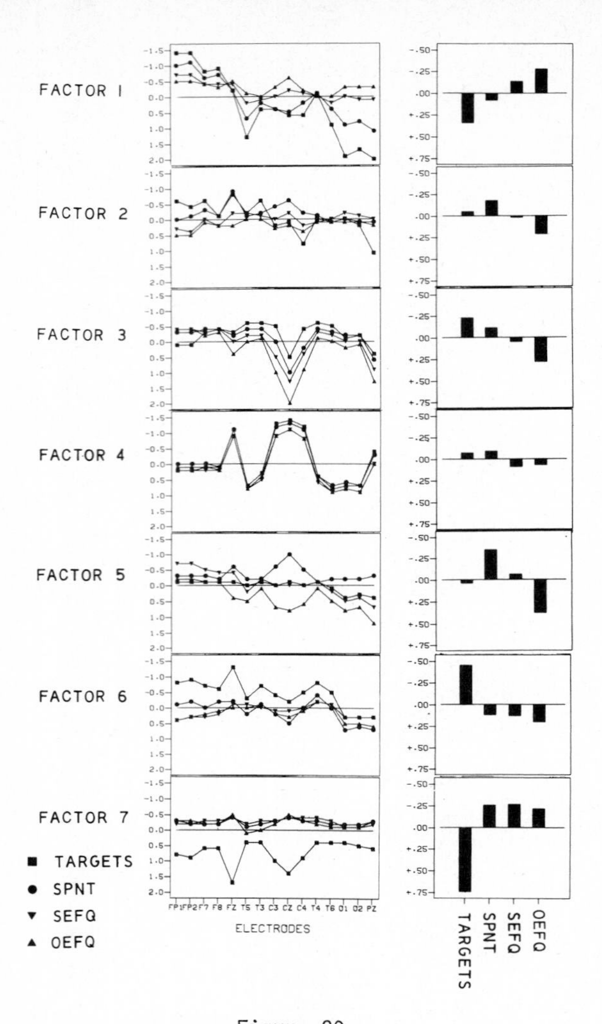

Figure 20
The left column illustrates the mean value of the weighting coefficients as a function of stimulus class and scalp location. The right column illustrates the overall class values obtained by averaging across electrodes.

to all 4 classes of stimuli. Over all recording sites the mean factor scores are inversely related to the proximity to the 'target'. This factor is smallest to 'targets' and largest to 'OEFQs'. All the paired comparisons between classes are significant. The factor (4) corresponding to the N96 peak also shows a near identical distribution for each of the stimulus classes. The mean values for both the two rare and the two frequent stimuli are similar. There are however significant differences ($p \leq .003$) between either rare and either frequent stimulus class. The distribution of basic wave 5, the long latency fronto-central N240, or 'mismatch negativity', varies markedly as a function of stimulus class. Over all electrodes the factor is most prominent in the 'SPNTs' and least prominent in the 'OEFQs'. Apart from 'target' vs 'SEFQ', all the other paired comparisons are highly significant ($p \leq .0001$). Basic wave 6, the later N1 peak coupled with a 500 msec slow wave, is primarily a target factor. At all but the 3 most posterior electrodes factor 6 is much larger for 'targets'. This effect is especially obvious at the Fz recording site. Over all electrode placements basic wave 6 is significantly ($p \leq .0001$) more prominent to targets than to any of the non-targets. There are no factor score differences among the 3 classes of non-targets. Basic wave 7 is even more dramatically a 'target' factor. At all electrode sites the 'targets' display much larger positive values. Averaging across the electrodes factor 7 is significantly ($p \leq .0001$) different from each of the non-targets. There are no significant differences between any of the non-target classes.

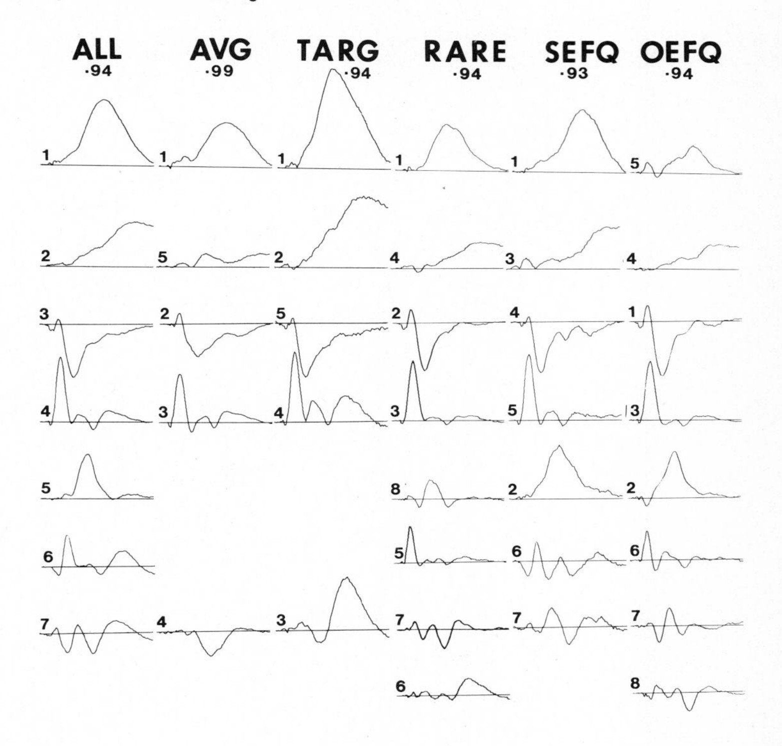

Figure 21
Comparison of the rotated basic waveforms obtained from PCAs of the full data set (ALL), of each of the class subsets (TARG, RARE, SEFQ and OEFQ) and of the across-subject averaged data (AVG).

Figure 21 compares the basic waves obtained from PCAs performed on the full data set, on the across-subject averaged data and on each of the 4 class subsets of the full data set. Each of the PCAs was performed on the covariance matrix and the resulting factors were subjected to varimax rotation. The basic waves in the first column are from the PCA of the full set. These are the factors that have already been presented. The basic waves in the column labelled 'AVG' were extracted from an analysis of the across-subject averaged data (removing subject variance). These 5 factors accounted for an impressive 99% of the total variance. The first 3 'AVG' factors correspond closely to the original basic waves 1, 3 and 4. The two remaining 'AVG' basic waves are similar to, but not identical with 2 of the 'ALL' factors. The PCAs on the class subsets extracted basic waveforms similar to the 'ALL' set for all but the 'target' data. The PCA of the 'target' subset produced only 5 factors which accounted for the same percentage of the variance (94%) as did the 7 or 8 factors from the PCAs of both the full data set and the other subset analyses. This reduction in the dimensionality of the factor structure of the 'target' data is primarily a reflection of the dominance of the P315 and slow wave in 'target' ERPs. These two features are of such large amplitude that in a covariance analysis they swamp some of the lower amplitude features. Although the stability of the factor structure in this experiment is not quite as good as in the previous study, there is still a close correspondence between the various sets of basic waves. Furthermore the observed differences are readily explicable in terms of the original data.

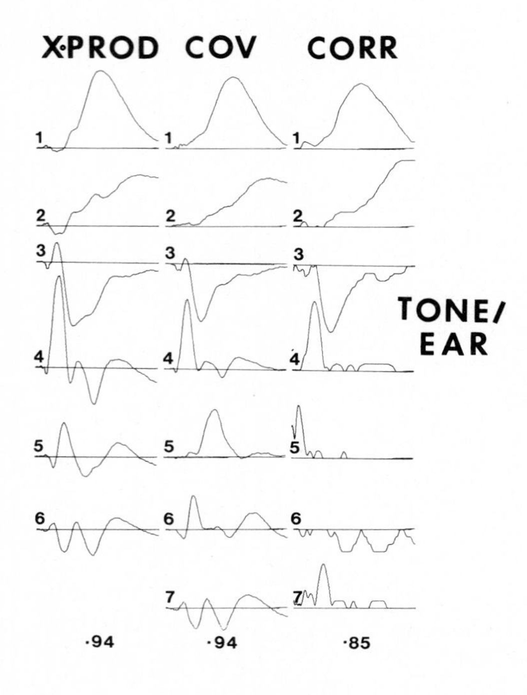

Figure 22
Comparison of the rotated basic waveforms obtained from PCA of the cross-products, covariance and correlation matrices.

Figure 22 presents the results of PCAs performed on the cross-products, covariance and correlation matrices. The illustrated basic waves have all been rotated to the varimax criteria. As in the previous two examples of factoring on the different matrices, the factor structure of the cross-product and covariance matrices are very similar. Both PCAs accounted for 94% of the total variance. The cross-product PCA extracted 6 factors. The PCA on the covariance matrix produced 7 factors. This discrepancy resulted from one of the covariance factors being shared over several of the cross-product factors. The first 7 factors of the correlation matrix PCA accounted for only 85% of the total variance. Furthermore only the first 4 basic waves correspond to the primary features in the data. In general it seems that PCA on the correlation matrix is of much less utility in both describing and understanding the data.

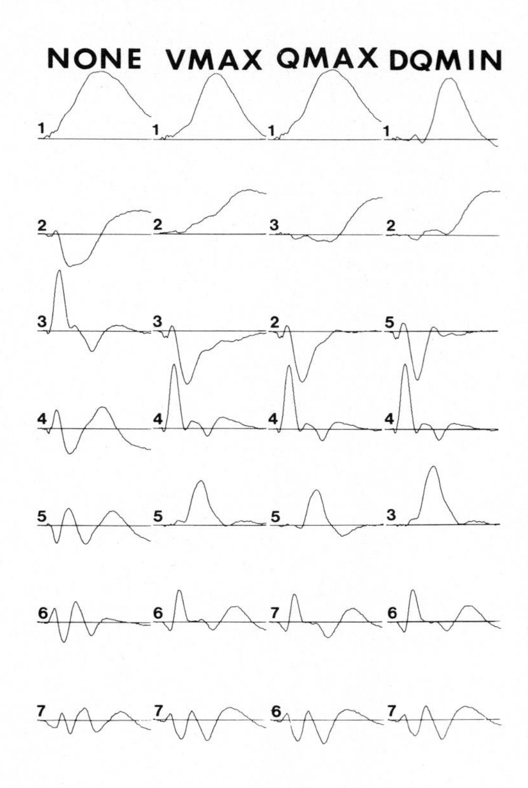

Figure 23
Comparison of various rotations performed on the same original basic waves (VMAX=varimax, QMAX=quartimax, DQMIN=direct quartimin, an oblique rotation).

The next figure (23) compares the effects of rotation on the basic waves obtained from a PCA on the covariance matrix of the full data set. In the left column are the unrotated waveforms. In the next 3 columns respectively are the loadings resulting from objective rotation according to the varimax, quartimax and direct quartimin criteria. Comparison of the unrotated factors with any of the rotated factors demonstrates the ability of rotation to simplify the pattern of factor loadings. The rotated loadings are generally simpler and much easier to identify with a particular data feature. As before the 3 types of rotation all produced an essentially identical set of basic waveforms. Thus it seems that at least with certain types of data the underlying structure is sufficiently strong that it cannot easily be altered.

DISCUSSION OF PCA

From the results of the two studies presented there seem to be several clear pragmatic statements that can be made about the use of PCA in the analysis of any one set of ERP data. The first point is that the factor structure will differ appreciably if the correlation matrix rather than the cross-products or covariance matrix is used. In the examples presented here the correlation matrix PCAs produced less desirable results in terms of both the amount of variance explained and the ease with which the factors could be related to the data waveforms. These empirical results coupled with our theoretical view that signal amplitude is of importance lead us to recommend that either the cross-products or covariance matrix should be used. One of the most important points is that using either a cross-products or covariance PCA produces a set of basic waveforms that provide a very good description of the underlying structure of the data. Each of the PCAs accounted for between 84 and 99% of the total variance. These basic waves were generally quite easily interpretable in terms of the original data. In many instances the PCAs were able to separate temporally overlapping components that were known previously to exist. In fewer instances the factors actually revealed significant features of the data that had escaped detection by mere visual inspection of the traces. The third point is that assuming the data set is sufficiently stable the factor structure seems to be quite robust. In the examples presented here the various PCAs performed on each full data set, the across-subject averaged data and on the class subsets all produced remarkably similar factors. Furthermore the sort of objective rotation employed did not seem to have much effect on the resulting factors.

It should be noted that each of the data sets analyzed here was normalised to the mean of the pre-stimulus activity prior to PCA. At present the effect that normalising has upon the resulting factor structure is undocumented. However it seems reasonable to expect that the differences between PCAs on normalised and non-normalised data will be substantial. Preliminary work on this problem suggests that this is indeed the case. PCAs of non-normalised ERP data tend to produce fewer significant factors because the first factor, which corresponds primarily to fluctuations of baseline, accounts for the bulk of the overall variance. An interesting observation concerning the effects of normalising is that the artificial reduction of variance at and near the normalising period seems to ensure that the shorter latency data features emerge as later, lower-order factors. This is quite possibly a rather misleading, and undesirable state of affairs.

To highlight the unique benefits and pitfalls of applying PCA to ERP data it will be useful to discuss the results from each of the data sets. At this point the discussion will be primarily methodological rather than subtraction. In the analysis of the 'fast' data from Experiment I, 7 of the 8 obtained basic waves were easily related to primary data features. As such it might reasonably be argued that PCA was non-contributory. After all PCA did not reveal any new data features. An analysis based on either the amplitudes of peaks and troughs (see figure 2) or difference waveforms (see figure 3) would have produced the same sort of conclusions. We would agree that many forms of analysis would have revealed an early selective attention effect in this data. However, the inferred nature and mechanism of this effect would have differed depending upon the method of analysis. For example, analysis of peak amplitudes and latencies would have produced effects on a whole series of peaks. Analysis of difference waveforms would have produced one or possibly two sustained shifts related to task parameters. As it turned out PCA revealed both peak and sustained shift effects. At this point in time it seems unwise to categorically state that any one analysis method is better than another for revealing differences in ERP data. However, there are other advantages of PCA that were useful in the analysis of this data set. First of all the cursor measurement of the shorter latency peaks and troughs on individual subject/ condition averages is extremely difficult and error prone. We believe the weighting coefficient to be a more accurate and objective matrix in situations where the signal-to-noise ratio is low. Secondly several of the basic waves were loading over the same period of time suggesting that a significant degree of overlap among components was present. This had been seen earlier in the difference waves illustrated in figure 3. A close examination of the basic waveforms (figure 4) reveals that both factors 1 and 6 load over a substantial part of the epoch. In fact the difference waveforms of figure 3 seem to be a composite of factors 1 and 6. This would suggest that the early shift over the first 30 msec is not identical with the later shift over the N100 period of time. It should also be pointed out that not all the early attention effects are accounted for by the basic waves relating to any sustained shift. Factor 7, the P26, is smaller for 'target' stimuli. This appears to be independent of any overlapping negativity.

A further benefit obtained by the application of PCA to this data set was the separation of two basic waves (1 and 3) that correspond to different elements of the N1 complex. This finding confirmed previous reports by McCallum and Curry (1980) describing multiple N1 peaks with differing scalp distributions and task sensitivities.

In general there are some problems with PCA that make its application less than straightforward. Most of these non-specific difficulties have been mentioned previously. We have considered briefly the choice of matrix, the type of rotation, the number of factors and the stability of the factor structure. In addition there are general difficulties with the identification of the basic waves and the analysis of the weighting coefficients. The difficulty with identifying and relating the factor to the data stems from the loss of polarity and the complexity of some of the basic waves. A specific example will be described below. The analysis of variance of factor scores is problematic because their statistics distribution is such as to increase the probability of a Type I error. At present there are unfortunately no hard and fast rules for the selection of appropriate alpha levels or suitable transformations to effectively reduce the probability of Type I error. Therefore we use analysis of variance of the factor scores in a rather conservative and descriptive fashion. For our purposes we need an alpha level of .001 or less to be reasonably convinced of the strength of the effect.

In relation to the PCA of the 'fast' data from experiment I there are two problems to be discussed. One of the problems with PCA as applied here on the covariance matrix is that data features with large covariance (amplitude) will dominate the factor structure. Furthermore as stated above, normalisation of the data for the pre-stimulus baseline reduces the variance over the first part of the epoch ensuring that shorter latency features have less effect on the analysis. Thus small amplitude short latency components such as the early sustained negative shift are not as clearly revealed as they may be with suitably constructed difference waveforms.

A second example illustrates the general problems of relating the basic waveforms to the data. As stated above the first 7 factors were easily related to the data. However, basic wave 8 was of sufficient complexity to cause difficulty (figure 4). This factor accounted for only 2% of the variance so it could reasonably have been dropped on those grounds. However, the weighting coefficients were significantly different depending upon the stimulus class. This was of sufficient interest for us to retain that factor in the analyses. We then had to determine what the factor was representing in the data. Eventually by close examination of the data, the basic wave and the weighting coefficients we were able to understand what this factor was related to. Basic wave 8 turned out to be inversely related early and late N100 components reflecting primarily scalp (both hemispheric and anterior-posterior) differences in the plateau of the N107 (see figure 1). These scalp differences held for all stimulus classes so this factor was more prevalent to 'targets' than 'non-targets'.

The primary advantage of applying PCA to data such as the 'slow' data from experiment I or experiment II is the ability to separate temporally and spatially overlapping components. The seriousness of the overlap problem in these data sets can be seen in the data figures 8, 10, 16 and 17 as well as in the difference waveforms presented in figures 11 and 18. Analysis based on the traditional cursor measurement of peaks and troughs can be very problematic in such cases of substantial overlap. The difficulties are two-fold. First of all there may be instances when one component overlaps another to the extent that the first becomes immeasurable. This is the case for the N232 peak at most scalp locations for the 'target' ERPs illustrated in figure 8. Secondly even if the various data features are all measurable the effects on any particular set of measures may very well be due only to fluctuations in the superimposed activity. Thus the amplitude differences in the P180, N240 and P315 peaks between SEFQ and OEFQ stimuli (figure 17) may be due as much to a superimposed shift (see figure 18) as to independent amplitude modulation of 3 peaks. PCA is able to separate the overlapping components and examine each for systematic relation to the experimental structure. In the example cited PCA revealed that the amplitude differences are due primarily to superimposed activity but that there is also some degree of independent peak modulation.

An additional but related benefit obtained by the application of PCA to these data sets was the appearance of basic waves not directly related to the primary data peaks and troughs. Factor 3 in figure 12 and factors 2 and 5 in figure 19 are good examples. These patterns of loadings had not previously been visualized as significant contributors to the data structure. It is interesting that these basic waveforms correspond only approximately to the sustained shifts observed in the difference waves (see figures 11 and 18). This suggests that the difference waves computed over the whole

768 msec epoch are reflecting differences due to the action of multiple overlapping activities. If so, these multiple activities are necessarily confounded. It is maintained that the degree of confounding with PCA should be substantially reduced.

The PCA analysis of these two data sets on the longer time base did not reveal any problems or pitfalls not previously discussed.

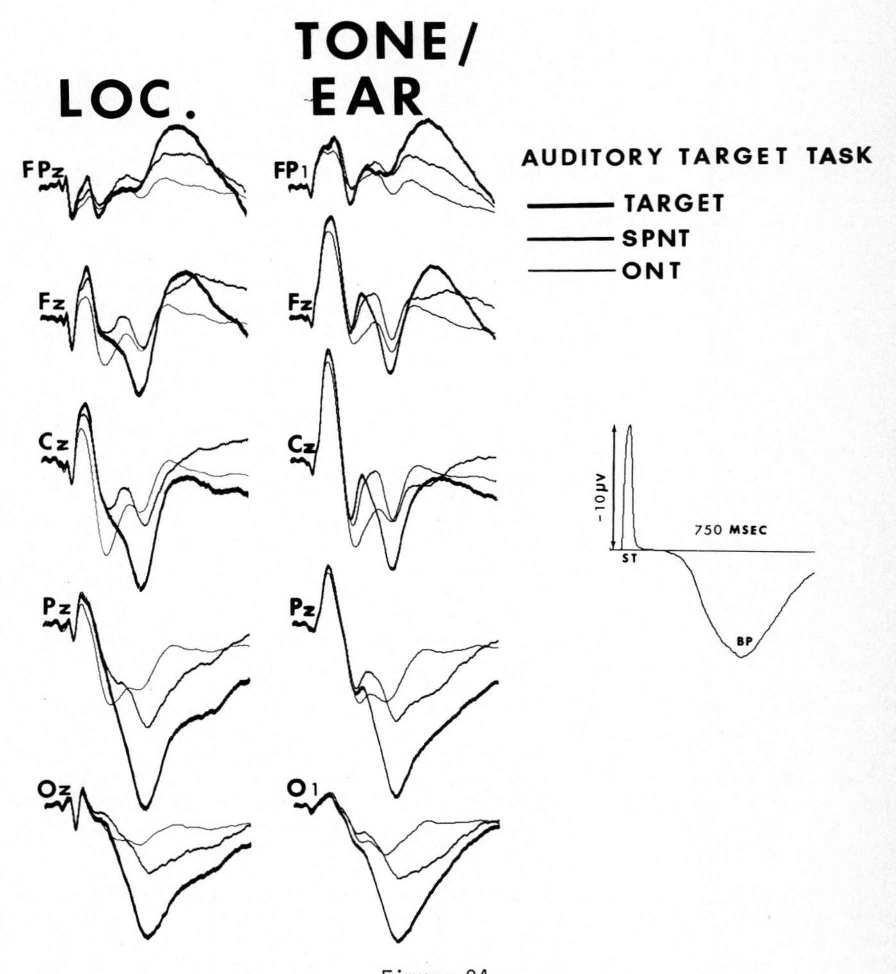

Figure 24
Comparison of the grand-averaged midline data from the two studies. In both columns the 'targets' are illustrated with the thickest line, the 'rares' (SPNT) with the intermediate line and the 'frequents' (ONT) with the thin line.

The remaining crucial question about the use of PCA concerns the stability of the basic waveforms from experiment to experiment. Are the principal components obtained from a PCA of one experiment going to be similar to those obtained in another experiment - or even in a replication of the first experiment? The answer to this question must necessarily be 'no' if the experiments are wildly different or if the recording montage is substantially altered. However, in the two experiments presented here the paradigms are at least superficially similar as being both 'auditory target detection tasks' with an extensive scalp coverage. Figure 24 presents a side by side comparison of the grand-averaged data obtained in the two studies. In the left column is the midline 'slow' data averaged by class. In the right column are the corresponding waveforms from the tone/ear study. As can be seen from the figure the main features of the two data sets are very comparable. The same series of peaks are prominent with only minor latency differences. What does differ in the two sets is the amplitude of the N1 complex and the variation of some of the peaks to the stimulus class. The N1 complex is much smaller in amplitude in the location study. This finding can be due to any one or any combination of three factors. First of all the ISI in the first study was 600 msec shorter. Secondly the stimuli were 'white noise' bursts as opposed to tone pips. Lastly the location task was much more demanding as evidenced by the performance data. Thus the amplitude difference is at present uninterpretable. Despite the lower N1 amplitude the amplitude differential between the classes of stimuli is greater in the location study. At frontal and central locations there is a clear ranking of N1 amplitude such that 'targets' are largest, 'rares' are intermediate and 'frequents' smallest. For the tone/ear study the N1 amplitude differential was only between rare and frequent events. There was no difference between 'targets' and 'SPNTs'. The N240 peak also responded differently to classes in the two experiments. In the location study the N240 amplitude was largest to 'rare non-targets' and almost vanished to 'targets'. In the tone/ear study the N240 is earlier to 'targets' but equal in amplitude to the N240 to 'SPNTs'. The 'frequent non-targets' produced the smallest amplitude N240 at most electrode sites. The final difference between the two data sets is the suggestion of an additional late slow wave at posterior scalp locations in the ERPs from the first study.

Thus the differences between the two data sets are rather small in comparison with the overall pattern of similarity. Figure 25 compares the rotated basic waves obtained from similar analyses of the two data sets. As can be seen from the figure the two sets of factors are very similar in shape. It should be noted that the fifth basic wave from the tone/ear study is inverted in this figure. This factor is not the P315 as is the second factor from the other set that it is plotted with. This is actually the crucial difference between the factor structure of the two data sets. In the location study the PCA extracted separate slow wave and P315 basic waves. In the tone/ear study the first basic waveform is a combined slow wave and P315 factor. At present it is unclear what is responsible for this relatively large change in the factor structure. The two obvious possibilities are the differing ISIs and differing levels of task difficulty. Another not quite so likely possibility is the differing scalp coverage in the two studies. The second study employed 4 additional recording channels which included FP1, FP2, O1 and O2 in place of FPz and Oz. This change may have been sufficient to alter the apparent relationships between the frontal negativity and posterior positivity.

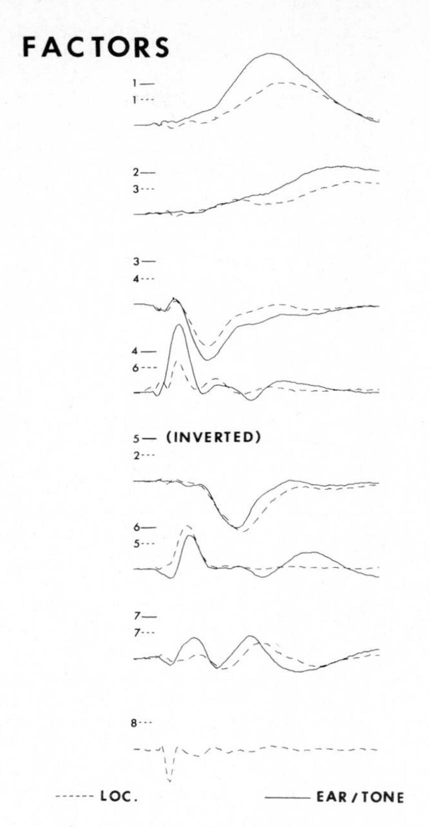

Figure 25
Overplotted basic waveforms for 768 msec
data from Experiments I and II.

The existence of multiple slow waves is demonstrated quite clearly in the factor structure of each of the data sets. The second factor of the tone/ear study and the third from the location study are very similar in shape and time course. Both of the factors have an onset latency of about 150 msec and peak very late in the epoch. This very late slow wave is quite clearly independent of the prominent frontal negative slow wave peaking at about 420 msec. This earlier slow wave emerges as factor 1 in both analyses. In the tone/ear study this frontal negativity was inseparable from the late positive activity at posterior recording sites.

It should be noted that in both studies the PCA extracted 2 negative basic waves peaking in the 80 to 120 msec bracket of time. The waveforms were separated on distribution and sensitivity to the task parameters. These findings support our previous findings on the multiplicity of N1 components (McCallum and Curry, 1980).

Based on the comparison of the factors obtained in these two task situations, it appears that the factors are in each case providing an accurate and parsimonious description of the original waveforms. Despite the similarity of the pattern of factor loadings the description of each of the data sets is however in a slightly different metric. Factor 1 in the first study is not equivalent to factor 1 in the second study despite the superficial similarity of the weighting coefficients. Therefore it seems unlikely that the basic waves have any necessary degree of generalizability to other data sets. In other words the factors are principal components only for the data set from which they were derived. They are not the principal components of auditory ERPs in target detection tasks. That is not to say that the factors must necessarily be idiosyncratic. In fact, as has been shown here, several of the factors have direct equivalents in the other data set.

DISCUSSION OF DATA

The short epoch data of Experiment I demonstrated that the effects of directed attention can manifest themselves earlier than the 60 or 70 msec onset of the auditory N1 complex. In this experiment there is considerable evidence to suggest that by about 10 or 15 msec stimulus relevance and rarity influence the scalp recorded ERP. It appears that in part the changes are due to the superimposition of a negative shift on the early components: There is however evidence from the PCA that some of the peaks such as the P26 are modulated in amplitude to an extent not accounted for by the superimposed activity. Further research is required to elaborate the nature and mechanisms of these early selective attention effects.

The differences in the ERPs to F/B and L/R stimuli are fascinating. In this experiment the subjects were effectively performing two very different tasks. In the conditions in which either left or right-sided sounds were designated 'target' the performance was excellent. This task is obviously an easy one for the human listener. This is to be expected because the neural mechanisms for determining left from right are robust and automatic. In contrast there does not appear to be any such automatic neural structure for determining front from back. Since the sounds arrive at both ears simultaneously there are no inter-aural differences. The only differences will be of sound intensity or composition. In either case the front-back discrimination requires the subject to 'match' the sound with some previously established template. The timing of this sort of matching process is likely to be extended - and error prone. The behavioural data lend support to the view. Both reaction time and error rate were greatly increased over those for L/R discrimination. This lack of early neural mechanisms for discrimination of front from back almost certainly accounts for the lack of differences between the early ERPs to 'targets' and 'rares'. It is likely that the point of discrimination was reached much later in time - perhaps as late as 100 msec after onset.

The slow data of Experiment I produced some interesting results. As expected the long latency ERPs varied dramatically as a function of stimulus class. For example the 'target' ERPs were dominated by large amplitude P300 and slow wave activities. What was not expected was the strong interaction between plane of localization (anterior-posterior or lateral and stimulus class (see figure 10). It is clear from this figure that the ERPs to the F/B and L/R 'rare non-targets' differ to a much greater extent than the ERPs to F/B and L/R 'targets' or 'frequents'. These differences between 'rare' ERPs are almost certainly due to the difficulty of the task producing uncertainty as to the origins of front or back sounds. One of the performance features apparent in the discrimination of front from back is the large number of errors of both omission and commission. Confidence levels were unfortunately not obtained on a trial-to-trial basis but each subject reported a great deal of uncertainty in the performance of this discrimination. It is argued that this uncertainty resulted in extended processing of F/B 'rare non-targets' and the ERP differences are reflecting this additional load.

The similarity of the F/B and L/R 'target' ERPs is surprising given the performance differences. The error rates were greatly increased and the reaction times were about 90 msec longer for targets from either the front or back. Accepting that reaction time and P300 latency are not directly coupled there are nevertheless certain aspects of the task structure reported to affect both equally. Recent work by McCarthy and Donchin (1980), adopting an additive factors approach, demonstrated that stimulus degradation led to both reaction time and P300 latency shifts. In contrast response incompatability altered only RT. The interpretation of these results was that P300 latency was indexing the time necessary for stimulus evaluation. Degradation of the stimuli required addition stimulus evaluation resulting in a longer latency P300. In the experiment reported here it seems likely that stimulus evaluation processes would have been extended in the discrimination of front from back. Nonetheless P300 latency remained constant. Stimulus evaluation is but one example of myriad hypothetical constructs alleged to relate P300 to behaviour. This data appears problematic to the vast majority.

In both 768 msec data sets there is evidence of multiple overlapping patterns of brain activity. These instances of overlap all appear to be related to, or at least augmented by, the processing of task relevant stimuli. In these data sets there appear to be three somewhat distinct types of overlap. First of all there is overlap of readily identifiable peaks such as the N240 and P315 in the 'target' ERPs. Secondly there is the overlap of an identifiable peak with a more amorphous waveform. The overlap of the P300 and the late slow wave in Experiment I is one example. Lastly there are instances of a sustained shift overlapping one or more known peaks. For example the differences between the ERPs to SEFQ and OEFQ stimuli in Experiment II appear to be due to a sustained shift over the 100 to 300 msec period (see figures 17 and 18).

The sustained shifts observed in these data sets are interesting in that they relate to the task structure, vary between the experiments and appear superficially similar to those that have been recently described by Hillyard (1981) and Näätänen (Näätänen and Michie, 1979). In both data sets there appear to be at least 2 negative shifts although the relationship of these to the task parameters differ. In Experiment I the early sustained negativity overlapped the 50 to 200 msec period. In the ERPs to L/R stimuli this shift was largest to 'targets', intermediate to 'rares' and smallest to 'frequents'. In contrast the ERPs to both F/B 'targets' and 'rares' displayed similar sustained activity. This lack of differentiation between F/B 'targets' and 'rares' is presumably due to the inability to make as rapid a discrimination between the two as is possible for L/R stimuli. In this regard it is interesting to note that this early shift has a shorter onset latency for the L/R 'target' ERPs than for the similar F/B ERPs (see figure 11). It is likely that the onset of this shift indexes a crucial stage in the recognition of input as possessing some 'target' criterion. This shift is probably equivalent to the 'processing negativity' of Näätänen and the 'Nd' of Hillyard. Following the early shift there is a second period of sustained activity over the 200 to 300 msec period observed only in the ERPs to F/B 'rare non-targets'. This shift is no doubt related to the additional processing of the F/B 'rares' due to the task difficulty and resulting lack of certainty. Whether this activity is merely an extension of the early shift or a different type of activity is difficult to resolve. The PCA results would support the latter. Basic wave 3 (see figure 12) loads highly over that period of time as well as over the last 200 msec. The relationship of this later shift to either Näätänen's 'mismatch negativity' or Hillyard's second 'Nd' component is uncertain. However the specificity of this effect suggests that this is not directly comparable with 'mismatch negativity' which should appear to all rare stimuli.

In the second experiment the pattern of sustained activity is even more complex. There appears to be a small amplitude and short latency (onset of 40-50 msec) shift that overlaps the N100 complex. This shift is of equal amplitude in the ERPs to both 'targets' and 'rares'. There is no additional 'target' increment in this data. This early negativity is not apparent in the difference waves of the two frequent stimulus classes (SEFQ and OEFQ). Following this early shift is a sequence of sustained activity over the next 250 msec. This appears as either one or two types of shift depending upon the particular subset of data examined. From the waveform differences illustrated in figure 18 the Target-SEFQ pattern appears to be two separate shifts separated by a positive difference. These features correspond with the N100, P180 and N240 peak differences. In the Rare-OEFQ differences the activity in the interval over 100 to 300 msecs is now merged although it is possible to see the two sub-components in the traces. The later activity is larger in amplitude. In the waveforms constructed by differencing the ERPs to SEFQ and OEFQ stimuli the major difference appears as one broad sustained shift over the 100 to 300 msec period. The peak of this activity is about 220 msec. The PCA results also indicate that there are at least two types of sustained activity. The second basic waveform (see figure 19) is a very broad component with an onset of about 30 msec but loading over the whole epoch and not peaking until about 500 msec. The fifth basic wave closely resembles the difference waves obtained both between the 'rare' and OEFQ ERPs and between the SEFQ and OEFQ ERPs. This factor loads heavily over the 100 to 350 msec part of the epoch with a peak at about 220 msec. This basic wave is most prevalent in the 'rare' data as can be seen from figure 20. The correspondence of these shifts to those previously described by Näätänen and Hillyard is even more problematic than it is for the ones obtained in Experiment I. Despite similarities of distribution the different pattern of class relationships rules against a simple equation. It is apparent that further work is required to elucidate the number and nature of negative shifts in auditory target detection paradigms.

CONCLUSION

In conclusion it seems to us that the ability of PCA to provide an accurate, objective and robust description of a data set recommends its use as a tool in the analysis of ERP data. Comparison of factor results from different studies is unfortunately problematic. This difficulty can probably be minimized by ensuring that the structured variance in the data sets is as similar as possible. This implies that the recording montage, the number of conditions and probably even the number of subjects should all be identical. In these circumstances it is expected that PCA will be an extremely powerful tool to assess changes in ERPs as functions of different experimental manipulation.

The substantive conclusions concerning the data presented here are first of all that there are systematic task-related changes in auditory ERPs as early as 10 or 15 msec post-stimulus. These changes are due to a combination of superimposed negativity and peak amplitude modulation. Several of the data features were differentially sensitive to 'targets' and 'rare non-targets' indicating that the salient stimulus property was 'target' and not just rarity. These data are at odds with most previous studies attempting to demonstrate early effects of selective attention.

Both studies on the longer time base have again demonstrated the sensitivity of the scalp-recorded ERPs to task parameters. This sensitivity is seen not only in the ERP differences between stimulus classes such as 'target' and 'frequent' but also in the differences between the F/B and L/R stimuli in Experiment I. In this study the difficulty of discriminating front from back as opposed to left from right meant that the subjects were performing two distinctly different tasks. In general the performance measures and ERP data support this task duality. However there are some interesting anomalies such as the P335 component - the primary feature of 'target' ERPs - which did not vary in amplitude or latency between the two sub-tasks despite the 90 msec reaction time difference.

In general both data sets give evidence of multiple overlapping activities which are often difficult to separate. As demonstrated here the overlap of brain activity can appear in several forms. First of all there are clearly instances when one peak overlaps another by virtue of a shorter onset latency for the second. The swamping of the N240 by the following P315 component in 'target' ERPs is one example. Secondly there are other instances when two components such as the P300 and following slow wave activity merge because of similar scalp distributions and task sensitivities in contrast to an obvious shortening of slow wave onset latency. Lastly, there are instances of sustained negative shifts overlapping the usual series of peaks and troughs. Several of these shifts were observed in each of the studies and all appear to relate to different aspects of the processing of task relevant stimuli. It appears that these negative shifts are only somewhat comparable to similar activities reported recently by Näätänen (Näätänen and Michie, 1979) and Hillyard (1981).

NOTES

We wish to thank Tony Gaillard, Risto Näätänen, Norman Loveless and Dan Ruchkin for their help in improving this chapter.

[1]A more extensive data-oriented report of each of these studies will be published in the near future.

Tutorials in ERP Research: Endogenous Components
A.W.K. Gaillard and W. Ritter (eds.)

5

THE ORIENTING REFLEX AND THE N2 DEFLECTION OF THE EVENT-RELATED POTENTIAL (ERP)

R. Näätänen
Department of Psychology
University of Helsinki
Finland

and

A.W.K. Gaillard
Institute for Perception TNO
Soesterberg
The Netherlands

This paper focusses on the N2 deflection of the ERP which the authors suggest is closely related to the orienting reflex. The available literature suggests that there are two negative components in the N2 time range which both contribute to the total N2 deflection. The earlier component, the so-called mismatch negativity (MMN), reflects an automatic pre-perceptual cerebral mismatch process, which occurs after a stimulus change in a repetitive homogeneous stimulus stream. This negativity is modality-specific and probably generated in the secondary sensory areas. This process is not dependent on, or modified by, the direction of attention. The second negative component, 'N2b', is modality unspecific and is superimposed on the MMN. The N2b occurs when the stimulus input is attended to. The N2b forms a wave complex together with the positive component 'P3a'. Both negative components appear to bear some relationship to the orienting reflex. The MMN reflects the neuronal mismatch process elicited by stimulus change but this does not appear as a sufficient condition for OR elicitation. The classical full-scale OR is only likely to emerge when the N2b-P3a complex occurs.

INTRODUCTION

The orienting reflex (OR) is one of the most important concepts in cognitive psychophysiology. The OR-theory has received a new impulse by the use of event-related brain potential (ERP) techniques to study brain activity during information processing. In this way it is possible to obtain information about the spatio-temporal activity pattern of the brain for different stimuli and tasks.

The OR is an old concept developed in the context of conditioning experiments with animals (see Pavlov, 1927) and it has been mainly studied by using measures of the activity of the autonomous nervous system such as the galvanic skin response, the vascular dilatation and constriction, and heart rate responses.

These measures are problematic, because rather slow responses and quite non-specific responses are involved, which makes it impossible to develop a detailed analytical view of the different phases of the OR. The consequence is that some basic problems relating to the initiation and determinants of the OR are still unsettled.

The basic paradigm in OR research involves the presentation of a repetitive sequence of identical stimuli. Usually, only the first stimuli in the sequence elicit 'the OR', a complicated pattern of organismic responses in different systems such as the autonomic nervous system lasting for a few seconds. A few repetitions of the same stimulus at not too long inter-stimulus intervals (ISI) attenuate and finally eliminate the OR. The OR is said to be habituated. On the other hand, when some feature of the stimulus is changed, the OR is re-established. The OR to stimulus change (change OR) appeared very similar to the OR elicited by the first stimulus (initial OR); it is often thought that the same response is involved in both cases (see O'Gorman, 1979; Bernstein, 1979). However, ERP research using the OR paradigm seems to suggest that different response patterns are involved in initial OR and change OR. In the former, a conspicuous feature is the large vertex N1 deflection to the first stimulus[1] which is reduced to approximately a half by one or two repetitions of the same stimulus with constant ISIs of the order of 1-2 seconds (Fruhstorfer, 1971; Öhman and Lader, 1977; Fruhstorfer, Soveri and Järvilehto, 1970; Chapman, 1965; Chapman and Bragdon, 1964; Ritter, Vaughan and Costa, 1968; Wastell, 1978). Stimulus change only partly restores the N1 amplitude and small stimulus changes have almost no effect on N1 (e.g. Butler, 1968; Näätänen, Hukkanen and Järvilehto, 1980a). Even inserting a stimulus of another modality brings the N1 amplitude only half-way between the unhabituated and habituated levels (Fruhstorfer, 1971).

On the other hand, the brain-response pattern to stimulus change is best identified on the basis of a negative component in the N2 time range called 'mismatch negativity' (MMN) (Näätänen, Gaillard and Mäntysalo, 1978). The positive deflections cannot be used in discriminating the two types of OR because both types tend to elicit rather similar slow positive components. Similarly, it appears that slow frontal negativities (for reviews, see Loveless, 1979; Rohrbaugh and Gaillard, this volume) are also common to both types of OR.

Hence it appears possible to distinguish the 'initial OR' and the 'change OR' from each other on the basis of the vertex N1 and the MMN. This suggests that there indeed exist two different forms of ORs which involve different brain mechanisms and might possess different functions. The present paper will be confined to the change OR with a particular focus on its initial stage and conditions of occurrence.

The classical OR theory suggests that the repetitive stimulus leads to the formation of the so-called neuronal model of the stimulus in the brain. That theory further suggests that it is the match of the incoming input to the features of the model that inhibits the OR. When a different stimulus is given, the sensory input does not match with the neuronal model and the OR is elicited (Sokolov, 1960, 1975).

In the current literature, there is a strong controversy as to whether stimulus change as such is sufficient to elicit an OR as Sokolov's theory suggests (see O'Gorman, 1979; Bernstein, 1979, 1981; Maltzman, 1979a; Siddle, 1979). Several contemporary investigators hold the opinion that not each stimulus deviance elicits an OR but that the deviance has to be experienced as somehow 'significant' before the OR emerges (e.g. Bernstein, 1979; Maltzman, 1979a). The slowness and diffuseness of the usual measures of the OR hinder an analysis of the different phases of the response complex and make it impossible to resolve several basic issues in OR research.

In the following we will review the ERP research relevant to the initial phases of the OR to stimulus change.

MISMATCH NEGATIVITY AS A REFLECTION OF THE MISMATCH PROCESS IN THE BRAIN

Recently, ERP research in humans has revealed a component which appears a good measure for the neuronal mismatch process. This component has been called 'N2' or 'N200' because it was the second major negative deflection, peaking approximately at 200 msec from the stimulus onset. The characteristics of the component have been investigated in a large number of studies during the seventies (e.g. Roth, 1973; N. Squires, K. Squires and Hillyard, 1975; Snyder and Hillyard, 1976; Ford, Roth, Dirks and Kopell, 1973; Ford, Roth and Kopell, 1976b; K. Squires, Donchin, Herning and McCarthy, 1977; Renault and Lesèvre, 1978, 1979; Ritter, Simson, Vaughan and Friedman, 1979; Vaughan, Ritter and Simson, 1980). For a review, see Donchin, Ritter and McCallum (1978). Of particular importance was the study of Snyder and Hillyard (1976) which provided evidence that the 'N2 component' is a specific brain response to stimulus deviance. This deflection was not observed in response to single loud clicks separated by long ISIs. However, when an occasional physically deviant auditory stimulus was interspersed in a sequence of repetitive auditory stimuli, a wave was elicited which they called the 'N2-P3a complex'. This occurred even when the deviant stimuli were merely of lower intensity than the 'standard' stimuli. Therefore, these authors associated N2 with Sokolov's OR theory "... the N2-P3a complex reflects the operation of a mismatch detector which signals any change in an ongoing background stimulus" (Snyder and Hillyard, 1976, p. 329). The authors seem to assume that stimulus deviance *per se* is sufficient to elicit the 'N2-P3a complex'.

There are other data which disclose an N2 deflection to a single click after a very long ISI. In a study of Ritter, Vaughan and Costa (1968) this component was elicited by the first stimulus of a sequence after a break of 2-5 min. In our view, this negative shift is not the MMN but rather represents another ERP component in that time range (to be discussed later).

Näätänen and his co-workers suggest that in the N2 time range there is a negative component which is a reflection of physical stimulus deviance *per se* i.e., which is insensitive to meaning, significance, and other psychological factors and, therefore, this compoent was called *mismatch negativity* (MMN) (Näätänen et al., 1978; Näätänen, Gaillard and Mäntysalo, 1980b; Näätänen and Michie, 1979a, 1979b; Näätänen, 1979, in press a; Näätänen, Simpson and Loveless, in press a; Näätänen, Gaillard and Varey, 1981; Näätänen, Sams, Järvilehto and Soininen, in press b).

In one of these studies (Näätänen et al., 1980b), the subject performed a dichotic-listening task, in which he detected and counted silently rare changes in pitch ('deviant stimuli') in one ear and ignored the input to the other ear. With constant ISIs of 800 msec, tones were delivered either to the left or the right ear in random order. The left-ear 'standard' and deviant stimuli were of 1000 Hz and 1150 Hz, respectively, and the respective pitches for the right ear were 500 Hz and 575 Hz. The probability of a deviant stimulus for each ear varied from 0 to 9%.

The results are presented in Fig. 1. The standards and deviants elicited a rather similar N1 deflection (peak latency approximately 100-120 msec). The

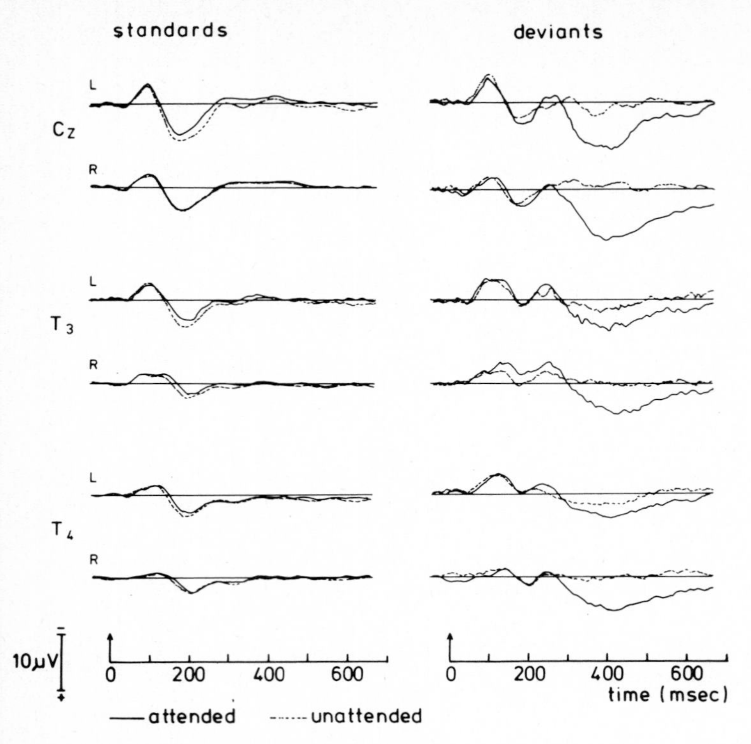

Fig. 1 Temporal and vertex EPs to standards and deviants for the left (L) and right (R) ear (from Näätänen et al., 1980b). Negativity is upwards in this and all subsequent figures.

deviant stimuli elicited a remarkable N2 deflection (peak latency 250-300 msec) which was not seen in the traces for the standard stimuli. The amplitude of the N2 deflection to the deviant stimuli in the attended and unattended ear was equal, demonstrating no significance or target effect. On the other hand, the large late positive deflection P3 or P300 (Sutton, Braren, Zubin and John, 1965; Wilkinson and Morlock, 1966; for reviews, see Pritchard, 1981 and Roth, this volume) was much larger for target than for non-target deviants.

Figure 2 presents subtraction curves in which the ERP to the standards was subtracted from the ERP to the deviants. In comparison to the ERPs to the standards, ERPs to the deviants disclose a continuous negative shift, with an onset around the peak latency of N1 and with a duration of ca. 200 msec. In the data recorded from the lateral electrodes, this negativity is very similar for the attended and unattended stimuli, followed by a large positive shift only to deviants in the attended input (target effect). An interesting feature of these data was that the lateral amplitudes of this negativity obtained at the T3 and T4 locations were significantly larger than those over the vertex. This suggests that this negative shift is modality-specific.

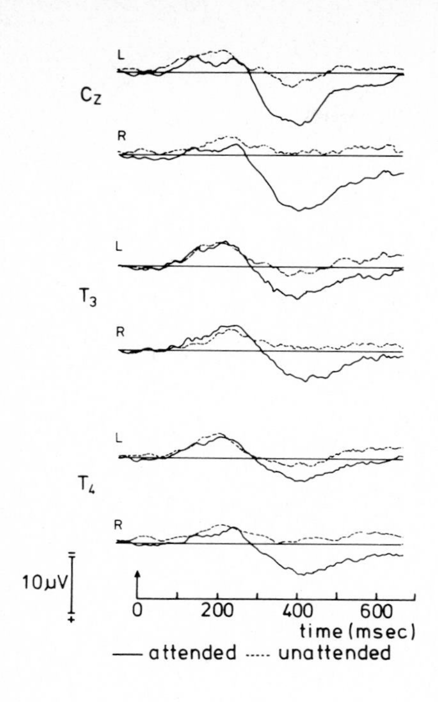

Fig. 2 The difference waveforms of the EPs presented in Fig. 1. The EPs to standards are subtracted from the EPs to deviants (from Näätänen et al., 1980b).

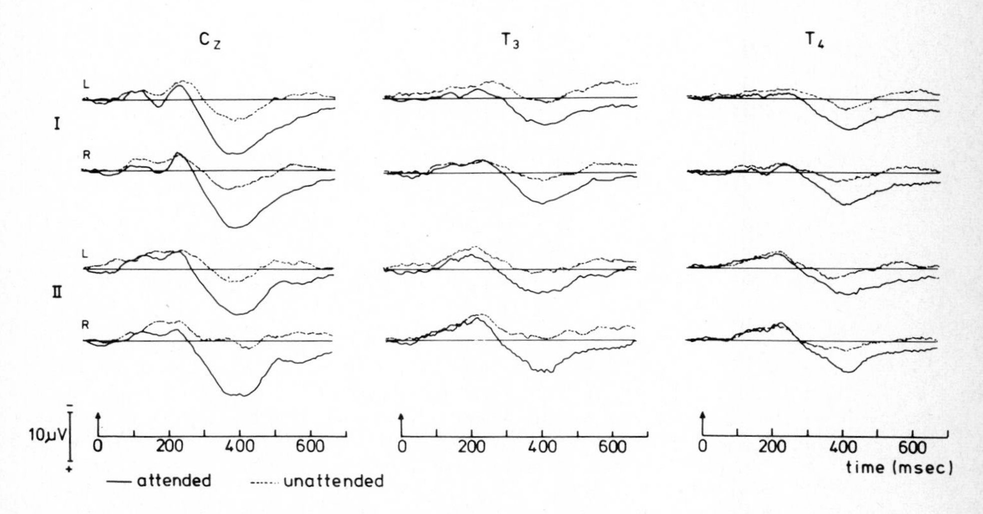

Fig. 3 Temporal and vertex difference waveforms for the left (L) and right (R) ear, separately for two experiments (I, II). The EPs to standards are subtracted from the EPs to deviants (from Näätänen et al., 1978).

In Fig. 3, difference waveforms from two similar experiments show the generalizability of these results. In contrast to Näätänen et al., 1980b, in this study (Näätänen et al., 1978) there was no pitch difference between inputs to the two ears. The deviants differed from the standards by increased intensity in Experiment I and by a higher pitch in Experiment II.

In all three experiments the MMN was elicited by deviants, whether they were attended or not. This insensitivity to stimulus signficance was taken as evidence that the MMN reflected a pre-perceptual physiological representation of stimulus deviance in a repetitive, homogeneous stimulus sequence,

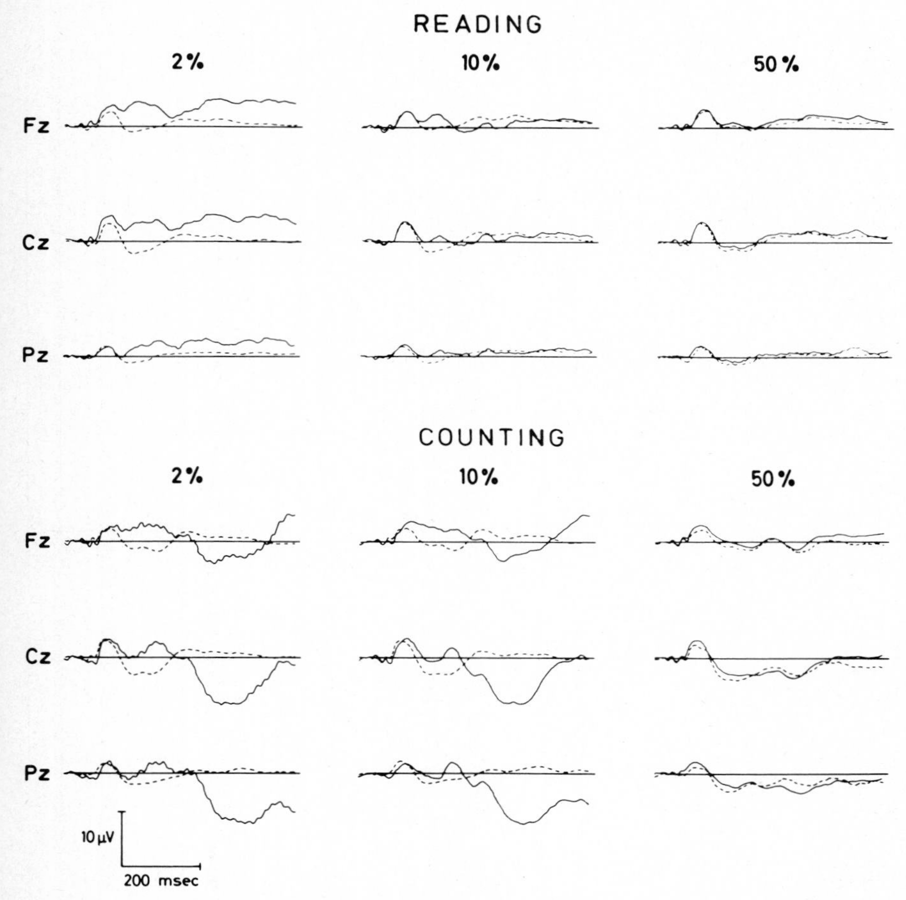

Fig. 4 Frontal, vertex and parietal EPs to deviants (solid line) and standards (broken line), as a function of the probability of the deviants (2, 10, or 50%). Stimulus onset is 30 msec after the beginning of the trace. The horizontal line is drawn at the average of this 30 msec period. Top panel: Reading condition; Bottom panel: Counting condition (from Näätänen et al., in press b).

i.e., a brain process which itself does not involve cognitive stimulus discrimination. This process was rather viewed as forming a gradual automatic build-up of a central representation of environmental change preceding such a discrimination process (e.g. Näätänen et al., in press a). Hence it was thought that the process which generates the MMN is the internal mismatch signal of the brain in the case of an abrupt change of a steady environmental stimulus. That generator process would constitute a necessary but not a sufficient condition for a conscious detection of the change. That generator process was regarded as an entirely *exogeneous* process and the codes (Uttal, 1965) of perception and discrimination should consequently be found in other processes.

In addition to the MMN, there appears to be a second component in the N2 time range, mostly superimposed on the later part of the MMN. This is illustrated by figures 4 and 5 which are from a study in which the a-priori probability of deviant stimuli was varied (Näätänen et al., in press b). This probability was either 2%, 10%, or 50%. The standard stimulus was a tone of 1000 Hz and the deviant stimulus 1044 Hz. For each probability, there was a reading condition, in which the subject was instructed to read a book, and a counting condition in which the subject counted the number of the deviant stimuli given in each block.

It is shown in Fig. 4 that in the reading condition the N2 negativity has a frontocentral distribution and a larger amplitude for the small probability (2%), whereas this negativity is absent in the 50% condition; presumably because in this condition the 'deviant' stimulus is not deviant anymore.

It can be seen in figure 4 that in the N2 time range the frontal data are similar for the reading and the counting condition. However, the central and parietal waveforms between the two conditions are different. In the

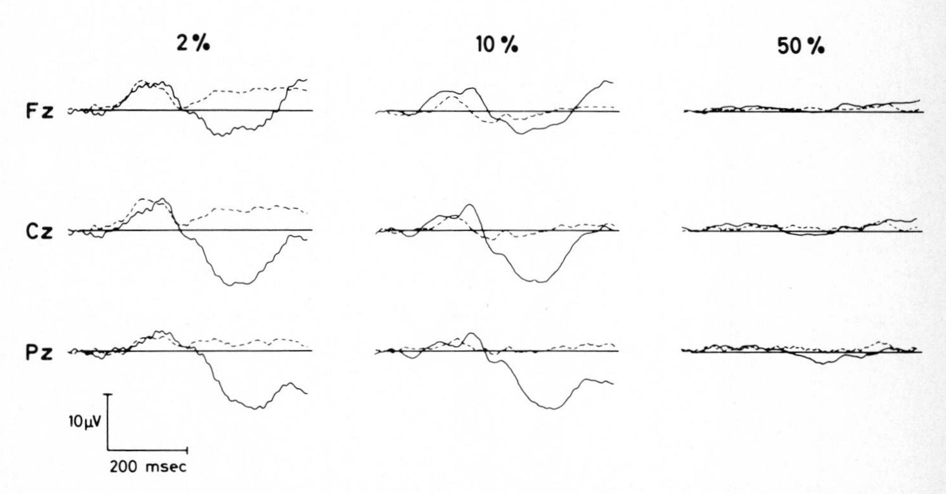

Fig. 5 The difference waveforms of the EPs presented in Fig. 4. Reading condition (broken line), counting condition (solid line). The EPs to standards are subtracted from the EPs to deviants (from Näätänen et al., in press b).

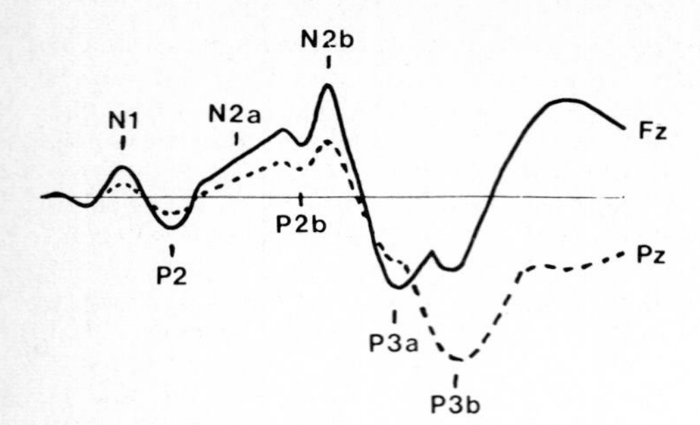

Fig. 6 A schematic illustration of the suggested component structure of the EP to a deviant stimulus under detection conditions. The time range involving the greatest overlap of components (N1-N2b) is stretched for illustrative purposes. N2a refers to MMN and P2b to the P165 component (from Näätänen et al., in press a).

counting condition there is a rather sharp negative peak, which occurs somewhat later (ca. 220 msec) and has a centro-parietal distribution. A careful examination of the topography of the subtraction waveforms for both conditions (Fig. 5) suggests that there are in fact two negative components in the N2 time range. An early component located anteriorly with a ramp-like form and a later peak which appears to be superimposed on the earlier component. This can be deduced from the fact that in the reading condition the topography of the negative deflection remains the same throughout the N2 time range, whereas in the counting condition the topography shifts posteriorly in the later part of N2 time range, when the second abrupt shift can be observed (see Fig. 5, in particular for the 10% probability).

This is illustrated schematically in Fig. 6 for the counting situation. The first component, the N2a is the 'mismatch negativity' (MMN) and the second component is the N2b (see Renault and Lesèvre, 1978, 1979; Renault, this volume).

The N2b is succeeded by a relatively sharp positivity, called P3a and by a large late parietal positivity ('slow wave'). Several studies have shown that the N2b and P3a are closely associated and form a complex together (e.g. Snyder and Hillyard, 1976; N. Squires, K. Squires and Hillyard, 1975; see also Picton and Stuss, 1980). It is even possible that this complex has an early positive component which is likely to be the 'P165' described by Goodin, Squires, Henderson and Starr (1978c). Picton and Stuss (1980) suggest that there is an endogeneous wave complex associated with signal detection, consisting of P165, N2, P3a, LPC (late positive component), and slow-wave components. They propose that the P165 and N2 components are associated with processes underlying signal identification while later components might reflect stimulus evaluation.

An N2b-P3a complex appears to be present in the vertex subtraction data of figures 2 and 3 for the attended input but not for the unattended input. In the lateral data (T3, T4), the complex is absent for both input channels. These data are consistent with the interpretation that the complex is triggered by the deviant stimuli in the attended input and that the complex does not extend laterally as far as the MMN which is similarly elicited by deviant stimuli within the attended and unattended input. Whereas the MMN has a modality-specific topography, that of the N2b, and hence of the N2b-P3a complex, appears unspecific.

It is tempting to postulate that the MMN is generated by a mismatch process, which may serve as a signal for further processing in the counting conditions. These further, task-dependent, brain processes may in turn be reflected by N2b.

With the distinction between the two N2s in mind, we will now systematically examine the main characteristics of both negativities. In the existing literature it is often difficult to separate these two effects, because in many cases the data were analyzed and interpreted in terms of a unitary N2.

In the present review the total negativity in the N2 time range is referred to as the 'N2 deflection'. The 'mismatch negativity' (MMN) component refers to the suggested contribution to the ERP waveform from the generator process which we assume reflects an automatic cerebral response to physical stimulus deviance (N2a in figure 6). N2b is a second negative component in the N2 time range. In contrast to the MMN, it appears to be task-dependent, occurring mainly in attention conditions, and belonging to a complex together with the P3a. Therefore, in ignore conditions (e.g. reading) only the MMN component will occur, whereas both components will be present in response to physically deviant stimuli under attention conditions. This gives us the possibility to differentiate between the two components and to study their characteristics.

CHARACTERISTICS OF THE TWO N2 COMPONENTS

1. Morphology

It is difficult to determine the onset latency of the MMN because the stimulus change also activates fresh afferent fibers which may enhance the N1 component. In most situations the onset latency appears to be ca. 100 msec (see figures 1 and 2 for typical examples). In the study of Simson, Vaughan and Ritter (1977a), the N2 deflection in their subtraction data started at 130 msec (± 25 msec) in an auditory task and at 170 msec (± 30 msec) in a visual task.

The time course of the MMN is dependent on the magnitude of the physical stimulus deviance. The MMN generally reaches its peak latency at 200-300 msec post-stimulus but the peak may occur later for very small, and earlier for very large, stimulus changes. This is illustrated in a reading condition with a standard tone of 1000 Hz (see Fig. 7). For the 3020 Hz tone the N2 peak occurs at 140 msec and for the 1020 Hz tone at 225 msec.

The duration of the MMN is also dependent on the magnitude of stimulus change. With very small stimulus changes the MMN may last for several hundreds of msecs, whereas its duration may be 100 msec when a very large stimulus change is involved (see the reading condition in figure 7).

The time course of the N2b component appears to be more rigid than that of the MMN. This is suggested by a study of Renault, Ragot, Lesèvre and Remond (in press; see also Renault, this volume). They analysed the ERP associated with stimulus omission in a regular visual stimulus sequence on a single-trial level. The subject's task was to flex his or her finger as fast as possible after detecting a stimulus omission. At about the moment of omission, a negative shift peaking at the parieto-occipital region (peak latency 220 msec) started to develop whose duration strongly correlated with the reaction time to the stimulus omission. Superimposed on the latter part of this rampllike negativity was a sharper negativity, which was centrally

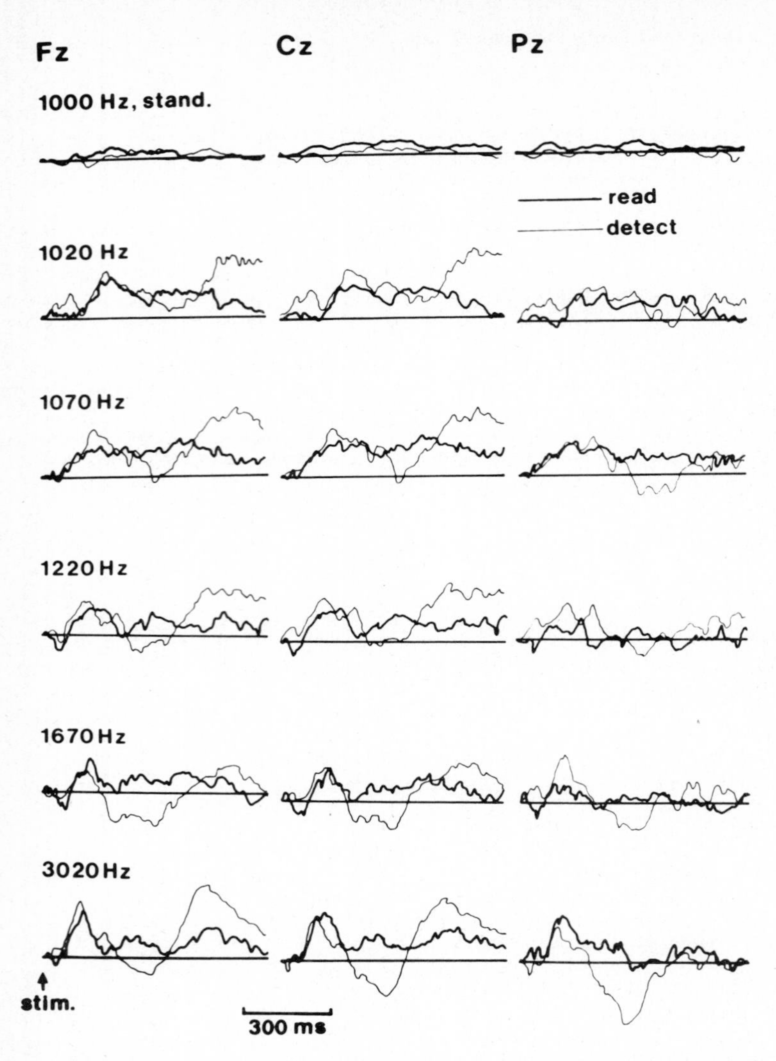

Fig. 7 Frontal, vertex, and parietal EPs to standard stimuli (probability 90%) and to each of five deviant stimuli (probability of each at 2%) randomized in the same stimulus block. Thin line refers to a detection condition where the subject counted all the deviants together, the thick line refers to the reading condition in which the subject ignored all the auditory input. Data of one subject (unpublished data of Loveless, Simpson and Näätänen).

distributed (peak latency 265 msec). The duration of the latter negativity, however, was quite constant and did not correlate with the reaction time. This negativity, which appears to be N2b, was always linked to a sharp positivity, presumably P3a with a similar central topography. On the other hand, the earlier, ramp-like negativity, which was modality-specific, cannot be a MMN because its onset occurs on average at the moment of stimulus omission. If it were, then there should be some latency as in the case of a physical deviance. A similar result was obtained in a study of Simson, Vaughan and Ritter (1976), in which the missing-stimulus negativity also started around the moment of stimulus omission. Thus, this missing-stimulus negativity can be regarded as a reflection of the mental representation of the regularly appearing stimulus triggered by the strong temporal expectation. Simson et al. (1976) interpreted their data similarly.

2. Topography

The data shown in figures 2 and 3, as already discussed, suggest that the MMN is modality-specific. In these dichotic-listening studies, the temporal subtraction waves were larger than, or as large as, those from the vertex data. (However, in a monaural stimulus situation this result was not found; Näätänen et al., 1980a.) The subtraction curves in figures 2 and 3 suggest that it is possible to distinguish the two N2 negativities on the basis of the lateral topography. In response to the deviant stimuli in the attended input, the N2b-P3a complex is present in the vertex data but not in the temporal data, where the MMN is prominent.

Simson et al. (1977a) have provided evidence for a modality-specific topography of the N2 deflection. In the auditory task they presented 'standard' tone bursts of 2000 Hz at 0.5-sec ISIs whereas the deviant tone bursts were of 1000 Hz. The topographical distribution of the subtraction waveforms was explained by the authors as reflecting "a source within the supratemporal plane projecting to the cortical surface in the central region, which sums with activity generated by a secondary source on the surface of the posterior portion of the superior temporal gyrus" (Simson et al., 1977a, p. 533). Later they concluded that this negativity appears to be generated principally within secondary auditory cortex, overlapping the N1 and P2 components of the auditory evoked potential.

In the preceding section it was shown that in the counting condition in the study by Näätänen et al. (in press b) the early part of the subtraction waveform had a frontal distribution and the later part centro-parietal one. In the reading condition, the topography of the subtraction waveform remained frontal or fronto-central throughout its duration suggesting that in that condition only the MMN occurred. However, it is not always possible to carry out a similar decomposition of the two negativities. For example in Näätänen et al. (in press a) both negativities had a frontal or fronto-central distribution.

Simson et al. (1977a) observed a modality-specific (parieto-occipital) topography in response to deviant stimuli in a visual discrimination task analogous to their auditory discrimination task. This subtraction waveform appeared to include a second component which is consistent with the present suggestion involving the two negative components in the N2 time range; Comparison of the subtraction waveforms obtained from frontal derivations with those from the posterior derivations disclosed "a difference in the waveshape, with the negative wave beginning and reaching its peak considerably later in the midfrontal region than elsewhere on the scalp" (Simson et al., 1977a, p. 533). This additional component seems to have occurred both at frontal and central electrode locations in their data (their figure 1) and it might consequently be an N2b component.

Further evidence for the existence of an N2b component in visual tasks was provided by Renault and his colleagures, reviewed in the previous section (see Renault, this volume). The available results obtained with two modalities are consistent with the view that physical stimulus deviance in a repetitive homogeneous stimulus sequence is associated with the modality-specific MMN, which is accompanied only in attention conditions by a second negativity with a sharper waveform and a non-specific topography (N2b).

3. Association with other ERP components

Early reports on 'N2' or 'N200' already suggested a close connection between that negativity and a frontal or fronto-central positivity (N. Squires, K. Squires and Hillyard, 1975; Snyder and Hillyard, 1976). However, it was shown by Näätänen et al. (1978, 1980b) that that negativity can appear even without a succeeding positivity, as is observed in ERPs to deviant stimuli in an unattended input in a dichotic listening situation. This negativity was described as the MMN and the N2b component was not present in this situation (figures 1-3). On the other hand, the deviant stimuli in the attended input elicited an N2b and this was accompanied by a P3a. Moreover, a slow wave (a late parietal positivity) seemed to emerge. Hence, it appears that the MMN is not intimately linked with any positive wave, whereas N2b never occurs without the P3a, that is, they form a unitary, inseparable ERP component: the N2b-P3a complex. In Renault and Lesèvre's (1978, 1979) data on omitted visual stimuli, there was usually first a ramp-like modality-specific negativity, which they called 'N2a'. This was followed by a sharp negative peak 'N2b' which had a central or fronto-central topography, like the positive peak 'P3a' which followed it. Their data are particularly convincing in linking N2b and P3a together to form a complex. For instance Renault and Lesèvre (1979) did not find one ERP to stimulus omission in which those two were dissociated. Further evidence was obtained by Näätänen et al. (in press a). They additionally observed that the N2b-P3a complex was preceded by a sharp positive wave of small amplitude with a similar midline topography. That positivity was identified as the 'P165' of Goodin et al. (1978c). In retrospect, a similar deflection appears to be present also in the vertex subtraction data of figures 2 and 3. In comparing vertex ERPs to the deviant stimuli in the attended and non-attended input in Fig. 2, the P165-N2b-P3a complex appears to be superimposed on the MMN in response to the deviant stimuli in the attended input. The complex is confined to the central electrodes whereas the MMN is large even at the lateral electrodes (T3, T4).

Hence, we conclude that the MMN is not linked with any positivity whereas the N2b is intimately linked with a subsequent positivity which has a similar topography and, perhaps, with a preceding positive component reflected by the 'P165'.

4. Dependence on stimulus deviance

The MMN seems to be elicited only by physical stimulus deviance and probably not by other stimulus conditions. This would qualify the MMN as a specific reflection of the neuronal mismatch process elicited by a deviant stimulus presented in a repetitive homogeneous stimulus series. As already mentioned, Snyder and Hillyard (1976) did not find any kind of N2 deflection in response to single loud clicks presented at very long ISIs. However, in the Ritter et al. (1968) data, an N2 appears to be contributing to the waveform (e.g., their figure 4) in response to the first auditory stimulus after a long interval of 2-5 min between stimulus sequences. In our view, the negativity observed in this and in other studies which do not involve physical stimulus deviance, is the N2b rather than the MMN.

Ritter and co-workers (see Ritter, Vaughan and Simson, this volume) presented slides representing words belonging to two semantic categories. 80% of the words were male names and the rest female names. When the subject was instructed to press a button each time a slide belonging to the rare category was delivered an N2 deflection was observed. A much larger N2 deflection with a different topography was obtained in response to a physically deviant stimulus in an otherwise similar experimental situation. However, when the subject was instructed to press to all the slides, no N2 deflection was elicited by slides belonging to the rare semantic category but an N2 deflection was still elicited by physically deviant stimuli. The MMN was evidently involved in the latter instance, and the N2 deflection observed in the semantic discrimination task may be the N2b. The small magnitude of the N2 negativity in this task lends credence to this suggestion, - the contribution of the MMN was apparently missing. Moreover, the N2 topography in the semantic discrimination task was different from that found in the physical discrimination task.

The authors themselves interpreted the N2 deflection as a specific reflection of the occurrence of a discrimination process under each of their conditions. The N2 topography in the semantic discrimination task indeed disclosed some features which suggest that it cannot, at least entirely, be regarded as a non-specific N2b. These topographical features were the laterality and hemispheric asymmetry lending credence to the authors' interpretation in terms of a discrimination N2. It is possible that this negativity is related to the processing negativity (Näätänen et al., 1978; Näätänen, in press b).

In summary, the MMN is only elicited by physical stimulus deviance whereas the occurrence of the N2b is not dependent on stimulus deviance.

5. Type of stimulus deviation

The MMN can apparently be elicited by any kind of physical stimulus deviance such as an increase or decrease in intensity of an auditory stimulus (N. Squires et al., 1975; Snyder and Hillyard, 1976; Towey, Rist, Hakerem, Ruchkin and Sutton, 1980; Näätänen et al., 1978, 1981), change in pitch (Ford et al., 1976b; Ford, Roth and Kopell, 1976a; Goodin et al., 1978c; Vaughan, Ritter and Simson, 1980; Näätänen et al., 1978, 1980b, in press a, b), or the shape of a visual stimulus (see Ritter et al., this volume). The MMN appears to be elicited even by an occasional change in ISI duration (Ford et al., 1976a; Ford and Hillyard, 1981).

As reviewed in the previous section, in contrast to the MMN, the elicitation of the N2b appears not to be dependent on physical stimulus deviance; physical stimulus deviance elicits the N2b mainly in attention conditions (see § 7 below).

6. Magnitude of stimulus deviance

The sensitivity of the N2 deflection to stimulus change is crucial in attempts to understand that deflection. It is even elicited by undetected deviant stimuli in a situation in which the subject's task is to detect slightly deviant stimuli in a homogeneous stimulus stream (Näätänen et al., in press a). This question was further explored by these authors in a study in which the standard stimulus was of 1000 Hz and the deviant stimulus usually of 1010 Hz individually determined for each subject according to his or her discrimination accuracy. When the subject was certain that a deviant stimulus was presented the response was to be given with one hand and when he was not certain the other hand was to be used. All the subjects produced a notable N2 negativity, even to undetected deviant stimuli (Fig. 8). In the latter case the N2 amplitude was mostly about the same magnitude as that for (correct) detections and 'non-sure' responses. In contrast, the late positivity was only elicited by detected deviants (see Pz in figure 8) and 'non-sures'. This supports the suggestion involving an automatic cerebral process which is a necessary, but not a sufficient condition for the conscious perception of stimulus deviance. In terms of the present hypothesis, the N2 negativity observed is the MMN.

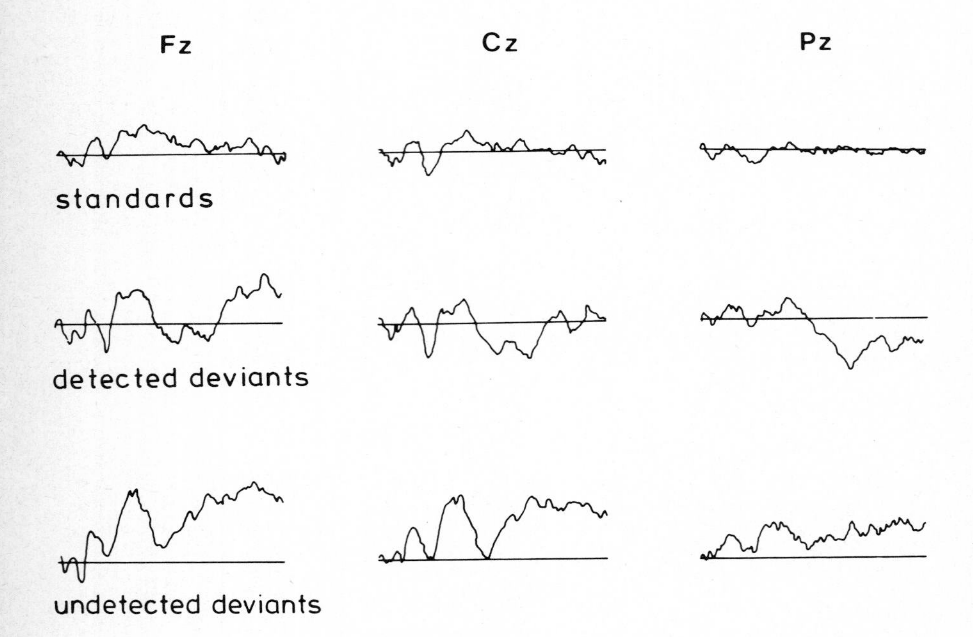

Fig. 8 Frontal, vertex, and parietal EPs to standard stimuli, detected deviant stimuli and undetected deviant stimuli. Standard stimulus 1000 Hz, deviant stimulus 1010 Hz. Trace length 768 msec. Data from one subject. (Unpublished data of Loveless, Simpson and Näätänen).

The amplitude of the N2 deflection is increased and its latency shortened when the magnitude of the physical stimulus deviance is increased (Ford et al., 1976b[2]; Ritter, Simson, Vaughan and Friedman, 1979; Ritter et al., in press; Towey et al., 1980; Lawson and Gaillard, 1981; Näätänen et al., 1980a; in press a). This effect can be illustrated by a study of Näätänen et al. (1980a), in which the standard tone was 1000 Hz and the two deviant stimuli had a pitch of 1044 and 1500 Hz (5% probability of occurrence each, randomized in the same block), which had to be counted. It can be seen in Fig. 9, that the N2 deflection is larger and occurs earlier for widely deviant stimuli (solid line), than for the slightly deviant stimuli (dotted line).

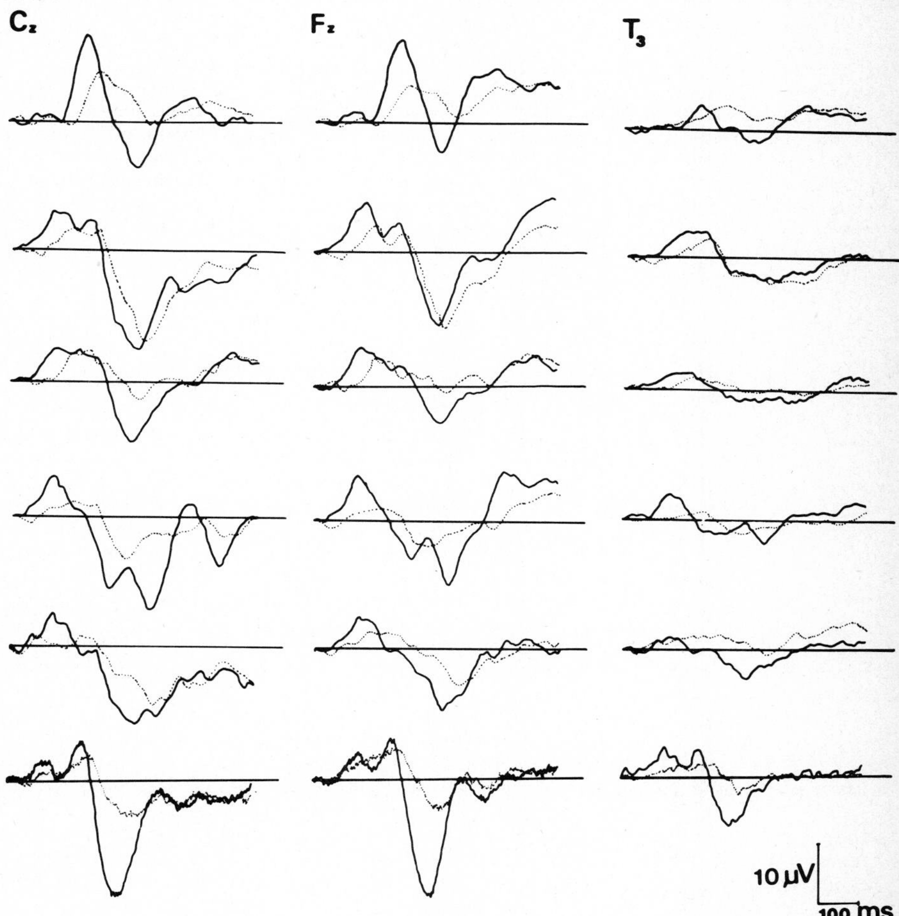

Fig. 9 Vertex, frontal, and temporal difference curves separately for each subject (row 1 for subject 1, row 2 for subject 2, etc.). The EPs to standards (1000 Hz) were subtracted from EPs to widely deviants (1500 Hz) - solid line; and from EPs to slightly deviants (1044 Hz) - dotted line (from Näätänen et al., 1980a).

In most experimental situations it is not possible to examine separately the MMN and N2b as a function of the magnitude of stimulus deviance. However, the study of Näätänen et al. (in press a) suggests that both are increased in amplitude and occur at shorter latency when this magnitude is larger.

It appears that the increase in the amplitude of the N2 deflection as a function of the magnitude of stimulus deviance reaches a ceiling at quite moderate stimulus changes (see Fig. 7) whereas the peak latency and duration of the N2 deflection may continue to change as the magnitude of stimulus deviance increases (Fig. 7). Both become shorter with greater deviance.

Corroborating results were obtained in a study by Lawson and Gaillard (1981) in which stimulus deviance was manipulated by varying the number of dimensions in which the stimuli differ. In their study two phonemes were presented in random order, one at a probability of 15% and the other of 85%. In different runs the phonemes differed in one (place of consonant or vowel), two (place and voicing of consonant) or three dimensions (place of consonant and vowel + voicing of consonant). The subject's task was to press a button to the infrequent stimuli. These stimuli elicited prominent N2 deflections which were larger in amplitude and earlier in latency when the two stimuli differed in more dimensions.

7. Sensitivity to attention

In the preceding section, it was shown that an N2 deflection can be observed in response to missed deviants (Näätänen et al., in press a; figure 8). This is very strong evidence that the MMN can be regarded as a reflection of a pre-perceptual mismatch process in the brain. In our view the mismatch process is not modified by perceptual and attentional factors. Fig. 8 shows that the N2 deflections to detected and missed deviants have approximately the same amplitude. When task-related differences in the N2 deflections exist, we attribute them to the presence/absence of the N2b (see also Näätänen, in press a, and Näätänen et al., in press a). There is of course a serious risk of circularity here but independent evidence is provided by (a) the unitary waveform of the N2 deflection in ignore conditions (such as reading), whereas in task conditions there are often two peaks (see figures 4, 5 and 7); (b) in ignore conditions, the topography of the N2 deflection remains the same throughout the N2 time range, whereas in attention conditions, a shift in the midline topography toward a more central focus occurs (figures 4 and 5), suggesting an overlap by a later negative component; (c) when this topographical shift occurs it is preceded by an early portion of negativity with a topography very similar to that in the ignore conditions.

Hence it is concluded that the MMN occurs independently and unmodified by attentional and perceptual factors, whereas the N2b is strictly dependent on such factors. It is possible, however, that some of the N2b is elicited in the ignore conditions when the stimulus deviance is so large as to be obtrusive (Näätänen et al., in press a).

Similarly, when the probability of the deviant stimuli is very small, the N2b-P3a complex will also be elicited by deviants, even when they are not designated as a target. In the study of Näätänen et al. (in press a), there were 3 deviants (within the same stimulus sequence), of which only one was a target, i.e. had to be counted. The amplitude of the complex was the same for target and non-target deviants.

However, when two easily separable input categories (inputs to the left and right ear) are concurrently used and the deviants in one input category are defined as targets and the other input category has to be ignored, the N2b deflection occurs only to the attended-input deviants (Näätänen et al., 1978, 1980b; figures 1-3 of the present paper).

A consistent result was obtained by Donald and Little (1981; see also Donald, this volume). They presented two independent stimulus sequences dichotically, instructing the subject to detect occasional deviant stimuli (of slightly different pitch) in one ear and to ignore the input to the other ear. Both inputs included deviant stimuli and their probability (which was equal for both ears) was varied from 0.25 to 0.0625 between different blocks. The probability had no appreciable effect on the ERP to the deviant stimuli in the unattended input but a drastic effect for the attended input. From their published data, it is impossible to separate the contributions of the two negative components to the total N2 deflection but it appears evident that the N2b-P3a complex only occurred to the deviant stimuli in the attended input and was strongly enhanced by decreased probability.

8. Probability of stimulus deviance

Within a certain probability range, the N2 deflection is larger the smaller the probability of the deviant stimulus (N. Squires et al., 1975; K. Squires, Donchin, Herning and McCarthy, 1977; Näätänen et al., in press b). The latter study investigated this effect separately for count and ignore conditions and the results obtained suggest that the MMN is more probability dependent than the N2b. In comparing the 2% and 10% conditions there was a probability effect on MMN in the reading condition whereas the N2b was about equal for these two probability levels (figures 4 and 5). On the other hand, there was no N2b peak in the 50% counting condition suggesting a probability effect on the N2b. The study of Donald and Little (1981) reviewed above also argues against the probability independence of the N2b, since their N2 deflection in response to deviants in the attended input showed great sensitivity to probability.

9. Sequential and temporal effects

The probability effects described in the previous section can be effects of either the a-priori or the sequential probability (see K. Squires, Wickens, N. Squires and Donchin, 1976; K. Squires, Petuchowski, Wickens and Donchin, 1977a). When the probability of a deviant stimulus is small, the standard has often been presented many times in succession prior to a deviant stimulus. This presumably strengthens the neuronal model of the standard stimulus and therefore a deviant stimulus would elicit a vigorous cerebral mismatch reflected by a large MMN. On the other hand, when two deviant stimuli happen to occur consecutively, the MMN to the second may be small or absent irrespective of the constant over-all probability in the stimulus block. This question was investigated by Sams, Alho and Näätänen (in preparation). The probability of a 1150 Hz deviant tone was 10% and the standard stimulus was a similar tone of 1000 Hz. In Fig. 10, ERPs are averaged

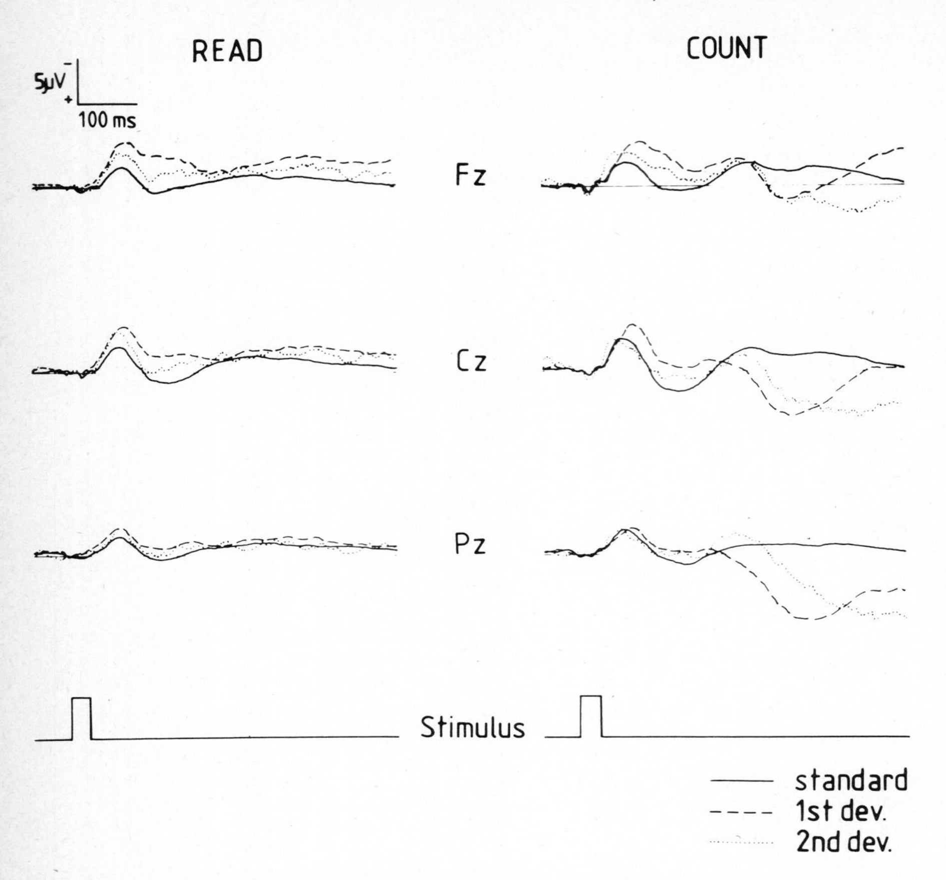

Fig. 10 Frontal, vertex and parietal EPs to the standard stimuli (solid line), to the first deviant stimuli (dashed line), and to the second successive deviant stimuli (dotted line) separately for reading and counting conditions (from Sams, Alho and Näätänen, in preparation).

separately for the first-deviant stimuli (after at least two consecutive standard stimuli) and the second-deviant stimuli (only every 100th stimulus on average was a 'second-deviant'). The amplitude of the MMN is larger for the first-deviant stimulus. However, the N2b-P3a complex is roughly the same for the first and second deviant (see the frontal and vertex data of the counting condition), which could be taken as another indication of the separability of the MMN and the N2b. The slow frontal negativity is absent in the ERP to the second deviant whereas the slow parietal positivity remains.

If the MMN is a reflection of an automatic cerebral mismatch process occurring when the sensory input does not match with the prevailing neuronal

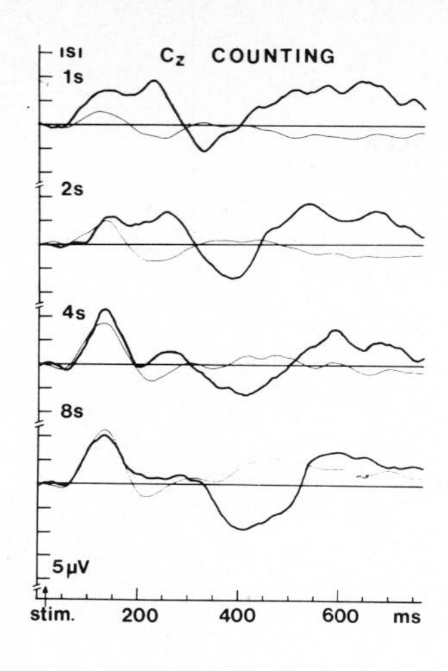

Fig. 11 Vertex EPs to standard stimuli (thick line) and to deviant stimuli (thin line) for ISIs of 1, 2, 4, and 8 sec in a counting condition; data from one subject (from Mäntysalo, Alho and Näätänen, in preparation).

model, then the MMN could be used to study the development and decay of the neuronal model. This was attempted in a study by Mäntysalo, Alho and Näätänen (in preparation) by varying the constant ISI from 1 to 8 sec between stimulus blocks. In this study the standard stimuli were tones of 950 Hz and the deviant stimuli to be counted were similar tones of 1150 Hz. The N2 deflection was much larger for the shorter ISIs than for the longer ISIs but there was some N2 deflection even in the 8 sec-ISI condition (Fig. 11). Tentatively, the MMN appears to be more sensitive to ISI duration than the N2b.

Fitzgerald and Picton (1981) also varied the constant ISI between stimulus blocks, from 0.25 sec to 4 sec. Standard tones were 1000 Hz and deviant tones (p = 20%) 2000 Hz or vice versa. The subject's task was to count the deviant tones. An N2 deflection was not readily observable for the two shortest ISIs and was maximal for the ISI of 2 sec (Experiment I). When the probability of the deviant stimulus is high, and the ISI duration short, as was the case in their study, only a short time will elapse on average between the delivery of two deviant stimuli. This high temporal probability may eliminate the MMN. Consistent results were obtained in their third experiment in which the N2 deflection was larger with decreasing time uncertainty. Consequently we suggest that if they had used a smaller probability they would have found the largest amplitudes with shorter ISIs than 2 sec. Therefore there is not necessarily any conflict between their results and those of Mäntysalo and colleagues. Because Fitzgerald and Picton (1981) did not have an ignore condition, it is impossible to disentangle possible differential effects on the MMN and the N2b in their data.

If the neuronal model involved in producing the MMN has indeed a short life time, then the MMN should disclose no long-term habituation. This is illustrated by figure 12, from an unpublished study from Loveless' laboratory. Figure 12 represents ERPs to the first eight and last eight deviant stimuli in a long sequence of tones. It can be seen that the N2 deflection for all the four subjects in each electrode is almost identical at the beginning and end of the sequence.

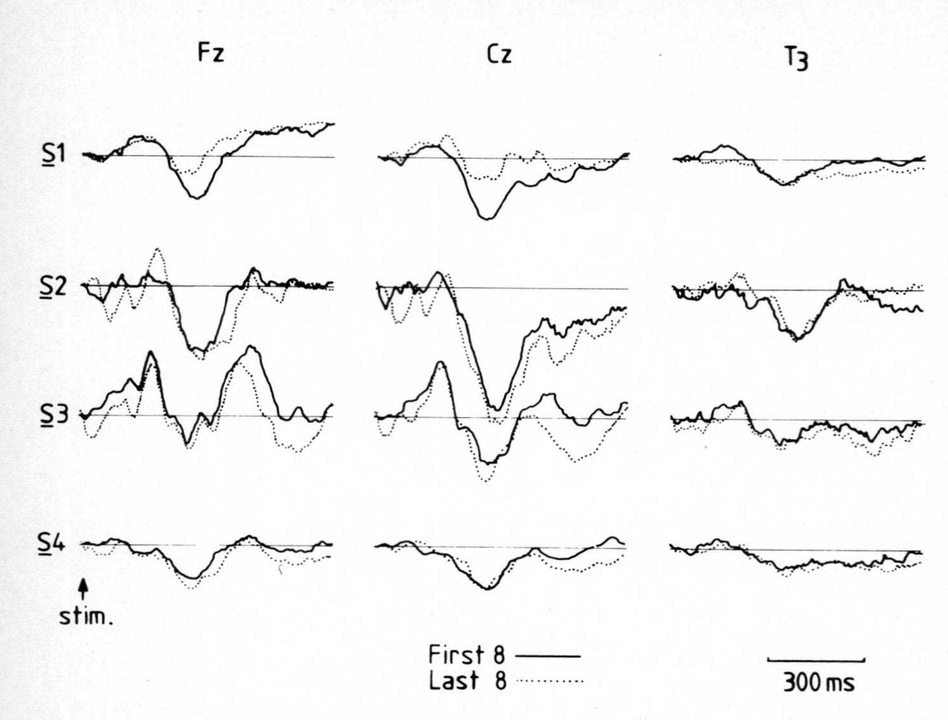

Fig. 12 Frontal, vertex and temporal EPs to the first 8 widely deviant stimuli (solid line) and last 8 widely deviant stimuli (dotted line) in a long sequence, separately for 4 subjects in a counting condition. The standard stimulus (probability 90%) was 2030 Hz, the widely deviant stimulus (probability 2%) 2999 Hz. (Unpublished data of Loveless, Simpson and Näätänen.)

GENERAL DISCUSSION

The results on the N2 deflection reviewed in the previous section suggest that any physically deviant stimulus in a homogeneous repetitive stimulus sequence elicits the MMN. The results suggest that the mismatch negativity is a scalp reflection of the mismatch process in the brain, and that this mismatch process develops automatically, irrespective of what the organism is engaged in and of the perceptual or behavioural demands. It also appears that the characteristics of the MMN such as its amplitude, locus, and time course are not modified by perceptual and behavioural factors.

The automatic cerebral mismatch process suggested here on the basis of ERP research is presumably of great interest to modern cognitive psychology where one of the main issues presently involves parallel vs. serial information processing. The reviewed ERP evidence involving the cerebral mismatch process suggests that highly accurate sensory analysis can take place automatically, without attention and conscious effort, at a high level of the central nervous system. This can be taken as evidence for parallel processing at the cortical levels. Particularly important in this respect are data disclosing a MMN in the unattended input in a selective dichotic listening situation.

An indication of the pre-perceptual nature of the mismatch is the fact that detected and missed deviants elicit a similar MMN (Fig. 8), which is interesting in view of general principles of brain functioning. The MMN is the outcome at a cortical level of a purely physiological process which is automatically triggered by the deviant stimulus and which is unaffected by cognitive or other psychological factors.

The mismatch process is of course a key postulate in the OR theory (Sokolov, 1960). Sokolov indirectly inferred the occurrence of a mismatch process from a variety of physiological data obtained from OR paradigms and suggested that that process initiates the OR. Now the question arises whether the mismatch process reflected by the MMN and the mismatch process postulated by Sokolov are identical. At the present stage of research, it appears that the two postulated processes have very much in common.

This question warrants a more detailed examination: The classical OR theory (Sokolov, 1960) claims that an OR is elicited by any stimulus for which a corresponding neuronal model does not exist. Therefore the first stimulus of a stimulus sequence tends to elicit an OR as does a stimulus change in that sequence of repetitive homogeneous stimuli. In the Introduction, we have referred to the OR elicited by the first stimulus as the initial OR. We suggest that in its elicitation, no comparison mechanism is involved. This is supported by the results which show that the first stimulus in a stimulus sequence does not elicit a MMN. Here the ERP research shows that the cerebral mechanism underlying the initial OR must be different from that underlying the change OR. The classical theory does not make this distinction (e.g. Bernstein, 1979) implying that in both cases the same neuronal-mismatch mechanism is activated, because an appropriate neuronal model is not available. Since the OR theory assumes that a cerebral mismatch process is elicited also by the first stimulus of a sequence, the mismatch process reflected by the MMN cannot be identical to the mismatch process postulated by the OR theory.

The mismatch process to stimulus change postulated by the classical OR theory appears closely related to the mismatch process reflected by the MMN. In particular, the sequential and temporal conditions of the elicitation of the MMN need further clarification. The OR as a function of various factors is quite well known, which provides the basis for inferences about the relationship of these factors and the neuronal mismatch process of the OR theory. Obtaining the same information for the MMN would presumably provide crucial information to determine how similar its generator process is to the mismatch process postulated by the OR theory.

An important step toward identifying the two processes is provided by the results demonstrating insensitivity of the MMN to attentional demands and stimulus significance. Sokolov (1975) has stressed that only physical and temporal stimulus factors affect the development of the neuronal model and the neuronal-mismatch process.

The topographical distribution of the mismatch negativity is modality-specific, which might be problematic in trying to identify the underlying mismatch process with that suggested by Sokolov (1975; and personal communication, November 1981). According to him, the mismatch process is generated by the novelty neurons of the hippocampus. The activity of these neurons ceases with repetition of an identical stimulus but is re-evoked by any change in physical or temporal features of stimulation (Vinogradova, 1975). This hippocampal activity presumably generated large non-specific negative fields on the scalp and it is hard to imagine how that hippocampal activity could be associated with modality-specific activity which would explain the modality-specific scalp topography. However, it is important to bear in mind that the hippocampal activity was recorded from rabbits and the MMN from humans. The prominence of the MMN above the specific sensory areas is in line with the assumed location of the mismatch process in these areas. However, there are probably no 'plastic' cells in these areas capable of preserving a neuronal model of a stimulus for a sufficiently long period of time (Sokolov and Vinogradova, 1975).

As mentioned in the Introduction, there has been a continuous controversy in the literature as to whether stimulus change is enough to elicit an OR or whether the change must be experienced as somehow significant by the organism before the OR emerges. The results on the MMN suggest that a mismatch process develops irrespective of psychological factors, such as 'attention' and 'significance'. Hence stimulus change in any case has a representation at the cortical level whether or not it is attended or perceived consciously. It is another question whether an OR always follows. In the present paper, we have observed that under some conditions the MMN was accompanied by the N2b-P3a complex, under others not. It is indeed tempting to suggest that when this complex (or any positivity) accompanies the MMN, the classical OR would be present, and that when the MMN is not accompanied by the N2b-P3a complex or any positivity, no OR is elicited by that stimulus change[3]. The conditions of the occurrence of the N2b-P3a complex are hardly known - but tentatively, they appear to be consistent with the conditions known to enhance the OR and to make it more resistant to habituation, e.g. directing attention to the stimulus category in question. As reviewed, the N2b-P3a complex does not occur in ignore conditions but it only elicited in attend conditions (when the magnitude of stimulus change is increased so as to make it 'obtrusive', the complex or some positivity may be elicited, even in ignore conditions). Similarly, in a selective dichotic-listening task, the deviant stimuli in a stimulus stream to be attended to (e.g. stimuli delivered to the left ear) elicited the N2b-P3a complex whereas the complex was not present to deviant stimuli in the ignored input (right ear).

It would be premature to speculate about the function of the generator processes of the N2b-P3a complex in OR elicitation but a conspicuous feature is the non-specific topography of the complex which strongly resembles that of the vertex N1 component. This suggests that they have the same generator mechanism. A parsimonious possibility is that the activation of that mechanism triggers the OR. This would explain the observed similarity of the initial OR and the change OR when automatic or peripheral measures are used. At the level of brain events the two forms of OR would, however, be quite different: The initial OR would be generated without any comparison mechanism whereas the elicitation of the change OR would be strictly dependent on the activation of a comparison mechanism. The MMN is a reflection of the activation of the latter mechanism.

NOTES

Supported by the Academy of Finland. The authors gratefully acknowledge the constructive criticism of E.N. Sokolov, N.E. Loveless and W. Ritter.

Send requests for reprints to A.W.K. Gaillard, Institute for Perception TNO, P.O. Box 23, 3769 ZG Soesterberg, The Netherlands.

FOOTNOTES

[1] The N1-deflection probably does have more than one generator, which may or may not be modality-specific. Only the non-specific effects are assumed to be related to the OR.

[2] Their amplitude effect failed, however, to reach statistical significance - though the tendency was quite apparent - which might be due to cancellation by a slow positivity which had a very early onset presumably because of the very large magnitude of stimulus deviance used. Their extreme deviant was 500 Hz when the standard was 1000 Hz and vice versa.

[3] However, elsewhere (Näätänen, in press a) it was suggested that even MMN alone would be associated with the kind of rudimentary stereotyped short-term OR with a function to facilitate perceptual processing of that stimulus. It was suggested that the MMN might be automatically linked with a transient spinal facilitation in a similar way as the vertex N1 component is known to be. The wide-spread negativity called sensitization negativity (Näätänen et al., in press a, b) might be another automatic concomitant of the MMN but when the probability of the deviant stimulus is increased the sensitization negativity appears to vanish before the MMN does.

Tutorials in ERP Research: Endogenous Components
A.W.K. Gaillard and W. Ritter (eds.)

6

ON RELATING EVENT-RELATED POTENTIAL COMPONENTS TO STAGES OF INFORMATION PROCESSING*

Walter Ritter, Herbert G. Vaughan, Jr. and Richard Simson
Departments of Neuroscience and Neurology
Albert Einstein College of Medicine
Bronx, New York
U.S.A.

The successful attempt to relate ERP components to specific stages of information processing would provide a means of understanding some aspects of how the brain works, as well as yield data relevant to theories of cognitive processing which have been based on behavioral observations. The ERP and behavioral sources of data both have inherent limitations, but taken together they might allow for converging operations to test theories common to both areas.

In this chapter, we describe recent experimental results obtained in our laboratory on two ERP components which we believe may reflect two stages of processing, namely, pattern recognition and stimulus categorization. Although RT data were obtained, the manner by which stages were identified was different than in behavioral studies. The first step was to identify a class of variables which affect the amplitude or latency of a particular component, or complex of components, and on the basis of the nature of the variables infer the functional significance of the associated physiological activity. We have relied mainly on latency because it can be related to the timing of behavioral responses. If the inferred function appears to be related to a general process, such as pattern recognition, then it was considered appropriate to hypothesize that the component is associated with a stage. An essential test as to whether such a component is associated with a processing stage would be whether a generic component is found with a variety of stimuli. A component related to stimulus classification, for example, should occur whether stimuli being classified are letters, digits, words, brackets or angles. In addition, the component should occur irrespective of the sensory modality of stimulus presentation. The term "generic component" is used because the precise locus of the brain activities, and the specific cellular processes associated with a particular processing stage, are likely to differ as a function of the nature and modality of stimulation. Thus, the scalp distribution of ERPs associated with a particular stage of processing may vary.

*This work was supported by grants HD 10804 and MH 06723 from the USPHS. The first author is also at the Department of Psychology, Lehman College, City University of New York.

If another component is found to be related to a different stage, then it is possible to determine whether variables which affect one component have an affect on the other component as well. Whereas it is unlikely that scalp recorded ERP components reflect all the neural processes associated with a given stage, it is likely that ERPs can index relative changes in particular stages. In this manner, ERP data may be relevant to models of stages of information processing.

Review of Pertinent ERP Data

A first stage of processing of stimulus input is reflected by the sensory ERPs that are obligatory responses and are modality specific in their scalp distributions. Depending on the sensory modality and nature of the stimulus, a number of sensory components have been identified which range in latency from about 1 to 300 msec. Relatively little has been established about the functional significance of specific sensory components, but a reasonable assumption is that they reflect the registration of the physical attributes of stimuli. The latencies of these components vary inversely with the intensity of stimulation, but otherwise remain essentially constant for a particular subject in a given state for physically identical stimuli. Alterations of stimulus intensity in a simple RT task are associated with changes in the latency of sensory ERPs which covary with reaction time (Vaughan, Costa and Gilden, 1966). Although never systematically studied to our knowledge, changes in latency of the sensory components due to variations in intensity are most likely associated with comparable changes in later task-related ERP components, such as N2 and P3 (P300). It has been shown, however, that the sensory components reflect a stage of processing independent of the subsequent stages to be discussed. For example, the long latency sensory component P2 (roughly 180-280 msec in latency) exhibits no change in timing during discrimination, regardless of the difficulty of the task. On the other hand, the latency of subsequent components can vary considerably as a function of task difficulty (with intensity held constant). Thus, the registration of the physical characteristics of stimuli appears to constitute a stage reflected by sensory ERPs which is independent of subsequent processing.

In the initial study that showed auditory P2 latency was not affected by the difficulty of a discrimination task, it was found that P3 was longer in peak latency for a more difficult than an easier pitch discrimination (Ritter, Simson & Vaughan, 1972). Since P3 latency also covaried with RT, it appeared that it might reflect the stimulus evaluation process necessary for the differential behavioral responses. These data, however, raise a critical issue. The circumstance that the peak latency of a component is differentially affected by changes in a particular processing demand does not distinguish between the possibility that the component reflects the relevant underlying process or an outcome of that process. At the present time we realize that P3 reflects an outcome of the stimulus evaluation process that covaries with RT rather than the process itself. In part this is because more recent studies have shown that N2, which precedes P3, also correlates with RT in discrimination tasks, and partly because investigators have noted that discriminative motor responses are often initiated in motor cortex prior to the occurrence of P3 (Donchin, Ritter, and McCallum, 1978). Although the data concerning components which occur prior to P3 will now be reviewed, it is clear that the question of determining whether alterations in a component reflect modifications of a process being studied, or an outcome of that process, will not always be resolved by identifying relevant earlier components, but requires some other solution.

A number of studies have found that N2 varies in latency as a function of the degree of physical differences between stimuli during discrimination (Ford, Roth and Kopell, 1976a; Towey, Rist, Hakerem, Ruchkin and Sutton, 1980; Näätänen, Hukkanan and Järvilehto, 1980; Lawson and Gaillard, in press) and correlates in latency with RT (Ritter, Simson, Vaughan and Friedman, 1979; Renault and Lesevre, 1979; Renault, Ragot and Lesevre, 1980).

In a recent study (Ritter, Simson and Vaughan, in press), we set out to determine whether an analogous N2 component is present for semantic discriminations. Three levels of task difficulty were used. In the easiest task, on 80% of the trials two brackets faced each other (i.e., []), and on 20% of the trials both brackets were reversed (i.e.,][ ). A similar arrangement was presented for two angles. In a harder task, four brackets, all facing in the same direction (see the bottom of Figure 1), occurred on 80% of the trials, and on 20% of the trials one of the four brackets was reversed (e.g.,]][]), subjects not knowing which of the four would reverse. Four angles were presented in a similar manner. The hardest task was a semantic discrimination. Two stimulus sets were employed: in one, all stimuli were names, with female names occurring on 20% of the trials and male names on 80% of the trials; in the other, 20% of the stimuli were animal words and the rest non-animal words. In all conditions the stimuli were presented in random order, and the subjects responded as quickly as possible to the 20% stimuli. The ERPs associated with the stimuli which occurred on 80% of the trials were subtracted from the ERPs associated with the stimuli which occurred on 20% of the trials, separately for each subject and condition, in order to delineate N2, which is larger for less probable stimuli (Simson, Vaughan and Ritter, 1977). Table 1 presents the relevant latency data averaged across subjects. The peak latencies of N2 and P3 progressively increased from the easiest to the hardest tasks, as did RT. Thus N2 was found for semantic discrimination, and it had about the same time relationship to P3 and RT for all tasks. However, the onset of N2 also occurred later across tasks, suggesting that the processes underlying N2 were dependent upon the outcome of a previous stage of processing.

TABLE 1

Mean Latency Obtained in Various Discrimination Tasks (in msec)

Condition	NA Onset	N2 Onset	NA Peak	N2 Peak	P3 Peak	RT
[]	134	180	225	251	422	425
<>	146	167	237	250	422	435
]]]]	143	214	271	307	453	448
>>>>	154	210	266	291	478	453
Names	144	266	299	385	593	562
Animals	167	285	298	372	570	560

In the study under discussion, subjects also performed a simple RT task to the stimuli employed on 80% of the trials during the discrimination tasks. For the physical stimuli, the brackets or angles on the bottom of Figure 1 were presented, in separate runs, on 100% of the trials. For the word stimuli, the name "DOUG" and the word "WINE" were presented, on separate runs, on 100% of the trials. Grand Mean ERPs averaged across subjects at the T5 recording site are presented in Figure 1. The top row displays the ERPs obtained in the simple RT conditions. In the left column, it can be seen that the same stimuli elicited a larger N1 (peaking at about 180 msec), and a smaller P2 (peaking around 240 msec), when they occurred on 80% of the trials of the discrimination task (middle row) compared to when they occurred on 100% of the trials of the simple RT task. In fact, P2 peaked on the negative side of baseline during discrimination. These opposite polarity effects on N1 and P2 suggested the presence of a negative displacement spanning both components, which was enhanced during discrimination. The difference waveforms on the bottom row of Figure 1 shows this negative displacement, termed NA, which was present in all conditions. In the experiments discussed in this paper, NA was generally largest in amplitude at the T5 recording site.

The NA component appears to reflect an earlier stage of processing upon which the processes underlying N2 might depend. There were no significant fluctuations in the onset of NA, but its peak latency increased as a function of the complexity of the stimuli employed in the discrimination tasks (Table 1). Thus, the progressively greater delay in the onset of N2 across conditions was associated with progressively longer durations of NA. On the basis of the relative timing of NA and N2, it was inferred that the

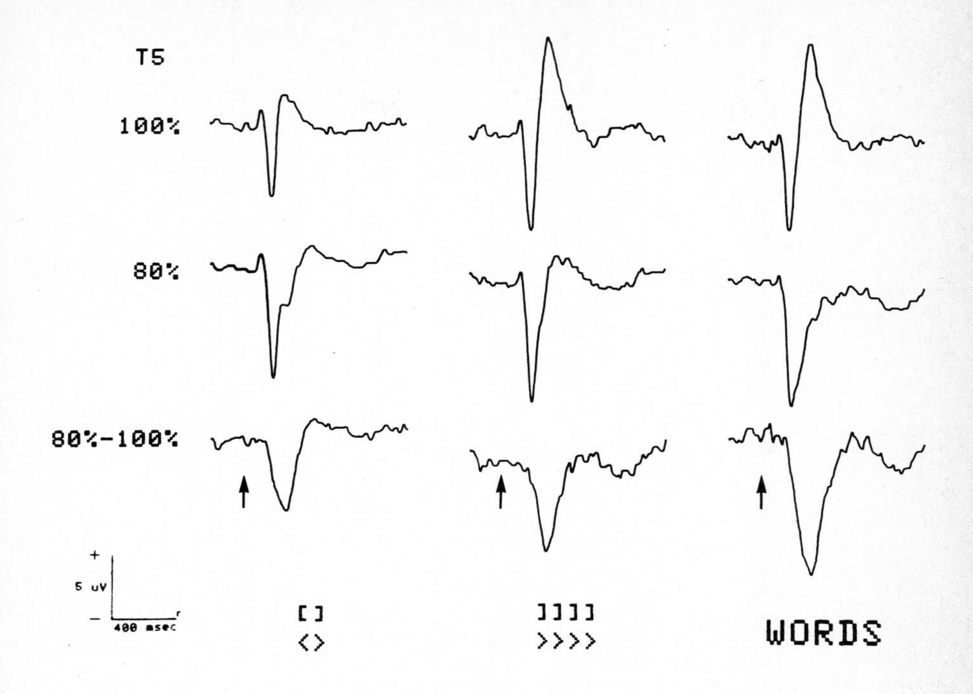

Figure 1. Grand Mean waveforms obtained at T5 during Simple RT when a given stimulus was presented on 100% of the trials (top row), for the stimuli which occurred on 80% of the trials during the discrimination tasks (middle row), and (bottom row) difference waveforms obtained by subtracting the ERPs obtained during simple RT from the ERPs immediately below them. Stimulus conditions indicated at the bottom of the figure.

physiological activity underlying NA might be associated with pattern recognition. N2 appears to reflect a subsequent stage of processing related to stimulus categorization.

Our hypothesis as to why NA is smaller during simple RT than during discrimination is that the extent of pattern recognition processes are less when the same stimulus is presented on all trials and the subject has been so informed.

If the hypothesis were correct, then NA should be enhanced if the subject thought the stimulus would be changing during a run, even though there were no changes. In order to test this idea, we are currently conducting an experiment in which the same stimuli (brackets) are presented on 100% of the trials of a simple RT task and in a condition where the subject is told that there are very infrequent stimulus changes to which differential responses are required (i.e., that this is a choice RT task). Preliminary results indicate that when the ERPs associated with the simple RT task are subtracted from the ERPs obtained when the subject is told to perform a choice RT task, the difference waveform contains an NA component. The latter is similar in amplitude to the NA obtained in another condition in which the subject performs a choice RT task and infrequent stimulus changes actually do occur. In other words, the ERP waveform obtained to the 100% stimuli when the subject expects stimulus changes is similar to the ERP waveform associated with the frequent stimuli when infrequent stimulus changes do occur. Moreover, RT to the 100% stimuli was short when the simple RT task was performed, and comparable to the choice RT task when the subject was falsely told there would be infrequent changes. Thus, the identical (i.e., 100%) stimuli elicited a larger NA and longer RTs when the subject expected stimulus changes, than when no changes were expected.

In Figure 1, a P1 component peaked at about 110 msec in all the waveforms of the upper row (simple RT) and middle rows (discrimination tasks). This component had a similar latency in all of the discrimination tasks for the ERPs associated with the stimuli which occurred on 20% of the trials. Using P1 as an index, arrival time at the cortex was roughly constant (luminance of the stimulus characters was the same across conditions), and the pattern recognition processes hypothesized to be associated with NA could begin at about the same time across conditions.

The amplitude fluctuations which can be seen in Figure 1 are consistent with the hypothesis that sensory ERP components and NA reflect different stages of processing. Looking across the top row, it can be seen that the number of stimulus characters used in simple RT had a small effect on the amplitude of N1 and a dramatic effect on P2. Both components were larger when four characters were used compared to two characters, but there was virtually no difference between four physical characters and four letter words (middle and right columns). By contrast, there were only small, non-significant differences in the amplitude of NA between the discrimination conditions which used two and four physical characters (bottom row, left and middle columns). On the other hand, NA associated with words which averaged about four letters (bottom row, right column) was almost twice the amplitude as for the physical discriminations. Thus, the amplitude of sensory components was affected by the number of stimulus characters, whereas the amplitude of NA was affected by the nature of the stimuli.

Changes in the duration of NA, as measured from onset to peak (Table 1), or from onset to return to baseline (bottom row of Figure 1), constitute an important reason for inferring that NA reflects pattern recognition processes rather than the outcome of these processes. By contrast, if NA had increased in peak latency across conditions, but maintained a constant duration, that would suggest it reflects the outcome of some other brain activity which varied in duration across conditions. The circumstance that NA did change in duration does not necessarily mean that it reflects the psychological processes hypothesized to underlie it, but it is consistent with such an interpretation.

It is possible that NA had the same duration on the individual trials of all the conditions but, by varying more in its latency from trial to trial in one condition than another, yielded a longer duration in the waveforms averaged across trials. In previous research, increased variability across single trials of the latency of a component, e.g., P3, has been associated with a smaller, broader morphology of the component. The Woody filter technique has been used to adjust for these effects of latency variability in order to obtain more accurate estimates of the amplitude of P3 (e.g., Ruchkin and Sutton, 1978a). As can be seen in the bottom row of Figure 1, however, increases in the duration of NA were accompanied by increases in its amplitude. Only the amplitude increase for words was statistically significant, but in no instance was an increase in the duration of NA associated with a decrease in its amplitude. Nevertheless, it is possible that a component could increase in amplitude and in latency variability from one condition to another such that the greater latency variability is compensated by the enhanced amplitude, resulting in averaged waveforms which may be equal in size in the two conditions, or even larger in the second condition. Indeed, NA was larger in amplitude in the averaged waveforms for the word than non-word conditions in Figure 1.

If the processing underlying N2 were dependent on the processes associated with NA, then an increased latency variability of NA would be expected to produce greater latency variability of N2 and thereby a lengthening of the duration of N2. Such was not the case (Table 1). Aside from a single trial analysis, however, there is no definitive way of demonstrating that longer durations of a component measured in averaged data are due to longer durations of that component on single trials. This issue is discussed further below.

To test the hypotheses concerning the functional roles of the activities associated with NA and N2, an experiment was designed in which it was predicted that the latencies of NA and N2 would be differentially manipulated (Ritter, Simson, Vaughan and Macht, in press). If NA reflects pattern recognition processes, then degrading stimuli should increase its peak latency, but not affect its onset. Differential manipulation of N2 timing was attempted by varying the time to perform a classification task. In Task A, subjects responded with one hand for digits and with the other hand for letters of the alphabet. In Task B, subjects memorized four letters of the alphabet and responded with one hand if a letter was in the memory set and with the other hand if it was not. In Task A, digits were presented on 20% of the trials, and in Task B letters in the memory set occurred on 20% of the trials. In both tasks, each trial consisted of one item, and all possible single digits and letters were used which were not confusable with one another. The stimuli were degraded by using a mask which consisted of a 4 x 3 matrix of small x's. Two simple RT Tasks were also employed, one with and one without the mask, in which the letter "F" was presented on 100% of the trials.

Figure 2 presents the Grand Mean waveforms for the simple RT tasks (left column), for the stimuli which occurred on 80% of the trials of the classification tasks (middle column), and the difference waveforms (80-100%) used to delineate NA (right column). The onset of NA was similar across conditions, but it had a longer peak latency when the stimuli were masked (lower two waveforms) compared to when no mask was employed (upper two waveforms).

Figure 3 plots the latencies averaged across subjects of the onsets of NA and N2, the peaks of NA, N2, and P3, and the RTs to the stimuli which occurred on 20% and 80% of the trials in each classification task, with and without the mask. The horizontal bars representing NA and N2 depict their relative durations, as estimated by the time from onset to peak. There was no significant difference in the onset of NA across conditions, but the duration of NA was affected by whether the stimuli were masked or not. On the other hand, the duration of N2 was unaffected by the mask, but differed significantly between the two classification tasks. The changes in the durations of NA and N2 were jointly related to RT.

In that the measures of component duration, based on the averaged difference waveforms of the subjects, are vulnerable to the arguments concerning latency variability, the results of a more conservative analysis based on peak latencies are presented. There was a main effect of stimulus quality (i.e., whether the stimuli were masked or not) on the peak latency of NA, but no main effect of classification task, and no interaction between stimulus quality and task. There were main effects of stimulus quality and task on the peak latency of N2, but no interaction. The interpretation of the main effects of stimulus quality on both NA and N2 is that these components reflect sequential processes, the activity underlying N2 being dependent on the output from the activity underlying NA. Consequently, an increase in the peak latency of NA would be associated with an increase

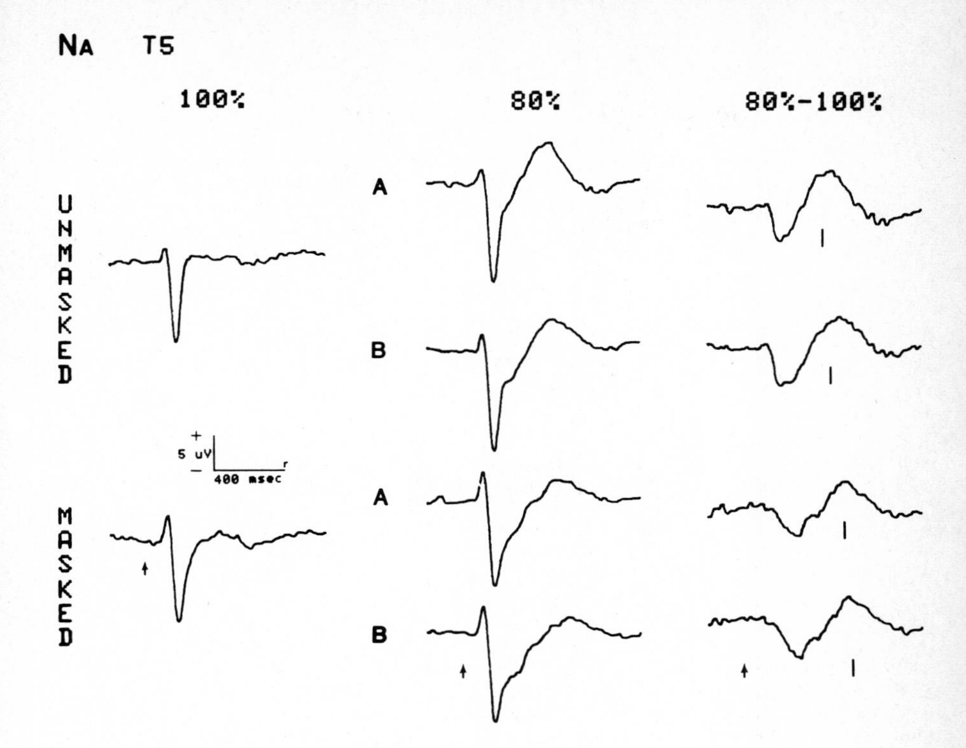

Figure 2. Grand Mean waveforms obtained at T5 for unmasked (top half of figure) and masked (bottom half) stimuli. ERPs associated with simple RT when the stimulus was presented on 100% of the trials (left column), for the stimuli which occurred on 80% of the trials of the classification tasks (middle column), and difference waveforms obtained by subtracting the ERPs obtained in the left column from the ERPs obtained in the middle column, separately for unmasked and masked conditions. A and B refer to the classification tasks described in the text. Arrow: stimulus onset. Vertical line: mean RT for 80% stimuli.

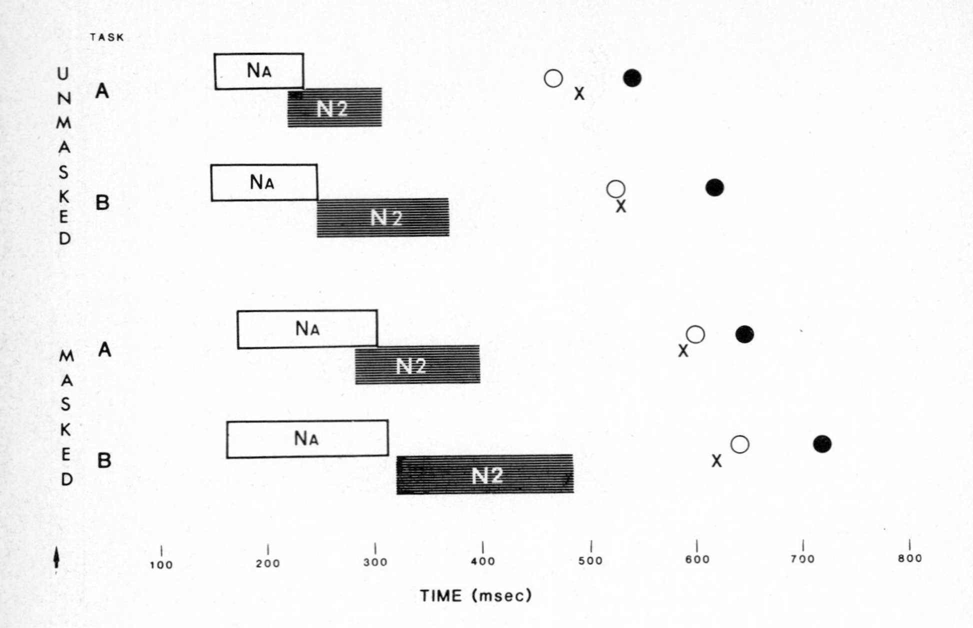

Figure 3. Mean onsets of NA and N2, peak latencies of NA, N2 and P3 (X), and RT for the stimuli which occurred on 80% (open circles) and 20% (closed circles) of the trials of Tasks A and B, obtained with unmasked and masked stimuli. NA and N2 measures taken from T5. P3 measures taken from Pz for 20% stimuli. The bars labelled NA and N2 connect their respective onset and peak latencies in the various conditions. Arrow: stimulus onset.

in the peak latency of N2. There was a main effect of classification task on the peak latency of N2, and no interaction between stimulus quality and task. As for RT, there were main effects of stimulus quality and task, and no interactions, for the stimuli which occurred on both 20 and 80% of the trials. Main effects of stimulus quality and task on the peak latency of P3 were also obtained, with no interaction.

The analysis of peak latencies leads to a weaker, but less vulnerable, inference that NA and N2 are at least associated with the outcomes of two stages of processing, one affected by stimulus quality and the other by the nature of the classification task. The data indicate that the timing of these two components can be independently manipulated and suggest that the variables which affect NA and N2 peak latencies affect the durations of the stages with which they are associated.

As with the experiment which used physical and semantic discriminations, if the brain activity underlying N2 were dependent on the processing associated with NA, and the increased duration of NA were due mainly to greater latency variability, then such increases would also be expected to be related to increases in the duration of N2. There was, however, no main effect of stimulus quality on the duration of N2 (measured as the time from onset to peak), despite changes in the duration of NA associated with the use of the mask. Moreover, there was a main effect of task on the duration of N2, while the nature of the task had no significant affect on the duration of NA. Though not conclusive, this pattern of results is consistent with the inference that the durations of NA and N2 were affected by the experimental variables rather than merely their trial to trial variability in latency.

Support for the latter conclusion comes from an important study of Renault, Ragot, Lesevre and Remond (in press), and discussed by Renault (this volume). They measured the duration of N2 in an omitted stimulus paradigm, from onset to return to baseline, on a single trial basis. It was found that the duration of N2 was correlated with RT for individual subjects. When the ERP data were sorted by RT quartiles and pooled across subjects, the N2 in the averaged waveforms increased in duration from the first through the fourth quartiles. These results demonstrate that increases in the duration of N2 in averaged waveforms can be associated with increases in the duration of N2 on individual trials, rather than latency variability. The theoretical significance of the study by Renault and colleagues is that it establishes that it is possible, in principle, for an ERP component to reflect a process under investigation, and not just an outcome associated with the process.

Some Theoretical Considerations and Implications

We have proposed that the ERP components under discussion, while overlapping in time, reflect sequential processing of task relevant information. Alternative interpretations of the data are possible, just as is true of models of processing stages based on behavioral data. For example, NA and N2 could be associated with physiological processes entirely unrelated to one another, and the experimental manipulations of their timing reported above due to effects on processes which operate independently and in parallel. Scalp recordings by themselves can not entirely rule this out, but an accumulation of data collected across experiments which is consistent with the sequential hypothesis can lend it greater credence. Moreover, experimental manipulations can produce ERP data which are inconsistent with a sequential interpretation. For example, Hansen and Hillyard (1980) found that the latency of the processing negativity in a selective attention task varied as a function of the degree of separation between channels, but RT to targets in the attended channel did not (nor did the latency of P3, Hillyard, personal communication). These results are inconsistent with the sequential hypothesis proposed by Hillyard and colleagues (Hillyard and Picton, 1979), wherein the activity underlying P3 was thought to be dependent on the operations associated with the prior processing negativity.

There are several lines of evidence which suggest the existence of sequential dependencies among certain ERP components. Courchesne (this volume) has found that P3 and Slow Wave both decrease in latency for the same task, from four years of age to the early twenties, but P3 precedes the Slow Wave by a constant interval of about 200 msec. His conclusion that Slow Wave is associated with processing which is influenced by the earlier processing related to P3 is consonant with the view of Ruchkin and Sutton (this volume) based on an entirely different set of experimental results. As a general rule, studies which have manipulated N2 latency have found similar changes in the subsequent P3 (e.g., Ritter, et al., 1979, and Table 1 of the present paper). Renault (this volume) has presented data which support the sequential interpretation of these components. In our masking experiment, the finding that P3 latency was affected by joint changes in the timing of NA and N2 (Figure 3), provides further support for the sequential nature of these ERPs.

There are several implications for theories concerning stages of information processing which follow from our analysis of the NA and N2 data summarized here. The idea of a series of subprocesses intervening between stimulus and response in RT tasks is supported by physiological data in a way which approximates the concepts of molar subprocesses or stages used by Donders (1969) and more recent theorists (Smith, 1968).

The data on NA and N2 also provide grounds for some comments on Donders' idea of the insertion of stages. In the experiment described above in which stimulus complexity was varied (Figure 1), subjects were run on additional simple RT tasks in which the stimuli were identical to those employed during the discrimination conditions. In other words, instead of presenting the same stimulus on all trials, as is usually done in simple RT, the stimuli changed during a run in the same way as during the discrimination tasks. The stimuli of the second condition (four angles or brackets) and third condition (words) were used. Mean RT, of course, was much shorter than for the discrimination tasks, but slightly longer than for the simple RT tasks in which the same stimulus was presented on 100% of the trials (cf. Hannes, Sutton and Zubin, 1968). NA components were obtained which had similar timing as given in Table 1 for the relevant discrimination tasks. An N2, also of comparable timing as during discrimination, was obtained for the runs which used changing angles or brackets, but not for the changing words. The circumstance that RT changed, but not the timing of NA or N2, when subjects responded in the same way to all stimuli in these simple RT tasks, indicates that the changes in NA and N2 latencies reported elsewhere in this paper are not due to changes in motor potentials. Instead, the results imply that pattern recognition was performed on the stimuli, even though unnecessary to the task, and that the physical changes, which occurred on 20% of the trials when angles or brackets were presented, were registered by the brain. Subjects reported that they perceived the identity of the stimuli (e.g., perceived the words as words, not just a blur), and were aware of the physical changes, but did not take note of the changes in the semantic categories of the stimuli. The NA and N2 components so observed appear to reflect processing of stimulus information irrelevant to the task. Since in these conditions RT was unrelated to the timing of NA or N2, a further conclusion can be drawn that the activities associated with these components can be functionally dissociated from the processing related to task performance.

When subjects were required to discriminate among these same stimuli, on the other hand, then the timing of NA and N2 were related to RT. In the case of the angles and brackets, the data suggest that the meaning of Donders' idea of insertion is that the stages of pattern recognition and stimulus classification (as reflected by NA and N2), which are activated by the stimuli regardless of the task, are inserted into the sequence of processes which are required to perform the discrimination task. In the case of the verbal stimuli, insertion refers to the inclusion of pattern recognition (NA), which is activated regardless of the task, and the inclusion of stimulus classification (N2), which in these conditions is only activated when the class membership of the words must be determined, into the sequence of processes required for differential responding. On the basis of the preliminary results reported above, in which the same stimulus was presented on 100% of the trials but the subject was led to think there would be infrequent stimulus changes which required differential responses, insertion occurred solely on the basis of the subject's belief in the need to discriminate to perform the task. The change in the nature of the stimulus processing upon which performance was based between simple RT and discriminative RT to the same (i.e., 100%) stimuli was associated with ERP as well as behavioral changes. At the present time, no ERP component has been uniquely identified with stimulus detection in simple RT, so it is not possible to know whether detection plays a role in discriminative RT tasks (e.g., whether pattern recognition processes are dependent on those involved in detection). However, Parasuraman, Richer and Beatty (in press) have used ERP data to support the view that the processes associated with detection and recognition occur in parallel, at least in part, during signal detection tasks.

The criticism directed against Donders, that it is not possible to demonstrate that the insertion of a stage has no effect on other relevant stages, cannot currently be answered with ERP data with regard to simple versus discriminative RT tasks. But it is possible that ERP investigations could provide relevant data in the future. In the meantime, ERPs can be used in the analysis of stages across some tasks. The differential effects of stimulus quality and categorization, reported above, is one example. Another is the use of P3 latency as a measure of relative stimulus evaluation time in comparing stimulus-response compatability and incompatability (McCarthy and Donchin, this volume). The experiments reported above provide tentative grounds for comparing Choice RT and Go No-go tasks (Donders' b and c reactions). Donders had inferred that the faster responses in Go No-go compared to Choice RT are at least in part due to differences in response preparation. The first experiment described above (manipulation of stimulus complexity) was a Go No-go task, whereas the second experiment (masking) was a choice RT task. Although these experiments were not designed to compare the Donders b and c reactions, the time from the peak of N2 to RT could be used as an estimate of response selection and execution time.

Table 2 presents the mean difference scores between RT to the stimuli which occurred on 20% of the trials of the two experiments and their respective N2 peak latencies. In every instance, the difference scores were less for the Go No-go tasks (left column) than the Choice RT tasks (right column), thus providing tentative support for Donders' idea concerning relative differences in response preparation in these two kinds of RT tasks. (If P3 latency were used as an estimate of stimulus evaluation time, comparable results would have been obtained: see Table 1 and Figure 3).

The data we have presented on NA and N2 are relevant to the additive factors approach of Sternberg (1969). Taylor (1976) has argued that additivity obtained by crossing two levels of two variables is uninterpretable. In our masking experiment, however, the lack of interaction between stimulus quality and the nature of the stimulus classification task on RT appears to be associated with independent effects of the

TABLE 2

Mean Time From N2 Peak to RT (in msec)

Experiment I		Experiment II		
Condition			Task	
[]	174	unmasked	A	233
<>	185		B	249
]]]]	141			
>>>>	162			
Names	177	masked	A	247
Animals	188		B	233

two variables on the relative durations of the pattern recognition and stimulus classification stages. In the masking experiment, the ERP data helps additivity become more interpretable. In other experimental conditions, where interactions occur between two variables, ERPs may be useful in identifying which stage or stages are affected in common.

The overlap in time of NA and N2 (Table 1 and Figure 3) does not support models of information processing based on discrete, serial stages. The ERP data also fail to support the cascade model of McClelland (1979), in which all stages begin to operate on stimulus information at about the same time, but accrue information at different rates. NA and N2 did not begin at the same time in any of the conditions reported, the onset of N2 being delayed with respect to the onset of NA as a function of stimulus complexity (Table 1) and stimulus quality (Figure 3). Thus, it appears that the activity underlying N2 does not begin to operate until the information processing associated with NA has reached some criterion pertinent to the task. Whereas the temporal overlap of NA and N2 may be associated with concurrent processing of information of two different stages, the data are also consistent with serial processing of task relevant information. The latter interpretation of the ERP data is compatible with the Asynchronous Discrete Stage model of Miller (in press), in which information on the various codes processed by a given stage is passed on to subsequent stages at varying points in time, depending on the completion time of a given code. Thus, in Miller's model, there is temporal overlap in the processing of sequential stages, but the overlap is consistent with serial processing of the information relevant to the performance of a given task.

The material reviewed in this chapter and the chapter by McCarthy and Donchin (this volume) indicates that ERPs can provide complementary data for assessing models of stages of information processing based on behavioral results. In our chapter, we have used RT to provide data which complement changes in ERP waveforms in order to test hypotheses about the cognitive processes associated with particular components. As more studies are conducted which provide sound grounds for interpreting the psychological significance of ERP components, ERP data may be expected to play an important role in the development of theories concerning stages of information processes.

ACKNOWLEDGEMENTS

We wish to thank Gregory McCarthy for helpful comments on the paper, and to acknowledge the technical assistance of Chester Freeman, the writing of computer programs by Alan Legatt, and the preparation of the manuscript by Anita Levine.

Tutorials in ERP Research: Endogenous Components
A.W.K. Gaillard and W. Ritter (eds.)

7

THE VISUAL EMITTED POTENTIALS : CLUES FOR INFORMATION PROCESSING

Bernard RENAULT

CNRS
Hôpital de la Salpêtrière
Paris, FRANCE

Three kinds of endogenous brain event-related potentials are described : an N200 wave peaking in the parieto-occipital region, the duration of which increases with perceptual processing ; a biphasic N2-P3a complex peaking in the central areas, probably related to "active" orienting and overlapping in onset with the terminal phase of the parieto-occipital N200 ; a parietal P3b wave generated mostly after the motor response. These findings suggest that the process reflected by N2-P3a is dependent upon cognitive decision indexed on line by the parieto-occipital N200 and develops concurrently with the end of stimulus categorization, the parietal P3b being generated after these stages.

INTRODUCTION

Several results have clearly established that the potentials associated with expected but missing stimuli are made up of at least two components, one negative, the N200, followed by a positive one, the P300 (Klinke, Fruhstorfer and Finkenzeller, 1968 ; Picton, Hillyard and Galambos, 1974 ; Simson, Vaughan and Ritter, 1976 ; Renault and Lesèvre, 1978 ; McCallum, 1980). Since the amplitude of this positive component varies as a function of the probability of the omitted stimulus in the same way as the P300 obtained to task relevant stimuli (Ruchkin, Sutton and Tueting, 1975) and has the same scalp distribution (Picton and Hillyard, 1974 ; Simson, Vaughan and Ritter, 1977a), it can be considered the same phenomenon. Moreover it has been demonstrated that the topography of the negative component changes according to the sensory modality of the stimulus, peaking in the preoccipital region when a visual stimulus is omitted and in the vicinity of the vertex in the case of an auditory one (Simson et al., 1976, 1977 ; Renault, Ragot and Lesèvre, 1980b). On the other hand, the topography of the positive component does not change with stimulus modality.

Taken together, these studies seem to establish that the modality specific negative potential (N200) and the positive non modality specific one (P300) are produced by at least two different intracranial sources. In fact, the actual description is still more complex since in response to an unpredictable task-relevant omitted stimulus, two different types of positive components have been distinguished according to topographical criteria : the vertex type and the parietal one (Renault and Lesèvre, 1978, 1979 ;

Ruchkin, Sutton, Munson, Silver and Macar, 1981). These are similar to the two types of P300 (P3a and P3b) first described by N. Squires, K. Squires and Hillyard (1975). The P3a was said to be preceded by a negative wave. This N2-P3a complex was related to mismatching and orienting processes (see also Snyder and Hillyard, 1976). Studies of the averaged visual omission response showed that two types of negative waves preceded both P300s : a central one and a parieto-occipital one (Simson et al., 1976 ; Renault and Lesèvre, 1975). These successive peaks could not be due simply to the latency variability of N200-P300 since they had different locations on the scalp. Actually, this variability has been shown to be more important in the omission paradigm than in the classic "odd-ball" one, due to the subject's temporal uncertainty concerning the expected stimulus time (Ruchkin and Sutton, 1979). This tends to wash out the averaged potentials. Therefore, in order to better indentify these various foci of activity, we analysed emitted potentials trial-by-trial.

The existence of these different foci of activity is a central issue in the field of event-related potentials and cognitive psychology insofar as the study of their relations to behavioural responses can shed some light on their functional significance and may help in the formulation of psychological models of information processing in humans. Indeed, the visual omission response may provide important clues since it contains the various waves described above. Here we shall summarize only those results concerning the relations between the task and the various types of N200s and P300s (Renault and Lesèvre, 1978, 1979 ; Renault, Ragot, Furet and Lesèvre, 1980a). In addition, we have recently demonstrated (Renault, Ragot, Lesèvre and Rémond, in press) that the duration of both N200 waves of different topography are differentially related to the duration of perceptual processing. However, for the sake of brevity, we were unable to extensively report all the results. The present chapter contains these additional results along with an attempt to relate each of the emitted waves to information processing stages.

METHODOLOGICAL REMARKS

The brain response to a relevant stimulus includes the exogenous N100-P200 potentials. Usually, this P200 component is so large and has such a wide distribution, that it not only conceals the endogenous N200 but sometimes also interacts with the P300. When the relevant stimulus is the omission of an expected stimulus, no exogenous potentials are recorded ; therefore the N200 and P300 components can be observed per se. However trial by trial studies of such emitted potentials remain difficult due to the low signal to EEG noise ratio. The spatio-temporal mapping method of Rémond (1961) was of great help for such an analysis. This display represents amplitude variations in the form of equipotential lines as a function of time (on the abscissa) and electrode location (on the ordinate) ; potentials between two successive electrode sites are obtained by a second order interpolation using a sliding window over three recorded potentials, thereby providing a better signal to noise ratio for potential variations between electrodes. In turn, these displays make the visual analysis of single trial EEG waveforms easier. On these maps the peak location of each wave was measured as a percent of the nasion-inion distance with respect to the inion. The baseline was determined by the averaged voltage level of the whole map. Only

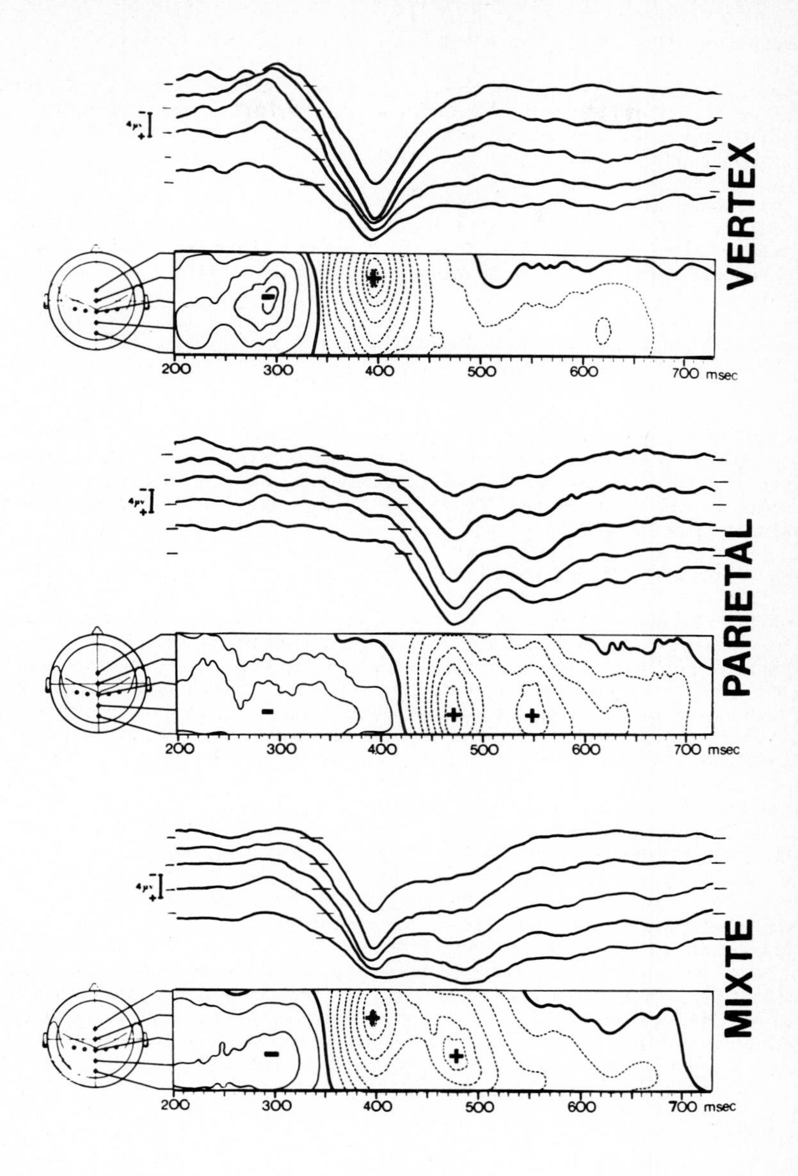

Figure 1 : Grand means (across subjects and situations) of the various types of emitted responses. Potential baselines are represented at the beginning, at the end and at the N200-P300 zero-crossing of each trace. On the maps, the potential increases of 2 µV between two isopotentials lines (see note 1 for complementary comments).

the responses free of eye movements and artefacts were plotted.

In the following experiments, only the visual emitted potentials were studied. Each run consisted of 450 pattern onsets for 22 msec, presented at a rate of 1 per sec. Ten percent of these stimuli, but never two in succession, were omitted randomly.

When a motor response was required from the subject, the movement consisted of a finger displacement towards a photo-electric cell. This kind of movement which required very little strength was chosen in order to minimize scalp motor-related potentials (Kutas and Donchin, 1977).

HOW MANY N200-P300S, THEIR TASK RELATIONSHIPS

Different types of emitted responses, differentiated on the basis of single trial analysis, were reported by Renault and Lesèvre (1978) and Renault et al. (1980a). In these studies three conditions were used : 1) execution of a motor act in response to the omission (GO) ; 2) withholding of the motor response to the omission and reaction to the visual stimulus ; 3) mentally counting of the omissions.

Across situations, three types of emitted potentials were differentiated on the basis of their spatio-temporal organization (Fig. 1) : first, the "mixed type" (64% of the trials) consisting of one negative wave beginning in the parieto-occipital region and a later, overlapping negative wave in the central region followed by two positive waves, the first one peaking at Cz, the second in the parietal region ; second, the "vertex type" (20% of the trials), composed of two successive waves of short duration (first negative, second positive) located in the central region ; third, the "parietal type" (16% of the trials), made up of a negative wave followed by a positive wave, both of long duration and both located in the parieto-occipital region. Latencies, topographies and amplitudes of these different types supported the hypothesis that the "mixed type" resulted from the addition of the vertex and parietal types of emitted responses. The vertex types appeared more frequently during the GO and NOGO tasks whereas the parietal types were more frequent during the counting task. In addition, in the GO condition, both the number of parietal types and the ratio between the parietal and the vertex positive waves amplitude increased with the mean value of the reaction time. These results demonstrated the existence of different kinds of emitted potentials. Moreover, they suggested that a shift in the brain activity, from the central to the parietal region, was linked to the motor (GO and NOGO) or sensory (mental counting) kind of task and, in a motor task, to the level of performance.

RELATION WITH THE MOTOR ACT

In the previous experiments, the N200s were the only waves which always occurred before the motor act. Whatever their topography, their peak latency showed a higher correlation with reaction time (.73) than did P300 latencies (.65 and .59). These relations were more extensively studied in a

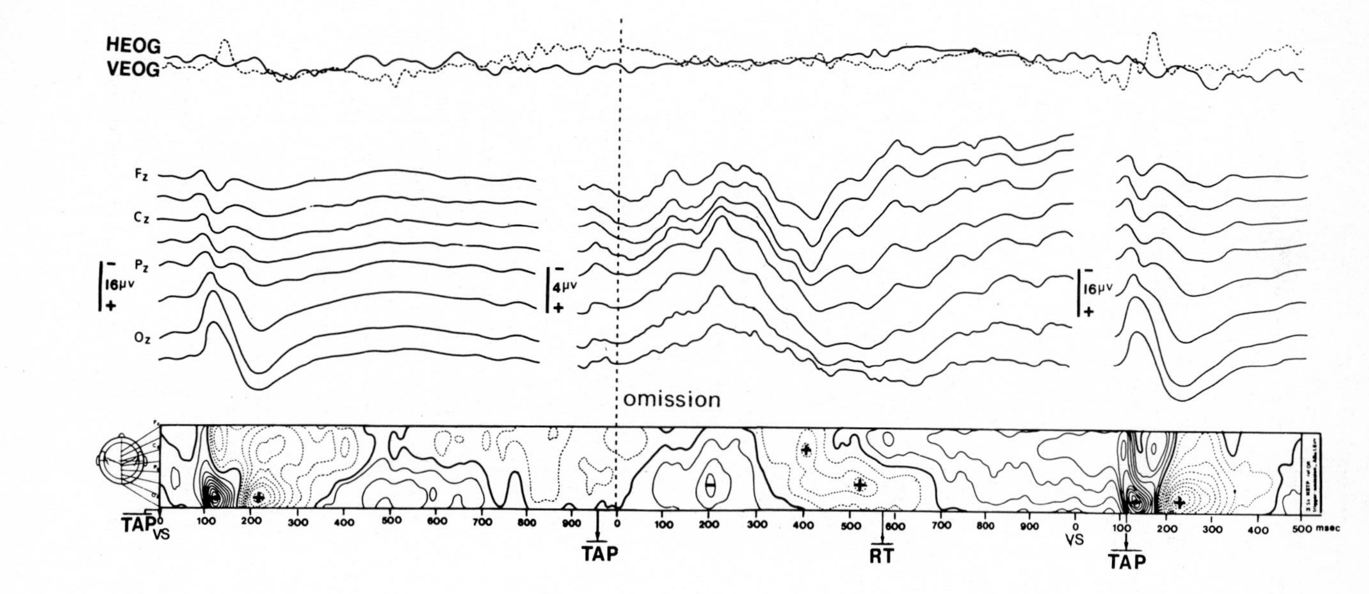

Figure 2 : Averaged potentials and corresponding spatio-temporal map of the emitted response in condition 1 (294 trials). The stimulus omission is preceded and followed by a visual stimulus (VS). The two curves above (HEOG and VEOG) represent horizontal and vertical eye movements. Note the change in the scale of potentials from 16 to 4 μV in order to better visualize the emitted response. However, on the spatio-temporal map, the scale does not change, the potential increasing of 1.6 μV between two isopotential lines.

second experiment (Renault and Lesèvre, 1979). Indeed, such a high correlation between N200 and the reaction time was not necessarily the sign of a causal relation since these phenomena could both depend upon a third internal event, e.g., the subject's time estimation of the moment the omission should have occurred. Therefore, in order to test this possibility, subjects were asked (condition 1) : to tap the rhythm at the same frequency as that of the visual stimuli, and whenever a stimulus did not occur, to give an additional motor response as quickly as possible. Each tap in the rhythm was considered an index of when the subject expected the stimulus to occur, with the additional motor response being a measure of the reaction time (RT) to the omitted stimulus. The rhythm tap and the RT response were performed by the same finger. In addition, two control conditions (conditions 2 and 3) were recorded in order to estimate the effects of : 1) the tapping of the rhythm (subjects were asked to give a motor response only after detection of an omission) and 2) the potentials related to the motor act (subjects were asked to perform self-paced finger displacements at the rate of approximately 1 per sec without any visual stimuli). The order of the three conditions was counterbalanced across subjects.

Basically, the results of this study were the same as those obtained by Ritter, Simson, Vaughan and Friedman (1979) and by Towey, Rist, Hakerem, Ruchkin and Sutton (1980). Not only did the N200 wave always precede the motor response, but its latency was also more correlated with the RT (.62) than was the parietal P300 latency (.56). In addition, this study demonstrated that neither the RT nor the peak latency of the N200 were highly correlated with the rhythm tap. This, therefore, supported the idea that the high correlation between N200 and RT was due to a causal relation between these phenomena (see also later on).

Two negative waves labelled Na (parieto-occipital) and Nb (central) were described in this study. However, they were not clearly differentiated. The results also suggested that the increase of the latencies of all peaks of the omission response (when RTs increased) was in fact due to an increase in the duration of the parieto-occipital Na wave and not to a delayed time estimation of the moment the stimulus should have occurred.

In order to confirm these results, the data were analysed a second time. The following assumptions were tested : 1) a relation between the onset of the emitted potentials and the expected stimulus time should be found insofar as these endogenous potentials reflect a process triggered by this time ; and 2) as the peak latencies of P300s and N200s are known to correlate with perceptual processing, their durations should increase when the behavioral output is delayed to the extent that their onset would covary with the expected stimulus time.

These assumptions prompted us to measure the onsets and offsets of the emitted waves (and therefore their duration) recorded in the Renault and Lesèvre's 1979 experiments, trial by trial. Detailed results follow.

AVERAGE RESPONSES

Whatever the omission condition (with or without the rhythm tap) for 5 out of 7 subjects the average spatio-temporal organization of the visual omis-

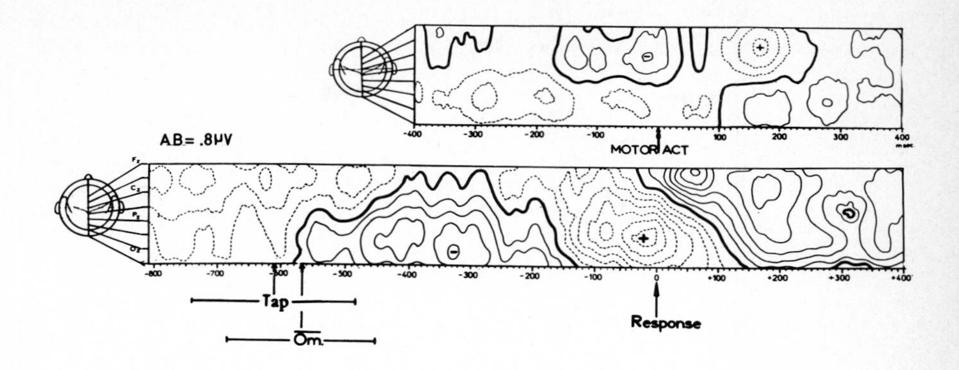

Figure 3 : Spatio-temporal maps triggered by the movement ; 7 subjects, 294 trials. Note the difference in the spatio-temporal organization of potentials in condition 1 (bottom) and 3 (top). .8μV between two isopotential lines.

sion response was similar to that reported previously. It consisted of a negative wave beginning around the moment of the omission (0 msec) in the parieto-occipital region and extending later (200 msec) towards the fronto-central region, followed by two positive waves, the first peaking (400 msec) in the central region and the second (500 msec) in the parieto-occipital region. The average responses of the two remaining subjects showed a similar spatio-temporal organization but the vertex components were not as clear. The Grand mean response (averaged across all subjects) is depicted in Fig.2. The averaged responses synchronised to the rhythm tap were similar. On the contrary, when the reaction time was the trigger, the central negative and positive waves disappeared. However, the topography and the timing of this averaged response was quite different from those obtained in the self-paced movement condition. Indeed the motor response potentials were less than 3 μV and peaked before and after the motor act whereas both waves of the omission response were of much higher amplitude and peaked before the motor response (Fig. 3).

SINGLE TRIAL ANALYSIS

For each subject, the averaged responses were utilized as templates for visual identification of each component of the single trial responses to the omission. These components were identified independently by two scorers. The scoring was done with regard to the point in time of the stimulus omission, but blind with regard to the rhythm tap and to the RT. Only those trials accepted by both scorers and for which at least two components (a negative and a positive one) were measured identically were used in subsequent analysis. Waveforms observed after the omission were considered as the superposition of relatively slow-rising (and slow-decaying) waveforms and EEG. Therefore, values which seemed to be the most likely to represent the underlying slow waves were taken as onsets and offsets. These values were always zero-crossings except for the central N200 onset (see later on).

In both conditions, with and without the rhythm tap, 294 and 242 maps were studied respectively, each of them corresponding to 2.5 sec of non-averaged EEG, including an omitted stimulus. From this total, respectively, 55% (162) and 58% (140) were accepted by the scorers. Except in 79 cases, the remaining maps did contain observable responses, but, on account of the low signal to noise ratio, no precise measures could be taken for at least two of the components.

The accepted single trial omission responses were divided into three different types according to the location of their peaks. For each subject, the most frequently observed pattern (81% of the accepted trials) showed two distinct foci of negative-positive activity : a parieto-occipital one and a fronto-central one. This pattern was therefore labelled "mixed type".

The existence of two active regions, each producing a negative-positive response to an omitted stimulus, was confirmed (as in previous studies) by the remaining 19% of the accepted trials. Indeed, the maps of the responses corresponding to these trials could also be divided into two groups : the first one (13%) contained responses made up of a negative-positive complex peaking at the vertex, the second one responses for which both components peaked in the parieto-occipital region (6% of the cases). Thus a second analysis of the mixed types was performed in order to better differentiate both negative and positive waves. The peak (latency, amplitude and location), the onset and the offset of each wave were measured when this was possible. Therefore 107 trials for condition 1 and 86 trials for condition 2 were selected for subsequent analyses. This reduction in the selected trials was due mainly to the difficulty in measuring the central N200 onset.

TABLE I : Overview of the results of the single trial analysis. Columns 1 and 2 refer to conditions 1 and 2 ($\bar{x}$ and s represent mean and standard deviation)

		Onset (msec)		Peak Lat. (msec)		Peak Amp. (μV)		Peak Loc. %		Offset (msec)	
		1	2	1	2	1	2	1	2	1	2
Parieto-occ.	$\bar{x}$	16	-5	218	207	19.8	20.4	22.3	24.4	321	333
N200	s	112	85	109	112	8.6	7.2	7.4	6.5	95	06
Central N200	$\bar{x}$	198	213	265	274	20.3	20.1	52.5	54.8	310	325
	s	91	109	93	102	7.7	10.2	10.4	8.6	92	104
Central P300	$\bar{x}$	310	325	381	384	21.6	22.1	53.3	58.4	456	467
	s	92	104	95	105	9.6	10.9	10.3	13.5	96	119
Parieto-occ.	$\bar{x}$	321	333	499	508	20.3	19.2	23.3	28.4	642	656
P300	s	95	106	107	114	9	7.2	8.9	6.8	128	139

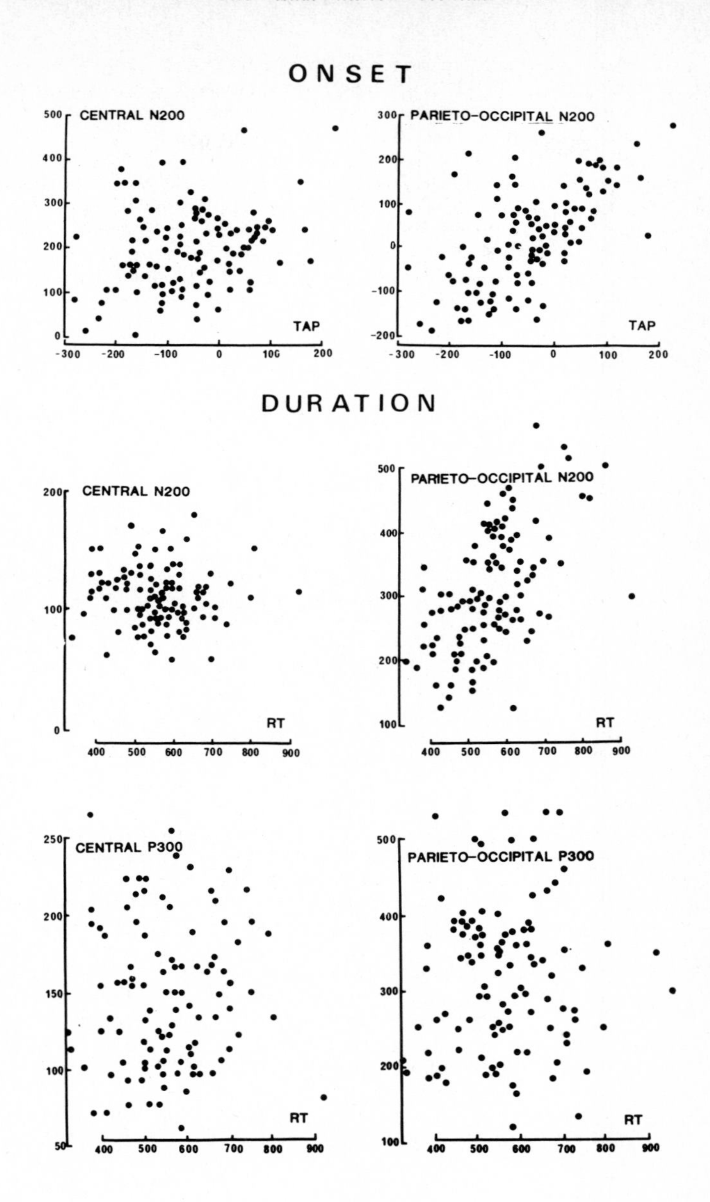

Figure 4 : On the top, onset of the parieto-occipital N200 and of the central N200 plotted against the rhythm tap (on the abscissa). On the bottom, duration of each wave plotted against the RT. Note the covariation of the parieto-occipital N200 with the rhythm tap and with the RT.

Indeed, the rapid potential changes between N200 and P300 provided an easy zero-crossing measure for N200s offsets (and P300s onsets) whereas the central N200, developing concurrently with the end of the parieto-occipital N200, partly added up with the latter, and the longer the duration of the parieto-occipital N200 the further the onset level of the central one was shifted away from the baseline. This sometimes resulted in different measures from the two scorers which therefore were not used. The onset of the parieto-occipital N200 and offsets of both P300s were easier to recognize and were usually scored in the same way.
As seen in table I, the characteristics of these endogenous brain waves remained the same whatever the condition and are similar to those of Fig.2 except for amplitude values (see note 2). For each condition, onset, peak latency, peak location and duration of the central N200 differed significantly from those of the parieto-occipital N200 at a minimum level of $p < .005$ (paired t test). Peak latency and location, offset and duration were also significantly different ($p < .005$) for both P300s. Thus, these results confirm the existence of two negative and two positive waves.

PERFORMANCE AND ERP TIME RELATIONSHIPS

The onset of the parieto-occipital N200 occurred approximately at the same time as the stimulus omission ; the mean differences were 16 msec (condition 1) and -5 msec (condition 2). These onsets varied from -182 msec before to 286 msec after the stimulus omission time in condition 1 and from -270 before to 190 msec after in condition 2. In condition 1 the tap of the rhythm was highly correlated only with the onset of the parieto-occipital N200 (.61, see fig.4 and Renault et al., in press). In contrast, neither the onset of the central N200 nor the onsets of the central and the parieto-occipital P300s were highly correlated with the tap of the rhythm (.33, .31, .33 respectively). The product moment correlations between the onset of each wave were computed. All these correlations were significant at the level of $p < .001$. In addition the correlations between the onsets of the parieto-occipital N200 and onsets of the other waves were significantly lower ($p < .001$) than those computed between the central N200, the central P300 and the parieto-occipital P300 onsets, these latter ones being nearly equal to 1 (table II).

TABLE II : Product moment correlations between onsets of each wave for condition 1 (107 trials) and 2 (86 trials)

	CONDITION 1			
	Par. Occ. N200	Central N200	Central P300	Parietal P300
Par. Occ. N200		.57	.57	.61
Central N200	.46		.97	.84
Central P300	.52	.96		.87
Par.Occ. P300	.53	.94	.97	
	CONDITION 2			

The offsets of both P300s were also highly correlated with their respective onsets (.88 and .78, situation 1 ; .98 and .77, situation 2).

These results demonstrate that the duration of the parieto-occipital N200 determines the electrogenesis of the other waves since only a variation in this duration (the duration of the other waves remaining constant) can explain the above observed correlations. More precisely, since the onset of the parieto-occipital N200 was correlated with the tap of the rhythm, it can be stated that the estimated stimulus time initiates the electrogenesis of the parieto-occipital N200, the underlying processes reflected by this wave leading in turn to the generation of the central N200-P300 complex and of the parieto-occipital P300. It must be noted (see table 1) that the processes reflected by the central N200-P300 are concurrent with those reflected by the parieto-occipital N200 and P300, since the central complex developed at the same time as the end of the parieto-occipital N200 and the beginning of the parieto-occipital P300.

Taking these results, and the fact that the RT was not highly correlated with the rhythm tap, into account (.29), the onset of the parieto-occipital N200 was used as a starting point for measuring the reaction time. Product moment correlations between RT's thus calculated, and the trial by trial duration of each negative and positive wave were computed. Results are shown on Fig. 4. For all subjects, and all single trials pooled, the correlations were .57 and .09 respectively for the parieto-occipital N200 and

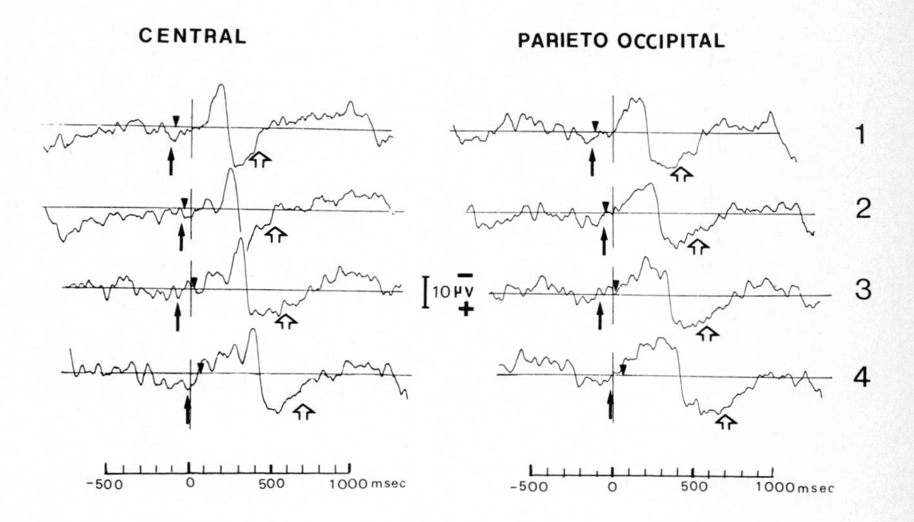

Figure 5 : Lines 1 to 4 represent latency corrected average potentials obtained for each RT quartiles in the central and the parieto-occipital region. The rhythm tap is indicated by the thin black arrow, the reaction time by the wide white arrow and the time of the omission by the black triangle. (Reprinted from Renault et al., Science, in press).

for the central N200 in condition 1 (.73 and .16 in condition 2) ; .01 and .25 respectively for the central and for the parieto-occipital P300 in condition 1 (.22 and .05 in condition 2). Thus, only the duration of the parieto-occipital N200 appeared as a determinate of reaction time, since it was the only wave whose duration was consistently correlated with the RT (Fig. 4). These results are emphasized on figure 5, where the pool of trials (condition 1) were divided into quartiles with respect to RT's. For each electrode location the trigger was the zero-crossing between N200 and P300 since those easily recognized points were the most reliable. A time scale translation was computed, the parieto-occipital N200 onset being taken as a "zero". This figure clearly shows that the duration of the parieto-occipital N200 increases across quartiles whereas only the peak latency of the other waves increases while their duration remains uncorrelated with the RT.

The mean reaction time was 559 msec (sd=108) and 494 msec (sd=125) in condition 1 and 2 respectively. Its range of variation in relation to the latency characteristics of the four waves is depicted in table III ; the N200s were the only waves which always preceded the motor response.

The RT was significantly shorter (65 msec) in condition 2 compared to condition 1 ($t=4.34$; $p<.001$) ; this difference was most probably due to the mechanical inertia of the finger since the finger displacement lasted about 60 msec and the tap and the motor response (RT) were given by the same finger. Therefore, if one takes into account the data reported by Ritter, Simson and Vaughan (1972), the central initiation of the motor act would take place from 165 msec before the registered photo-electric signal, i.e., at the same time as the mean offset of the negative waves. When the RT is short (see table III) the central initiation of the movement would take place from 103 msec to 123 msec before the offset of the parieto-ocipital N200 in condition 1 (or from 144 msec to 161 msec, in condition 2). In other words, in the case of the shortest RT's, the central initiation of the motor act, the process reflected by the central N200 and those reflected by the end of the parieto-occipital N200 would be concomitant.

TABLE III : Occurrence (msec) of the motor response in relation to the endogenous components (RT minus wave characteristics) ; min, $\bar{x}$ and max represent minimum, mean and maximum values.

	Condition 1			Condition 2		
	min	$\bar{x}$	max	min	$\bar{x}$	max
Offset of the par.-occ.N200	62	238	630	24	161	526
Offset of the central N200	104	249	620	4	169	526
Peak of the central P300	-12	178	604	-56	110	440
Peak of the par.-occ.P300	-124	60	460	-172	-14	318
Offset of the central P300	-60	103	540	-127	27	402
Offset of the par.-occ.P300	-334	-83	288	-320	-162	180

DISCUSSION

1. Spatio-temporal organization.

These results clearly establish that the potentials associated with missing visual stimuli are made up of four distinct waves, two of long duration, peaking in the parieto-occipital regions and two of short duration, peaking near the vertex. These distinct foci of activity have already been partially suggested by Simson et al. (1976), Renault and Lesèvre (1978 and 1979), Harter and Guido (1980), Ruchkin et al. (1981). Furthermore, these trial by trial studies have shown that this kind of "mixed response" is observed quite frequently, even in the case of subjects who had an average response (triggered by the moment of the omission) with only one focus of negative-positive activity ; indeed, the trial by trial examination of the omission responses of these subjects showed that the two foci of same polarity seen on the averaged maps of the other subjects were in that case so closely located in space and time that they mingled in the averaged data and thus appeared as one focus.

The disappearance of the central N2-P3 waves when the average was triggered by the motor response whereas the parieto-occipital N200 and P300 waves did not disappear (Fig.3) is not surprising. Indeed, these central waves were washed away because of their short duration and their important latency variability. The same kind of phenomenon is also more or less observed when the trigger is the omitted stimulus time and does not exist in latency corrected averages (compare Fig. 2, 3 and 5).

On the basis of their waveform and scalp distribution, the central N200 and P300 waves are quite similar to the N2-P3a complex, the parieto-occipital N200 to the modality specific N2 and the parieto-occipital P300 to the P3b. Two points must be emphasized concerning this latter wave. First, its real location is most probably parietal (as seen in counting and NOGO conditions) rather than parieto-occipital (GO condition). In the latter case, this wave overlaps, at least partially, the negative pre-motor potential and thus appears more occipital (see Banquet, Renault and Lesèvre, in press, for similar results) ; this was confirmed by the fact that its location became more and more parietal as the difference between its peak latency and the RT increased. Second, this wave terminates at the same time as appears across the fronto-central regions a late negative wave, increasing in amplitude until the appearance of the next visual stimulus (see the map of Fig. 2). Thus the last part of this parietal P300 (P3b) is quite similar to the "slow-wave" (S.W.) described by N. Squires et al. (1975). Nevertheless, the spatio-temporal organization of this S.W. suggests that its positive part may be independent of its negative one as argued by Kok (1978). Further studies comparing the negative part of the slow wave obtained in this experiment with the "O wave" (Rohrbaugh, this book) and/or with the CNV (which can also be triggered by an omission : McCallum, 1980) are necessary before drawing conclusions. However, our data concerning topography and temporal evolution suggest that the negative part of this slow wave is similar to a CNV. Indeed, under our experimental conditions a CNV could be generated by the regular train of visual stimuli, could habituate, and be reactivated by an omission. Figure 1 strongly supports such an assumption : no negative wave can be seen between the preceding visual stimulus and the omission whereas a fronto-central negativity, similar to a CNV, develops from 500 msec after the omission until the next visual stimulus. Furthermore, this result underlines the differences between the spatio-temporal organization and reactivity of the CNV and those of the omission response.

2. Functional significance.

The timing of these different waves and their relationships with the estimated time of the stimulus and with the motor response provide some clues to their functional significance. Only the onset of the parieto-occipital N200 was strongly dependent upon the rhythm tap ; moreover this wave began at the same time as this tap, that is before the occurrence of the omission. This supports the suggestion by Simson et al. (1976) that the onset of this negative wave is linked to some anticipatory event activated by an internal representation of the stimulus delivery rate. Furthermore, this wave is the only one whose duration increases when increases the reaction time, the motor response being initiated centrally approximately at the same time as its offset. Therefore, as we have already argued (Renault et al., in press), this parieto-occipital N200 can be considered as reflecting real time perceptual processing previous to P300 generation. Since the central N2-P3a complex and the P3b wave were highly time-locked to the offset of this wave (and thus to its duration) this demonstrates that their electrogenesis depends upon the processes reflected by the parieto-occipital N200. In that sense, the P300s may also index information processing (Kutas, McCarthy and Donchin, 1977 ; McCarthy and Donchin, 1981 ; Duncan-Johnson, 1981). However, P300s are delayed indices of the first stages of perceptual processing (Ritter et al., 1979 ; Renault and Lesèvre, 1979 ; Towey et al., 1980) and probably reflect the "context updating" (Donchin et al., 1978) or the inhibition of an arousal reaction that follows a decision (Desmedt and Debecker, 1979). In fact, the existence of two distinct P300s may well be a first step in validating both these interpretations of the P300 (Banquet et al., in press).

A very interesting issue concerning the modality specific N200 is to determine when it starts in relation to information processing stages. Recently, Sanders (1980) suggested six different stages on the basis of results of the Sternberg's additive factor method (1969). First, the preprocessing depending upon the signal contrast ; second, the feature extraction depending upon the signal quality ; third, the identification depending upon the signal discriminability ; fourth, the response choice depending upon the stimulus-response compatibility ; fifth, the response-programming depending upon the response specificity and sixth, the motor adjustment depending upon the muscle tension. In the omission paradigm, the subject's internal representation of the stimulus delivery rate determines these successive processing stages. As the onset of the modality specific N200 occurred about 70 msec after the rhythm tap it can be assumed that this onset should at least be concomitant with the identification stage. Actually, Harter and Previc (1978) and Harter and Guido (1980) showed that a wave identical to this modality specific N200 (named "selection negativity") varied according to spatial frequency and orientation of visual stimuli ; therefore, feature extraction would play a role in the electrogenesis of this wave. Furthermore, Ritter, Simson, Vaughan and Macht (in preparation ; see also Ritter and Vaughan, this book) demonstrated that two negative components (NA and N2), in the latency range of the modality-specific N200 described here, were differentially modified by signal quality and discriminability. Consequently, assuming these NA and N2 are parts of this modality-specific N200, this wave would most likely start with the feature extraction stage (as does the Na wave). However, experiments manipulating both signal contrast and quality would be necessary before drawing definitive conclusions.

Regarding the question : "when does the modality-specific N200 finish?", converging results provide some clues. Duncan-Johnson (1981), McCarthy and Donchin (1981) and Ragot and Renault (in press) agree that the P300 is generated when full stimulus categorization occurs. However, in relation to Sander's stages it seems important to test whether the stimulus-response (S-R) compatibility affects the N200 duration (and thus, P300 latency) or not. Ragot and Renault manipulated the S-R compatibility and found that P300 latency was longer for incompatible than for compatible cases whereas it did not seem to vary with motor programming. Consequently, it can be suggested that the modality specific N200 should finish with the response choice stage and then determine the electrogenesis of the parietal P300. However, McCarthy and Donchin did not find similar result concerning the effect of S-R compatibility upon P300 latency. In fact, as Sanders underlines (1980, p.339), compatibility in an "ill-defined variable"... and therefore "comparisons between studies on S-R compatibility are often difficult since the operational meaning of 'compatible' and 'incompatible'varies across experiments". The lack of compatibility effects found by McCarthy and Donchin seems therefore to be due to a matter of labelling of variables and stimulus presentation (see Ragot and Renault, in press, for a more detailed discussion of this point).

The modality specific N200 wave has often been called "mismatch negativity" in opposition to the "processing negativity" (see Näätänen and Mitchie, 1979b). More recently, Näätänen (in press) discussed the relations between N200 and orienting and concluded that the N200 "could be used as a measure of the neuronal-mismatch process postulated by the orienting theory". However, in such a view, the N200 reflects a process which is dependent only upon the automatic comparison between the on-going stimulus and its neuronal model and thus most probably taking the same amount of time whatever the vigilance level. The modality-specific N200 described in our studies increases in duration with the decrease of vigilance (i.e. increase of the RT). Moreover, its onset is dependent upon the rhythm tap which is known to be a cognitive task (Fraisse and Voillaume, 1971 ; Wing and Kristofferson, 1973), and its offset probably occurs with the response choice, i.e., at the end of a series of cognitive processes. Therefore it is unlikely to reflect only a neuronal mismatch in the sense of Sokolov (1975) and is more likely to be related to "active" perceptual processing including template matching.

On the other hand, the central N200 of the N2-P3a complex, whose duration is constant, has often been related to orienting following a mismatch (N. Squires et al., 1975 ; Snyder and Hillyard, 1976 ; Ford, Roth and Kopell, 1976a ; K. Squires, Donchin, Herning and McCarthy, 1977 ; Näätänen, Gaillard and Mäntisalo, 1978 ; Renault et al., 1980b). Näätänen and Gaillard (this book) basically agree with this interpretation, pointing out a possible interpretation of this complex in terms of "significance orienting reflex" (Näätänen, Simpson and Loveless, in press). These authors advance the view that the elicitation of the N2-P3a complex depends upon the N200 wave which has to reach some threshold amplitude in order to trigger this complex. Our interpretations based on the trial-by-trial analysis reported here support their suggestion in part. Indeed, the time occurrence of the N2-P3a complex appeared to be dependent upon the duration of the modality-specific N200 since this complex was generated concurrently with the end of this wave (cf. also the correlation analysis). However, this was the case only for the "mixed types" of omission responses since the N2-P3a ("vertex types") and the parieto-occipital N200 and P300 ("parietal types") could each be seen alone. Therefore, trial-by-trial data do not suggest that the N2-P3a is dependent upon a threshold amplitude level of the modality specific N200. On the contrary, it seems that these two kinds of ERPs reflect concurrent processes, each of them being able to trigger the behavioral response. However, most often ("mixed type" of responses) these two parallel processes are coupled and time dependent. This coupling would possibly induce a balance between the motor and sensory aspect of the task performance or, in other words, favor either speed or accuracy (Kutas et al., 1977). More precisely, when the motor aspect is emphasized only a N2-P3a complex, reflecting an orienting reflex which would facilitate the motor reaction (Näätänen, in press), would often be observed. By contrast, when the sensory analysis is emphasized a modality-specific N200 followed by a parietal P300 would be recorded more often (see Renault et al., 1980a for a compatible view). In fact, most of the time, these two concurrent processes can be evidenced together ("mixed type"), therefore suggesting that orienting, in these cases, is dependent at least upon stimulus identification and its template matching. Such an interpretation is in-keeping with the cognitive theory of "active" orienting as being linked to experienced significance of the stimulus (Bernstein, 1979). Furthermore, this interpretative line suggests that orienting, response choice and (when the RT is short) central initiation of the movement coexist concurrently and with mutual interdependences since the waves described above i) overlap in part, ii) are linked together and iii) determine the reaction time.

CONCLUSION

These trial-by-trial studies of ERP spatio-temporal organization and in particular of their durations were able to differentiate the modality specific N200 from two other events : the N2-P3a complex and the parietal P3b. This former wave is currently quite important since its time course seems to reflect on line perceptual processing until the response choice stage. Actually, other recent advances in the field underline the importance of the duration of brain events for studying mental chronometry (Ritter and Vaughan, this book) and implement Woodworth's suggestion (1938) of using "brain waves as indicators of the beginning and end of a mental process". Other negative waves have been described and related to information processing such as the "selection negativity" (Harter and Guido, 1980) and the so-called "N1 effect" first observed by Hillyard, Hink, Schwent and Picton, 1973, and reinterpreted as due to a "processing negativity" (Näätänen and Mitchie, 1979a ; Hansen and Hillyard, 1980 ; Okita, 1981). Although these various waves were revealed by substraction, and observed in various experimental paradigms manipulating different psychological variables, further experiments seem necessary before concluding that they reflect different functions and are much different from this modality specific N200. In any case, from a conceptual point of view, it must be emphasized that these endogenous negative waves related to perceptual processing and selective attention clearly demonstrate that mental chronometry is possible today. Furthermore, their timing, which is consistent with the proposition that the brain processes information in a parallel manner (Turvey, 1973 ; McClelland, 1979 for examples), should be of great help to develop psychological models of information processing in human.

NOTES

This work was supported by the "Centre National de la Recherche Scientifique" (CNRS). I wish to thank Colette Gaiche for the typing and presentation of this article and Tony Gaillard, Marta Kutas, Nicole Lesèvre and Walter Ritter for their helpful comments and discussions on earlier drafts. Adress requests for reprints to : Dr. Bernard Renault, GR9 du CNRS, LENA, Hôpital de la Salpêtrière, 47, Bd de l'Hôpital, 75651 Paris Cedex 13, France.

1 - After trial-by-trial recognition of each type, the averaging trigger (on Fig. 1) was the peak latency of the first positive wave. For this reason the amplitude of the N200 waves are reduced. Since, in this study, windows from only zero to 1 sec after the omission have been analysed, the averaged maps presented start at 200 msec and finish at 700 msec. This is due to the latency variability of the emitted responses which had the effect of erasing out early and late part of the averaged window.

2 - Amplitude values are much larger in table I than in Fig. 2 since : 1) these mean values are computed on the basis of trial by trial measurements whereas latency variability reduces the averaged potential of Fig. 2 and 2) Fig. 2 contains all the trials whether they had the four waves or not, whereas the table is based on selected trials where the four waves were measured by the two scorers.

Tutorials in ERP Research: Endogenous Components
A.W.K. Gaillard and W. Ritter (eds.)

8

A COMPARISON OF P300 AND SKIN CONDUCTANCE RESPONSE

Walton T. Roth

Veterans Administration Medical Center, Palo Alto
and
Stanford University School of Medicine
California, U.S.A.

P300 and skin conductance response (SCR) show many similarities. Empirically, both are elicited by stimuli of any modality that are surprising, task-relevant, or intrinsically salient. Theoretically, both P300 and SCR have been explained in terms of neuronal, probabilistic, cognitive, or computer models of the orienting reflex (OR). However, the OR is not a unitary reflex, but a group of related processes that presumably correspond to different psychological functions. Although P300 and SCR are very different biologically, there is as yet a lack of evidence for their having different psychological correlates.

INTRODUCTION

Contemporary researchers in psychophysiology tend to be divided into those studying autonomically controlled responses and those studying event-related potentials (ERPs). There is amazingly little communication between these two groups, though they may brush shoulders at scientific meetings. The results of this are shown most vividly in the failure of each group to cite the papers of the other, although both groups are trying to understand the meaning of responses observed under similar psychological conditions. There are several reasons for this split: for example, evoked potential researchers are impressed with the proximity of the scalp to the seat of the mind as compared to the distance of the skin of the hand. But probably the most important reason is the vast scientific literature in the two areas. Few feel able to spend the time to read papers in both areas in a critical way.

An integration of the findings of classical psychophysiology and ERP research is beyond the scope of any paper. Here, I will outline a comparison of two variables: SCR and P300. These variables were chosen because the experimental conditions for eliciting them are strikingly similar, though the nature of the responses has led to somewhat different paradigms and analysis methods. First, the temporal characteristics of SCR and P300 are quite different. SCRs to simple stimuli have an onset latency of more than 1 sec, peak around 3 sec, and may take 7 sec or longer to return to baseline. P300s to simple stimuli have peak latencies between 200 and 400 msec and return to baseline in about 100 msec. Their temporal recovery and dimensions are such that several distinct P300s can be elicited in 1 sec (Woods et al. 1980), while 4 sec are required for a single SCR. Thus, interstimulus intervals (ISIs) are usually much longer in SCR experiments (typically 15 to 60 sec) than in P300 experiments (typically 1 to 3 sec). A stimulus train with 1 sec ISIs is considered unsuited for SCR elicitation because of SCR overlap and doubt as to which stimulus elicited the SCR. The long ISIs in SCR paradigms are inefficient for investigating P300s in that fewer P300s are available per unit time for signal averaging or other kinds of data analysis. Also the amplitude of certain earlier positive components, such as the auditory P180, increases with ISI, and these components may begin to merge with P300, making separate measurements of these components uncertain. This is part of the general concern in P300 experiments about overlap of different positive and negative potentials from various locations in the brain synchronized to the same stimulus that elicits P300.

Second, less elaborate methods of data analysis are required for SCRs than for P300s. SCRs are recorded with much less noise than P300s and with a better baseline reference. SCRs require no averaging or special methods to detect single responses in the presence of noise. "Non-specific fluctuations" or "spontaneous" SCRs that are not synchronized with stimuli are easily detected, though sometimes such SCRs occur in a latency range that makes them impossible to differentiate reliably from evoked SCRs. Tonic basal levels of conductance can be reliably measured. In constrast, signal averaging or special filtering methods are usually required for reliable evaluation of P300. While template matching and adaptive filter techniques have enabled P300 latency estimation in single trials, it is questionable whether these methods are adequate for detecting spontaneous occurring P300s, if such exist. Finally, reports of accurately measuring tonic, direct current (DC) brain potential baselines from the scalp are viewed with skepticism.

These differences have contributed to obscuring the essential fact that the general experimental conditions for eliciting SCRs and P300s are identical. These conditions are that stimuli must be surprising, i.e., violate the subject's expectations, and/or task-relevant, and/or intrinsically salient, e.g., intense. SCR researchers usually discuss all three conditions in the context of the orienting reflex as conceptualized by Sokolov (e.g., Sokolov 1963a). ERP researchers have concentrated on the first two conditions and have used ideas of information processing taken from current cognitive psychology as their theoretical framework.

In the following section, I will review the evidence that P300 and SCR are both controlled by the same stimulus properties. In doing so, I will assume a basic unity among positive peaks in the 200 to 400 msec latency range, an assumption that could be questioned. A single average ERP waveform may contain multiple positive peaks in this range, which might be due simply to latency variations in a single component in the underlying trials. Although topographic differences between different P300s suggest different generators, various other components might overlap with a single P300 and change its apparent distribution. Multiple positive peaks with different distributions are especially prominent when subjects are required to process stimuli in complicated ways (Stuss and Picton 1978; Friedman et al. 1978; Friedman et al. 1981). Unless the contrary is explicitly stated, it may be assumed that either P300 or SCR was measured in a given study, not both.

STIMULUS PROPERTIES ELICITING P300 AND SCR

1. Surprise

A stimulus is surprising when it is not expected. Surprise is enhanced when a specific expectation is built up and then violated by the unexpected event. Expectation encompasses whether an event will occur, what the event will be, and when it will occur. The experimenter varies expectations and surprise by varying the probability of stimulus events and their similarity. Paradigms can be classified by how surprise is achieved.

a. Occurrence uncertainty. In the seminal experiments of Sutton et al. (1965), when the modality of the second stimulus of a pair of stimuli was uncertain, the second stimulus elicited a P300. Tueting et al. (1970) showed that P300 amplitude was inversely related to event probability in a paradigm where subjects guessed whether the next click would have a high or low pitch. Duncan-Johnson and Donchin (1977) presented subjects with a random series of high or low pitched tones at a fixed ISI of 1.5 sec. The event probabilities ranged from 0.1 to 0.9 in steps of 0.1. Subjects either counted the occurrences of the high pitched tone or performed a distraction task to which the tones were irrelevant. In the counting task, P300 was inversely proportional to the a priori probability of the eliciting stimulus. By contrast, in the distraction task, P300 was not elicited. The amplitude of P300 was also dependent on the sequence of stimuli, with the alternations of stimuli evoking larger P300s than repetitions. This finding is consistent with the observation that the longer the sequence of repetitions of a stimulus preceding its discontinuation, the larger the P300 (Squires et al. 1976).

Grings and Sukoneck (1971) performed a guessing experiment similar to those of Sutton and his colleagues, but one in which SCRs were measured. Four different cue lights indicated the probability of a shock 10 sec later. The probabilities were 0.0, 0.25, 0.75, and 0.1, and the subject had to indicate during the 10-sec interval his guess as to whether a shock would be given. The size of the SCR to the shock was inversely related to its probability. In a somewhat different paradigm, Öhman et al. (1973) found a U-shaped function, with 25 and 100% probable shocks giving the largest SCRs. The aversive properties of the stimuli and preparation for them may have modified the probability effect in this experiment.

Sequential probability effects for the SCR have been studied under the rubric of "below-zero" habituation. Thompson and Spencer (1966) coined this term to refer to a hypothetical continuation of the habituation process when stimuli continue to be presented after they fail to elicit detectable responses. Evidence for additional habituation would be a delay in spontaneous recovery of the response or increased difficulty in dishabituating it. Stephenson and Siddle (1976) gave passive subjects 70-dB SPL, 1000-Hz, 3-sec tones with a random ISI between 20 and 40 sec. Different groups received "test" stimuli of either 1400-Hz or 670-Hz after either 5, 10, or 15 no-response trials. The number of repetitions of the 1000-Hz stimuli in the three cases averaged 16, 20, and 26. For both test stimuli, SCRs were proportional to the number of repetitions, which is contrary to Thompson and Spencer's theory, but is consistent with the interpretation that the subject builds an expectation that the series will continue without change.

One version of the event uncertainty paradigm is to create the expectation that an event will occur at a certain point in time, and then not fulfill the expectation. Both EEG and autonomic responses occur to this kind of "missing" stimulus. Sutton et al. (1967) first showed that P300 can be elicited by missing stimuli. In an experiment of Simson et al. (1976) trains of 200 auditory or visual stimuli were given at an ISI of 1 sec. Five percent of the stimuli were omitted at random. Subjects signaled missing stimuli after the next stimulus. Missing stimuli elicited parietal P300s with latencies around 500 msec which were preceded by negative peaks around 250 msec.

The omission of expected stimuli also elicits SCRs. This is observed in experiments in which conditioned stimulus (CS) - unconditioned stimulus (UCS) pairs are followed by extinction trials in which the CS is given alone. During extinction, SCR may appear at two latencies, one a few seconds after the CS, and another a few seconds after the missing UCS. The first SCR is regarded as the recovery of an orienting response that habituated during conditioning. The second is a conditioned response to stimulus omission. When CS - UCS intervals are 1 or 2 sec, these two responses overlap in a way that makes their origin uncertain (Furedy and Poulos 1977). With longer intervals, unambiguous SCRs to missing UCSs can be demonstrated. Williams and Prokasy (1977) used a random reinforcement paradigm in which the CS (a 68-dB tone) was followed after 6 sec by the UCS (a 115-dB, 200-msec white noise burst) with a probability of 0.33, 0.67, or 1.0. "Third interval UCS omission responses (TORs)", measured in a 7 to 9.5 sec latency window from CS, were much more likely to occur in groups where CS and UCS were paired than in control groups where CS and UCS were never paired.

In experiments more analogous to that of Simson et al. (1976), Siddle and Heron (1975) presented 70 or 90-dB, 3-sec, 1000-Hz tones at fixed ISIs of 12 or 21 sec. The 7th, 10th, 13th, 16th, and 21st stimulus were omitted, but no reliable omission SCR was obtained. The long ISIs used may have prevented the formation of accurate expectancies, especially when the stimuli were not task relevant.

b. Stimulus quality deviation. The degree of deviation of stimulus qualities from what is expected, influences P300 amplitude. Ford et al. (1976) did a parametric study in which 50-msec tone pips were presented at ISIs of 1, 0.5, or 0.25 sec. One out of every 12 stimuli deviated in pitch from the frequent stimulus type by 5%, 25%, or 100%. The amplitude of P300 to mismatches was proportional to the degree of discrepancy. This effect was larger when subjects responded to the mismatches by quickly pressing a button than when they were reading a novel.

A similar effect was found by Siddle and Heron (1976) for SCRs in a "below-zero" habituation paradigm. SCRs were larger to a pitch change of 500 mels than to a change of 250 mels, after SCRs had been habituated to repetitions of standard stimuli. Siddle et al. (1979b) tested semantic deviation gradients. A single word was visually presented 12 times, followed by a second test word. When the standard words had been "quilt" or "pillow", the test words "pillow" or "quilt," respectively, evoked smaller SCRs than when the standard words had been "hair" or "peace." It is of interest that the authors do not mention instructions to the subjects in describing their methods and procedure. The tradition of OR experiments is to consider cognitive processing as a kind of reflex, even when the stimuli are verbal.

Conditioning procedures evidently give different deviancy gradients depending on whether the CS is complex or simple. When words or shapes are paired with aversive UCSs, closely related words or shapes elicit larger SCRs than distantly related words or shapes (see Grings and Dawson 1973, for a review). However, if the pitch of the CS is varied, the opposite result may occur (see Prokasy and Kumpfer 1973, for a review). For example, Öhman (1971) used 70-dB, 3000-Hz, 10-sec tones as CS and a 50 msec shock immediately following tone offset as UCS. After 24 conditioning trials, of which 67% were reinforced, the subjects received four 70-dB, 10-sec test tones of 200, 500, 1200, and 3000-Hz in a random order balanced across subjects. Intertrial intervals varied from 30 to 60 sec. SCRs in the interval 1-4 sec after CS onset were proportional to the deviation of the pitch of the test tone from the pitch of the conditioning tones. A similar finding emerged in a sensitization group for which UCSs were randomly scattered in the intertrial interval, but never following or preceding CS by less than 10 sec.

c. Temporal uncertainty. This factor tends to confound other factors. First, it is a part of occurrence uncertainty since, even though the timing of infrequent events within the trains of stimuli is known in terms of ISI, their timing is not known in the sense of when in the train the infrequent event will occur. Second, temporal uncertainty is a part of ISI effects in that time estimation becomes less precise as intervals increase beyond a few seconds.

For both P300 and SCR, temporal uncertainty effects are difficult to demonstrate. McCarthy and Donchin (1976) compared ERPs to stimuli initiated by the subject pressing a button and therefore temporally certain, and stimuli presented at identical intervals by a computer. In one set of conditions, the button or computer triggered a 1500-Hz tone with a probability of 0.10 and a 1000-Hz tone with a probability of 0.90. In both cases, subjects counted the high pitched tones. These tones evoked a P300 which differed in distribution in the two conditions, but differed little in amplitude. Control conditions in this experiment took into account the influence of the ERPs related to button pressing per se. Mean ISI in various conditions were between 4 and 10 sec with standard deviations between 1 and 2 sec.

Pendergrass and Kimmel (1968) reported that SCRs to 90-dB tones habituated more quickly when the ISIs were fixed at 40 sec compared to when the ISIs varied with a mean of 40 sec. After the first two trials, SCRs to corresponding trials of the variable ISI group were always larger than to those of the fixed ISI group. However, a follow-up study in the same laboratory using shocks did not replicate this result (Kimmel and Ray 1979).

Fitzgerald and Picton (1981) gave 55-msec, 65-dB SPL tones at 1000 or 2000-Hz. In one set of conditions the sequential probability of one pitch was kept at 0.20 while the ISIs were fixed in different runs at intervals from 250 msec to 4 sec. Subjects counted the less probable stimuli. P300 amplitude was proportional to ISI. In conditions where temporal and sequential probability were varied, only temporal probability, i.e., the probability of a target occurring in a given time interval, affected P300. I am unaware of a good analogy to this experiment in the SCR literature. Ray (1979) compared habituation to 50-dB, 1000-Hz tones presented at mean ISIs of 40 secs with those at mean ISIs of 120 sec. Habituation was slower for the group receiving the longer mean ISIs, which means that average SCRs would probably be larger for this group. However, the ISIs in Ray's experiment are 10 times or more those used by Fitzgerald and Picton.

d. Temporal deviation. When a subject is led to expect that a stimulus will occur at a certain point in time, stimuli occurring at different times may produce P300s or SCRs. Ford and Hillyard (1981) presented 50-msec, 62-dB SPL noise bursts at ISIs between 300 and 1200 msec in various conditions. In one experiment, the ISI was usually 1200 msec (p =.91) but occasionally was 600, 450, or 300 msec (p = .03 for each). Even when subjects were reading a book, a N130-P215 complex was elicited by the early tones. Its size did not vary with the degree of earliness. Although the P215 is earlier than many of the other late positive waves considered in this paper, there are reasons for considering such waves to be members of the P300 family (Roth 1973; Squires, N.K. et al. 1975).

Grings (1965) observed time disparity effects in a conditioning paradigm. After a tone CS and shock UCS had been presented together with fixed ISIs of 5 sec, different subject groups were tested with trials having intervals of 1.0, 2.5, or 4.0 sec. After 8 reinforcements (pairings), the greater the deviation of the test interval from 5 sec, the larger the SCR. As the author acknowledged, these ISIs are so short that there is possible overlap between SCRs to the CS and to the UCS, which makes interpretation of the results more equivocal than if longer intervals had been used.

2. Task Relevance

There are numerous ways in which stimuli can become relevant to the subject. Two general concepts associated with stimulus relevance are signal value and direction of attention.

a. Signal Value: Giving stimuli signal value by having the subject count them or press a button at their occurrence increases the amplitude of the P300 they elicit (e.g., Duncan-Johnson and Donchin 1977; Roth et al. 1978a; Roth et al. 1982; Fitzgerald and Picton 1981). The effect of the motor response of the button press per se does not account for the P300 enhancement (McCarthy and Donchin 1976).

Requiring a button-press to a stimulus also enhances the SCR it elicits. Siddle et al. (1979a) found that when subjects had to press quickly to tone offset, SCR to tone offset was larger than when no response was required. The stimuli were 70-dB SPL, 1000-Hz, 5-sec tones given at random ISIs between 20 and 50 sec. Subjects pressing the button had SCRs twice as large as those not pressing after the first two trials, for which SCRs were equally large in both groups.

Bernstein and Taylor (1979) required a rapid pedal press to tones of certain pitches, and measured SCR. In certain conditions of the experiment 10 trials of a 3-tone sequence were presented at intervals between 30 and 60 sec. Each of the three pitches 1000, 1400, and 1850-Hz, was chosen once in random order to form the sequence. Each tone was 7 sec, with no separations between tones. Different pitches were targets for different groups of subjects. Two factors determined the amplitude of the SCR. The first was whether the stimulus was relevant to the pedal press. Target stimuli and the stimuli preceding them elicited larger SCRs than stimuli following the targets. Second, the more the event uncertainty of relevant stimuli or the earlier they appeared in the sequence, the larger the SCRs. In any sequence the sequential probability of the first stimulus was 0.33, the second 0.5, and the third was 1.0.

A method for giving signal value to a stimulus without requiring an immediate motor response is to have subjects discriminate stimuli and report their judgment after the relevant ERP has occurred (e.g. Hillyard et al. 1971). This results in larger P300s than in conditions where no discrimination is required. Discriminative judgements also enhance the SCR. Greene et al. (1974) gave subjects visual stimuli of three hues and three durations (10, 15, and 20 sec). Intertrial intervals varied randomly from 30 to 40 sec. Different groups of subjects judged either hue or duration, and reported their judgement 3 sec after stimulus offset. SCRs were much larger to stimulus offset in subjects reporting duration than in subjects reporting hue.

Sometimes paradigms employ pairs of stimuli, in which the signal value of the first depends on the second. Johnston and Holcomb (1980) presented a digit in a train of irrelevant digits that indicated, with various degrees of certainty, which imperative stimulus in a two-choice reaction time task would be given. The sequential probabilities between the predictor stimuli and the imperative stimuli were 0, 0.25, 0.5, 0.75, or 1.0. After about 300 trials, when the subjects had thoroughly learned the signal values of the predictor digits, those digits that indicated the imperative stimulus with less certainty elicited larger P300s. In addition, the monetary value of very short reaction times was zero, $1.00 or $2.00 in different trials. When monetary values were higher, P300s to the predictor digits were also higher.

Öhman et al. (1973) used as a warning stimulus a visual display that informed the subject of the probability of a shock to follow after 8 sec (0, 0.25, 0.50, or 1.0) and its intensity (strong or weak). The SCR to the visual display was larger when higher probabilities or stronger shocks were indicated. The probability effect was largest in the first block of 10 trials, and replicated a previous result of Grings and Sukoneck (1971). The SCR finding for shock strength is comparable to the P300 finding for monetary value by Johnston and Holcomb (1980) in that, in both cases, one can imagine that the signal value of stimuli predicting strong shocks and high gains is greater than the signal value of stimuli predicting the opposite. The findings for probability were the opposite in the two experiments, perhaps because the shocks were aversive and the imperative stimuli were not.

Another way to give stimuli signal value is have them deliver feedback about performance. Johnson and Donchin (1978), in an extension and replication of a study by Adams and Benson (1973), showed that the more useful the feedback stimuli are, the larger the P300 they elicit. Subjects were asked to press a button exactly 1 sec after a cue light came on. Whether their press fell within an acceptable time range was indicated by feedback tones of two different intensities. The difference between these intensities varied between runs. When the difference was less, P300s were smaller and time estimation worse, since the tones were difficult to distinguish and, therefore, of less informational value. That the tones close in intensity were harder to distinguish was corroborated in a separate RT paradigm.

Stuss and Picton (1978) presented slide pictures that could be classified by any of five criteria. After a slide was presented for 1.2 sec, the subject chose which criterion he thought was relevant on that run by pressing the appropriate button. A feedback tone 1.5 sec after the offset of the picture told him if the classification was right (1000-Hz pitch) or wrong (4000-Hz pitch). P300s to the feedback tone were largest in trials immediately prior to when the subject became confident that the correct criterion had been found, presumably the trials for which the feedback was of greatest interest.

De Swart and Das-Small (1979) gave pictures that could be classified by any of four criteria. The pictures were presented for 12 sec, and immediately afterwards, the subject indicated by button press whether he thought they were an example of the relevant category or not. In addition, he indicated the certainty of his choice. Seven seconds later a slide came on informing the subject if the choice was correct or incorrect, or the slide gave <u>no</u> information. SCRs were larger for confirming feedback in the early trials when he was uncertain of the category than in later trials when he was more certain. SCRs to disconfirming feedback had the opposite pattern. Noninformative feedback tended to elicit smaller SCRs than informative feedback. In general, SCRs were larger when the feedback gave more information, i.e., violated expectancies.

<u>b. Direction of Attention</u>. Giving a stimulus signal value is usually thought of as directing attention towards the stimulus. In some experiments subjects are simply told to attend or to ignore all or some of the stimuli. Instructions to ignore sometimes have paradoxical effects, as illustrated by Kohlenberg (1970). He presented 75-dB tones of 500, 600, and 700-Hz in a fixed sequence of pitches (low, medium, high) for a total of 18 trials. The ISI was 20 sec. Different groups of subjects were instructed to ignore each of the pitches. SCRs to the ignored tones were largest in each group. Probably the instructions in effect give signal value to the "ignored" stimuli.

Broadbent's ideas about channel selection and "pidgeon-holing" (Broadbent 1971) suggest a distinction between attention to categories of stimuli (e.g. all stimuli of a certain modality) and attention to target stimuli within a category (e.g. tones of a slightly higher pitch than the majority of tones presented). This distinction sharpens the meaning of signal value by implying that three levels of task relevance may be created in certain paradigms. In ascending order these are: task-irrelevant, task-relevant, and target.

3. Salience

Salience is used here to designate the qualities of stimuli that make them stand out regardless of their probability or of their relevance to any task assigned to the subject. Intensity is the aspect of salience that is easiest to scale, although reinforcing qualities of stimuli (e.g., pleasurable or painful qualities) may also affect salience in a way that cannot be wholly described in terms of intensity. There is a semantic overlap between salience and the other stimulus qualities eliciting P300 and SCR. It is natural to call salient stimuli surprising and to consider them as having intrinsic signal value and being innately task relevant to the survival of the organism. However, for scientific purposes, it is desirable to use different terms for factors that can be distinguished operationally. Thus, we have considered under the heading "Surprise," only experiments that manipulated probabilities and expectations, and under "Task relevance," only experiments that explicitly manipulated this variable. However, reinforcing qualities of stimuli (e.g. pleasurable or painful qualities) may also affect salience in a way that cannot be wholly described in terms of intensity.

Parametric variations in auditory stimulus intensity were used in two experiments that demonstrated the importance of this factor in eliciting P300. Roth et al. (1980, 1982) presented 50-msec white noise pips at a fixed ISI of 4 sec. Intensities ranged from 55 to 115-dB SPL in a 10-dB steps. Stimuli were delivered under three attention conditions: counting the stimuli, ignoring them, or performing a visual task. Principal Components Analysis was used to decompose the waveforms. P280 and P340 components were both affected by intensity regardless of the attention condition. The intensity effect was monotonic for P280 in the 115 to 85-dB range. Johnson and Donchin (1978) observed a linear relationship between a P300 area measure and the intensity of 50-msec, 1000-Hz nontarget tones over a range of 10 to 50 dB SPL. The subject's task was to count 30 dB tones, which had a probability of 0.5 in each series of stimuli. ISIs were irregular with a mean of 6.75 sec.

Raskin et al. (1969) demonstrated intensity effects for several autonomic variables including SCR in a parametric experiment. Stimuli were 0.5 or 5.0-sec white noise bursts given at ISIs of 15 or 45 sec and at 40, 60, 80, 100, or 120-dB SPL. Subjects were assigned to a single combination of durations, intervals, and intensities. They were given no task involving the sounds. Stimulus intensity effects were greatest after the 10th trial and persisted to the 30th and last trial. Regardless of duration or ISI, SCR amplitudes were ordered by intensity, except that SCRs to 40 and 60-dB stimuli did not differ.

STIMULUS NONSPECIFICTY

A discussion of the experimental conditions for eliciting P300s and SCRs should also identify conditions which are not necessary for their elicitation. Neither P300 nor SCR are dependent on specific physical properties of the stimulus. Stimuli of various modalities and qualities can elicit identical P300s or SCRs, if other conditions are matched (see Pritchard (1981) for P300 examples, and Sokolov (1963) for SCR examples). Furthermore, a given stimulus may or may not elicit a P300 or SCR depending on whether other conditions are met. Sutton et al. (1965) referred to ERPs with these properties as "endogenous" responses and included P300 in this class. Such responses were contrasted with "exogenous" responses that are closely related to physical parameters of the stimulus (see Donchin et al. 1978 for further discussion of this distinction). Brainstem ERPs to auditory stimuli are an example of "exogenous" responses. Sokolov, in developing criteria for the orienting response (OR) to be used in identifying its neural substrates, emphasized the same stimulus nonspecificity (Sokolov 1963a). Identical ORs can be elicited to stimuli of a wide variety of physical characteristics. ORs can occur to both the onset and offset of stimuli, and to expected stimuli that fail to appear.

As I have discussed elsewhere (Roth et al. 1980), stimulus nonspecificity is rarely absolute for any response. For example, stimulus intensity, although it is a simple physical parameter, can have a strong influence on P300 and SCRs that cannot be annulled by changing other antecedent conditions for these responses.

HABITUATION

One of the best known features of SCR is its tendency to decrease in amplitude as stimuli are repeated. The study of Raskin et al. (1969) showed the usual SCR decrease over trials. SCRs to more intense stimuli declined more slowly. Generally, stimulus properties that enhance SCR - more unexpectedness, more signal value, and higher intensity - retard habituation. It is not unreasonable to assume that at least the first two properties change as the stimulus is repeated, in a way that completely accounts for SCR decline. At first the stimulus is unexpected: but, with repetition, its temporal and physical charateristics become known. The subject may also regard it as having signal value at first, since the experimenter would not be presenting stimuli if they were not important to someone. In addition, the subjective aspect of the third property, intensity, may decline with stimulus repetition. Schell and Grings (1971) gave 20 trials in which a 5-sec red light was followed after 5 sec by a 2-sec 100-dB white noise. The perceived intensity of the noise was assessed by having the subject match it to a tone of variable intensity after trials 1, 6, 10, 15 and 19. Both SCR amplitude and perceived intensity declined markedly after the first four trials. Oddly, however, neither declined in a control condition in which the stimuli were not paired but given separately.

Habituation of P300 is often not considered in the analysis of P300 experiments. Habituation may be slow or absent if stimuli are task relevant. When stimuli are task irrelevant, any P300 that might have been elicited by the first few stimuli will tend to have habituated so rapidly that it will not be apparent in the usual average based on 30 or more trials. However, if task-irrelevant stimuli are sufficiently salient and occur at long enough ISIs, they will elicit a P300 which can be measured in averages based on only a few trials or even single trials. Under these conditions, P300 habituation occurs at a slow enough pace to be demonstrated.

We have shown repeatedly that if stimuli are not task-relevant, P300 declines over trials. Roth and Kopell (1973) used trains of 30-msec tone bursts of either 1000-Hz or 2000-Hz pitch. The ratio of the occurrence of the two pitches was 20/1, and the ISI was 2 sec. Subjects were told to ignore the stimuli. P300s could be seen in averages of the first 15 infrequent tone trials in 8 of 12 subjects, and these P300s declined in averages of subsequent groups of infrequent trials. In a second experiment (Roth 1973), a similar paradigm was used except that there were three different levels of infrequent stimulus probability. A P210 wave to the infrequent stimuli was observed, which declined in amplitude over three consecutive blocks of 10 infrequent trials.

Presumably because of temporal uncertainty, a non-signal tone is especially effective in eliciting P300s when ISIs are on the order of 12 sec or more, though the P180 component is larger at these ISIs and can merge with P300 and complicate its measurement. Under these conditions Lutzenberger et al. (1979) found P300s between 270 and 350 msec in the first 10 trials that declined over the first 20 stimuli. Schandry and Hoefling (1979), in a similar experiment, found slower P300 habituation with irregular ISIs. Other examples of P300 habituation in this type of paradigm are Becker and Shapiro (1980), Roth et al. (1982), and Verbaten (this volume).

Using complex visual patterns as stimuli, Courchesne et al. (1975) examined P300 in single trials. P300s were largest (about 30 μV) to the first presentation of the patterns. The second instance of this type of stimulus evoked a P300 50% smaller than the first, and in the third instance, a P300 60% smaller than the first. Further decline was not noted in the next three trials. Kok and Looren de Jong (1980b) presented geometric figures that were made task relevant by requiring the subject to recognize them later. Stimuli were given in trains and ordinal averaging, that is, grouping the stimuli by their ordinal position in the train and averaging across trains, demonstrated a decline in P300 and two later positive waves over the first five repetitions of infrequent stimuli.

The classical demonstration of habituation requires OR recovery to the introduction of a test stimulus and to the stimulus following the test stimulus. In that way, true habituation can be distinguished from effector fatigue or receptor adaptation. These recovery properties have been established for SCR (e.g., Sokolov 1963), but the evidence for their applying to P300 is more indirect. P300 experiments show either amplitude decline but not recovery (experiments cited in this section) or what might be interpreted as amplitude recovery without evidence of prior amplitude decline (experiments cited under Surprise). The exception to this generalization is the flawed experiment of Megala and Teyler (1979), who tried to test P300 and other ERPs for true habituation by means of ordinal averaging. They presented two blocks of 30 repetitions of trains of 11 stimuli. At one of the later positions in the train (6-10), a test stimulus more or less intense than the others in the train was given. Subject instructions were to "attend" to all stimuli. Averages were calculated across the trains: of stimuli at each ordinal position, of the test stimuli, of the stimuli preceding them, and of the stimuli following them. These averages showed an initial decline in P300 amplitude, followed by enhanced P300s both to test stimuli and to standard stimuli immediately following test stimuli. The effects were seen in both the auditory and visual modalities. Unfortunately the leads used - F3, T3, and O1, referenced to linked ears - are far from optimal for measuring P300. The waveforms of a "representative subject" in their Figs. 2 and 4 do not reassure the reader that a reliable component was being measured.

P300 AND SCR DIFFERENCES

The argument so far has been that the stimulus properties that elicit P300s and SCRs are identical. If this were true, it would be expected that P300 and SCR would covary when measured in response to the same stimuli. The amount of data available for testing this prediction is quite meager, since few attempts have been made to record P300s and SCRs concurrently.

Roth et al. (1978b) presented 50-msec, 95-dB SPL chords at random ISIs. The probability of the chord occurring after any 1-sec interval was 0.1. In some conditions, a 65-dB noise burst was presented at 1-sec intervals between occurrences of the chords. Instruction conditions included one in which subjects pressed a button quickly when they heard the chord, and another in which they read a book. Chord trials were sorted on the basis of the amplitude of SCR elicited in a way that controlled for time effects so that covariations of P300 and SCRs could not be attributed simply to their parallel course of habituation. The only P300-SCR covariation found was an inverse one: larger SCRs were associated with smaller P300s in the read condition. The ERP waveform suggested that this effect might be due to a negative shift that was larger when SCRs were larger, rather than due to changes in P300 itself.

In our most recent study (Roth et al. 1982) there were significant correlations in single trials between the amplitudes of blink and P300, and blink and SCR, but not between P300 and SCR. The blink-P300 correlation was no longer significant when the effect of trial number was removed by partial correlation.

Becker and Shapiro (1980) compared the habituation of alpha blocking, SCR, and P300 at Cz to 60 1-msec, 115-dB clicks given at a fixed ISI of 15 sec. The shortness of the clicks made their intensity seem moderate. One group of subjects counted the clicks silently, and another was told to ignore them. Blocks of 10 trials were averaged, and P300 was measured as the maximum positivity in the 250 to 370 msec latency range. Alpha blocking and SCRs were not affected by instruction condition, while P300 was 10 μV larger in the count condition than in the ignore condition. Alpha blocking and SCR habituated at the same rate, both being completely habituated by the 25th click. P300 was not measured in individual trials, but its amplitude in blocks of 10 decreased over at least the first 4 blocks. Since there is ample evidence from experiments cited in this paper that SCR amplitude can be altered by instructions giving a stimulus signal value, one can only conclude that P300 was more sensitive to the instructions used by Becker and Shapiro than was SCR.

Verbaten (this volume) recorded ERPs, SCRs, and visual fixation time to visual stimuli of varying complexity. These stimuli could be made task relevant by requiring subjects to recognize them afterwards. A number of dissociations were observed between the response measures, suggesting to the author that they might be functionally differentiated. For example, for task-irrelevant stimuli, visual fixation time and P300 at Fz and Cz were more affected by the information content of the stimuli than were SCR or P300 at Oz.

Thus, these four experiments suggest that, although the general conditions for eliciting P300 and SCR are the same, the two responses can be dissociated. At the present, the meaning of these dissociations is unclear. It is my impression that there is a difference between the rates of amplitude decrease of P300 and SCR to task-relevant stimuli of moderate intensity. While P300 can continue with little decrement over hundreds of trials if the behavioral response to the stimuli is held constant, SCR is likely to decline to near zero levels long before a hundred trials are reached. However, this impression has not, to my knowledge, been explicitly tested.

THEORETICAL CONVERGENCE: THE ORIENTING REFLEX

The OR has consistently been the theoretical point of departure for discussing SCR, while theories of P300 have been drawn from less general concepts of information processing. Only in a few of the earlier papers was P300 associated with the OR (Ritter et al. 1968; Roth 1973; Roth and Kopell 1973). A short review supporting a relationship between P300 and the OR was written in 1976 (Friedman 1978). Recently, however, there has been an upsurge of interest in the OR on the part of ERP researchers as evidenced by concern with the OR in the 1980 Presidential Address of the Society for Psychophysiological Research (Donchin 1981) and the Orienting Reflex Panel at the 1980 Sixth International Conference on Event Related Slow Potentials of the Brain. The publication of a book based on the 1978 NATO Conference on the Orienting Response in Humans has also been a stimulus to reexamine the implications of the OR for ERPs (Kimmel et al. 1979).

1. Nature of the OR

Pavlov (1927) described how his dogs would react to unexpected stimuli by physically orienting the appropriate sensors towards them. After a few repetitions of the same stimulus, this reaction ceased. Pavlov referred to the OR as an investigatory or "what-is-it" reflex, and pointed to its obvious evolutionary advantage for the organism. This reaction is so fundamental and universal in the animal kingdom that it can hardly be asserted that Pavlov discovered something that had never been noticed before. Rather, Pavlov drew our attention to this phenomenon and demonstrated that it could be an object of scientific investigation. The OR is an innate reaction, built in as part of the "hardware" that regulates the interaction of an organism with its world. As such, its importance cuts across academic disciplines such as comparative biology, neurophysiology, and psychology.

2. Criteria for OR components

Many behavioral and physiological reactions occur when the organism encounters unexpected stimuli. Which of these reactions qualify as OR components? The OR, as originally conceptualized, was based on elements of overt behavior: the dog turns his head towards a new sound and pricks up his ears. Or if a new smell wafts into the room, he sniffs the air. But, in addition, many covert changes have been postulated to be parts of the OR: for example, sympathetic discharge in the autonomic nervous sytstem or the firing of certain single units in the brain. Sokolov (1963a) has proposed specific criteria for deciding if a response is a component of the

OR: stimulus nonspecificity and selective extinction. Stimulus nonspecificity refers to the independence of the response from physical properties of the stimulus, a characteristic of both P300 and SCR, as discussed earlier in this paper. Selective extinction means that a change in any stimulus property can temporarily reverse habituation. This is certainly true of SCR, and there is some evidence that it is true of P300 also, as discussed earlier.

There is a second way of arguing that a response is an OR component. If the OR is a functional entity based on an integrated biological substrate, the various OR components will tend to appear together to a given stimulus. Hence, temporal association between a new candidate for membership in the OR such as P300 and an old established member such as SCR is a recommendation for acceptance in the club. This argument was used by Friedman et al. (1973), who found that concurrently measured P300 and pupilary diameter both varied inversely with stimulus probability. They considered this evidence for P300 being an orienting component since the pupilary response was already an established component, or at least was considered as such in the Russian literature. As this review has documented, the stimulus conditions for eliciting P300 and SCR are virtually identical, so there is good reason to expect that both will occur to the same stimuli. Unfortunately, as discussed under <u>P300 and SCR Differences</u>, these components often appear to be dissociated in the few experiments where they have been measured simultaneously. These dissociations need not disqualify P300 as an orienting component if, for example, the two responses habituate at different rates or are elicited by OR stimuli of different potencies. Dissociations between OR components have long been recognized in the concept of general vs. local ORs (Sokolov 1963). Potent or unhabituated stimuli can elicit ORs with a wider spectrum of components than weak or more habituated stimuli. For example, an unexpected visual stimulus may initially elicit SCRs, vasomotor responses, and generalized alpha blocking. With repetition, SCRs and vasomotor responses will successively disappear, and alpha blocking will be confined more and more to the occiput. Thus, covariation of putative OR components must be observed under a number of conditions before firm conclusions can be reached. Furthermore, individual differences in the relative responsivity of physiological response systems must be taken into account.

3. Models of the OR

OR theory as developed by Sokolov and others has used models of several kinds. These models, or at least their terminology, have been applied to ERPs, sometimes explicitly in the context of the OR and sometimes without reference to it.

a. <u>Neuronal models</u>. Sokolov has proposed neuronal models for the OR both at the level of cell aggregates such as the reticular formation and its connections with various parts of the brain and at the level of single cells (Sokolov 1975). Bringing together P300 and SCR in a common neurophysiological scheme in our present state of knowledge would be a work of science fiction. Even the concept of a "neuronal model" of stimuli (Sokolov 1963) seems more metaphorical than concrete in the case of higher animals and man. The stimulus properties abstracted to create an expectation of future stimuli can be so complex, that assignment of the neuronal model to specific interconnections of neurons is impossible. Nevertheless, ERP researchers have been attracted by at least the language of neuronal models. Paradoxically, P300 has been said to occur when the current stimulus matches the neuronal model or template (Hillyard et al. 1971; Squires et al. 1973a; Thatcher 1976), or when it fails to match the template (Ford 1979; Pritchard 1981). Apparently, templates of task-relevant stimuli give rise to P300 when matched, while templates built up from recent environmental events give rise to P300 when mismatched.

b. Probabilistic models. The fact that stimulus probability is a crucial variable in determining the size of the OR that a stimulus elicits, led to theoretical formulations using concepts of information taken from the theory of communication. The template or model of the environment that the organism builds up is a representation of high probability events. Improbable stimuli that do not match the template convey more information than probable, matching stimuli. Sokolov (1966; 1969) created a formal probabilistic model that described the occurrence of the OR in the context of the organism forming hypotheses about the environment. He included in his equations such terms as the probability of a hypothesis Ai given a datum Kj , the uncertainty or entropy remaining after Kj , and thresholds of uncertainty for eliciting an OR. In essence, an OR is elicited when a datum fails to reduce entropy adequately. Velden (1974), on the basis of SCR ORs to words that make a sentence, disputed Sokolov's assertion that residual entropy determines the magnitude of the OR. Instead, in Velden's experiment, the amount of information given by the individual stimulus appeared to be the determining factor.

Sutton in his early formulations associated P300 with delivery of information to the subject (Sutton et al. 1967). Ruchkin and Sutton (1978b) extended this idea by noting that the information in a signal is a function of both the a priori probability that the signal will be presented and the a posteriori probability that the signal was presented. The less certainty a subject has about having correctly perceived the signal, the less information value the signal has. Ruchkin's review of P300 literature proved the usefulness of information theory concepts for understanding P300 amplitude. There is an essential limitation to this theory, however. Mathematical definitions of information do not take into account whether the information is psychologically meaningful, and a recognized metric for psychological meaningfulness is lacking (Velden 1978). A failure of probability measures to predict P300 amplitude is seen in studies where stimuli with low global probability continue to elicit larger P300s than stimuli with high global probability, even when the sequential probability of the former kind of stimulus is made certain by informing the subject which stimulus will occur next (Friedman et al. 1973; Sutton 1969; Tueting et al. 1970). Along the same line, Campbell et al. (1979) plotted P300 as a function of bits of information in one experiment and found that increases in information beyond 2 bits were not always accompanied by increases in P300.

c. Cognitive models. The development of cognitive psychology has led to new models of attention, memory, and other kinds of information processing. Cognitive theorists have at times included the OR in their formulations. Kahneman (1973) described the OR as an involuntary reallocation of attention. Posner (1978) performed RT experiments that demonstrated what he called activation and orienting.

There is a striking parallel between recent cognitive theories of OR and of P300. Öhman (1979) has described the OR as a call for processing in a central channel. The call is initiated by a preattentive or automatic process involving short-term memory store (STS). He identified the function of the central channel with focal attention, a controlled process. Thus, the OR represents a shift from automatic to controlled processing (Schneider and Shiffrin 1977; Shiffrin and Schneider 1977). A call on the central channel can be initiated either by a stimulus that disconfirms the expectancies built up in STS or by a stimulus that matches primed items (that is, targets) in STS. Posner similarly proposed that P300 reflected invocation of limited capacity conscious processing (Posner et al. 1973; Posner 1975). In a recent review, Rösler (1980) concluded that P300 arose when additional processing resources had to be allocated to the stimulus, that is, when automatic processing had to give way to controlled processing.

d. Computer models. The appeal of computer models of thought is no longer limited to scientists engaged in, or even familiar with, research in artificial intelligence. Pyschophysiologists use computers so extensively nowadays that computer analogies are a natural way of thinking for them. For example, Öhman's theory that the OR is a call on the central processor suggests an analogy with turning on a computer interrupt that diverts the execution of commands from the main program to more or less specific subprograms designed to process various contingencies. The strength of the OR would be analogous to the priority of the interrupt. Donchin (1981) sees the goal for a theory of P300 to be specification of the "subroutine" that it represents and the role of that subroutine in the program in which it was called.

FUNCTIONAL DIVERGENCE

The previous section described an empirical, biological, and theoretical unity, the OR. Now it is time to question if such a unity really exists. An alternative approach to understanding the diverse phenomena that have been brought under the OR umbrella is to ask what functions these phenomena represent (e.g., Verbaten, this volume), group these functions into more coherent entities, and assign physiological and behavioral components to the appropriate functions and entities.

1. Multiple functions

A multitude of processes must be presumed to contribute to the phenomena that have been called the OR. These processes are at the same time physiological and psychological. They can be divided into processes that initiate the OR and processes that follow initiation.

Initiating processes may be complex. Disconfirmation of expectancies - Sokolov's mismatch detector - requires pattern recognition and stimulus encoding, retention of encoded items in STS along with expectancies of their future occurrence, comparison of newly encoded items with these expectancies, and updating of expectancies. In the case of ORs to primed items, a control process must have occurred that enabled STS to make certain matches quickly. On arrival of a new item, a comparison is made with the primed item or items. Intense or salient stimuli may engage special sensory mechanisms which react to abnormal sensory inputs.

If expectancies are violated, target items are detected, or stimulus overload occurs, a number of futher processes may ensue. General arousal or activation may increase to facilitate sensory, central, and motor processes. Total processing resources may temporarily increase. More specific postural and sensory adjustments may increase sensitivity to stimuli of the kind eliciting the reaction. Whatever is going on in the central channel is interrupted, and more processing is devoted to the recent event. Increased processing can mean looking up relevant information in LTS, alternate or "deeper" categorization of the item, symbolic transformations of it, as well as all that belongs to the common-sense realm of thinking and planning. Some specific motor processes serve to prepare the organism for fight or flight. In the case of stimulus overload, various protective response such as blinking and flexor contractions appear.

2. Multiple reflexes

In view of the variety of the antecedent conditions for eliciting the OR and the number of functions involved, it is not surprising that its theoretical unity has been questioned, and divisions into several more or less distinct reflexes have been proposed.

a. Voluntary vs. involuntary ORs. ORs to task-relevant stimuli have been hard to fit conceptually with ORs to intense or surprising stimuli. When a subject is looking for a target, detection of this target is part of a unitary searching activity rather than an instance of deflection of attention by a salient distractor. Of course, the target sometimes may appear in the periphery of the visual field or at a time when the the subject has momentarily stopped searching, but why should the OR associated with deflection of attention occur in cases when the stimulus is already at the focus of attention? Furthermore, with target detection, an OR is elicited by a template match rather than a mismatch, as in the case of surprising stimuli. These contradictions, along with certain empirical findings, have resulted in a proposal that two kinds of ORs with different properties must be distinguished: voluntary and involuntary (Maltzman 1979; Näätänen 1979).

The OR to targets has been called voluntary, since an act of will, or in the less every-day language of cognitive psychology, a control process, is directing a search for a target. Voluntary ORs are contrasted to involuntary, automatic ORs to surprising or intense stimuli. The use of the words "voluntary" and "involuntary", however, can be misleading. If the item to be detected is primed in such a way that it resides in STS before the stimulus arrives, the matching process may be almost as rapid and automatic as the mismatch or overload detection process is. From a mechanistic standpoint, voluntary and involuntary responses can be viewed as depending on the number of possible outcomes that can follow an initial event. Voluntary or control processes can preset processsing parameters before a stimulus arrives in various ways resulting in various outcomes , or they can alter outcome in various ways after a stimulus has arrived. However, control processes take time to operate, and the interval between initial stimulus processing and the triggering of the OR is probably so short that the response cannot be modified very much by post-stimulus control processes. In this sense, all ORs are involuntary, although the will can influence the OR by its actions before the stimulus arrives.

b. ORs vs. defense and startle reflexes. That stimuli with higher absolute intensity should be more potent in evoking ORs is not logically entailed by theories of expectancy mismatching or target matching. Furthermore, high intensity stimuli may elicit responses that are qualitatively different from responses to lower intensity stimuli, yet retain some OR properties (Sokolov 1963b). Abrupt auditory stimuli give rise to eye blinks and other muscle contractions that similar stimuli of a slower onset do not elicit. These considerations have stimulated attempts to separate response complexes called defense and startle reflexes from ORs (Graham 1979). Such attempts have only been partly successful, since either the component response of these three reflexes overlap, or a single stimulus can elicit more than one reflex. That could mean either that the same basic processes are evoked under different antecedent conditions resulting in similiar physiological responses, or that different processes elicit overlapping physiological responses.

3. Components and Functions

A radical functional approach implies that it is meaningless to try to decide whether a response is an OR component, since the OR, insofar as it can be considered as existing, is only a grouping of functions. There is much more profit in assigning responses to function, and in the best of all possible psychophysiological worlds, each distinct physiological response corresponds to a distinct function.

P300 and SCR are only two of a score or more responses that could be considered for possible assignment to functions associated with the OR. At least five ERP components have appeared in paradigms that might be expected to produce ORs, and attempts have been made to assign these ERPs distinct roles in information processing. Their different latencies has generally been taken as indicating how early the psychological operation they represent can occur. Very briefly, they are the following:

a. Processing negativity (Nd). A negative component that can begin less than 100 msec after stimulus onset. It is associated with further processing in a selected channel (Näätänen, this volume) and has been called an OR since it is elicited in response to stimulus matching to preset target criteria (Näätänen 1979).

b. N200. A negative component at about 200 to 250 msec reflecting a stage of template matching or mismatching that is later than that of Nd (Simson et al. 1976; Ritter, this volume).

There are a number of possibly distinct late positive waves, classification being based on latency, topographic distribution, and experimental conditions (for reviews, see Roth 1978; Pritchard 1981). Only two types will be mentioned here, early P300 and late P300.

c. Early P300. Positive components like the P3a of N.K. Squires et al. (1975) and the P280 of Roth et al. (1982) may reflect the same function as late P300, but follow a simpler and less thorough stimulus evaluation than the late P300. Pritchard (1981) associates P3a with updating of neuronal models of the physical properties of the stimuli, while late P300s are associated with updating more cognitive models.

d. Late P300. Positive components like the P3b of N. K. Squires et al. (1975) and the P340 of Roth et al. (1982) are the category into which most P300s in the experiments cited in this paper fall. Since P300 latency depends on the difficulty of stimulus categorization but not on response compatibility, it has been concluded that P300 coincides with or follows a stimulus evaluation stage (Donchin et al. 1978). P300 in a secondary task is more affected by the perceptual than by the motor difficulty of the primary task, indicating that P300 shares resources with perceptual processing (Donchin 1981). Yet, Donchin concludes that P300 is more likely to represent the updating of schema in STS or LTS than merely to represent some final stage of stimulus evaluation. Assigning P300 to the updating of schema or "context updating" (Pribram and McGuiness 1975) gives it a function central to the classical OR.

e. Slow wave. A broad parietally positive peak sometimes associated with simultaneous frontal negativity begins at about the same time as the late P300 but can extend for seconds. The slow wave appears in most P300 paradigms and after the warning stimulus in Contingent Negative Variation paradigms. Since it seems to represent prolonged stimulus processing and is proportional in amplitude to the relevance of the stimulus, it has been called an OR component (Loveless 1979).

Lists of non-ERP physiological responses relevant to the OR paradigms could be constructed and their possible functional correspondences explored, but here I will only mention that SCR is one of many manifestations of sympathetic autonomic discharge, which are usually thought of as part of preparation for physical action.

P300, SCR, AND THE OR: A CRITICAL FORMULATION

1. The Nature of the OR

Neither taking the OR for granted as a unified biological entity nor defining it arbitrarily as a certain group of functions is satisfactory. The fact that three different situations - expectancy mismatch, target match, and sensory overload, elicited by the stimulus properties of surprise, task relevance, and intrinsic salience - can result in the same physiological responses, namely P300 and SCR, must be revealing something about the innate organization of the brain. However, further questions must be asked about just what processes are involved. I would like to call the common processes immediately and automatically invoked by extremes of these three stimulus situations, the primary OR. In all of these situations, more resources need to be devoted to that kind of stimulus than to others. We can imagine various processes being called upon by primary OR stimuli: reallocation of attention or increase in processing resources, deeper analysis of the stimulus by symbolic transformations or searches for relevant information in STS and LTS, updating of expectancies, enhancement of sensory sensitivity, and non-specific preparation for motor response.

Whereas subprocesses of the primary OR are elicited together by any adequate OR stimulus, other processes may have a close but more optional relationship to the OR. These latter secondary group of OR processes may be neither necessary nor sufficient for an OR, but under certain conditions occur before, after, or during a primary OR. The antecedent processes differ in the three OR situations, so no single one is necessary for an OR. They are not sufficient, since apparently some threshold must be exceeded before an OR occurs. For example, target selection and categorization are part of target matching but do not necessarily lead to a measurable P300 or SCR. Only when targets are especially hard to identify or of special interest do these responses occur. Likewise, mismatches can probably be perceived without evocation of an OR. In a number of experiments, mismatching stimuli failed to elicit SCRs although subjects later reported that they had noticed the mismatch (for a review, see O'Gorman (1979). Processses that may or may not follow or occur in parallel with the primary OR are behavioral orientation towards the stimulus (the classical OR), blinking (the startle reflex), or stimulus avoidance (the defense reflex).

The subprocesses of the primary OR need to be operationally defined in a way that will allow empirical testing of how they are associated. It may turn out that not all of these subprocesses occur to every OR stimulus, but they must be correlated more highly with each other than with other processes like behavioral orientation if the distinction between the primary OR and other related reflexes is to have any validity. On the other hand, it is not necessary or even likely that basic OR processes will only occur in OR situations. Reallocation of attention may take place without regard to immediate stimuli. To consider this reallocation an OR to internal events, would be a major extension of the conventional notion of the OR.

2. Classification of P300 and SCR as ORs

The concept of the primary OR implies a modification of Sokolov's criteria for OR components. The new criteria for classifying a response as a component of the primary OR are the elicitation of the response by each of three different kinds of stimuli: ones that mismatch expectancies, ones that are targets, and ones that are innately salient, e.g., intense. These criteria extend the idea of stimulus nonspecificity beyond modality and beyond physical characteristics, except those that contribute to salience, to a cross-situational nonspecificity. Selective extinction or habituation is no longer an OR criterion, since it may be absent if stimuli continue to be task relevant or if they are intense enough. The habituation expected to task-irrelevant stimuli is a consequence of their diminishing surprise value.

The presence of P300 and SCR in the stimulus situations reviewed in this paper is evidence, therefore, for their being components of the primary OR. Whether other ERPs also meet these criteria or whether P300s of only certain latencies or distributions meet them are important questions that cannot be adequately discussed here, except to say that, with respect to the first question, I disagree with some statements of Näätänen (1979). He asserted that Nd originated from voluntary orienting and a N200 from involuntary orienting. First, Nd or N200, cannot be considered ORs _just_ because they are part of target detection and categorization. As argued above, there is no reason to believe that these functions necessarily lead to any kind of OR. Second, Nd is an involuntary response in the sense discussed under _Voluntary vs. involuntary ORs_, since it is an automatic reaction to stimuli preset by a controlled process.

3. Functions of P300 and SCR

Classifying P300 and SCR as primary OR components does not assign them to specific subfunctions of that OR. What function or functions do these responses reflect? Donchin et al. (1978) and Donchin (1981) answer this question using a military metaphor: SCR is a "tactical" component of the OR, a specific response to the current stimulus. P300, on the other hand, is a "strategic component", updating the context which rules the tactics used on the next trial.

In evaluating this answer, it is useful to consider what the biological nature of a response can tell us about it. Moderately moist hands have more friction than dry hands when grasping tools or climbing trees, so SCR has been considered to have the function of preparation for motoric action (e.g., Kuno 1930). Unfortunately, there is no way to assign a biological role to P300 or any other ERP. Too little is known about the brain and the anatomic sources of P300 to make valid inferences about its function from its polarity, waveform, latency, topographic distribution, etc. (But see Ritter and Vaughan, this volume; Renault, this volume). Not being able to assert that a physiological process like P300 is the actual substrate of a psychological process, we are inclined to use the more distancing metaphor of mirroring - P300 "reflects" this or that process. Here we must be satisfied with the establishment of a correlation between a physical and a psychological process by manipulating experimental parameters in a way that affects both the inferred psychological process and the physical response.

The correlative assignment of a function to SCR could be quite different from the biological assignment. Sweaty palms are an indicator of very widespread peripheral and central nervous system activity governed by the reticular system. Reticular system activation influences the efficiency of many aspects of information processing, beginning with initial sensory processes and ending with response execution. It is likely that context updating is promoted by activation as well. Since activation can be the result of the most sophisticated stimulus categorization that man is capable of, it derives from the same centrally located brain mechanims which underlie ERPs. Thus, it is not impossible that on a correlative basis, a 3-sec latency, peripheral response like SCR will be assigned to the same psychological function as the 300-msec latency, central P300. It might be argued that such an assignment of P300 and SCR to the same process, say, context updating, only would be a sign that motor preparation of the SCR type is closely correlated with context updating of OR stimuli. However, it should be clear that the same might apply to P300: P300 could be a process whose biological significance in the brain has nothing to do with context updating per se, but which correlates with it under certain circumstances.

From correlative experiments, theoretical inferences are made that suggest new experiments. Donchin (1981) proposed, for example, that the context updating hypothesis for P300 can be tested by examining the consequences of P300 evocation for the memorability of events, and certain evidence from verbal learning experiments suggests that P300s are indeed larger to words that will be remembered later (Karis et al. 1981). The prospect that these findings will give validity to the distinction between the strategic P300 and the tactical SCR is dimmed by likelihood that SCR also predicts the memorability of items. Bagshaw et al. (1965) postulated that SCR was related to the registration of events in memory on the basis of amygdala ablation in monkeys. Ablated monkeys lost SCRs but maintained a behavioral orientation to stimuli that was slow to habituate, apparently because expectancies were not being updated. Öhman (1979) asserted that his model of the OR implies that the OR is a prerequisite for storage in LTS (see also Siddle et al., in press). Items that elicit ORs should be remembered better since a call on the central channel means that these items will be processed more thoroughly, and greater depth of processing promotes recall. Also, the surprise associated with OR-eliciting items may serve as an added association or cue that facilitates remembering them.

A considerable literature already exists, suggesting that in humans, arousal in general and SCRs in particular, are greater to items that later will be retrieved from LTS (for a review, see Craik and Blankstein 1975). As long ago as 1937, Brown reported that visually presented nonsense syllables that were learned faster elicited larger SCRs than did syllables that were learned more slowly. In a more modern study, Corteen (1969) read 21-word lists to subjects over a loudspeaker without telling them they would be asked to recall the words later. Subjects were divided into three groups: one was tested for free recall immediately, another after 20 min, and the third after two weeks. For each retention interval, correlations showed that words that elicited bigger SCRs were more likely to be recalled. Other OR components have the same properties. In another incidental word learning paradigm, skin potential response and alpha blocking predicted free recall immediately after the words were presented (Warren and Harris 1975).

Thus, P300 and SCR are again expected and observed under the same conditions. This expectation derives from simply conceptualizing them both as OR components. Contrary to the proposal of Donchin, SCR has the same attributes of a "strategic" context-updating component as P300. However, my notion of the primary OR implies that P300 or SCR may be correlated with any or all of the processes associated with that OR - phasic arousal, reallocation of attention, deeper stimulus analysis, memory search, as well as memory or context updating. Since these processes tend to occur together as part of the OR, it may be impossible to dissociate them sufficiently in an OR context to be able to assign P300 or SCR to specific subprocesses. Of course, the postulated functions of the primary OR are only meaningful insofar as they can be operationalized; otherwise, these theoretical distinctions will end up in the wastebasket of empty words. I think that unlikely, however. The experimental literature of cognitive psychology gives us reason to believe that these internal processes can be objectified and teased apart as others have been.

CONCLUSIONS

The original, prototypic experimental conditions for eliciting P300 and SCR were so different that the essential similarities of these responses were often missed. P300 was typically elicited by stimuli that gave the subject information relevant to the task the experimenter had assigned. P300 could be elicited in hundreds of trials of this kind with little decline in amplitude. SCR, on the other hand, was typically elicited by task-irrelevant stimuli, and habituated within a few trials.

Yet, as this review has documented, their antecedent conditions are generally the same, and much of what seemed different about them could well be a consequence of the data analysis required by the poor signal-to-noise ratio of P300 and the historical interest in involuntary, reflex phenomena by students of the OR. The burden of proof should lie on those who contend that P300 and SCR have different psychological meanings rather than on those who consider these responses equivalent.

Theoretical statements about P300 have emphasized establishing a correspondence between P300 and contructs from cognitive psychology, while such statements about SCR have regarded SCR as a component of an integrated, biologically-based reflex. These theoretical approaches are complementary, and the experimental programs they entail should be applied to the other response. SCR should be investigated in experiments that untangle processing stages, and P300 should be studied where its relation to other physiological responses can be observed across subjects and within individual subjects over time.

The OR as an overarching theoretical structure has been slowly cracking under theoretical and empirical pressure. Yet, there is reason to assume that certain processes constituting a primary OR are invoked when the antecedent conditions discussed in this review are present to a sufficient degree. Whether P300 or SCR are more closely related to one or the other of these processes is presently not known. In any case, it is premature to assign P300 specifically to context updating.

Finally, it is important for the reader to bear in mind that the restriction of this review to P300 and SCR was an expediency to lighten the burden on the reviewer. The evidence for other ERPs being OR components was not examined in any detail. The possibility that only certain types of P300s are OR was not explored here. Similarly, SCR was chosen rather arbitrarily from among other autonomic responses. Components of the evoked heart rate response, for example, have many of the same properties as SCR.

Notes

Preparation of this paper was supported by the Medical Research Service of the Veterans Administration and NIMH Special Research Center Grant MH 30854. I thank Connie C. Duncan-Johnson, Ray Johnson, Jr., Judith M. Ford, Margaret J. Rosenbloom, David Siddle, M.N. Verbaten, Walter Ritter, Adolf Pfefferbaum, and Karen S. Haney for their helpful suggestions. Send reprint requests to Walton T. Roth, M.D., Psychiatry (116A3), V.A. Medical Center, 3801 Miranda Ave., Palo Alto, California 94304, U.S.A.

Tutorials in ERP Research: Endogenous Components
A.W.K. Gaillard and W. Ritter (eds.)

9

THE INFLUENCE OF INFORMATION ON HABITUATION OF CORTICAL, AUTONOMIC AND BEHAVIORAL COMPONENTS OF THE ORIENTING RESPONSE (OR)

Marinus N. Verbaten

Department of Psychophysiology
University of Utrecht
Varkenmarkt 2, 3511 BZ Utrecht
THE NETHERLANDS

This chapter discusses some recent data which may contribute to an understanding of the orienting reaction (OR) (Pavlov, 1960; Sokolov, 1966). Experiments were designed to test the effect of stimulus information on strength and habituation rate of the OR in task relevant and non-task relevant conditions. Furthermore we investigated whether the so-called P300 component habituates in the same period of the time as the SCR. On the basis of these experiments and other data, current theories about the OR and its habituation (Bernstein, 1979; Öhman, 1979) will be evaluated.

During his conditioning studies Pavlov (1960) noticed that the conditioned behavior of dogs was sometimes interrupted when something happened that differed from the experimental situation, e.g. somebody entering the room. The dog then shifted its attention to the man or woman entering the room. Pavlov recognized the broader meaning of this reaction and called it the orienting reflex. The work of Sokolov (1966) was based on this notion. Still today, the model for this reflex he proposed is one of the most quoted models in the OR literature. The interesting fact, from a psychophysiological point of view, is that Sokolov described the OR as consisting not only of behavioral, but also of physiological reactions such as changes in electrical activity of the cerebral cortex, electrodermal changes, heart-activity changes etc. Sokolov's theory is a psychological theory in the sense that the explanation for the decrement of the OR as a function of trials is given in terms of psychological processes rather than neurological ones as found in the habituation theories of Kandel (1976) and Groves & Thompson (1973). The latter theories explain habituation in terms of changes in the activity of specific interneurons. Sokolov explains habituation by assuming that incoming afferent information is constantly compared with a memory model of the stimulus. When there is a match between the stimulus and the model, the OR no longer occurs and habituation of the OR is then complete.

So the OR has been described both in terms of behavioral, cortical and autonomic changes contingent on the presentation of a novel stimulus and in terms of a complex psychological process, consisting of several components,Sokolov (1969). It has been demonstrated (Barry,1979) that presenting a novel stimulus leads to various responses, but a problem arises when some responses show reliable habituation with repeated stimulus presentations and others do not. Some habituate quickly and others slowly. The pattern of physiological responses is complex. Therefore, the question of whether all these responses have the same function must be addressed. According to

Sokolov, all responses have the same function but differ in sensitivity. The SCR reflects, according to Sokolov, the properties of the OR accurately and has been extensively used as an OR index both in the Russian and in the Western literature.

Although Sokolov has changed various parts of his theory since 1963, his description of the OR has not changed. What has been changed are his ideas about the function of the OR. In more recent work (Sokolov, 1969) the function of the OR now includes the context of the situation in which the stimulus is perceived. The information which has to be gathered by the OR now not only includes physical aspects of the stimulus, like frequency or intensity, but also its probability of occurrence, complexity and spatial location.

Given the description of the OR as an information regulator (Sokolov, 1966), the OR should be stronger and habituate slower when stimuli with more information are presented. Another point about the OR theory is that it was formulated for the explanation of reactions to non-signal stimuli. The OR in reaction to signal stimuli is treated by Sokolov (1963) as a special case. Here where the OR to non-signal stimuli no longer occurs after 3-12 stimulus presentations, signal stimuli can induce ORs dozens of times. There is no empirical support for Sokolov's prediction that habituation of the OR to non-signal stimuli which contain more information is slower. Surprisingly enough, the most frequently used OR index, the SCR, shows the expected slower habituation only when the stimuli have signal value or task-relevance. Bernstein (1973) found that subjects, although able to report changes in stimulus conditions, did not show an OR, indexed by the SCR and concluded that stimuli have to be novel and task-relevant in order to induce an OR. This became the basis for his two-stage "significance" theory of the OR. It could be argued, however, that there is a circularity in Bernstein's reasoning. As was pointed out by O'Gorman (1979) "An OR is the result of a judgement of significance, but the only method proposed to date for determining whether a judgement of significance has occurred is the appearance of the OR". This circularity is caused by the fact that the OR is mainly indexed by the SCR. Bernstein could not exclude that his subjects had reacted in some other reaction system. The same remark may be made in regard to the absence of the effect of information in non-signal conditions.It can not be excluded that the effect of information may have registered in response systems other than the electrodermal system. This possibility was investigated in a first experiment (Verbaten, Woestenburg and Sjouw, 1979) in which we measured concurrently two indices of the OR, the SCR and the visual orienting reaction (VOR). The VOR is a measure of the direction of the gaze and corresponding fixation time. A method was developed for measuring concurrently the SCR and the VOR in a habituation paradigm. The VOR is one of the few externally measurable activities of visual stimulus processing (Didday, 1975) and consists of the turning of the eyes towards the source of stimulation. It has been shown that stimuli which contain more information induce longer spontaneous fixation times (Leeuwenberg, 1967) and that eye movements show habituation to repeatedly presented stimuli (Mackworth and Otto, 1970). The approach to the VOR in the experiments which will be reported here was based on the consideration that when stimuli are presented in the periphery of the visual field (so called "eye-field", Sanders (1963)) subjects have to make a saccadic eye-movement for optimal processing. We presented stimuli of different complexity in the right upper corner of a TV screen at an angle of about 28 degrees from the subject's sagittal line of view. Around that stimulus, two windows were constructed for measuring the eye movements towards the stimulus. One field was established in which the effects of the repeated stimuli and information were expected to occur (field 1) and a control field (field 2) in which these effects were not expected (see fig.1)

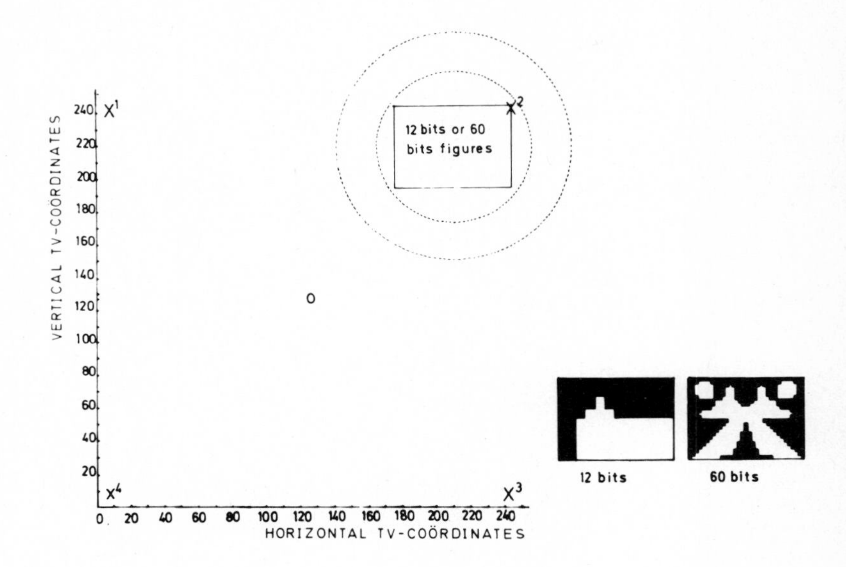

Fig.1. Distances of the four calibration stimuli (X1, X2, X3, X4) and the information containing stimuli in TV coordinates (256 x 256). The discontinuous lines indicate the two windows field 1 and field 2 around the information containing stimuli, which are presented during the habituation trials. Horizontal distance is 45 degrees of arc, vertical distance is 36 degrees of arc.

In the first experiment 28 subjects received two blocks of 14 trials. Half the subjects received first 14 presentations (trials) of a stimulus with 60 bits of information (Spinks and Siddle, 1976) where complexity was measured by Attneave's (1954) system and then 14 stimuli with 12 bits of information. The other half of the subjects received the reversed order. Interstimulus intervals (ISIs) were randomly chosen and varied between 15 s and 25 s. Subjects were told that some pictures would be presented on the TV-monitor, but that they did not have to react to these pictures in any special way. This is a standard instruction in non-signal stimuli habituation studies. On the basis of Sokolov's (1966) theory we expected larger ORs and slower habituation of the OR in the 60 bits condition. The VOR was both stronger ($F(1/26)=5.47$, $p<5\%$, one tailed) and habituated slower ($t(26)=2.15$, $p<5\%$, two-tailed)(in terms of number of trials to reach the habituation criterion of three consecutive trials with a zero reaction) in the 60 bits condition than in the 12 bits condition in the absence of any effect on SCR magnitude ($F(1/26)=0.62$, $p>10\%$)

or SCR habituation rate ($t(26)=0.61$, $p>10\%$). SCR habituation rate was defined in the same way as the VOR; three consecutive trials with a SCR less than 0.01 umho was the habituation criterion (see also fig.2).

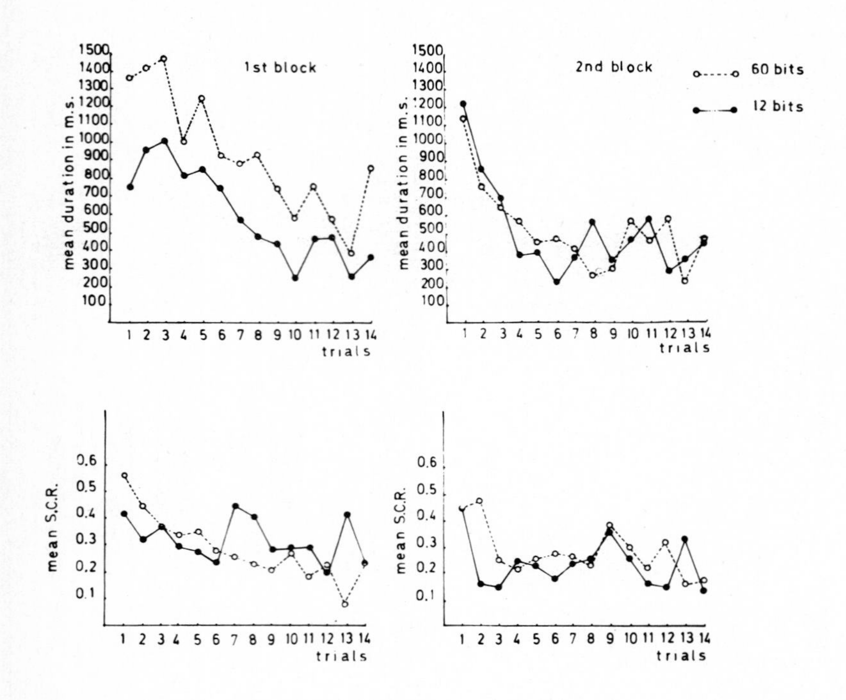

Fig.2: Mean fixation times in field 1 and SCR magnitude during block 1 and block 2. Solid line: 12 bits condition, discontinuous line: 60 bits condition.

These results demonstrated that in non-signal conditions the predicted effect of stimulus information was seen in the VOR response system, while there was an absence of any effect on the SCR. The latter results are in accordance with the results of e.g. Berlyne, Crow, Salapatek and Lewis (1963).

On the basis of these results we concluded that the two response systems may not have the same function, which is the basis for the so-called unitary theory of the OR of Sokolov.

In a second experiment (Verbaten, Woestenburg and Sjouw, 1980) we tried to replicate the effects of information on the SCR in signal studies found by several investigators (Berlyne et al., 1963; Spinks and Siddle, 1976; Fredrikson and Öhman, 1979). The conditions of the first study were again used in this study but the stimuli were made task-relevant. This was done by asking subjects to memorize the 60 bits and 12 bits stimuli and giving them a reward for correct recognitions at the end of the experiment. The design of the study was similar to the design of the first study. The dependent variables were the VOR, the SCR and the number of non specific fluctuations (NSFs) in skin conductance. The latter variable was measured in order to test whether the predicted effect of information on the SCR was the result of a more general sensitization of the electrodermal system, or a more specific effect related to stimulus processing. The VOR habituated quickly and was not significantly influenced by information (Fs(1/26) <1). SCRs were larger and SCR habituation slower to stimuli containing more information (respectively F(1/26)=5.60, p<5%, two-tailed, and F(1/26)=8.45, p<1%, two-tailed) (see fig.3).

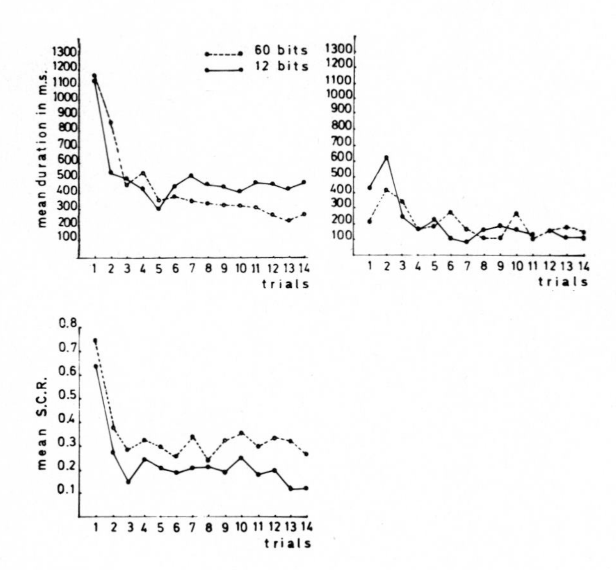

Fig.3: Mean fixation durations inside field 1 and mean SCR magnitudes in $\sqrt{\Delta C}$. Solid line: 12 bits condition, discontinuous line: 60 bits condition.

No effect of information was found on NSFs (F(1/26)=0.17, p>10%), which made it improbable that the effect of information on the SCR was caused by a more general sensitization. Furthermore, this supplied in another way evidence against an unitary theory of the OR. The hypothesis that all responses have the same function, but differ in sensitivity could not be ruled out on the basis of the first experiment. This was true, because it could be argued that the conditions of that experiment, in particular the reclining position of the subject, who lay on an examination table, were such that sympathic activation was low, thereby lowering the sensitivity of the electrodermal system. The results of this experiment did not support such an interpretation; the determinant factor for slower SCR habituation to more complex stimuli was the task-relevance condition.

The lack of effect of information on the VOR was unexpected and demanded an explanation. The similarity of the VOR habituation curves observed in the second block of the first experiment and the VOR habituation curves in the second experiment was conspicious. Subsequent analysis revealed that, without realizing it, we had in the second study also manipulated the uncertainty of the subject about the oncoming stimulus condition. The subjects had been able to deduce that only two stimuli would be presented. In the first study, the usual habituation instruction which implies very little about the oncoming stimulus situation had been given. The subject is told in such situations that a number of stimuli will be presented and that he/she does not have to react to these stimuli in any particular way.

In a third experiment (Verbaten, Woestenburg and Sjouw, 1981) the effects of stimulus uncertainty and stimulus complexity were investigated.According to Sokolov (1966) there is not necessarily a difference between the effects on the OR of information in terms of complexity or information in terms of stimulus uncertainty. Therefore we expected an additive effect of these independent variables on the VOR and, because we used non-signal stimuli, no effect on the SCR. One half of the 56 subjects which took part in this experiment were told that the same stimulus would be presented a number of times in the right upper corner of the TV-screen ("certain condition"). The other half was not informed about the stimuli ("uncertain condition") and were told that some pictures would be presented on the TV-screen and that they did not have to react to them in any special way. Within each of these groups, half the subjects received 14 stimuli with 12 bits information and the other half received 14 stimuli with 60 bits information. Stimulus complexity ($F(1/52)=5.84$, $p<5\%$, two-tailed) and stimulus uncertainty ($F(1/52)=8.01$, $p<5\%$, two-tailed) both led to slower habituation of the VOR. Neither factor, however, had a significant effect on the SCR (see fig.4).
The results of this study demonstrated not only that VOR habituation is not influenced by complexity in "certain" conditions, but also that it is actually quicker in such conditions. Fig. 5 shows the effects of the "certain" and "uncertain" instructions on the eye movement behavior during a base-level measurement (see fig. 4, trial B). No stimulus was presented but all the responses were measured in the same way they were during stimulus presentations. In a related study (Verbaten, 1981) in which the same "Uncertain" and "Certain" instructions were given to the subjects, we found indications that subjects process stimuli in a different way when given a priori instructions. Fig. 5 shows that the subjects in the "certain" condition obviously anticipated the stimuli in the right upper corner, while the subjects in the "uncertain" could not do so, because they had not received any a priori information about the stimuli.
The results of the experiments reported thus far do not support an unitary conception of the OR. They also do not support the assumption that the SCR is the best or most sensitive OR index. Instead, the various responses which occur when a novel stimulus is presented may be better understood by examining their possible functions. A more complete understanding of the OR requires more insight into central processes, among other things, because the SCR and the VOR should be seen as by-products of central processes or consequences of central processes. This point of view deserves further elaboration. Two experiments were done in which cortical responses (ERPs) were measured concurrently with the SCR using the same habituation design. A problem was however, that averaging usually occurs over a large number of trials, relative to the number of trials in which for example the SCR habituates. At least 10-15 stimulus presentations are used for

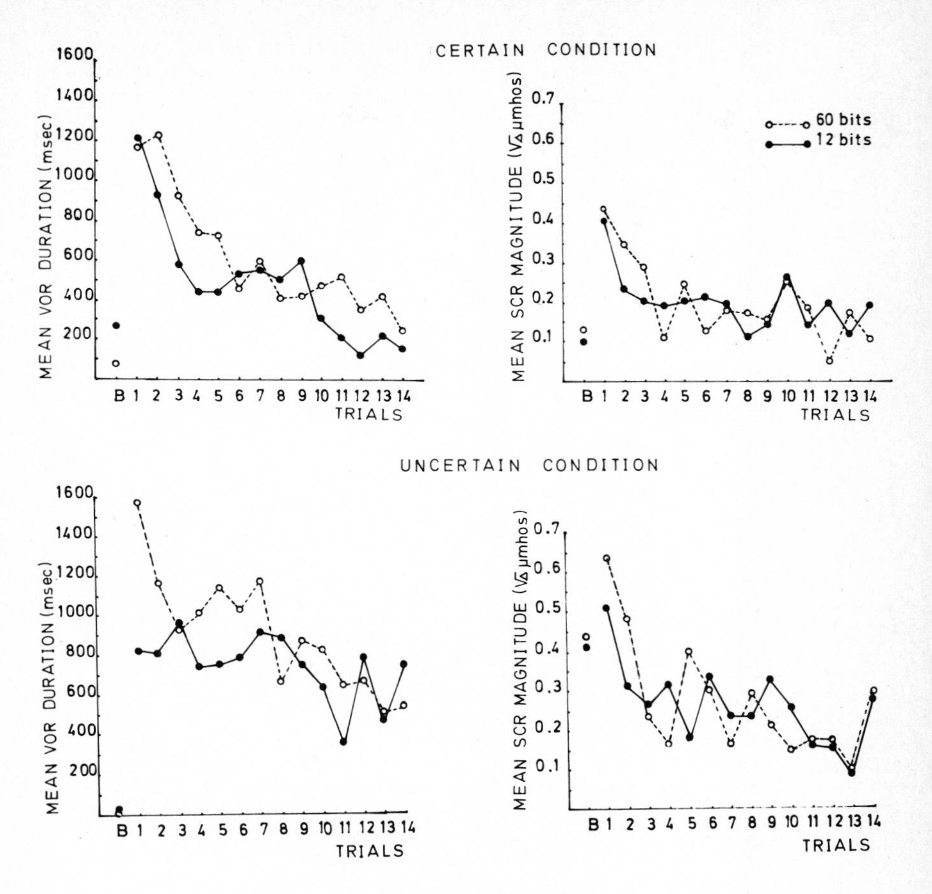

Fig. 4: Mean fixation duration in field 1 and SCR magnitude in the "Uncertain" condition (below) and "Certain" condition (above). Solid lines: 12 bits condition, discontinuous lines: 60 bits condition.

averaging out ERPs, which means that habituation can at best be studied over 30-45 trials. This is a period of time in which the SCR normally has already disappeared. The so-called fast habituation design (Ritter, Vaughan and Costa, 1968) is one solution because averaging occurs over ordinal positions, but this has the disadvantage of contaminating earlier and later responses in a habituation period (Roth, 1973). Another solution for this problem is to use other forms of averaging which need less trials. For this reason we used an improved Wiener filter procedure (Woestenburg, Verbaten, Sjouw and Slangen, 1981 a) which enabled us to determine ERPs on the basis of 4-6 trials. The SCRs were averaged over the same number of trials. In this way it was possible to directly compare changes in ERP parameters and the SCR over trials. In a fourth experiment (Woestenburg, Verbaten, Sjouw and Slangen, 1981 b) we investigated the influence of information on habituation of the visual ERP and the SCR, using non-

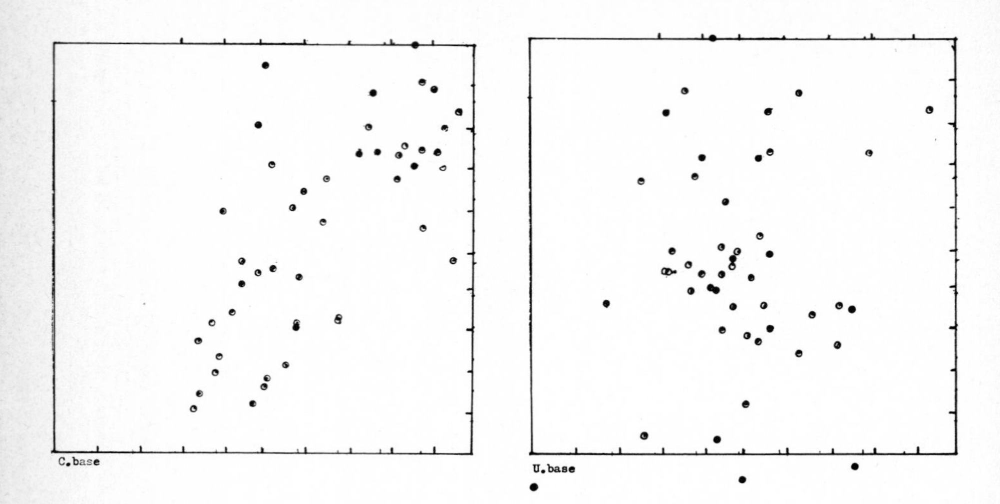

Fig.5: Distribution of fixation points over the TV-monitor during the base-level trial. Each point indicates one uninterrupted fixation of at least 100 msec. "C" indicates the "certain" condition and "U" the "uncertain" condition.

signal stimuli. The difficulty was that while we could not instruct subjects to look at the stimuli in order not to violate the non-signal character of the stimuli, we also had to be sure that the subjects looked at the stimulus when it was presented. For this reason we used the same eye movement measurement procedure as in the three former experiments and measured continuously where the subject fixated. When the subject fixated within a preset window, (see fig.6) a stimulus was presented. Half the subjects (13) received 6 ensembles (1 ensemble contains 6 stimulus presentations) of a 60 bits stimulus and the other half (13) received 6 ensembles of a 4 bits stimulus. The main aim of this study was the identification of ERP components which habituated in about the same period of time as the SCR, a major definitial requirement of the OR. Another property of an OR is that it will be larger when reacting to stimuli which contain more information. The investigation of this question pertaining to ERP components was the second aim of this study. In order to determine in which latency areas ERP components showed a linear decrease of amplitude as a function of stimulus repetitions, we did a multivariate linear regression analysis (Finn, 1976). The input for this analysis consisted of averages of 10 consecutive sample points, totalling to 20 datapoints per lead per information condition (in this way the original dataset of 200 datapoints was reduced to 20 datapoints). We found significant linear regression in the 240-480 msec latency area, most clearly so in the Fz and Cz leads. (see fig. 7).

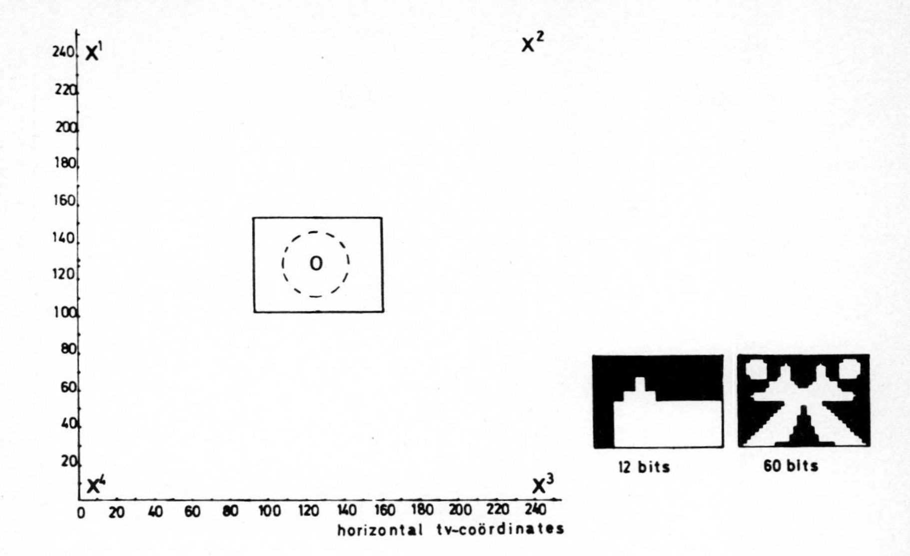

Fig.6: Distances of the four calibration stimuli (X1, X2, X3, X4) and the letter O in TV-coordinates (256 x 256). The discontinuous lines indicate the preset window A around the centre of the TV-screen. The position of the information containing stimuli is indicated by solid lines.

This latency are is conventially chosen as the P300 latency area. For subsequent analyses we therefore scored maximal amplitudes in this latency area. We found a significant linear decrease over the six emsembles of SCR amplitude ($F(1/23)=22.10$, $p<1\%$) and P300 amplitudes in the Fz ($1/23)=16.60$, $p<1\%$) and Cz ($F(1/23)=14.03$, $p<1\%$) locations.No such an effect was found for the P300 in the Oz location, indicating a difference in this respect between more anterior and more posterior leads. SCR habituation was faster than P300 habituation, because analysis over the first 4 ensembles showed still a significant linear decrease of SCR magnitude, but no longer such an effect of P300 amplitude. This difference in rate of decrease of response strength may mean that the SCR and fronto-central P300 are related to different aspects of the OR. This suggestion is also supported by the fact that fronto-central P300 amplitudes were larger in the 60 bits condition than in the 4 bits condition (Cz $F(1/23)=4.43$, $p<5\%$, one tailed) (Fz $F(1/23)=336$, $p<10\%$, one tailed) in the absence of such an effect on SCR magnitude and Oz P300 amplitudes (see fig.8).

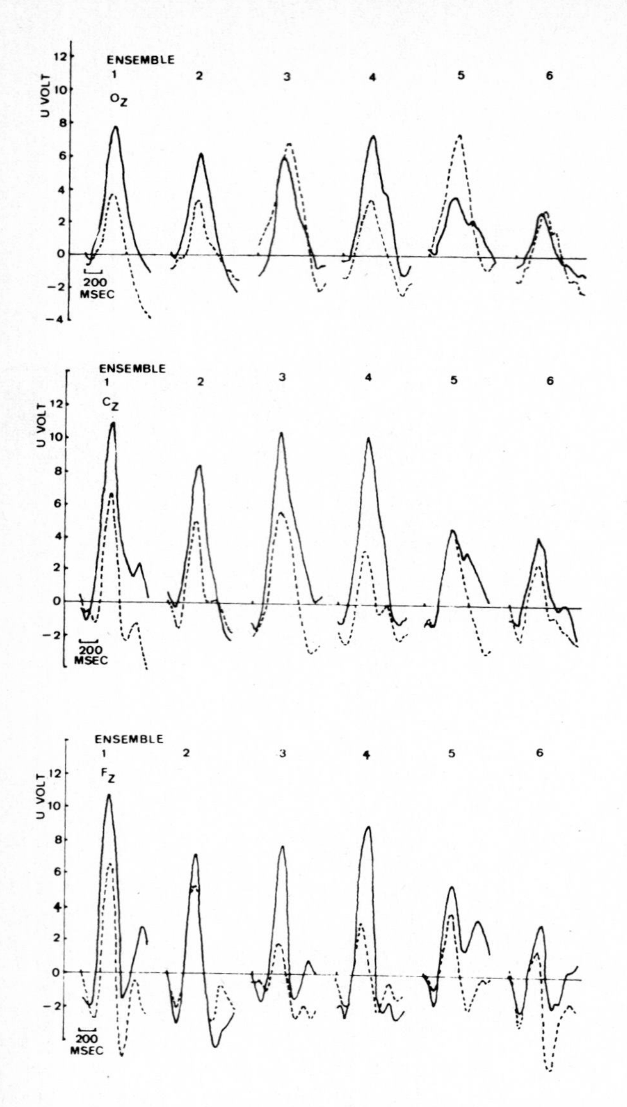

Fig. 7: Mean ERPs over all subjects in Oz, Cz and Fz leads, (positive is upwards). Solid line: 60 bits condition. Interrupted line: 4 bits condition.

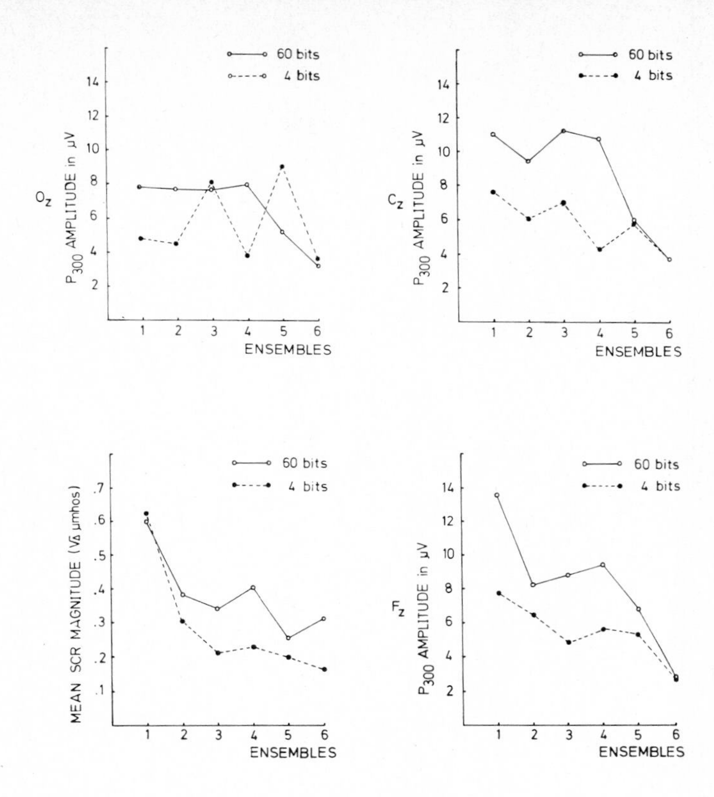

Fig.8: Mean SCR magnitudes and P300 amplitudes in the Oz, Cz and Fz locations as a function of blocks of 6 stimulus presentations. Solid line: 60 bits condition, discontinuous line: 4 bits condition.

The fronto-central P300 showed two characteristics of the OR, namely the decrease of response strength over trials and stronger responses to stimuli containing more information.
Task-relevance or irrelevance seems important in regard to the topographical distribution of the P300. Desmedt and Becker (1979) found large late positive components, which they called the P350, in more posterior leads (centro-parietal) in task-relevance conditions, while the fronto-vertex amplitudes were very small in such conditions. In a fifth experiment (Woestenburg, Verbaten, Sjouw and Slangen, 1981 c) we used the same habituation design as before but gave task-relevance to the stimuli. We instructed the subjects to memorize the stimuli because at the end of the experiment there would be a recognition test. The subjects were rewarded for correct recognitions. There were 28 subjects. Half the subjects received first 6 ensembles (one ensemble contains 6 stimulus presentations) of a 60 bits stimulus and then 6 ensembles of a 12 bits stimulus, the other half of the subjects received the stimuli in the reversed order. There was a pause of 1 minute between the two blocks of 36 stimuli. ISIs were

randomly chosen and varied between 6-10 s. P300 amplitudes were scored in the same latency areas (250-400 msec) as in the former experiment.We found a significant linear decrease of SCR magnitude ($F(1/24)=8.16$, $p<1\%$) and Cz and Fz P300 amplitudes (resp. ($F(1/24)=8.05$, $p<1\%$ and $F(1/24)=10.16$, $p<1\%$) in the absence of such an effect on Oz and Pz P300 amplitudes (see fig.9 and fig.10).

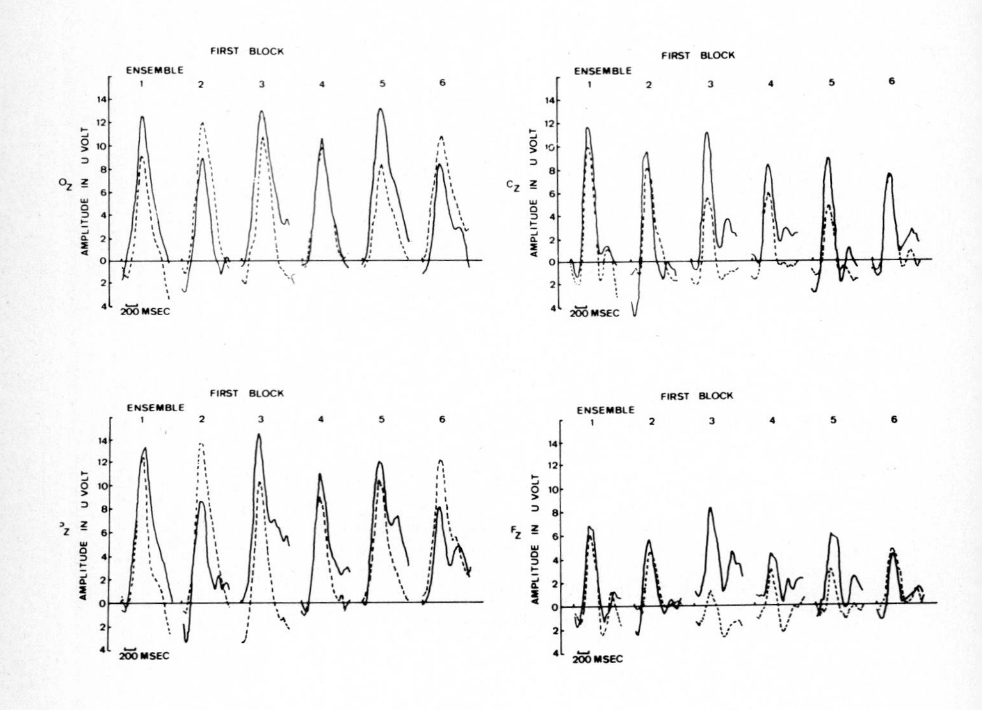

Fig.9: Averaged ERPs over all subjects in the first block. Solid line: 60 bits condition. Interrupted line: 12 bits condition.

During the first block we found significantly larger P300 amplitudes in the high complexity condition in contrast to the low complexity condition, but only in the Fz and Cz leads. No such effects were found, neither in block 1 nor in block 2, in Oz or Pz leads.

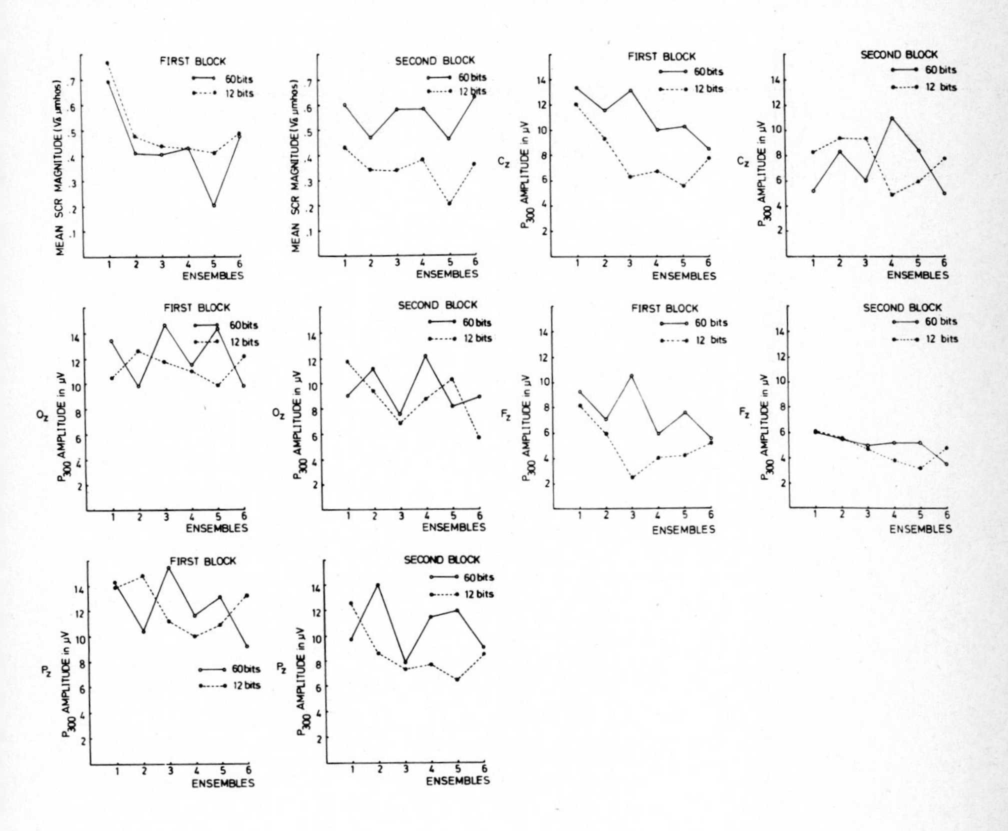

Fig.10: SCR magnitude and P300 amplitudes in Oz,Cz,Fz and Pz locations as a function of ensembles in block 1 and block 2. Continuous line: 60 bits condition. Discontinuous line: 12 bits condition.

The absence of an effect of stimulus complexity on SCR magnitude in this experiment was unexpected, because we have found in experiment 2 that in task-relevance conditions the SCR magnitude was larger in high complexity conditions. However, it might be that the difference in number of presented stimuli (72 in this and 28 in the second experiment) influenced the results.

DISCUSSION:

The results of the experiments thus far do not support an unitary conception of the OR. They also do not support the assumption that the SCR is the best or the most sensitive OR index. Instead the various responses which are contingent on a novel stimulus may be better understood by examining their possible functions. In this respect the SCR might be better viewed as a system's response which is activated when effort has to be invested in a given situation. The sympathic system supports the somatic system when some effort has to be made, whether that effort be physical, like running, or mental, like memorizing important information. From this point of view we can understand the effect of information in task-relevance conditions on the SCR. These conditions demand more effort, therefore they result in more electrodermal activity. This approach to the SCR-OR is not incompatible with Bernstein's (1979) two-stage model of the OR. The resemblance is that a SCR occurs when at the secondary stage the information is found (evaluated) to be important or relevant. According to Öhman (1979) the SCR might be seen both as a call for space in a central channel and probably as an answer to that call as well. The important aspect of Öhman's theory is the link it supplies between OR theory and selective attention theories like those of Sternberg (1969) and Shiffrin and Schneider (1978). Unfortunately, the identification of the SCR with a call for central processing space leads to some difficulties. On this assumption, we might expect that the VOR would always coincide with the SCR because it is not easy to understand how in the absence of a focal inspection of the stimulus (in the case of visual stimuli like in the above mentioned experiments), central processing, involving a controlled memory search for specific information in order to compare with the perceived stimulus, can take place. Moreover, according to Just and Carpenter (1976) eye movements reflect what is currently processed in the short term memory (STM), which is considered to be a part of the central processor. The most parsimonious explanation seems that pre-attentive processes identify mismatches (in the case of non-signal stimuli) or matches (in the case of signal-stimuli) with what is currently processed in the central channel, leading automatically to a call for central processing. In the experiments reported above we used visual stimuli effecting foveal inspection of the stimulus, and because the stimuli were presented in the periphery of the visual field, they induced a VOR. In such conditions a SCR might also occur, reflecting support of the somatic system and/or a general arousal increase thereby facilitating stimulus intake. When after an initial processing of the stimulus in the central channel, the stimulus is found to be of significance, more mental effort might be invested, leading to an increase in electrodermal activity (Kahneman, 1969). Given the VORs latency time (150-200 msec), this assumption requires a pre-attentive processing of stimuli, allowing at least a global matching of stimuli with what is currently present in the STM. ERP studies have supplied convincing support for such an assumption. It has been shown that the amplitude of the N1 component (80-150 msec latency time) is larger in reaction to stimuli which have been made task-relevant (or in other words, for which

a stimulus set had been induced) than for task-irrelevant stimuli (Hink, Hillyard & Benson, 1978). That pre-attentive processing influences the OR was shown in the third experiment reported above. Supplying subjects with a priori information ("certain" condition) might be seen as a procedure for inducing perceptual or attentional set. Fig. 5 shows that such a set had indeed been induced. During the base-level trials, subjects in the "certain" condition scanned the place on the TV-screen where the stimulus was expected. Inducing a set and thereby the pre-attentive processes clearly influenced the VOR and habituation of the VOR in the experimental conditions reported above. The VORs were less intense and habituated quicker, which is to be expected if we assume that in the case of non-signal stimuli more working space is needed when the mismatch with what is currently processed is larger. There are some convincing arguments in favour of the assumption that stimulus processing has its effect mainly on the perceptual side and not on the motor side of the stimulus processing chain (Posner, Klein, Summers and Buggie, 1973). Although with a different paradigm, Stanovitch and Pachella (1977) found,like we did in experiment 3, an interaction between complexity and uncertainty. The latter authors explained their results by assuming that in conditions of low stimulus uncertainty, stimulus processing might be more efficient because several (assumed serial) stimulus processing stages might overlap. Because we assume that the N1 component reflects pre-attentive processing of stimuli, and the OR represents an attentive response requiring processing in a central channel, we don't agree with hypotheses identifying the N1 as an OR component (Näätänen,1979). An additional argument in favour of such a position might be found in the fact that the N1-P2 component does not show reliable habituation as a function of stimulus repetition, one of the most basic properties of the OR (Lader & Öhman, 1976).
The direct comparison between SCR habituation and ERP habituation demonstrated that the SCR and the P300 habituated in about the same period of time. Moreover the fronto-central P300 was larger when stimuli with more information were presented. These results support the assumption that the fronto-central P300 is a component of the OR (Ritter e.a., 1968, Roth, 1973, Courchesne, 1975, 1978a). We found that the fronto-central and the more posterior P300 were influenced by different aspects of the stimulus situation. The fronto-central P300 was largest when unexpected, complex stimuli were presented and habituated when such stimuli were repeatedly presented. The more posterior P300 was largest when expected, task-relevant stimuli were presented and did not habituate. In such conditions the fronto-central P300 was smaller. The P300 seems an EEG aspect of an attentional response, different from the pre-attentional characteristics of the N1-P2. Hink, Hillyard and Benson (1978) showed that the P300 only occurs to target stimuli, while background stimuli,although inducing a N1-P2, did not lead to a P300. It has been shown before that uncertainty has a reliable effect on the vertex P300 (Friedman, Hakerem, Sutton and Fleiss, 1973), leading to larger P300 amplitudes when uncertainty is higher. It might be that the smaller fronto-central P300 amplitudes found in this experiment were the result of reduced uncertainty, because the experimental conditions here were quite similar to those in experiment 2 and there we suggested that the task-relevance instruction had reduced stimulus uncertainty. The results are in accordance with Becker and Desmedt (1979) who also found relatively small fronto-central and large posterior P300 amplitudes in task-relevance conditions. It might be, however, that fronto-central P300 amplitudes found by the latter authors were large in the beginning of the experiment and habituated quickly and that their method of averaging obscured this fact.

Both the fronto-central and the posterior P300 are connected with some aspect of post-stimulus processing, but probably in different ways as indicated by the different topographical distributions in different experimental conditions. One of the possibilities mentioned by Courchesne (1978) in relation to the function of the fronto-central P300 was that it is larger in conditions where stimuli are not easily categorized. The results of this study are in accordance with such an assumption. Uncertainty or stimulus set alone seems not sufficient for the explanation of the function of the fronto-central P300, because more complex stimuli lead to larger amplitudes both in uncertain and certain conditions. Although the VOR is, like the fronto-central P300, larger in the case of stimulus uncertainty and stimulus complexity there are indications that they are differently influenced by experimental conditions. In task-relevance conditions P300 amplitudes were still larger when more complex stimuli were presented, while such an effect was absent in the VOR in similar conditions.It seems reasonable to assume that the VOR is connected with some earlier information gathering phase, while the fronto-central P300 is connected with some later evaluative phase in which the stimulus is categorized. The posterior P300 seems related to other aspects of stimulus processing. It is large in task-relevance conditions independent of the complexity of the stimulus and stable in that it does not show habituation. It is interesting to note that the posterior P300 responds best to the description of the OR in signal conditions; even the SCR habituates within 10-20 trials in such conditions, while the posterior P300 does not habituate over 36 trials.

In summary the results of this study show that there are several responses contingent on the presentation of a novel stimulus; the VOR, the SCR, the fronto-central and the posterior P300. In some way or another, all these responses demonstrate some property of the OR as described by Sokolov, but there is no response which shows consistently all the properties of the OR. The unitary approach of the OR was not supported by the results of this study and a theory which assumes several phases of information processing reflecting different physiological response systems seems more adequate. In this sense, OR theory might profit from the selective attention theory. On the other hand, the OR paradigm might clarify some problems which are not adequately solved by the selective attention paradigm, e.g. the topographical distribution of the P300.

Tutorials in ERP Research: Endogenous Components
A.W.K. Gaillard and W. Ritter (eds.)

10

HUMAN ENDOGENOUS LIMBIC POTENTIALS: CROSS-MODALITY AND DEPTH/SURFACE COMPARISONS IN EPILEPTIC SUBJECTS

Nancy K. Squires
Department of Psychology
SUNY, Stony Brook, New York

Eric Halgren, Charles Wilson, and Paul Crandall
UCLA Neuropsychiatric Institute
Los Angeles, California

Field potentials can be recorded from the human limbic system under the same task conditions that produce endogenous event-related potentials at the human scalp. In this chapter, recent findings are described that further support the local generation of the endogenous limbic potentials, the distributions of the limbic potentials are compared in auditory and visual tasks to evaluate the possibility of multiple endogenous components in the depth potentials, and clinical data are reviewed that bear on the issue of the limbic system as the neural source of the endogenous scalp potentials.

INTRODUCTION

In a recent report we described field potentials from various limbic system structures in man, recorded under conditions known to produce endogenous event-related potentials at the human scalp (Halgren, Squires, Wilson, Rohrbaugh, Babb, and Crandall, 1980). These limbic field potentials occurred with latencies similar to the endogenous scalp potentials. Furthermore, they had similar functional correlates: they were larger to unexpected than to expected stimuli, they were larger when the subject actively attended to the stimuli than when the subject was reading, and they were evoked by unexpected stimuli in either auditory or visual modalities. Since the limbic field potentials met the major criteria used in identifying endogenous event-related potentials, we called them "endogenous limbic potentials" (ELPs).

Evidence was also presented that the ELPs are generated locally, in the region of the recording electrodes. The ELPs were large relative to the surface potentials recorded concurrently, even in bipolar recordings from electrode only a few millimeters apart. Polarity reversals were seen between different structures, and sometimes even between homologous structures in the two hemispheres (e.g., left and right hippocampi). The latter result was tentatively attributed to the termination of the electrodes in the two hemispheres at different points relative to the local dipole generator.

Perhaps most significantly, in several cases we were able to record multiple-unit activity concurrently with the field potentials from the same electrode. In most cases the pattern of unit firing corresponded both temporally and functionally to the ELPs. While these results clearly indicated that the limbic system was active under conditions that produce endogenous ERPs at the scalp, they did not resolve the question of whether the scalp ERPs are a volume-conducted reflection of the limbic potentials, or whether the two are simply correlated.

Similar recordings have now been made from a second series of six patients. Data from this second series will be presented here which focus on three issues: 1) ELPs recorded from multiple electrodes within a given structure are compared with the aim of clarifying differences previously seen between structures and between homologous sites in the two hemispheres; 2) Comparisons of depth recordings and surface recordings are presented showing substantial covariation between the two, and suggesting a functional relationship; 3) Comparison of the effects of different stimulus modalities on the depth and surface ERPs are presented to assess whether multiple endogenous components exist at the depth as well as at the surface.

METHODS

Electrodes were implanted, bilaterally, in hippocampus, hippocampal gyrus, and amygdala. The implantation was performed in order to locate the epileptic foci in patients for whom anterior temporal lobectomy was contemplated. (For a more detailed description of the patients and the recording techniques see Halgren et al., 1980). The electrodes were of two types, .2 mm diameter wire macroelectrodes, and 40 micron diameter Pt. alloy fine wire microelectrodes. In addition, stainless steel electrodes were placed in the outer table of the skull according to the 10-20 system. Of these, Cz and Fz were the most commonly used. All recordings were made to a nose reference. The EEG was amplified with Grass 7P511 amplifiers with a bandpass of .3 Hz to 5 kHz, and recorded on FM tape for off-line analyses. Averages were obtained with a PDP 11/10 computer which digitized the EEG at 2 msec/address. For the auditory stimuli 200 msec of pre-stimulus activity and 800 msec of post-stimulus activity were included in the average. For the visual stimuli no pre-stimulus baseline was obtained.

In the visual task the stimulus was either a backward "Z" (P = .8) or a backward "N" (P = .2). The two stimuli occurred in random order with a constant ISI of 1.2 seconds. The subject was asked to count the rare stimuli. In the auditory task a series of tones was presented, with low tones (1 kHz, P = .8) and high tones (1.5 kHz, P = .2) occurring in random order at a constant ISI of 1.2 seconds. The subject was asked to count the high tones. The fixed ISI in the auditory task represents a change in procedure from the previous experiment where the tones were presented with a long, variable ISI ranging from 3 to 6 seconds. The procedure was changed to increase the compatibility of the auditory and visual conditions for the cross-modal ELP comparisons. However, it also reduces the possibility of observing the very long-latency endogenous waves that are enhanced by long ISIs (Rohrbaugh, Syndulko, and Lindsley, 1978).

Evoked potentials were averaged separately for the rare and frequent stimuli. An automatic artifact-rejection system was used so that if the EEG exceeded either a criterion maximum or minimum at any point during the averaging epoch the data of that trial were excluded from the averages. All waveforms presented here represent the subtraction of the evoked potential to the frequent stimulus from the evoked potential to the rare stimulus unless otherwise noted.[1]

RESULTS AND DISCUSSION

A. Depth distribution of auditory ELPs

Figure 1 shows the difference waveforms in the auditory task from the vertex and from three right-hemisphere electrodes: hippocampal gyrus (RHCG), amygdala (RAm), and hippocampus (RHC). The vertex ERP shows a large positive wave with a peak latency of about 350 msec (P3). Concurrently at the depth a large positive wave is seen in the hippocampal gyrus and a large, almost mirror-image negative wave from the amygdala. The hippocampus shows a smaller negative potential at a slightly longer latency. These data similar to the data previously reported (Halgren et al., 1980), showing polarity reversal of the ELP between different libmic sites.

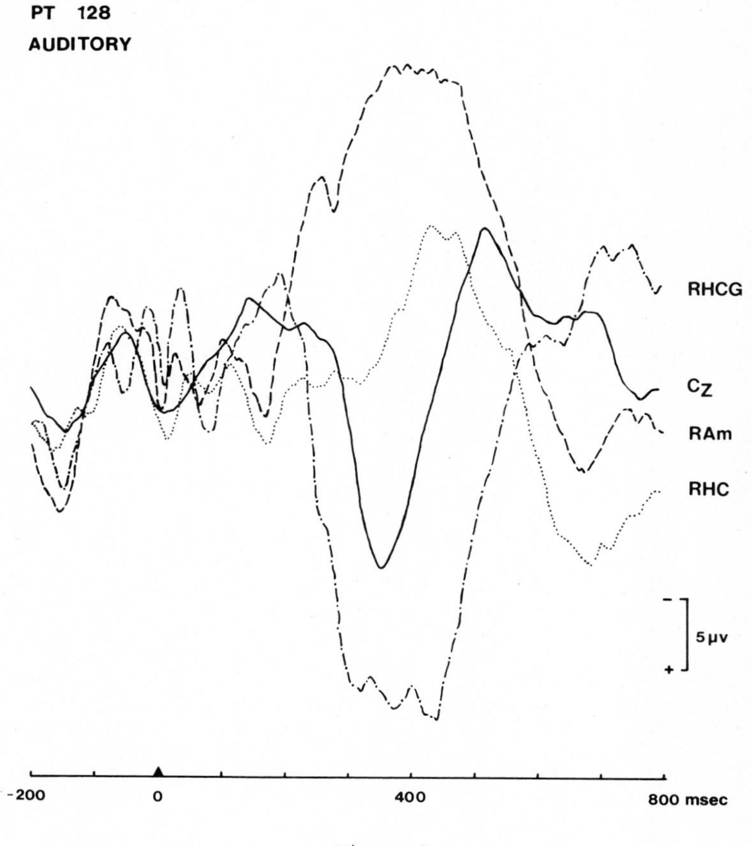

Figure 1
Patient #128: Rare-minus-frequent difference waveforms from Cz (solid line) and from the right hippocampal gyrus (RHCG), amygdala (RAm), and hippocampus (RHC) in the auditory oddball task. Tone onset is at time "0."

Figure 2 shows data from the same subject from the homologous left-hemisphere electrodes. Again there is a polarity reversal between the hippocampal gyrus and the amygdala. However, in these recordings the ELPs are morphologically more complex. Whereas the right-hemisphere potentials were essentially monophasic, the left-hemisphere potentials are biphasic or triphasic, and the peak latencies vary somewhat with respect other and to the peak of the surface P3. Nevertheless, the maximum difference in voltage between the depth potentials occurs in the P3 latency region.

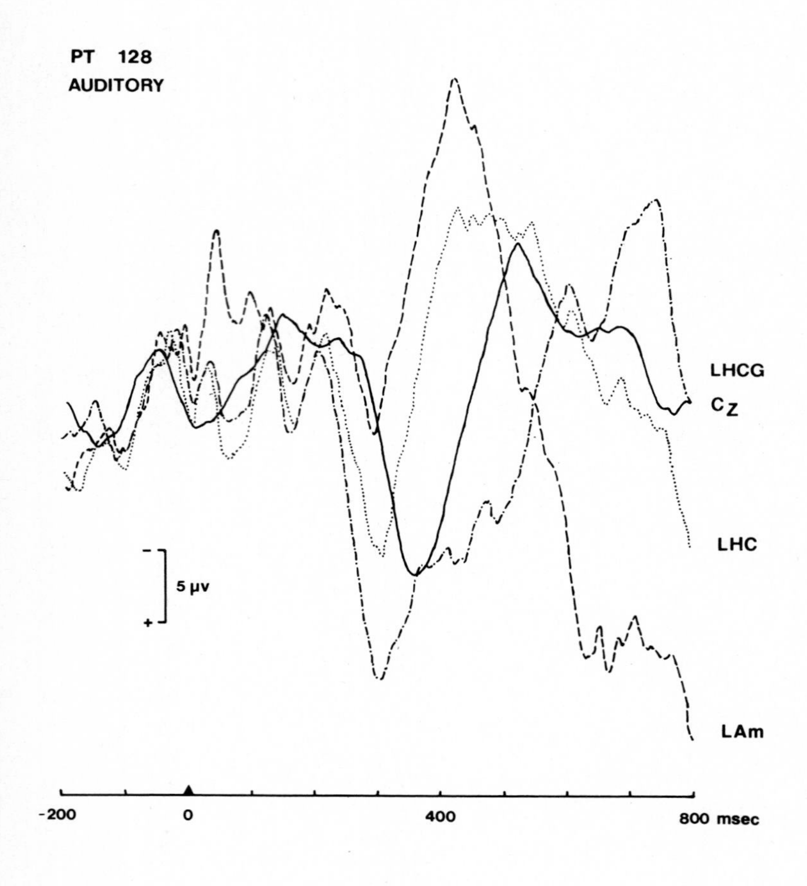

Figure 2

Patient #128: Recordings from the left hippocampal gyrus (LHCG), hippocampus (LHC), and amygdala (LAm). The Cz recording is the same as that shown in Figure 1.

Recordings from multiple electrodes in the same structure are shown in

Figure 3 for the same subject, where the ELPs from three left hippocampal microelectrodes are superimposed, along with the same surface waveform shown previously. The variations in waveform and polarity seen within the same structure are as great as the variations seen across structures in the same subject. (Compare, for example, Figures 2 and 3). This finding supports our earlier conclusion that polarity reversals between the two

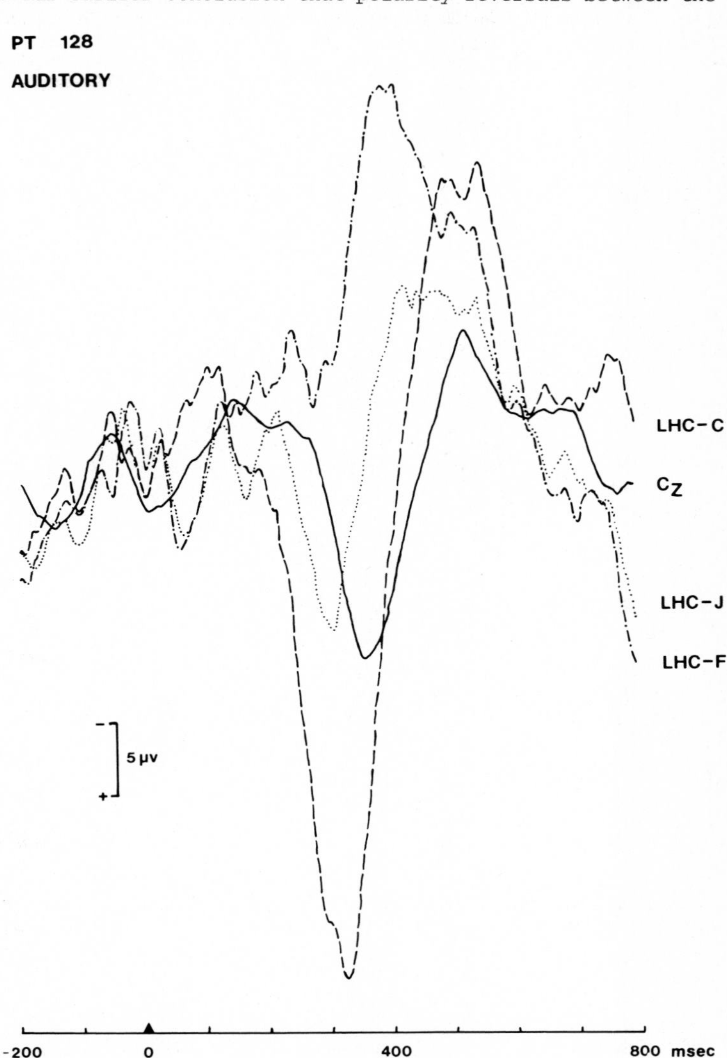

Figure 3
Patient #128: ELPs from three microelectrodes in the left hippocampus. The solid line represents the same vertex recording shown previously.

hemispheres, but from electrodes nominally in the same location, can be attributed to differences in the exact location of the electrode relative to the local dipole since, as shown here, small variations in electrode placement within a structure can result in almost complete polarity inversion. (The distance between the hippocampal microelectrodes was approximately 1-5 mm). Furthermore, the data of this subject suggest that the major voltage gradient, regardless of the recording configuration, occurs at the latency of the surface P3. In some cases (e.g., Figure 2) large variations in activity were also seen at longer latencies, but this was less common. The onset of the ELP always occurred at or before the onset of the surface P3.

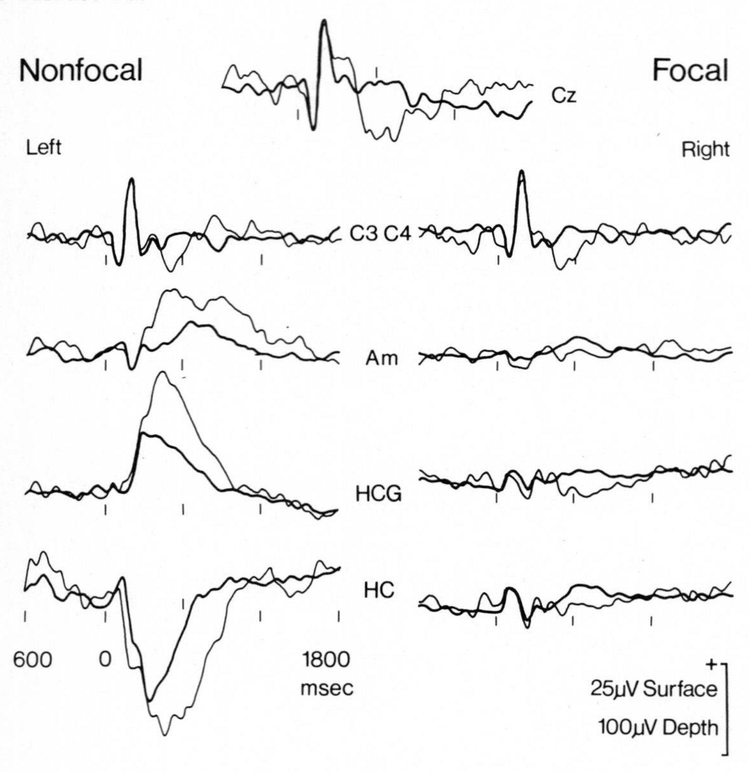

Figure 4
Patient #118: Comparison of the ELPs from the more epileptogenic side (the right) and the less epileptogenic side (the left) in the auditory oddball task. ERPs to the frequent tones are indicated by the heavy lines. Recordings are from surface electrodes Cz, C3, and C4, and from depth electrodes in amygdala, hippocampal gyrus, and hippocampus. These recordings were taken during the auditory task with long, variable ISIs, and the time base extends from 600 msec before the stimulus to 1800 msec after.

B. Depth activity and lateralization of pathology

Figure 4 shows ELPs to rare and frequent tones from a subject in the first series of patients. This subject's pathology was clearly lateralized to the right side: all seizure onsets were on the right, the inter-ictal spike rate was higher on the right than on the left, and the results of the Wada test showed disruption of memory on the left (intact) side. Furthermore, this patient subsequently underwent surgery and a marked sclerosis was observed in the tissue excised from the right side. Corresponding to the unambiguous lateralization of the temporal lobe pathology in this patient, there was a marked asymmetry in the ELPs, with essentially no evoked activity seen on the side of the seizure focus, to either the rare or the frequent tones.

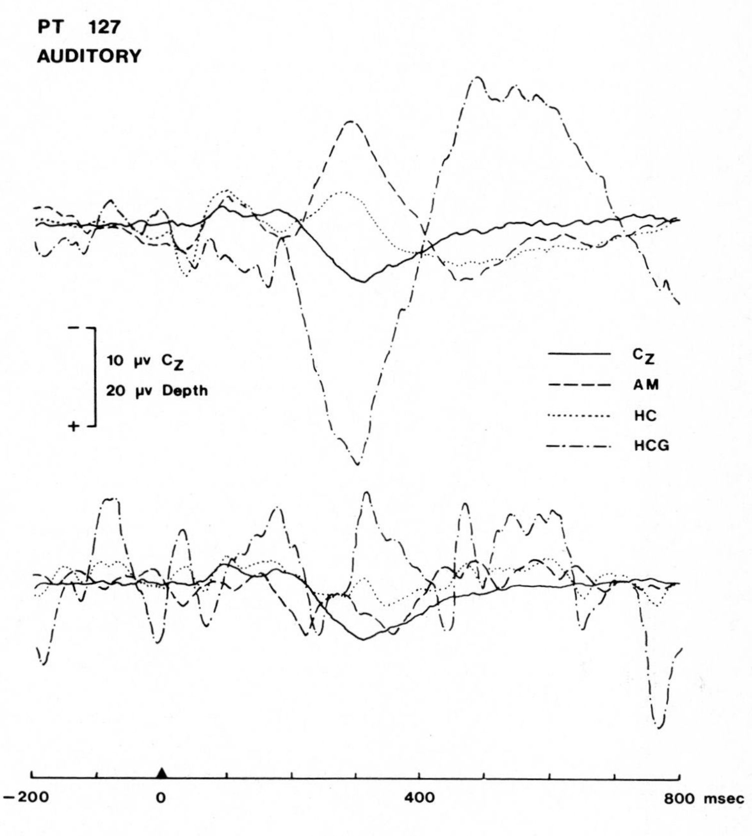

Figure 5
Patient #127: Rare-minus-frequent difference waveforms from Cz, amygdala, hippocampus, and hippocampal gyrus on the right (top) and the left (bottom). This patient's seizure onsets were lateralized to the left side.

Similar findings were obtained from one subject in the second series, in whom the seizures were lateralized to the left side. The difference waveforms for this subject are shown in Figure 5. Large ELPs are seen in the recordings from the right hemisphere (top), but on the epileptogenic side (bottom) there is no time-locked endogenous activity that exceeds the level of the background noise.

Although such clear lateralization of the seizure focus and the ELPs occurred in only one patient of each series, their data provide strong support for the local origin of the ELPs; if the endogenous potentials recorded from these electrodes were volume conducted from a distant source, they should be unaffected by the integrity of the structures in the neighborhood of the recording electrode. As shown here, however, there is a striking correspondence between the loss of the ELPs and the loss of function in the local area.

C. Correlations between depth and surface recordings

Over the total subject population there was considerable variability in the quality of the endogenous potentials. However, the quality of the depth and surface recordings in individual subjects showed good correspondence. Summarized across the two series of patients, six subjects showed good quality endogenous potentials at both the surface and the depth, where "good" refers to the unambiguous identification of these potentials on the basis of latency, waveform, and variation with experimental conditions. One subject had poor (ambiguous) results for both the surface and the depth, and in the remaining five subjects both ELPs and surface endogenous components were absent. No subjects showed clear dissociation of the depth and surface ERPs. These findings are consistent with the hypothesis that the surface endogenous ERPs are a direct (volume-conducted) reflection of limbic activity, since the endogenous components were missing at the surface in every case in which they were missing at the depth. An alternative explanation is that the P3 generator is activated in consequence of limbic activation, and therefore absent in patients with limbic system damage. A third possibility is that the absence of both the ELPs and the surface potentials are the result of a disruption at an earlier processing stage, upon which both are dependent. A trivial example of this might occur if the subject failed to pay attention to the task, or was insufficiently motivated, so that both the endogenous surface ERPs and the ELPs were absent because they share a common cognitive prerequisite, rather than because they are anatomically dependent. However, all of the subjects appeared to be adequately motivated, and did correctly perform the counting task. There was no obvious differences in the performance of the subjects who did and who did not have clear endogenous waves. Thus it is of some significance that the surface and depth potentials disappear as a unit since it suggests that they reflect the same cognitive operation.

D. Cross-modal comparisons

Simson, Vaughn, and Ritter (1977a) have extensively examined the scalp topography of two of the endogenous scalp components, N2 and P3, under task conditions similar to those employed here. Their data indicate that while the P3 component has essentially identical scalp distributions for rare auditory and rare visual events, the distribution of the N2 is modality specific, having an occipital focus in the visual condition and a broader distribution with a central maximum in the auditory task. The distributional data suggest that the P3 has the same generator source regardless of stimulus modality, while the generators of the auditory and visual N2 waves are different.

Since the cross-modal comparison has been shown to be an effective technique for the functional differentiation of scalp endogenous components, we reasoned that it might also prove useful in partitioning the ELP and/or further identifying its psychological correlates. Although the ELP tends to peak in the P3 latency region, as already noted it has an earlier onset and longer duration, thus overlapping several surface components. Furthermore, its morphology is sometimes complex (e.g., Figures 2 and 3) and like the scalp ERP might consist of several sub-components.

For the majority of recording sites the ELPs in the auditory and visual tasks had the same polarity and morphology. The peak latencies of the visual ELPs were somewhat later than those of the auditory ELPs, corresponding to the latency differences in the surface P3s. (The visual task appeared to be slightly more difficult than the auditory task, which probably accounts for the latency differences). However, clear dissociations between the auditory and visual ELPs did occur at some electrode sites, as illustrated in Figure 6. For this subject, of the 14 electrodes examined, the ELPs from the hippocampal electrodes showed the greatest

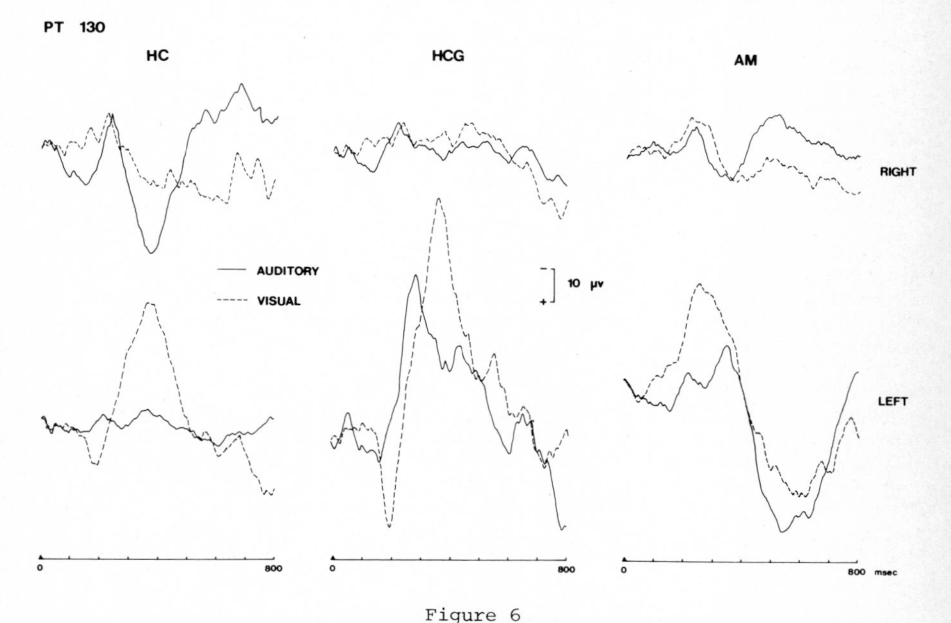

Figure 6
Patient #130: Difference waveforms in the auditory (solid lines) and visual (dashed lines) tasks, for the hippocampus (HC), hippocampal gyrus (HCG), and amygdala (AM) of each hemisphere. The surface P3 had a peak latency of 356 msec in the auditory condition and 400 msec in the visual condition.

cross-modal discrepancy: the right hippocampal ELP was larger in the auditory condition whereas the left hippocampal ELP was larger in the visual condition. In contrast to cross-modal differences in scalp recordings, however, it was not the case for the ELPs that part of the waveform varied with modality while part was modality independent; instead, the ELP at some electrodes was more sensitive to one modality than the other, while at other sites it was equally sensitive or equally insensitive to stimuli in both modalities. Smaller cross-modal differences were seen in the recordings from the same subject from three micro-electrodes in the right hippocampus (Figure 7). The ELPs in the auditory task showed amplitude variations across the three hippocampal electrodes while the ELPs in the visual task did not.

Vinogradova (1975), in summarizing her extensive work on the orienting response in rabbits, concluded that throughout the functional circuit that includes the hippocampus, hippocampal gyrus, and amygdala, there are certain areas that code stimulus novelty while ignoring stimulus modality and other stimulus characteristics. Hippocampal area CA1 is one example. Other areas (e.g., hippocampal area CA3) code both novelty and modality. Assuming that the differences shown in Figure 6 are attributable to differences in electrode placement within the hippocampus between the right and left hemispheres, our data are in agreement with the conclusions of Vinogradova. Thus the left hippocampal gyrus electrode would be in an area that responds to novel events regardless of modality, the left hippocampal electrode would be in an area that responds only to novel visual events, and the right hippocampal electrode would be in an area that responds only to novel auditory events.

While this formulation suggests certain discrepancies between the behavior of the ELPs and the scalp endogenous ERPs, it should be accepted only tentatively due to two possible confounding factors. First, as noted above, the visual task was somewhat more difficult than the auditory, producing longer P3 latencies; possibly it is the factor of task difficulty rather than modality that has produced the differences in the ELPs across tasks. Second, the auditory and visual conditions were presented on different days, so that factors such as subject state might account for the differences in the ELPs. Given the substantial co-variation of the ELPs in auditory and visual tasks at the majority of the electrode sites we consider these explanations less likely than true cross-modal differences in the ELPs at some sites. Assuming local differences in the auditory and visual ELPs, an important question to be addressed is whether these local differences would be visible at distant recording sites. Conceivably the local field potentials may sum in such a way to make them indistinguishable at a distance. For example, subtle differences in auditory and visual activation of hippocampal fields may well not produce distinguishably different waveforms at the scalp. This is even more likely to be the case when other areas are active at the same time and do not show modality differences in the ELP. Thus, while our data indicate that the limbic system, and particularly the hippocampus, differentiates between novel-auditory and novel-visual events, it is not possible to infer from these data that the corresponding scalp potentials will differ. Conversely, identical scalp distributions in two conditions need not imply identical activation at the depth. The subtle distributional difference observed cross modally at the depth, however, are probably not great enough to give rise to the substantial differences seen in the scalp topography of the N2 across modalities.

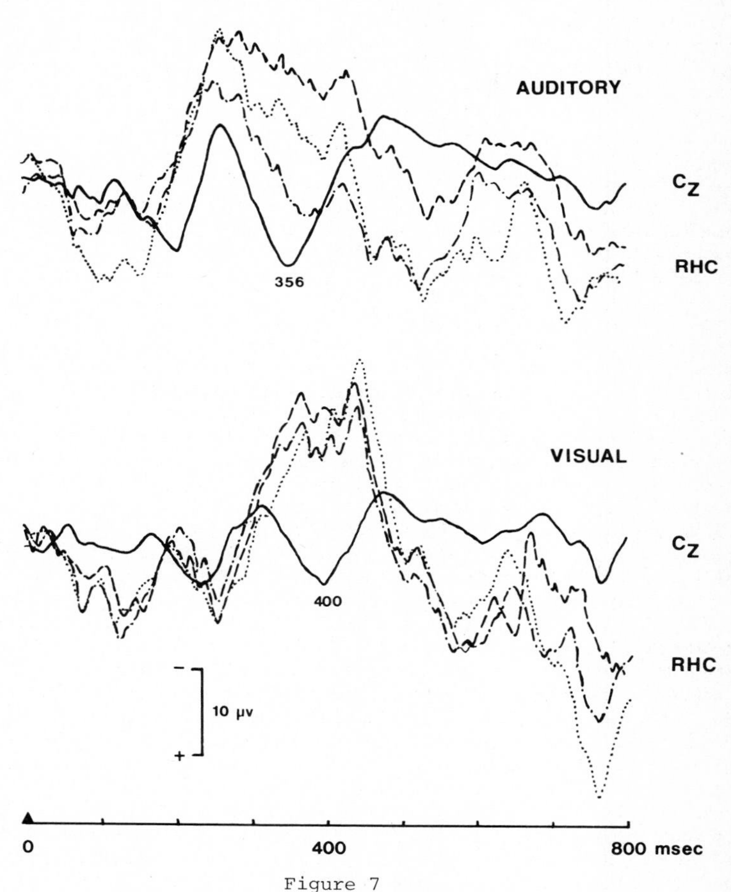

Figure 7
Patient #130: Difference waveforms from Cz (solid lines) and three microelectrodes in the right hippocampus in the auditory task (top) and the visual task (bottom). The morphology of the ELPs varied across electrodes in the auditory task but not in the visual task.

E. Endogenous ERPs in clinical populations

One implication of the current results is that the limbic system is intimately involved in the types of information-processing activities that result in the scalp-recorded endogenous ERPs. This may be inferred both from the number of functional parallels that have been drawn between the locally generated ELPs and the surface ERPs, and from the correspondence in the quality of depth and surface ERPs in epileptic subjects. This hypothesis is in accord with the theoretical formulations of hippocampal function based on animal research. For example, both Vinogradova (1975) and Pribram and McGuiness (1975) have stressed the role of the hippocampus in registering the mismatch of incoming information by comparison to a neuronal template of ongoing events, and have related these findings to Sokolov's (1963) theory of the orienting response. Similar parallels have been drawn between the endogenous scalp ERPs and Sokolov's model (e.g., N. Squires, K. Squires, and Hillyard, 1975; Hillyard, K. Squires, and N. Squires, in press). According to Vingradova, "different lines of experimental evidence help to restrict the hypothetical function of the hippocampus to a relatively limited complex: detection of novelty (orienting reaction) and its registration (memory)."

Furthermore, the hypothesized role of the limbic system in the generation of endogenous ERPs is consistent with the clinical data on endogenous (scalp) ERPs. Table 1 summarizes the data obtained with a variety of clinical populations. The findings have been divided into two general categories: populations in which the primary finding is a prolongation of the latency of P3, versus populations with normal latencies but decreased P3 amplitude. In the former category (latency prolongation) belong such diagnostic groups as presenile dementia, normal aging, head trauma, and mental retardation. In the second category (reduced amplitude) are schizophrenia, depressive illness, and hyperactivity. The categorization of chronic alcoholics is uncertain; while some studies report prolonged latencies in this group (Pfefferbaum, Horvath, Roth and Kopell, 1979), normal latencies have also been reported (Begleiter, Porjesz, and Tenner, 1980). This discrepancy may relate to variations in the degree of neuropsychological deficit (see below). Finally, for two groups the data are currently insufficient for even a tentative determination of the status of P3 latency (category III). In two studies on autistic children (Novick, Kurtzberg, and Vaughan, 1979; Novick, Vaughan, and Kurtzberg, 1980), P3 amplitudes in the clinical subjects were so severely reduced that reliable estimates of P3 latency could not be made. This was also the case in one study on reading-disabled children (Schaub, 1981). Dainer, Klorman, Salzman, Hess, Davidson, and Michael (1981), using a continuous-performance task with RD children, found reduced amplitudes of the late-positive component, and prolonged latencies.

Dichotomization of the clinical groups on the basis of amplitude versus latency abnormalities is consistent both with physiological evidence on the independence of latency and amplitude measures (see, for example, Buchwald, 1981), and with the effects of experimental variables on the human endogenous potentials. While attention (Isreal, Chesney, Wickens, and Donchin, 1980) and motivation (Sutton, in press) have been shown to modulate P3 amplitude, latency appears to be determined primarily by the time required to evaluate and categorize stimulus input (N. Squires, Donchin, K. Squires, and Grossberg, 1977; Kutas, McCarthy, and Donchin, 1977; McCarthy and Donchin, 1980).

Relating this to the clinical populations described in Table 1, it is those groups exhibiting abnormal P3 latencies for which the clearest evidence of medial temporal lobe pathology exists. Histopathological changes in the hippocampus have repeatedly been found in dementia of the Alzheimer's variety and Down's Syndrome mental retardation (Soltaire and Lamarche, 1966; Burger and Voegel, 1973; Ellis, McCulloch, and Corley, 1974; Crapper,

Table 1: P3 latency and amplitude abnormalities in clinical populations with diminished mental function. A "+" in the latency or amplitude column indicates an increase in that measure over the control group; a "-" indicates a decrease; and an "N" indicates normal values.

	Population:	References:	Latency:	Amplitude:
I.	Latency prolongation			
	Normal aging	Goodin et al., 1978a	+	-*
		Pfefferbaum et al., 1980a	+	-
		Pfefferbaum et al., 1980b	+	N*
		Ford et al., 1979a	+	N
		Ford et al., 1979b	+	N
		Porjesz et al., 1980	+	N
		Syndulko et al., in press	+	-
	Dementia	Goodin et al., 1978b	+	-
		K. Squires et al., 1980	+	-
		Syndulko et al., in press	+	-
	Head Trauma	K. Squires et al., 1980	+	
	Retardation	N. Squires et al., 1979	+	-
	Alcoholism	Pfefferbaum et al., 1979	+	-
		Begleiter et al., 1980	N	-
		Porjesz et al., 1980	N	-
		Skerchock and Cohen, 1981	+ and N	-
II.	Amplitude reduction			
	Hyperactivity	Prichep et al., 1976		-
		Klorman et al., 1979	N	-
		Loiselle et al., 1980	-	-
	Schizophrenia	Roth and Cannon, 1972	N	-
		Levitt et al., 1973	N	-
		Verleger and Cohen, 1978	N	-
		Pass et al., 1980		-
		Roth et al., 1980a	N	-
		Roth et al., 1980b	N	-
	Depression	Levitt et al., 1973	N	-
		Goodin et al., 1978b	N	
		Litzelman et al., 1980	N	-*
III.	Amplitude reduction, latency uncertain			
	Autism	Novick et al., 1979	?	-
		Novick et al., 1980	?	-
	Reading Disability	Schaub, 1981	?	-
		Dainer et al., 1981	+	-

*Scalp-distribution changes

Dalton, Skopitz, Scott, and Machinski, 1975; Ball, 1976, 1979a, b; Ball and Nuttal, 1980). Furthermore, the distribution of these abnormalities within the hippocampus and hippocampal gyrus are very similar in Down's Syndrome, Alzheimer's disease, and normal aging (Ball, 1979a, b; Ball and Nuttal, 1981). Temporal lobe atrophy (as seen in the CAT scan) is found in chronic alcoholism without Korsakoff's psychosis (Cala, Jones, Wiley, and Mastaglia, 1980). Information on the pathology resulting from head trauma is limited, but there are some indications that the temporal lobes are particularly vulnerable to damage in closed head injury (Hecaen and Albert, 1975).

Temporal lobe pathology has also been implicated in some of the groups that show severely diminished P3 amplitudes. Recent findings suggest that the medial temporal lobes are decreased in volume in children with early infantile autism (Hauser, DeLong, and Rosman, 1975), and parallels have been drawn between the neuropsychological deficits seen in the autistic with those of temporal lobe amnestics (Boucher and Warrington, 1976; DeLong, 1978). In our population of temporal lobe epileptics, there is an unusually high incidence of absent P3s (50%). Also, we have preliminary data on patient HM, collected in collaboration with Dr. Susan Corkin. HM underwent bilateral temporal lobectomy about 25 years ago and exhibits profound deficits in recent memory. He shows severe abnormalities in the endogenous ERPs, similar perhaps to those seen in the autistic subjects of Novick et al. (1979, 1980).[2]

We are suggesting, therefore, that prolonged P3 latencies or grossly abnormal P3 morphology may be indicative of medial temporal lobe pathology. Reduced amplitude in the presence of normal latency may indicate abnormalities in other brain areas, perhaps those subserving attentional or motivational functions rather than those related to orienting and memory. These speculations are admittedly premature in view of major gaps in the data base. However, a consideration of these gaps may ultimately lead to a clearer picture of the relationship of endogenous ERPs to neuropathology. In the first place, the temporal lobe neuropathological data relate only to subsets of each clinical population: e.g., dementia of the Alzheimer's variety, and Down's Syndrome mental retardation. There is a paucity of histological evidence on many of the other sub-groups that can be correlated with the electrophysiological data. More detailed correlations of the electrophysiological findings and the neuroanatomical evidence (e.g., CAT scans) in individual subjects would facilitate inferences about the anatomical substrates of the endogenous ERPs. Secondly, pathological changes in the brain are rarely limited to a single anatomical region in serious, mentally debilitating disorders; in chronic alcoholics, for example, abnormalities are found in the frontal lobes to even a greater extent than in the temporal lobes (Cala et al., 1980). Thus on the basis of electrophysiological/neuroanatomical comparisons in a single group it would be difficult to attribute ERP abnormalities to pathology of a particular brain region. However, comparison of ERPs across groups with different patterns of neuropathology may clarify the picture. Thirdly, the neuropsychological deficits of patients in all of these groups are variable; for example, chronic alcoholics may be divided into those with and without mnestic deficits (Skerchock and Cohen, 1981). Such heterogeneity indicates the importance of relating ERP measures to specific neuropsychological patterns if a coherent picture is to emerge. In the Skerchock and Cohen study it was found that chronic alcoholics with mnestic deficits exhibited prolonged P3 latencies while those without mnestic deficits did not. Finally, although most studies on endogenous ERPs in clinical groups now distinguish at least the N2 and P3 components and report measures of both amplitude and latency of each component, few studies demonstrate cognizance of the ever-growing list of endogenous ERPs and the ways in which they may overlap and interact. (For an exception see Pfefferbaum et al., 1980b, who discuss the possibility of slow-wave decrements in aging, and the way in which it may confound P3 amplitude measures). If these components prove to be psychologically distinct, increased sophistication in the choice of procedures for distinguishing among them may well be essential in making precise correlations between electrophysiology, neuropathology, and neuropsychological deficits.

In summary, if clinical findings are to prove useful in identifying the sources of the endogenous ERPs, greater attention needs to be given to the neuropathological and neuropsychological findings in individual subjects, and their relationship to more carefully defined patterns of ERP abnormalities. Inevitably, the analysis presented in Table 1 will be further refined. Nevertheless, we believe that the existing data provide support for the hypothesis that the limbic system plays a role in the generation of the endogenous ERPs.

F. Sources of ERPs - some procedural considerations

Even if all these problems in subject grouping, choice of experimental procedures, ERP measurement, and identification of pathology were to be solved, it must be emphasized that clinical evidence for the neural sources of ERP components will never be totally sufficient. To conclusively identify the generators of specific ERP components we also need direct recordings from a variety of brain areas during tasks that produce the endogenous waves, either in human or in animal subjects. The following points may be self-evident, but their enumeration will help to put the data presented here in the appropriate perspective. Depth recordings of the endogenous ERPs need to address three questions: 1) in which areas of the brain is there locally-generated activity that conforms to our definition of "endogenous"? 2) in which brain areas that appear to be likely candidates for generator sources is there <u>no</u> such activity? and 3) in the case of locally-generated endogenous activity, is this activity volume-conducted to the surface? By these criteria, the existing recordings from human subjects present a very incomplete picture. Our own data, restricted as they are to limbic structures, are informative only with respect to the first question. They do clearly demonstrate that the limbic system is active at the same time, and under the same circumstances as the surface endogenous ERPs. They do not bear on the question of other possible areas of activity, or on whether the limbic activity is volume conducted to the surface. Wood, Allison, Goff, Williamson, and Spencer (1980) recorded from multi-contact probes entering the brain at various cortical locations, and aimed at the temporal pole. This technique has the advantage over ours that the activity of extensive areas of the brain can be sampled, and to some extent the degree of volume conduction from active sites may be assessed. However, in their original recordings, although Wood et al. found the largest P3-like waves at subcortical sites, there was no evidence of local generation, and although similar waves could be recorded from depth to surface, latency changes in these waves rendered the data ambiguous as to whether the large subcortical waves were the same as those seen more superficially. In more recent recordings (Wood et al., in press) polarity reversals have been observed along the temporal probe which are consistent with a medial temporal lobe site of generation. The absence of polarity inversion trans-cortically suggests that these potentials are not cortically generated, but this is conclusive only for the specific areas through which the probes passed, and does not preclude the possibility of more distant cortical activation.

Three hypotheses have been presented in the literature as to the neural generator of the P3 component: hippocampus (Begleiter and Porjesz, 1980), parietal cortex (Vaughan and Ritter, 1970; Simson, Vaughan, and Ritter, 1976), and frontal-cortex/mesencephalic reticular formation (Desmedt and Debecker, 1979). The deficiencies in the existing data from depth recordings in humans are highlighted by the fact that they do not allow us to conclusively decide between these three very different hypotheses. However, both our own data and those of Wood et al. are probably most consistent with a limbic source for at least part of the endogenous activity. Final resolution of this issue awaits recordings from a wider variety of brain structures under the appropriate psychological circumstances, and may depend as well on greater attention to the precise definition of components in psychological terms, and to their more refined operational definitions.

Notes

This research was supported by National Science Foundation grant BNS 77-17070 and by Public Health Service grant NS02808. We wish to thank John Rohrbaugh for his helpful comments on an earlier draft of this chapter.

[1]The data of the earlier report suggested that in addition to endogenous activity that differentiated the ELPs to rare and frequent stimuli, there was activity common to the rare and frequent ELPs but which varied with attention. (See, for example, the waveforms labeled "LHC" in Figure 2 of that report). This activity might also be considered endogenous. However, in the current report we have focused on the aspects of the waveform related to stimulus probability. The use of rare-minus-frequent subtractions in isolating probability-related endogenous components has proven useful in previous studies on the scalp topography of the endogenous components (e.g., Simson et al., 1977).

[2]In all these cases where absent or grossly abnormal P3s were found (autism, temporal-lobe epilepsy, and temporal lobectomy) the subjects were able to perform the auditory-oddball task. Thus the generation of a P3 does not appear to be necessary for adequate task performance. This suggests that the P3 process reflects additional transformations on the information (e.g., orienting or registration in memory) that are not essential to the task itself.

Tutorials in ERP Research: Endogenous Components
A.W.K. Gaillard and W. Ritter (eds.)

11

POSITIVE SLOW WAVE AND P300: ASSOCIATION AND DISASSOCIATION

Daniel S. Ruchkin
Department of Physiology
University of Maryland, School of Medicine
Baltimore, Maryland 21201
U.S.A.

Samuel Sutton
Department of Psychophysiology
New York State Psychiatric Institute
722 West 168 Street
New York, N.Y. 10032
U.S.A.

INTRODUCTION

Slow Wave and P3b are members of a group of long latency, positive polarity, endogenous event-related potential (ERP) components now known as the late positive complex (LPC). In the earliest experiments in which Slow Wave was found, it appeared that Slow Wave related to experimental variables in much the same manner as did P3b. As a result, during the early period there was little focus on Slow Wave. However, in recent years evidence for a behavioral dissociation between Slow Wave and P3b has been accumulating. The purpose of this chapter is, in the light of current knowledge, to review and attempt to evaluate the similarities and differences between Slow Wave and P3b.

The initial reports of LPC activity described a single prominent component with a peak latency of about 300 msec (Sutton et al., 1965,1967). It was referred to as the late positive component or P3 or P300. Subsequent experiments established that P300 amplitude is generally largest over parietal scalp. P300 potentials have generally been elicited by events that are made relevant by serving one or more of the following purposes: (1) to provide a subject with feedback information concerning the outcome of a prior task; (2) to be the object of a discrimination or counting task; (3) to be an imperative signal requiring performance of a motor response. The utility of information extracted from an event, rather than its physical properties, is a key determinant of P300 behavior. For example P300 can be elicited by events that confirm or disconfirm a subject's prediction. When the task is changed so that the subject is no longer concerned with these events, then P300 is much smaller or absent. The subject's expectancy concerning an event, which can be manipulated by changing event probabilities, is also a determinant of P300 behavior. The more unexpected (i.e. informative) the event, the larger P300 amplitude (Sutton et al, 1965; Tueting et al, 1970; Duncan-Johnson & Donchin, 1977). However, relationship to probability is maintained even when the identity of the stimulus is known in advance (Sutton, 1979), although under these conditions P300 is considerably reduced in amplitude.

Experiments have been conducted in which discrimination between events is made difficult, so that subjects can extract relatively little useful information due to their a posteriori uncertainty of whether a particular event occurred, even though its occurrence was inherently informative. The information loss due to difficulty in forming a perceptual judgment is referred to as equivocation (Shannon and Weaver, 1949; Ruchkin and Sutton, 1978b). The more difficult the judgment, the greater the equivocation and the smaller is P300 amplitude (K. Squires et al, 1975b; Johnson and Donchin, 1978; Ruchkin and Sutton, 1978b; Ruchkin et al, 1980a; Johnson, in press).

With the accumulation of P300 data, it became apparent that there were other overlapping, endogenous late positive components that could occur separately or in conjunction with P300. Although these components are dissociable from P300, in some respects they resemble P300. In most earlier studies distinctions were not made between P300 and other members of the LPC. Some of the key steps towards recognizing that there was a complex of dissociable positive components were provided in reports by N. Squires et al, Donchin and co-workers, Courchesne et al, Roth et al, Renault and Lesevre, and Friedman et al. N. Squires et al (1975) observed, in addition to P300 (which they designated as P3b): (1) a shorter latency (240 msec) component with a fronto-central distribution, designated as P3a; and (2) a long duration, slowly varying wave that overlapped with and followed P3a and P3b. The long duration wave, designated Slow Wave, was maximally positive over parietal scalp and negative over frontal scalp. Both P3a and Slow Wave resembled P3b to the extent that they were largest when elicited by low probability events.

P3a is one of a number of LPC components with latencies shorter than or in the range of P3b that have been observed either in conjunction with P3b or in conditions which fail to elicit P3b. In general when the eliciting stimuli were not task relevant, no P3b was present (e.g. a non-attend condition or novel stimuli), and the LPC components appeared to relate to aspects of "orienting" behavior. When the eliciting events were task relevant P3b was also present. Under these circumstances the multiple positivities have been distinguished from P3b and one another primarily on the basis of scalp topography and latency. The functional significance of these short latency LPC components is not understood at this time.

In the present review of Slow Wave, we have for the most part limited ourselves to studies in which some technique such as principal components-varimax analysis (PCVA) has been used to cope with the temporal overlap between Slow Wave and other components. We have reviewed only a few of the large number of studies in which what may well be Slow Wave may be seen by visual inspection, but no attempt was made to cope with its overlap with other components.

INITIAL SLOW WAVE FINDINGS

The N. Squires et al paper (1975) was the first published report of Slow Wave findings. These authors used tone pips with unpredictable intensity shifts or unpredictable pitch shifts in different experiments under both attend and ignore conditions. No differences were found between P3b and Slow Wave in relation to experimental variables. Both of these components were larger when the subject counted target stimuli than when the subject ignored the stimuli, and larger at lower stimulus probability than at higher stimulus probability. Slow Wave and P3b differed only in waveshape and scalp topography. P3b had a well defined peak in the 300-400 msec latency range, whereas Slow Wave was a relatively broad wave with a duration of about 700-1000 msec. On the basis of visual inspection, Squires et al estimated that Slow Wave onset latency was about 180 msec. Unlike P3b, which had large positive amplitudes at Pz and Cz (about 10 microvolts) and a small positive amplitude at Fz, Slow Wave amplitude was about 5 microvolts positive at Pz, near zero amplitude at Cz and negative at Fz (about -2 microvolts). Donchin et al (1975) were the first to obtain overlap-free estimates of Slow Wave and P3b waveshapes via the use of PCVA. P3b began at about 150 msec and had a well defined peak at about 300 msec while Slow Wave began at about 250 msec, peaked at about 670 msec and had a duration of over 800 msec. Data concerning Slow Wave topography and its variation with experimental conditions were not reported by Donchin et al (1975).

N. Squires et al's (1975) Slow Wave findings were essentially replicated in three subsequent experiments that also employed auditory stimuli in counting paradigms (McCarthy and Donchin, 1976; K. Squires et al, 1977; Duncan-Johnson and Donchin, 1977). In these studies PCVA estimates of Slow Wave showed onset latencies of about 100-200 msec, peak amplitudes in the 400-700 msec latency range and durations greater than 500 msec. McCarthy and Donchin (1976) compared several conditions involving subject triggering and computer triggering of stimuli. The findings for the computer triggered condition were comparable to those of the N. Squires et al (1975) experiment. In contrast with previous reports, when the subject triggered the stimuli (with no delay between button press and stimulus occurrence), Slow Wave was negligible for the low probability, counted stimuli, and for the high probability stimuli it was positive at all electrodes, being largest at Fz. The data for the subject-triggered condition are difficult to interpret, partly due to the possible confounding with motor response related activity.

The K. Squires et al (1977) design, which was very much like that of N. Squires et al (1975), involved either silent counting of the loud stimuli in a Bernoulli series of loud and soft stimuli or engaging in another task. Both P3b and Slow Wave were small in the ignore condition. In the attend condition, P3b was larger to rare, counted stimuli than to rare, uncounted stimuli while Slow Wave was large for both rare counted and uncounted stimuli. In a more extensive investigation of the effects of probability, Duncan-Johnson and Donchin (1977) used a counting task in which probability of the counted stimulus varied in 10% increments for different blocks from 10% to 90%. In this experiment, in the attend condition, both P3b and Slow Wave were large for all rare stimuli, whether counted or not.

The results of the above experiments make clear that Slow Wave is an endogenous component that is influenced by some of the same behavorial variables that influence P3b, differing mostly in scalp topography and timing. These results led Donchin et al (1978, pg 356) to take the position in their review of endogenous ERPs that "...as Slow Wave is so closely associated with P300, it will not be further discussed..."

DISSOCIATION OF SLOW WAVE AND P300

The first inkling that Slow Wave and P3b may behave differently in relation to experimental condition may be seen in the plots of "repetition free" averages for the two components in Figs. 5 and 6 of Duncan-Johnson and Donchin (1977). The PCVA estimates of amplitude for Slow Wave at posterior electrodes increase at high probabilities as well as at low probabilities (Fig. 5, center panel). In contrast, at the same electrodes, P3b appears to decrease monotonically as probability increases (Fig. 6, center panel). However, the difference was not commented on by the authors and in fact may not be statistically significant.

Roth et al (1978) were the first to explicitly report evidence of dissociation between P3b and Slow Wave. They used auditory stimuli in a go/no-go reaction time (RT) paradigm. Slow Wave and P3b were affected in the same way by probability of stimulus occurrence (increased amplitude for decreased probability) but had an opposite relationship as a function of RT. Larger Slow Waves and smaller P300s were associated with longer RTs, and the reverse for shorter RTs. Baseline-to-peak amplitude measures were used. It should be noted however, that because of the opposite relationship between P3b and Slow Wave, this finding would hold even if there is confounding due to overlap of P3b and Slow Wave, since such overlap would tend to diminish the observed opposite patterns in relation to RT. This finding has been replicated (Friedman, 1981) in an experiment with a similar design in which PCVA measures of P3b and Slow Wave were obtained. Roth et al presented a number of arguments that their results were not due to movement potentials or movement related artifacts. The most compelling with respect to Slow Wave were that its anterior-posterior scalp distribution differed from that of the positive component of the movement potential and that Slow Wave was also observed in no-go trials (although its amplitude was lower). With respect to P3b, they noted that the response synchronized average P3bs had neither constant amplitude nor constant latency with respect to time of movement as a function of RT quartile, properties that movement potentials would be expected to have. A possible artifactual contribution to their findings not dealt with by Roth et al is that the lower amplitude of the average P3b for long RTs (from trials in the RT fourth quartile) in comparison with the short RT average P3bs (from trials in the RT first quartile) could, at least in part, be due to greater latency variability of P3bs in the RT fourth quartile. The range of RTs is largest in this quartile. Conceivably the latency variability of P3b, with respect to stimulation and response times, may also have been greater for fourth quartile trials.[1]

Ruchkin et al (1980a,b) reported further instances of behavorial dissociation and opposite relationships for Slow Wave and P3b. In order to reduce possible confounding of amplitude measures due to component overlap, PCVA derived estimates of Slow Wave and P3b components were utilized. Neither study involved contamination by motor factors.

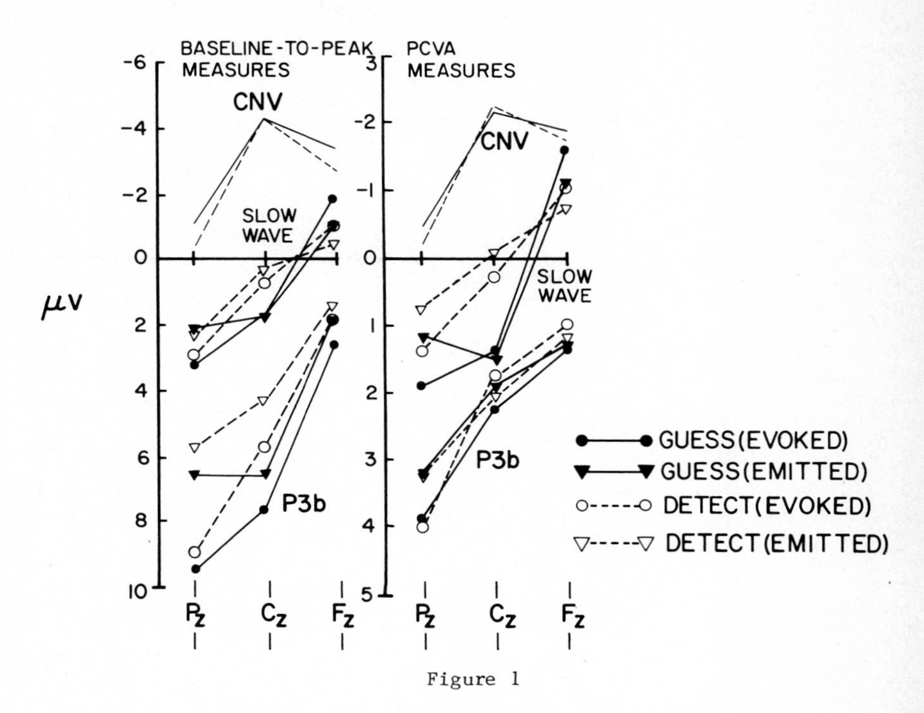

Figure 1

Left panel: Across subjects averaged baseline-to-peak measures of evoked (second click present) and emitted (click absent) P3b and Slow Wave amplitudes plotted against electrode site for Detect and Guess-Detect Tasks. The CNV baseline-to-peak amplitude measure was based upon the epoch preceding the time of the second click and thus is pooled across click present and absent conditions. Right panel: Across subjects averaged amplitude measures of CNV, P3b and Slow Wave obtained from a cross-products PCVA. The PCVA ordinate is in arbitrary units. (Adapted from Ruchkin et al, 1980b).

In the first study (Ruchkin et al, 1980b) a comparison was made between ERPs observed in two tasks. The stimulus conditions were identical in the two tasks, but the instructions to the subjects were different. The two tasks were (1) report at the end of a trial whether a second click following a warning click was present or absent (Detect task) and (2) predict prior to a trial whether the second click will be presented or omitted (Guess-Detect task). It should be noted that in order for subjects to discover whether their predictions were correct or incorrect in task 2, they also had to detect the presence or absence of the second click. The second click occurred on 50% of the trials and its intensity was adjusted so that it was detectable at approximately 80% accuracy for all subjects. For both tasks feedback was not given until the end of the session. Note that for detection alone the task is complete when the stimulus event has been identified. In the guessing task processing is more complex since the detected event must be compared to the prior prediction in order to determine whether the prediction was correct. In the PCVA analyses there were no differences in topography or amplitude of P3b due to tasks (Fig. 1, right panel). However, there was a significant task effect upon Slow Wave topographic profiles. In the detection task Slow Wave topography was essentially the same as in the counting and RT experiments cited above, positive over posterior scalp, negative over frontal scalp and near zero at Cz. In contrast, in the guess task Slow Wave was relatively large and positive at Cz, about equal in amplitude to the amplitude at Pz.

In the second study (Ruchkin et al, 1980a) Slow Wave and P3b were compared as a function of detectability in a signal detection experiment. Three levels of stimulus detectability were used, with corresponding average accuracy levels of 63%, 87% and 99%. Slow Wave and P3b varied in an opposite manner with detection difficulty (see Fig. 2.); Slow Wave amplitude increased and P3b amplitude decreased as difficulty increased. It is of particular interest that these relations held for correct rejection trials as well as for hit trials. In other words, absence of the physical stimuli at the three different accuracy levels also gave an inverse relationship for emitted P3b and emitted Slow Wave. Thus the findings obtained for P3b and Slow Wave are related to detectability rather than intensity.

Kerkhof (in press) observed relationships similar to those described above between P3b, Slow Wave and confidence ratings in a signal detection experiment at a fixed intensity level. The more confident the subject, the larger (and earlier) was P3b and the smaller was Slow Wave.

The P3b data in the above studies are consistent with the notion that P3b represents completion of an evaluation and categorization of an event, with P3b amplitude partly determined by the amount of information extracted from the event percept (the more unexpected and less equivocal, the greater the information and larger the amplitude). The Roth et al (1978), Ruchkin et al (1980a,b) and Kerkhof (in press) studies also indicate with respect to P3b that this evaluation may well be a relatively early stage of processing. Additional processing may yet take place in relation to the event under consideration, and Slow Wave may be a sign of such further processing activity. Thus the difference in Slow Wave scalp topography between detect and guess tasks may be due to the additional comparison operation that occurs in the guess task, and increased Slow Wave amplitude in the detection experiments may reflect greater mobilization of effort when stimulus conditions are such as to make a decision difficult.

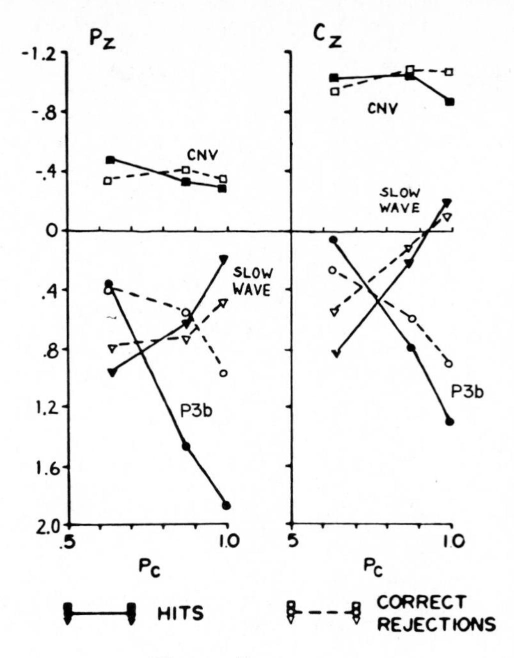

Figure 2

Across subjects averaged amplitude measures of CNV, P3b and Slow Wave at Cz and Pz, obtained from a cross-products PCVA, plotted against percentage accuracy (Pc) of detection. Accuracy was manipulated by varying signal intensity (Adapted from Ruchkin et al, 1980a).

A NOTE ON DATA ANALYSIS

In view of the temporal and spatial overlap of P3b and Slow Wave components, care must be used in the formulation and interpretation of amplitude measures. In some instances, when the recording epoch is sufficiently long, it may be possible to obtain baseline-to-peak amplitude measures of Slow Wave at long latencies that have little or no P3b contamination. Presumably, the Roth et al study succeeded in obtaining dissociation between P3b and Slow Wave for this reason. However, it is often (although not always) the case that P3b coincides with the initial portion of Slow Wave, so that there is no time during the P3b epoch when Slow Wave is minimal. In such a case, one still may be able to utilize topographic separation (e.g. if Slow Wave is minimal at Cz) to obtain an uncontaminated baseline-to-peak measure of P3b amplitude. However, the Ruchkin et al (1980b) data indicate that this maneuver is not always possible. In that study Slow Wave was large and positive at Cz in the guessing condition. As a consequence, it was not possible to tell from the baseline-to-peak amplitude analysis whether the ERP variations with task were Slow Wave or Slow Wave plus P3b phenomena. The baseline-to-peak amplitude analysis in the variable difficulty signal detection experiment (Ruchkin et al, 1980a) was also confounded by overlap. In that experiment increasing detection difficulty decreased the amplitude of P3b and increased the amplitude of Slow Wave. Therefore in baseline-to-peak measures P3b variation in relation to signal detectability appeared to be reduced. In both these studies a procedure such as PCVA was required to delineate the different relationships of P3b and Slow Wave to experimental variables. Discussion of a rationale for using PCVA to deal with P3b and Slow Wave overlap is presented in Ruchkin et al (1980b). More detailed discussions of the strengths and weaknesses of PCVA analysis of ERP data are provided by Glaser and Ruchkin (1976), John et al (1978), Donchin and Heffley (1978) and in a forthcoming review by Wood and Ruchkin.

RESERVATIONS

The Ruchkin et al (1980a,b) data provide what appears to be evidence for a late, positive Slow Wave that is distinct from P3b and is a sign of a later (or longer duration) stage of event evaluation. However, the experimental designs were such that other interpretations cannot be ruled out. In both experiments each condition (task or signal intensity level) was fixed over blocks of trials, so that subjects knew in advance what the conditions in a trial would be and thus could have prepared differently in the different conditions for a trial. Conceivably the Slow Wave components relate in part to termination of such prior preparation states and differences associated with differences in task type and task difficulty reflect effects of differential preparation. Note, however, that the effect of probability upon Slow Wave observed in the initial investigations cannot be readily attributed to differential preparation because subjects did not know in advance which stimulus would be presented.

Kok and Looren de Jong (1980a) have obtained Slow Wave data that overcome some of these difficulties. Kok and de Jong varied difficulty from trial to trial on a random basis. Thus there could be no systematic differential preparation effects. Visual stimuli (letters) were used in a delayed response discrimination paradigm. In 50% of the trials the letters were degraded and thus difficult to recognize. The ERP data were analyzed via PCVA. Even with differential preparation ruled out, P3b amplitude was smaller and Slow Wave amplitude was larger for the difficult to recognize stimuli.

P4, SLOW WAVE AND TASK DEMANDS

The above studies suggest that, in addition to timing and scalp topography, Slow Wave and P3b differ in their relationship to task processing requirements. There have been reports of other long latency, posteriorly distributed positive waves, sometimes designated as P4, that also appear to differ from P3b along the dimension of task processing requirements. Whether or not these P4 components correspond to Slow

Wave is an open question. However, at the least, P4 and Slow Wave appear to be members of the same sub-species of the LPC in that, on the basis of currently available data, they behave similarly in relation to experimental variables.

Kok and Looren de Jong (1980b) compared the effect of stimulus familiarity upon different members of the LPC. Subjects were instructed to memorize three visual patterns (polygons). The three patterns were presented in a randomly sequenced block of 50 presentations. The subjects were previously exposed to two of the patterns, so that these patterns were familiar. One of the familiar patterns was presented 5 times (infrequent-familiar) and the other was presented 40 times. The third pattern was new to the subject and was presented five times (infrequent-unfamiliar). PCVA measures of LPC components were obtained for the two kinds of infrequent stimuli. Stimulus familiarity had little effect upon P3b amplitude (peak: 384 msec) but P4 (maximum amplitude at Pz; onset: 250 msec, peak: 600 msec, duration > 750 msec) was clearly larger for the unfamiliar stimulus. Amplitudes of both components decreased with repeated presentations. A smaller, longer latency (onset: 500 msec, duration > 500 msec) slow positive component, largest over frontal scalp, was elicited by the first presentation of the unfamiliar stimulus and then decremented abruptly with subsequent repetitions (the authors designated this component as "Slow Wave"). There was no PCVA analysis of the ERPs elicited by the frequent stimulus. However, inspection of across subject average waveforms indicate that for the frequent familiar stimulus P3b was relatively small but still apparent while P4 and "Slow Wave" appeared to be negligible.

Stuss and Picton (1978) and Stuss et al (1980) obtained LPC data in experiments in which subjects learned by trial and error the rule for classifying visual or auditory patterns. The rule that specified the appropriate response to a given stimulus occasionally changed over sequences of trials. In effect the subject's task was to detect these transitions in rules. At the end of each trial an auditory feedback signal indicated whether or not the subject was correct. In baseline-to-peak measures, the feedback signal elicited a P3b component followed by a long latency (peak: 590 to 647 msec), broad duration P4 component, maximally positive over centro-posterior scalp. Both P3b and P4 were largest in those trials between the trial on which the rule transition occurred and the trial just prior to the subjects' discovery of the new rule. In trials in which a known rule was confirmed P3b was still apparent, although with a diminished amplitude, while P4 was negligible. This difference between P3b and P4 was not commented on by the authors but can be seen in their figures (e.g. Fig. 3 of Stuss et al, 1980).

Johnson (1979) had subjects count one of two tones presented in a random sequence. The probability of the counted tone shifted occasionally between .33 and .67. Subjects were instructed to report transitions in probability. Starting with the 4th trial prior to a report, a long duration Slow Wave appeared in parietal recordings and increased in amplitude over subsequent trials, reaching its maximum one trial prior to which the report was made. Slow Wave was negligible in other trials. In contrast with Slow Wave, a P3b was present in all trials. However, the P3bs were generally largest in trials leading up to and including detection of transitons.

Still another partial dissociation between P3b and Slow Wave is reported by Parasuraman et al (in press). Subjects had two tasks in each trial: 1) to detect whether or not a low intensity signal was presented (detection); and 2) to report which of two different pitch signals had been presented (recognition). They found that P3b amplitude was related both to correctness of detection and correctness of recognition, while Slow Wave amplitude was related only to correctness of recognition. The amplitude of the earlier N100 was related only to confidence rating.

A common element in the Kok and de Jong (1980b), Stuss and co-workers (1978, 1980) and Johnson (1979) experiments is that a P4 or Slow Wave component becomes apparent when the task appears to require more processing. The variation of P4 (or Slow Wave) with variation in processing requirements appears to be more marked than for P3b, i.e. even when there is an association between P3b and P4 (qualitative similarity), there is still a quantitative difference. Unfortunately in none of the studies was there a statistical assessment of this apparent difference between P3b and P4 (Slow Wave) behaviors, so that the significance of this difference is not known. The nature of the putative additional processing is not clear at this time. Conceivably it relates primarily to either or both quantitative and qualitative changes in uncertainty associated with the transition in rules or degree of familiarity of stimuli. Other possibilities are that it may relate to increased memory activity, either in utilizing past experiences to deal with transitions in rules or, in the case of the Kok and de Jong (1980b) experiment, in memorization of an unfamiliar stimulus.

SLOW WAVE AND TASK DEMANDS - II

The construct of increased Slow Wave amplitude with increased task demands may provide an explanation for what appeared to be an unusual finding reported by Jenness in 1972. In that experiment subjects performed a very difficult pitch discrimination, indicating their decision with a button press. Following the button press there was a feedback interval. If the subject was correct, a stimulus with the same pitch as in the discrimination phase of the trial was presented. If the subject was wrong, no stimulus was presented. Thus presence of the feedback stimulus served two roles: (1) it told subjects that they were correct, and (2) it provided further rehearsal in that the high or low pitch stimulus coincided with the identification they had just made. Following the feedback click there was a very extensive sustained positive wave at Cz (the only recording site) which looks larger and longer than the kind of activity one obtains in relation to motor movement per se. A tantalizing aspect of the Jenness data is that the Slow Wave was larger for the feedback click than for the task click. Since the recordings are from Cz, this could be consistent with the Guess-Detect topographic findings of Ruchkin et al (1980b).

Friedman et al (1981) have obtained data from adolescents that provide a further example of P3b/Slow Wave dissociation and limited support for the construct of increased Slow Wave amplitude with increased task demands. A continuous performance task that employed visual stimuli was used with two levels of complexity. Complexity was manipulated by specification of the target stimulus, as follows: Task A (simple), the target was the numerals 08, and non-targets were other numerals in the range of 02 to 19; Task B (complex), the target was the repetition of any immediately preceding numeral, and non-targets were non-repeating numerals. Subjects responded with a finger lift to targets and withheld response to non-targets. P3b (designated P450 by Friedman et al) was larger for targets than non-targets in both tasks and was larger in task A (simpler) than in task B (more complex). Slow Wave on the other hand was larger in Task B (i.e. where task demands were greater). Within each task non-targets elicited larger Slow Waves than targets. This latter finding does not appear to be entirely consistent with the task demand construct. Although the memorization requirements associated with non-targets in task B may well be a relatively demanding aspect of the subjects' task, the non-targets in task A do not appear to be more demanding than the targets. Given our current understanding of Slow Waves, we are not able to offer an explanation for the finding of a larger Slow Wave for non-targets.

SLOW WAVE LATENCY

Slow Wave onset latency has not been systematically reported. Available data come primarily from inspection of published PCVA derived estimates of Slow Wave component waveforms. Such data are available in most of the counting and detection experiments cited above. Generally onset latency is in the range of about 90-300 msec for auditory stimuli and 200-350 msec for visual stimuli, but longer latencies have been observed. In the Friedman et al (1981) study onset latency to visual stimuli was about 450 msec.

A Slow Wave with a very long onset latency occurred in a match/mismatch experiment conducted by Sanquist et al (1980). The subjects' task was to judge whether two words matched on one of the following criteria: (1) orthographic, (2) phonemic or (3) semantic. The criterion was fixed within a block of trials. In a trial the two words were presented in succession. Each word was displayed for 100 msec. The interval between words was 2 sec. Estimates of P3b and Slow Wave components were obtained via PCVA. P3b and Slow Wave components were elicited by both the first and second word. The onset latency of Slow Wave to the second word was about 800 msec. The Slow Waves elicited by the second word did not vary with judgment outcome but did vary with criterion. When the criterion was semantic or phonemic Slow Wave was significantly more positive than when the criterion was orthographic. Upon completion of the judgment trials, the subject was given a memory recognition test for items presented in the second word position. After sorting the second word ERPs into "subsequently recognized" and "subsequently unrecognized" categories, it was found that both P3b and Slow Wave were larger when elicited by items that were subsequently recognized. However, these data were sparse, and so there was no statistical analysis of this effect.

Comparison of onset latencies from Sanquist et al (800 msec) and the detection, discrimination experiments cited above (90-350 msec) suggests that Slow Wave onset may relate to the processing stage at which additional effort is required. Presumably the perceptual processing in the detection experiments occurs earlier than the conceptual processing involved in phonemic or semantic judgments. It is interesting in this context that within the perceptual stage no onset latency differences were found as a function of detection difficulty (Ruchkin et al, 1980a, unpublished results). These findings suggest that it is the stage of processing and not the level of difficulty which determines onset latency of Slow Wave. However, Slow Wave onset latency data are still sparse and so this formulation is tentative.

SLOW WAVE DURATION

There is little available information concerning Slow Wave duration. In many experiments the analysis epoch is not long enough to observe Slow Wave termination. On the basis of visual inspection of waveforms presented in figures from the studies reviewed above, Slow Wave duration generally appears to be greater than 600 msec and in some instances exceeds 1400 msec. There can be considerable variation in the time course of Slow Wave. In some cases it has a phasic appearance, with durations in the range of 700-1200 msec. In other cases it appears to be tonic, with duration exceeding 1400 msec. Examples of tonic and phasic Slow Waves are provided in Fig. 3a, right panel. In our experiments tonic Slow Waves were elicited by the final stimulus in a trial, where inter-trial intervals were long (6-20 sec) and trials were initiated by the subject. Phasic Slow Waves were elicited by earlier stimuli in the trial or by events that required a prompt response. We do not know of any instance where the same stimulus elicited both a phasic and a tonic Slow Wave. It is not clear at this time whether these waves are a single component of varying duration, or whether we are dealing with two or more components.

That the phasic and tonic Slow Waves behave similarly in relation to experimental variables and differently from the behavior of P3b in relation to the same variables is illustrated in Fig. 3. These data come from a study concerned with ERPs in an experimental paradigm in which information is transmitted to a subject by a combination of two successive events (a 'message'). Each event delivered essential information for understanding the message, but the subject could not determine the meaning of the message until occurrence of the second event (closure). The amount of information in each event was varied by varying its perceptual difficulty. Increasing perceptual difficulty causes an information loss, termed equivocation.

The data in Fig. 3 consist of PCVA basis waveforms (Fig. 3a) and weighting coefficients (Fig. 3b). Slow Wave (right panels) is negligible for both EVENT 1 and EVENT 2 when there is no equivocation (easy discrimination). When equivocation is present (difficult discrimination), Slow Wave (maximal at Pz) is large for both EVENT 1 (phasic Slow Wave) and EVENT 2 (tonic Slow Wave) with no significant difference at Pz between the peak amplitudes of the phasic and tonic Slow Waves. In contrast P3b (left panels) is absent in EVENT 1 regardless of level of equivocation but clearly present at both levels of equivocation in EVENT 2. Here P3b is larger for the no equivocation condition. In summary, these data indicate that the effect of closure upon P3b is to elicit a relatively large P3b when there is closure

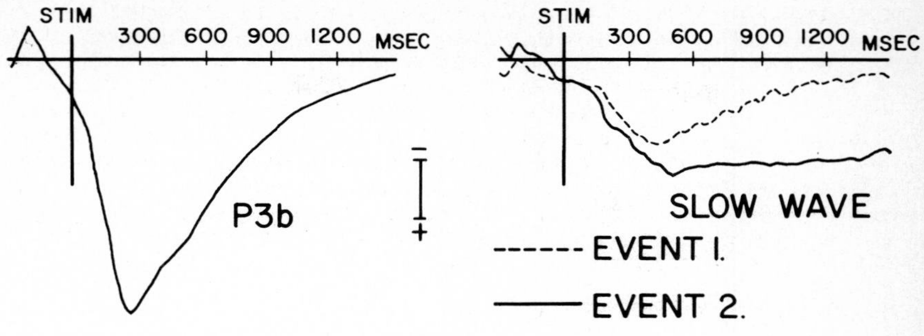

Figure 3a

Cross-products PCVA derived estimates of P3b and Slow Wave component waveshapes in an experiment in which a trial consisted of a sequence of two events with an inter-event interval of approximately 5.8 sec. The first of the two events (EVENT 1) elicited phasic Slow Wave activity but negligible P3b activity. The second event (EVENT 2) elicited both a P3b and tonic Slow Wave activity.

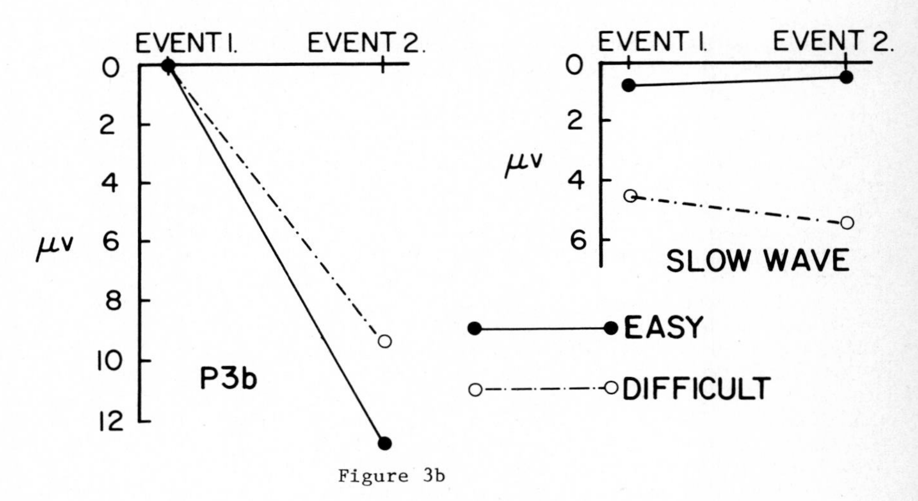

Figure 3b

Cross-products PCVA derived measures of P3b and Slow Wave peak amplitudes at P_z plotted against event for easy discrimination (low equivocation) and difficult discrimination (high equivocation) conditions.

(EVENT 2) and negligible P3b when there is no closure (EVENT 1). In contrast, there is no effect of closure upon the peak amplitude of Slow Wave. However, in these data a tonic Slow Wave was elicited when there was closure and a phasic Slow Wave when there was no closure.

These data replicate the findings that Slow Wave amplitude increases with increased task difficulty while P3b amplitude decreases with increased difficulty. The data also provide an instance where Slow Wave is obtained but P3b is not elicited. This latter finding provides evidence that the processing underlying Slow Wave is not necessarily dependent upon the processing underlying P3b.

SLOW WAVE, CNV AND TASK DEMANDS

There have been a number of reports of reduced contingent negative variation (CNV) (less negativity) when the subject's task in some way becomes more complex. Introduction of an ancillary, distracting task during the warning (S1) - imperative (S2) stimulus interval (Tecce and Cole, 1976) or increasing memory load (Roth et al, 1975; Ford et al, 1979) reduces CNV magnitude, and CNV shifts towards positive amplitude when warning stimulus intensity decreases (Loveless and Sanford, 1974b; Ruchkin et al, 1981). Increasing the difficulty of the task associated with the S2 stimulus also reduces CNV (Delse et al, 1972; Marsh and Thompson, 1973; Gaillard and Perdok, 1979,1980). It has been argued that task difficulty causes internal distractions away from the contingency that exists between warning and imperative stimuli, thereby resulting in reduced CNV magnitude (Ford et al, 1979). We have observed (Ruchkin et al, 1981) that when S1 intensity was decreased, so that it was not easily discerned, what appeared to be a positive Slow Wave rather than a CNV was present in the S1-S2 interval, as illustrated in Fig. 4. This finding raises the possibility that the slow ERP activity following S1 may sometimes consist of a mix of CNV and Slow Wave, and that increased Slow Wave may have contributed to some of the reported reductions of CNV magnitude. Decreased motor preparation in those CNV paradigms involving a motor response may also contribute to the observed reductions in CNV magnitude (Gaillard, 1978).

It has been argued that the CNV can be divided into two components: an "early CNV" and a "late CNV" (Loveless and Sanford, 1974; Rohrbaugh et al, 1976). Gaillard (1978) and McCarthy and Donchin (1978) have commented upon similarities between the early CNV component and Slow Wave, which can also be negative over frontal scalp. Similarly, Kok (1978) has suggested that the early CNV and the frontal Slow Wave are the same phenomenon. However, Ritter et al (1980) have found that the scalp topographies of the early CNV and Slow Wave are different, and, in contrast with the early CNV, Slow Wave topography is similar for visual and auditory stimuli.

Our use of the terms CNV, early CNV and late CNV is not meant to imply that a position is being taken in this chapter with respect to the complexities of slow negativities (Gaillard, 1978; Rohrbaugh et al, 1978). Rather, we are suggesting that positive Slow Waves may also occur in the time domain in which slow negativities are found.

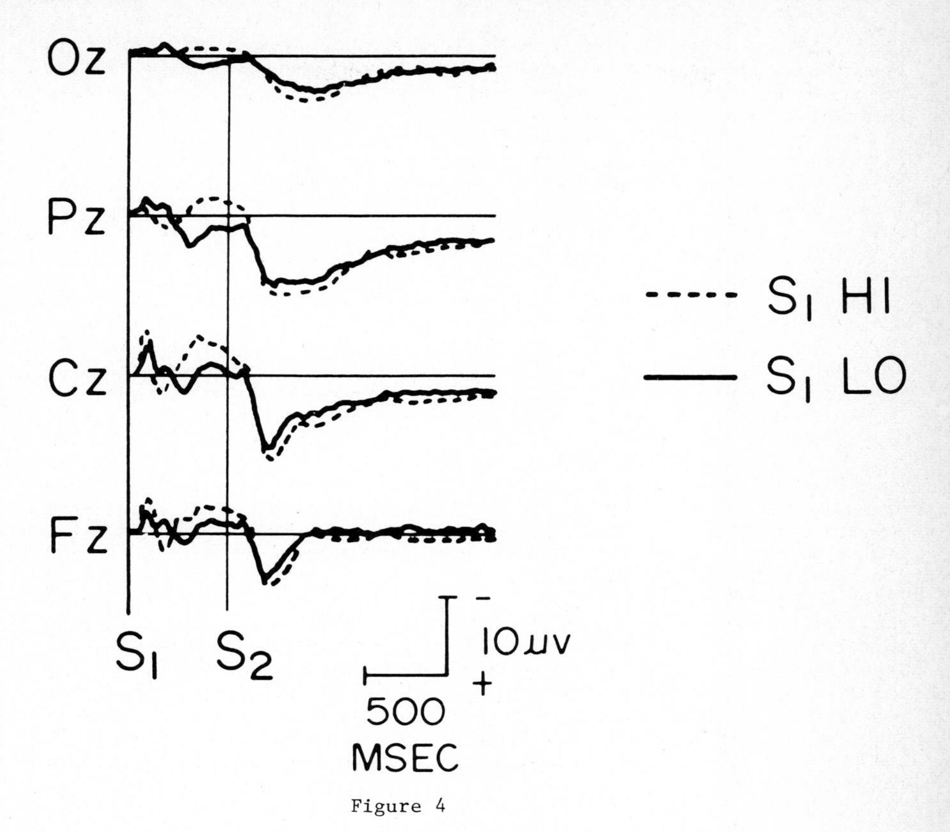

Figure 4

The effect of S1 intensity upon slow potential(s) in the S1-S2 interval. Averages are across subjects and conditions other than S1 intensity. S1 HI is a .1 msec click with a 50 dB SL intensity and S1 LO is a 22 dB SL click. The S1-S2 interval is 600 msec. The S2 stimulus conveyed task relevant information to the subject. There was no motor response following S2. (From Ruchkin et al, 1981).

There are few data available concerning possible relationships of pre-stimulus CNV with post-stimulus Slow Wave. Ruchkin et al (1980a,b) have observed dissociation between CNV and Slow Wave, the latter varying with experimental conditions while the former remained constant. In Ruchkin et al (1980b) Slow Wave amplitude at Cz varied as a function of the task requirements (detection vs. prediction) while in Ruchkin et al (1980a) Slow Wave amplitude at Pz and Cz increased as a detection task became more difficult.

SUMMARY AND GENERAL IMPLICATIONS

Even a casual inspection of the current and earlier literature indicates that Slow Wave appears in many studies in which its presence was neither analyzed nor even noted. Inspection of Fig. 1 in Sutton et al (1965) indicates that there were clear Slow Wave contributions in the very first report of P300. It should be remembered that the earlier studies used amplifiers with relatively short time constants (e.g. .3 Hz high pass in Sutton et al, 1965). Thus particularly for the earlier studies, visual inspection would tend to underestimate the relative contribution of Slow Wave. The presence of Slow Wave could have profound implications for our attempts to generalize about P3b (These implications have been developed at greater length in a forthcoming paper; Sutton & Ruchkin, in press). Here we simply note that the recent data show cases where Slow Wave and not P3b is related to the experimental variables, although the baseline-to-peak measures indicated that both were related to experimental variables (Ruchkin et al, 1980b). In still other cases, P3b and Slow Wave are related in an opposite manner to the same experimental variables (Ruchkin et al, 1980a). In a recent report by Daruna (1981), the difference between two populations is obscured in the baseline-to-peak measures because Slow Wave varies in an opposite direction for the two populations. In summary therefore, the recent findings for Slow Wave do not only add new information concerning this ERP component, but also have some potential for clarifying earlier findings with regard to P3b.

The degree to which the confound has been serious or relatively minor depends on three factors: 1) the relative amplitudes of the various components; 2) the relative onset latencies and latencies of the peaks of the various components, the confound being potentially greater the more they overlap in time; and 3) whether or not the components relate in the same or opposite manner to the same experimental variables. Thus, although the relationship to detection accuracy is opposite for P3b and Slow Wave, the baseline-to-peak findings are distorted but not invalidated by the overlap because of the smaller amplitude of Slow Wave in this paradigm. For cases where Slow Wave and P3b relate in the same manner to experimental variables, the confound is even less serious. For example, in a recent study by Steinhauer (1981), both a Cz maximum P300 and Slow Wave relate in a similar manner to various manipulations of the role of the stimulus. However, even where the confound does not invalidate baseline-to-peak findings, it should be remembered that the quantitative relations are distorted in baseline-to-peak measures.

In the Friedman et al (1981) and Sanquist et al studies (1980) where Slow Wave onset latency (relative to P3b) is fairly late, the overlap results in less of a confound.

There are several possible permutations in the relationship between P3b and Slow Wave to experimental variables:

1. Neither P3b nor Slow wave is related to a particular experimental variable.

2. P3b, but not Slow Wave, is related to an experimental variable.

3. Slow Wave, but not P3b, is related to an experimental variable.

4. P3b and Slow Wave are related in the same manner to an experimental variable.

5. P3b and Slow Wave are related in an opposite manner to an experimental variable.

To the degree that all of these permutations are found in different studies, it provides clear evidence that P3b and Slow Wave are independent phenomena. On the whole, it appears that cases which fit each of the five permutations have been found. However, some caution must be exercised in accepting this conclusion because it is not yet clear how many Slow Waves we may be dealing with. There appears to be considerable variation in Slow Wave onset latency, duration, and topography. It is not known at this time the extent to which Slow Wave may be affected by sensory modality, although few if any, differences have emerged between evoked and emitted Slow Wave (Ruchkin et al, 1980a). We have commented on the possibility that the more tonic and more phasic Slow Waves may be different components, or the same component with different durations. It has been suggested that the frontally negative portion and the parietally positive portion of Slow Wave may be different components (Picton & Stuss, 1980). There are certainly cases where most of the "action" in relation to an experimental variable is in the frontally negative portion (Fitzgerald & Picton, 1981; Friedman et al, 1981) while for the Slow Waves in our experiments most of the action was in the parietally positive portion. The fact that PCVA does not separate the frontal negative and parietal positive Slow Waves as different components cannot be taken as evidence that they are not two different components. The PCVA analysis could be hampered in separating them if they have an essentially identical time course. Still another topographic complication is illustrated in Ruchkin et al (1980b). In this study, the detect task has a parietally maximal Slow Wave while in the Guess-Detect task amplitude is relatively equal at central and parietal sites.

In summary, we now know that Slow Wave and P3b are functionally distinct. The major problems which remain are: (1) how many different slow waves there are, and (2) what are the unique functional correlates of each of them. Resolution of these problems may also contribute to a clarification of the functional role of P3b.

Notes

The preparation of this chapter was supported in part by a U.S.P.H.S. NINCDS grant, NS11199. We are indebted to Drs. Nancy Squires and Walter Ritter and Mr. Robert Munson for their comments on this chapter and to Ms. Tammy Patterson and Maria Tate for typing the manuscript. Address reprint requests to Dr. Daniel Ruchkin, Department of Physiology, School of Medicine, University of Maryland, Baltimore, Maryland, 21201, U.S.A.

[1]We are indebted to Dr. Walter Ritter for this observation.

Tutorials in ERP Research: Endogenous Components
A.W.K. Gaillard and W. Ritter (eds.)

12

CHRONOMETRIC ANALYSIS OF HUMAN INFORMATION PROCESSING

Gregory McCarthy
Neuropsychology Laboratory
Veterans Administration Medical Center
West Haven, Connecticut 06516
and
Depts. of Neurology and Psychology
Yale University
New Haven, Connecticut 06520

Emanuel Donchin
Cognitive Psychophysiology Laboratory
University of Illinois
Champaign, Illinois 61820

The covariation of P300 latency and reaction time was assessed in two sets of experiments designed to test the hypothesis that P300 latency reflects a subset of the information processes whose durations are reflected in RT. In the first experiments, P300 latency and RT were shown to covary as a function of the time required to categorize a stimulus. However, this relationship could be dissociated by requiring subjects to make highly speeded responses which also resulted in increased error rates. In the second experiments two factors, stimulus discriminability and stimulus-response compatibility, were shown to have additive and independent effects upon RT. The latency of P300 was strongly affected by stimulus discriminability but only minimally affected by S-R compatibility. These results are interpreted to support the hypothesis that P300 latency reflects primarily the durations of processes concerned with stimulus evaluation and is relatively unaffected by processes concerned with response selection and execution.

INTRODUCTION

Our topic is the joint analysis of reaction time (RT) and P300 latency in studies of human information processing. We will review two sets of experiments, both of which were performed at the Cognitive Psychophysiology Laboratory of the University of Illinois. The first set concerns the trial-by-trial covariation of P300 latency and reaction time in categorization tasks in which either the speed or accuracy of performance is emphasized. The second set concerns the relative changes in RT and P300 latency associated with manipulations of the durations of two stages of information processing. These experiments illustrate the manner in which variation in event-related potentials (ERPs) can be used to illuminate the operation of cognitive processes.

A few general points must be made prior to a discussion of the data. While most of contemporary Cognitive Psychology finds the analysis of information processing into component mental processes a profitable research strategy, there is no general agreement about the properties of such processes and how they interact. Does a given process provide continuous or discrete output? Can several processes be active simultaneously? Do processes compete for common resources? Different assumptions lead to different models with differing prescriptions for the analysis of reaction time.

Component mental processes are hypothetical entities which provide a valid functional description of the transformations of information in the nervous system. While these or analogous processes must take place, the manner in which they are realized by brain systems may be complex. Certainly we must not require that each process identified in our psychological models be assigned to a particular neuroanatomical structure or to a particular ERP component. It is premature to accept any particular functional model of information processing as a reality on which to impress our psychophysiological data. It is of interest, however, to examine the departures of psychophysiological data from the behavioral performance data on which a given psychological model is based.

A second point concerns measurement of the physiological signals which we record. It is quite common to parcel the ERP waveform into a small set of components, yet the criteria employed for identifying components vary greatly among investigators and are rarely made explicit. This is especially true for the endogenous potentials, where functional criteria often dominate the component's identification. It is difficult, in principle, to determine if a change in amplitude in a 'component' is simply that, or is due to the superimposition of other components. Similarly, an apparent change in the latency of a component may be due to the introduction of a new component or to a change in the relative amplitudes of overlapping components. When analyzing components strongly associated with functional processes, such as P300, we wish to interpret apparent changes in amplitude and latency of the physiological signal with changes in the strength or timing of the associated psychological process. It is clear, however, that alternative explanations for the ERP changes observed must be considered. With these caveats in mind, we will now consider data relevant to these issues.

THE CORRELATION BETWEEN P300 and RT

Ritter and Vaughan (1969) first observed that the late positive component (or P300) varied in latency depending upon the nature of the stimulus changes used to elicit it. These observations were followed by a systematic analysis of P300 latencies measured on single trial waveforms and RTs obtained concurrently in vigilance and discrimination tasks (Ritter, Simson, and Vaughan, 1972). P300 latency and RT were positively correlated across and within tasks.

Kutas, McCarthy, and Donchin (1977) examined the covariation single trial P300 latency and RT in categorization tasks. Three categories of stimuli were employed. In the 'fixed names' condition, subjects were presented on each trial with one of two names (DAVID or NANCY) with NANCY presented on 20% of the trials. In the 'variable names' condition, subjects were presented with different males' and females' names, but females' names were presented on only 20% of the trials. In the 'synonyms' condition, 20% of the words presented were synonyms of the word 'prod,' while the remaining words were unrelated. Three response conditions were used for each stimulus condition: subjects were required either to count the occurrences of the rare category stimuli, make a choice RT response with an emphasis on response accuracy, or make a choice RT response with an emphasis on speed. Large amplitude P300s were elicited by the rare category stimuli in all conditions. Estimates of P300 latency for the rare stimuli were obtained by a template fitting procedure (see Kutas et al., 1977, for procedural details).

The shortest RTs were obtained for the fixed names, while the longest RTs were obtained for the synonyms. Over all stimulus conditions, the RTs to the rare category stimuli were 97 msec longer than to the frequent category. The RTs obtained in the speed response conditions were about 90 msec shorter than the RTs in the accurate response conditions; however, incorrect responses to rare stimuli were made on 9% of the trials in the speed conditions as compared to 3% for the accurate conditions. The ordering of P300 latency was similar to that of RT. Figure 1 presents mean P300 latencies (as measured from the single trials) for the three stimulus and response conditions. As can be seen at the right of the figure, P300 latency was longer in the count conditions than it was in either of the RT conditions. P300 latencies were about 19 msec shorter in the speed than accurate RT conditions.

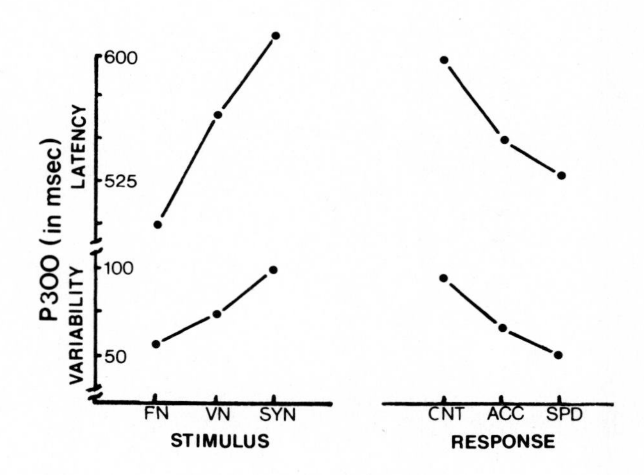

Figure 1
Mean latency and standard deviation of P300 latency (as determined from single trial measurement) for the three categorization tasks (fixed names - FN, variable names - VN, synonyms - SYN) and three response conditions (count - CNT, accurate-RT - ACC, speed-RT - SPD). (from Donchin and McCarthy, 1980).

The latencies of P300 were estimated for each single trial at the point of the best-fit between the template and record obtained on that trial. The correlation of RT and P300 latencies obtained in this manner for the accurate response condition (excluding error trials) was .66 while for the speed condition the correlation was .48. In the accurate condition, average P300 latency preceded RT by about 91 msec, while in the speed condition P300 latency was shorter than RT by only 26 msec. However, on trials in which errors were made (i.e., a 'frequent' response made to a rare stimulus), the overt response always preceded P300.

The error rate in this experiment was too low to permit a detailed examination of the error trials. We conducted a second experiment (McCarthy, Kutas, and Donchin, 1978; McCarthy, in press; see also McCarthy and Donchin, 1979) in which we used only the 'variable names' stimulus condition. However, many more trials were obtained from each of the 10 subjects for whom the necessity for speeded responses was emphasized. The RTs and estimates of P300 latencies were obtained in the same manner as the preceding experiment. Figure 2 presents the mean RTs and P300 latencies for the three response conditions.

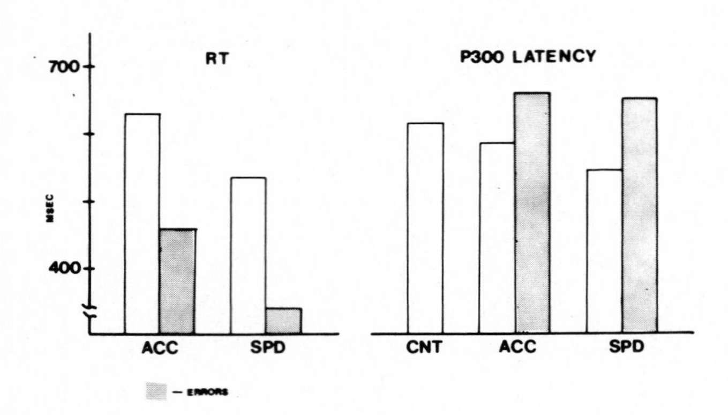

Figure 2
Mean P300 latencies and reaction times (RT) for the count (CNT), accurate-RT (ACC), and speed-RT (SPD) conditions. Shaded bars represent trials in which incorrect responses were made. (from McCarthy and Donchin, 1979, reprinted with permission).

The RTs in the speed condition were 95 msec shorter than in the accurate condition; however, in the speed condition incorrect responses were made to 40% of the rare stimuli while only 10% errors occurred in the accurate condition. The P300 latencies were longer in the count condition than in either RT condition. As before, the P300 latencies in the accurate condition were somewhat longer than in the speed condition. The P300 preceded RT in the accurate condition by 47 msec on the average. In the speed condition, RT preceded P300 by 7 msec.

The results for the error trials are given in the shaded bars. The RTs for the error trials are much shorter than they were for correct trials in both RT conditions, and faster for the speed than accurate conditions. Conversely, the P300 latencies are <u>longer</u> for the error trials than for the correct trials. This pattern of results is clearly apparent in the ERPs presented in Figure 3 in which ERPs recorded on correct and error trials have been overlapped for each subject. RT preceded P300 by an average of 286 msec for error trials.

The correlation of single trial P300 latency and RT was determined for each

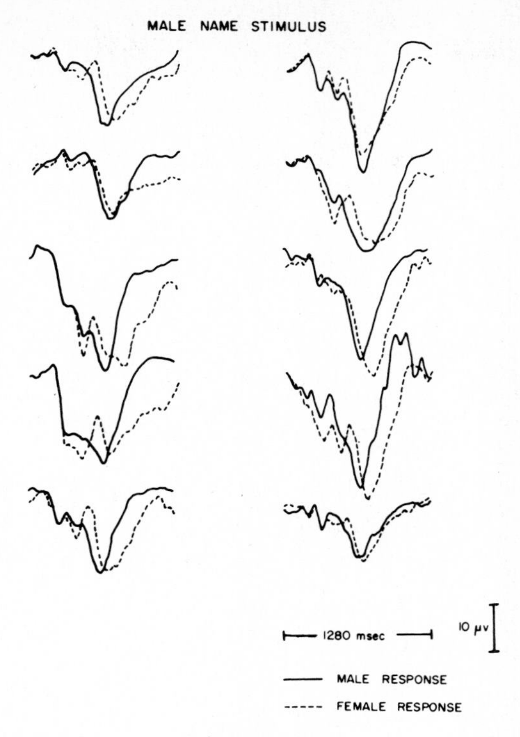

Figure 3
ERPs from ten subjects elicited by low probability (p=0.2) males' names. Waveforms drawn with solid lines are for correct responses in a choice-RT task, while superimposed waveforms drawn with dashed lines are for incorrect responses. (from McCarthy and Donchin, 1979, reprinted with permission).

subject for the two RT response conditions excluding error trials. The mean correlation for the accurate condition was .45 while for the speed condition the mean correlation was .33. The difference between these correlations was not statistically significant.

Taken together the results of these experiments indicate that the latency of P300 elicited in a choice response task can be dissociated from the response by varying the speed requirements of the task. In the first experiment, subjects correctly responded to the rare category stimulus (across all stimulus conditions) at 97% and 91% percent accuracy with correlations between P300 and RT of .66 and .48 respectively. In the second experiment (with only variable names) the categorization accuracy to the rare stimulus was 90% and 60% with P300-RT correlations of .45 and .33 respectively. The mean difference between RT and P300 latency (RT-P300) for these same accuracy levels were 91, 26, 47, and -7 msec. Thus, as speed increases (and accuracy decreases) the correlation between RT and P300 diminishes. The difference between RT and P300 decreases in magnitude and even reverses sign. Although we have noted that P300 latency decreases slightly for the speed as compared to the accurate RT condition, most of this difference is due to the effects of the speed instructions on RT. A dramatic dissociation occurs for trials in which the subject incorrectly responds to the rare category stimulus. The subjects are aware of their errors on these trials which are associated with very fast RTs, suggesting a fast guess or a response bias error. On these trials, the RT precedes P300 by more than 200 msec.

In these experiments, P300 precedes and closely covaried with RT when the accuracy of the response is emphasized. This result suggests that the latency of P300 reflects the categorization of the stimulus. However, as the speed of the response is progressively emphasized, RT tends to precede P300 latency and more errors are committed. If P300 reflects some terminal stage in the evaluation of a stimulus, this result suggests that subjects are responding prior to the completion of these evaluative processes. The pattern of results for the error trials implies that the evaluation process goes on to completion, even when an erroneous response had been made earlier.

In these experiments, P300 latency and RT were both strongly affected by the difficulty of the categorization tasks. However, P300 latency was relatively insensitive to the speed manipulation which strongly affected RT. If the speed manipulation had its primary effect upon response processes, then these data indicate that P300 latency is not dependent upon response processes and that P300 should be dissociated from RT whenever response processes are varied. A more direct test of this hypothesis is examined below.

STAGES OF INFORMATION PROCESSING

As stated in the introduction, many different models of information processing have been advanced, each making different assumptions about the properties of mental processes. One very influential approach has been Sternberg's (1969a, 1969b) additive factors method. In Sternberg's model, information processing occurs in independent, serial _stages_ whose durations sum to produce the composite RT. The output of these functional stages are presumed not to be affected by factors influencing their durations. Therefore, manipulating the discriminability of a stimulus will vary the duration of a stimulus-encoding stage, but it will not result in 'inferior' output which would affect the durations of subsequent stages. Thus, indirect interactions among processing stages are not allowed by the model. Stages are identified by manipulating many experimental factors and observing their patterns of interaction. Experimental factors which produce additive effects upon RT are presumed to affect the durations of different stages. Factors which have interactive effects upon RT are presumed to affect the same stage. The function of the processing stage is inferred from the nature of the experimental factors which influence its duration.

Two points must be emphasized. First, Sternberg's model is concerned with functional _stages_ and not with processes per se. Stages may comprise many processes, which may be concurrently active and interdependent. Factors which influence different processes within a stage will most likely produce an interaction, the form of which (underadditive or overadditive) may give insight into the underlying organization of the stage (Sternberg, 1969a). Second, the additive-factors method does not reveal the absolute durations of the stages or the order in which the stages are executed.

Additive-factors provides a powerful method for examining dissociations between P300 latency and RT. If experimental manipulations have independent and additive effects upon RT, the degree to which P300 latency is sensitive to each component of the RT can be directly assessed. On the basis of our speed-accuracy results for P300 and RT, we asserted that P300 latency is correlated with the time required to categorize a stimulus, but can be dissociated from response selection and execution. We tested this assertion in an additive-factors paradigm by varying two factors, stimulus discriminability and stimulus-response (SR) compatibility, which were presumed to affect stimulus encoding time and response selection time respectively. Stimulus discriminability and SR compatibility have been previously found to have additive effects upon RT (Bierderman and Kaplan, 1970; Frowein and Sanders, 1978; Shwartz, Pomerantz, and Egeth, 1977; Stanovich and Pachella, 1977; Sternberg, 1969a; cf., Rabbitt, 1967).

In a choice RT experiment, subjects indicated with a button press which of two target words (either "RIGHT' or 'LEFT') were presented on a given trial. There were two experimental factors. Stimulus discriminability was manipulated by presenting the target word alone in the center of a CRT screen ('normal') or surrounded by an array of noise characters ('degraded'). This manipulation was presumed to affect the encoding of the target word, thereby affecting categorization time, and should affect P300 latency and RT similarly. SR compatibility was manipulated by varying the relationship between the responding hand and the target word. As the target word had to be identified before the responding hand could be determined, this manipulation was presumed to primarily affect RT. In the 'compatible' condition, the target word 'RIGHT' required a right hand response, and 'LEFT' required a left hand response. In the 'incompatible' condition, the target word 'RIGHT' required a left hand response, and 'LEFT' required a right hand response. The compatibility condition was signaled to the subject by the case of the stimulus array. Upper case characters indicated the compatible condition while lower case characters indicated the incompatible condition. Figure 4 presents a sample of the eight resulting stimulus types. The characters composing the noise array were randomly chosen on each trial as were their positions around the target word. The target words, level of discriminability, and the SR compatibility were chosen independently, equiprobably, and randomly on each trial.

Ten subjects participated in this study, in which RTs and ERPs were recorded concurrently. The RT results are presented in Figure 5. The results for two conditions are shown. In condition 1 (circles), stimuli were presented as described above, but the subjects always made the compatible response. This condition served as a control to determine if the case of the stimulus array had a systematic effect upon RT. There was no effect of target word or case on RT; however, degrading the discriminability of the target word increased RT by a mean of 86 msec. In condition 2 (squares), the case of the stimulus array indicated a compatible (C) or incompatible (I) response. As in condition 1, there was no effect of the target word upon RT. The incompatible response increased mean RT by 123 msec while degrading the target word increased RT by 53 msec. These effects were additive, i.e., there was no interaction of discrimin-

RIGHT LEFT

right left

TIN Q VA
SRIGHTL
X UE S N

HQLM S
BZLEFTLR
CPADI R

eg k nz
d rightb
bml pfq

fctu lk
amleftqd
p ozm

Figure 4
Prototypic stimuli used in the additive factors experiment described in the text. On each trial subjects are presented with either the word 'right' or 'left,' presented in upper or lower case, with or without accompanying 'noise' letters in random surrounding positions. Displays in upper case indicate that subjects are to respond with the hand indicated by the target word (i.e., RIGHT = right hand response). Lower case displays indicate a crossed response mapping (i.e., right = left hand response). (from McCarthy, in press).

ability and compatibility. Note, however, that the overall RTs obtained in condition 2 were longer than those obtained in condition 1, and that the effect of stimulus discriminability was somewhat smaller in the second condition.

The P300 latencies were estimated for each trial using the template matching procedure described in Kutas et al. (1977). The results for conditions 1 and 2 are presented in Figure 6. In condition 1 (circles) there was no effect of target word or case on P300 latency. Degrading the stimulus array increased P300 latency by a mean of 50 msec. In condition 2 (circles) there was no effect of target word or SR compatibility upon P300 latency. Degrading the stimulus array increased P300 latency by a mean of 17 msec. Note, that as is true for RT, the overall P300 latencies were longer for condition 2 than for condition 1, and that the effect of stimulus discriminability was smaller in condition 2.

Our main prediction, that P300 latency would be relatively insensitive to manipulations of SR compatibility, was confirmed. We found no increase in P300 latency for the incompatible response despite a 123 msec effect on RT. P300 latency did appear to be sensitive to the discriminability of the target stimulus, especially in condition 1. In condition 2, however, this effect was only 17 msec. Given the imprecision in estimating single trial P300 latencies, a latency difference of this magnitude is suspect.

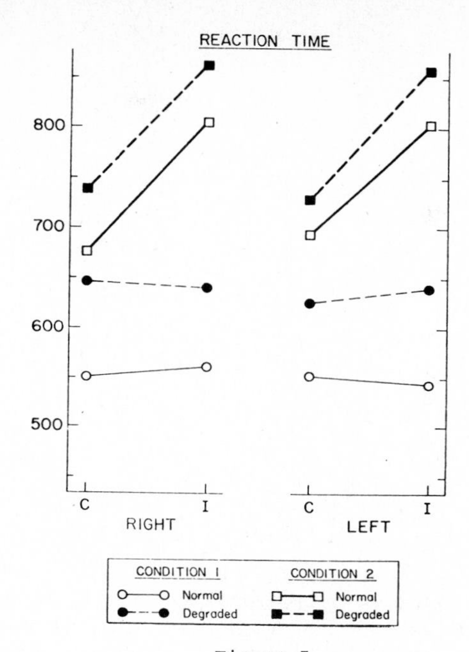

Figure 5
Mean reaction times obtained for two experimental conditions. In condition 1, subjects always responded with hand indicated by the target word, regardless of case. In condition 2, the case of the stimulus indicated either a compatible (C), or incompatible (I) stimulus-response mapping. (from McCarthy, in press).

For both conditions, however, the magnitude of the discriminability effect was smaller for P300 latency than for RT. The addition of the SR discriminability manipulation to condition 2 increased both RT and P300 latency relative to condition 1. For example, the time required to make a compatible response to a non-degraded target in condition 2 was about 130 msec longer than in condition 1. P300 latency, likewise, increased by nearly 50 msec.

In condition 2, we found that both stimulus discriminability and SR compatibility had additive effects upon RT. There was no measurable effect of SR compatibility on P300 latency, but the predicted effect of discriminability on P300 latency was small. We decided, therefore, to conduct another experiment in which stimulus discriminability was varied over a larger range (McCarthy, 1980; McCarthy and Donchin, 1981). A matrix of characters consisting of 4 rows and 6 columns were presented in the center of a CRT for 400 msec. Only one target word ('LEFT' or 'RIGHT') was present in the matrix on each trial. The targets were always written horizontally but could appear in any row. The rest of the positions in the matrix were either filled with # symbols in the high dis-

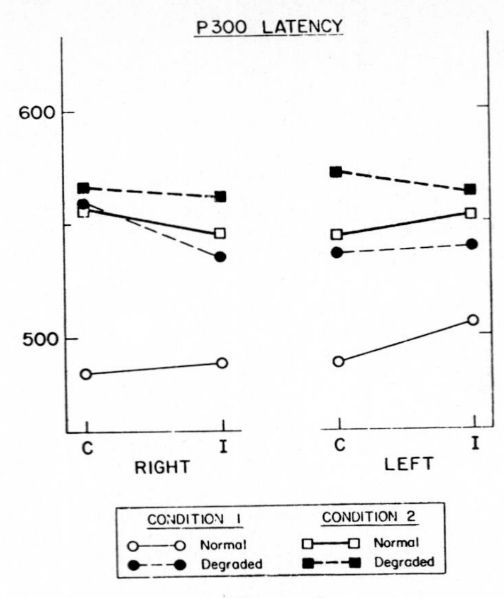

Figure 6
Mean P300 latencies from single trial estimates from the additive factors experiment described in the text. Conditions are as described for Figure 5. (from McCarthy, in press).

criminability (or 'no noise') condition, or with randomly chosen alphabetic characters in the low discriminability (or 'noise') condition. Upper case characters were always used. Four sample matrices are presented in Figure 7.

SR compatibility was manipulated by presenting a cue word 100 msec prior to the onset of the matrix. The cue SAME indicated a compatible SR mapping; a right hand response for the target RIGHT and a left hand response for LEFT. The cue OPPOSITE indicated an incompatible SR mapping; a right hand response for the target LEFT and a left hand response for RIGHT. Responses beyond 2000 msec were ignored and the trial was coded as no response. Feedback informing the subject of his/her RT and whether the response was correct was given on the CRT 2500 msec after the matrix onset.

Twenty subjects (10 males and 10 females) participated in an initial experiment to determine if our experimental manipulations would have additive effects upon RT. In this preliminary experiment, the matrix was exposed for only 250 msec and the cue preceded the matrix onset by 2600 msec. Responses were made within 2000 msec on 99.4% of the trials. Correct responses were made on 87% of the trials. The mean RTs for each target word, discriminability condition, and SR compatibility condition are presented in Figure 8. As in the previous experiment, stimulus discriminability and SR compatibility had additive effects upon mean RT as well as on RT variance. The 'noise' background increased RT by a mean of

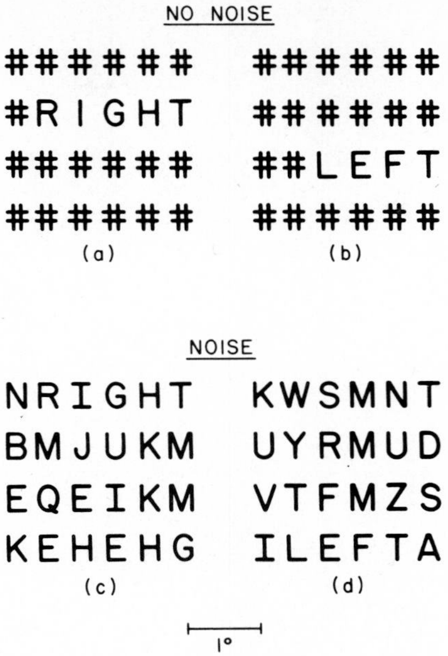

NO NOISE

```
######  ######
#RIGHT  ######
######  ##LEFT
######  ######
  (a)     (b)
```

NOISE

```
NRIGHT  KWSMNT
BMJUKM  UYRMUD
EQEIKM  VTFMZS
KEHEHG  ILEFTA
  (c)     (d)
```

1°

Figure 7
Four prototypic matrices in which the target words 'RIGHT' and 'LEFT' are shown presented against a 'no noise' (a and b) and 'noise' (c and d) background. (from McCarthy and Donchin, 1981, Science, Vol. 211, pp. 77-80, reprinted with permission. Copyright 1981 by the American Association for the Advancement of Science).

247 msec, while the incompatible SR mapping increased RT by a mean of 124 msec. Male subjects were faster than female subjects by an average of 167 msec. This sex difference was completely additive with all other factors. Stimulus discriminability and SR compatibility also had additive effects upon percent correct responses. There were no sex differences for this measure. Additional analyses of these data can be found in McCarthy (1980).

Fifteen male subjects participated in the next phase of the experiment in which both ERPs and RT were obtained. As in the preliminary behavioral experiment, discriminability and SR compatibility had additive effects upon RT. The noise characters increased RT by a mean of 266 msec while the incompatible response increased RT by a mean of 91 msec. The subjects responded correctly on 91.7% of the trials. Percent correct did not interact with any factor.

ERPs from five midline electrode sites are presented in Figure 9. These

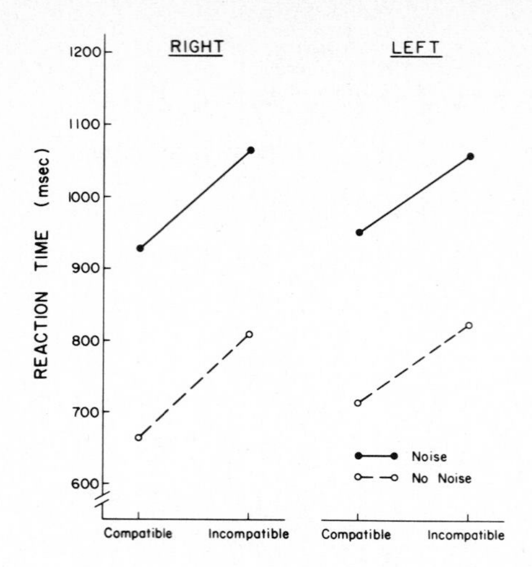

Figure 8
Reaction times presented for each target word (RIGHT and LEFT) as a function of stimulus discriminability (no noise = open circles, dashed lines; noise = filled circle, solid lines) and S-R compatibility (compatible or incompatible) collapsed across the twenty subjects participating in the behavioral experiment. (from McCarthy, 1980).

ERPs have been averaged across subjects and target words. The morphologies of the waveforms prior to the onset of the matrix are similar for all experimental conditions. These similarities contrast with the large differences seen in the region of the waveform following the presentation of the matrix. Most notable is a large positive potential recorded from Pz in the 'no noise' conditions which peaks at about 430 msec after matrix onset with an inflection in its rising negative slope at about 600 msec. In the 'noise' condition there are two positive peaks of smaller magnitude with latencies of approximately 360 msec and 700-800 msec. On the basis of a series of Principal Component Analyses (see McCarthy, 1980), we concluded that there are two major positive potentials overlapping in the 'no noise' condition. These potentials become temporally dissociated by the 'noise' manipulation as the second potential moves out in latency. In this regard, our data resembles that of Friedman, Vaughan and Erlenmeyer-Kimling (1978) who found that two overlapping positive potentials elicited by visual stimuli with amplitude maxima over posterior scalp became progressively temporally dissociated as a function of RT quartile. In their study the earlier positivity had a peak latency of 341 msec which did not change as a function of RT quartile. The second positivity increased in latency by 210 msec from quartile 1 to 4.

Single trial estimates of P300 latency were obtained for each single trial ERP from the Pz electrode. It was determined that template fitting was not an appropriate method to estimate P300 latencies in these data, as the template fit was biased by the positive going slope at the termination of the CNV. Instead we picked the most positive peak in the interval between

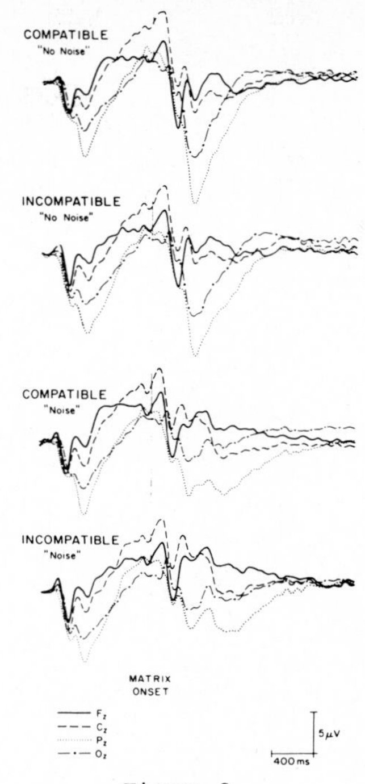

Figure 9

ERPs recorded from five midline electrode sites for each stimulus condition as described in the text. The recording epoch consists of a 50 msec baseline, a 1000 msec epoch between cue and matrix onset (vertical line), as well as 2000 msec of activity following matrix onset. Negativity is plotted upwards. (from McCarthy and Donchin, 1981, Science, Vol. 211; pp. 77-80, reprinted with permission. Copyright 1981 by the American Association for the Advancement of Science).

200 and 1500 msec post matrix onset (for details of this procedure, see McCarthy, 1980). Figure 10 presents the mean P300 latency estimates obtained by this procedure as well as the mean RTs. An ANOVA performed on these data indicated that only stimulus discriminability had a significant effect upon P300 latency.

The increase in P300 latency due to the 'noise' characters was 191 msec. The overall increase in latency associated with the incompatible response was 16 msec, but this difference was not statistically significant. None of the experimental manipulations significantly affected the P300 latency variance.

The subjects in the preliminary behavioral experiment reported that the matrix row in which the target word appeared affected their responses. In this experiment, row position was stored for each trial so that this effect could be investigated. Figure 11 shows the result of this analysis for both RT and P300 latency. Targets presented in the outer rows (top and

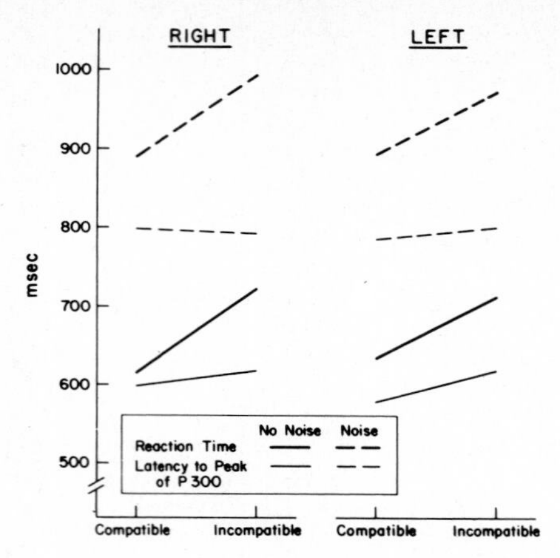

Figure 10
Mean reaction times and P300 latencies obtained across subjects for stimulus discriminability ('no noise' and 'noise'), S-R compatibility ('compatible' and 'incompatible'), and target word ('RIGHT' and 'LEFT'). (from McCarthy, 1980).

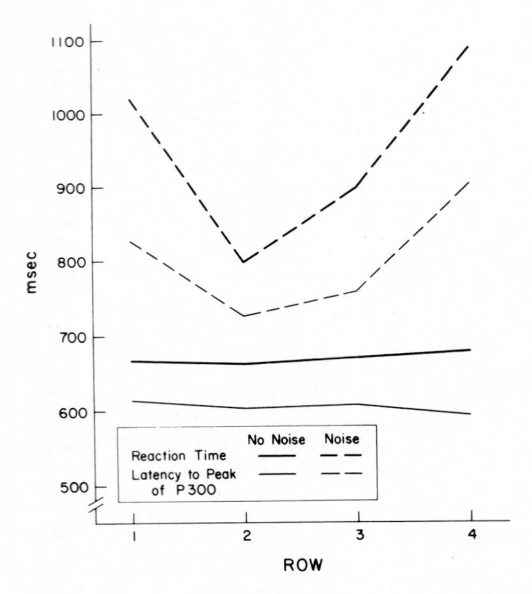

Figure 11
Mean reaction times and P300 latencies obtained across subjects for target words presented in each matrix row (1, 2, 3, 4 - from top) as a function of stimulus discriminability ('no noise' and 'noise'). (from McCarthy, 1980).

bottom) were responded to more slowly than targets in the middle rows. However, this effect was only obtained for the 'noise' trials. In the 'no noise' trials, the RTs were equivalent for each row. The 'row' effect strongly interacts with stimulus discriminability indicating that both variables affect an encoding processing stage. P300 latency, then, should also be affected by this variable. P300 latency was longer for targets in the outer than inner rows but, like RT, only for the 'noise' condition. This difference in P300 latency is readily apparent in the ERP waveforms. In Figure 12 are the waveforms from Pz averaged separately for the inner and outer rows from the compatible response conditions. ERPs from the incompatible response conditions are presented in Figure 13. In the 'no noise' conditions, the waveforms are very similar. However, in the 'noise' conditions the second positive potentials for the outer rows occurs much later than in the inner rows for both the compatible and incompatible response conditions.

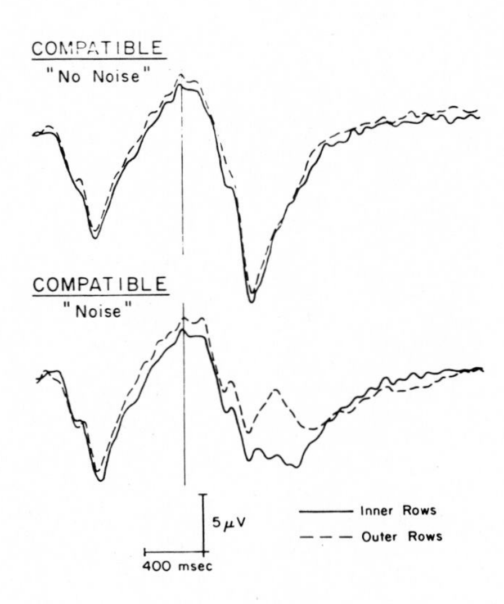

Figure 12
ERP waveforms obtained from the parietal electrode site (Pz) for the compatible S-R mappings for the 'no noise' and 'noise' trials. The waveforms from the inner rows (2 and 3 - solid line) and outer rows (1 and 4 - dashed line) are shown overlapped. Negativity is plotted upwards. (from McCarthy, 1980).

The single trial correlations of P300 latency and RT were also examined. For all manipulations, significant positive correlations were obtained, but of small magnitude. For the 'no noise' conditions, a correlation of .20 was obtained for the compatible responses and .10 for the incompatible responses. The correlations for the 'noise' conditions were somewhat higher, probably reflecting the greater range of P300s and RTs as well as the additional variance introduced by the row position of the target. For the compatible responses the correlation was .30 while for the incompatible conditions the correlation was .22. The decrease in correlation for the incompatible responses was significant and reflects the addition of RT variance by this manipulation without corresponding P300 latency variance.

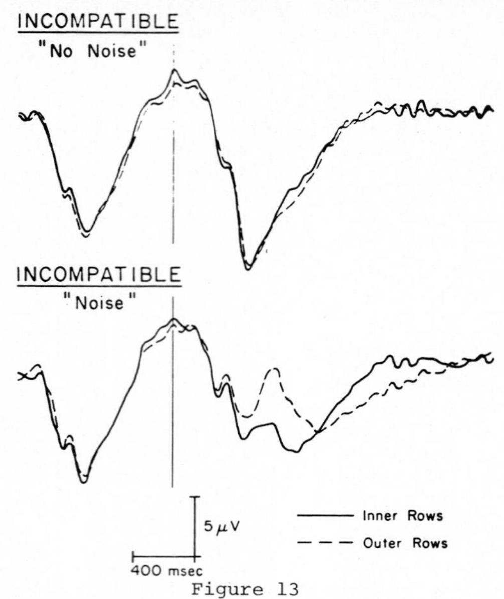

Figure 13
ERP waveforms obtained from the parietal electrode site (Pz) for the compatible S-R mappings for the 'no noise' and 'noise' trials. The waveforms from the inner rows (2 and 3 - solid line) and outer rows (1 and 4 - dashed line) are shown overlapped. Negativity is plotted upwards. (from McCarthy, 1980).

In these experiments, P300 latency was strongly affected by two variables, stimulus discriminability and row position, both of which are presumed to affect stimulus encoding. P300 was relatively insensitive to manipulation of SR compatibility which is presumed to affect response selection and execution. RT, however, was strongly affected by all of these variables. These data, then, support our hypothesis that the latency of P300 reflects a subset of the durations of processes which are reflected in RT, specifically those processes on which the categorization of a stimulus depends.

DISCUSSION

Our results indicate that both P300 latency and RT are affected by manipulations of stimulus discriminability and of the difficulty in stimulus categorization. Manipulations of speed-accuracy tradeoff and stimulus-response compatibility strongly influenced RT; however, P300 latency was minimally affected by these manipulations. Thus, our hypothesis that P300 latency primarily reflects a subset of information processing activities, specifically those related to stimulus evaluation, was upheld. If further processes are required to translate the stimulus category, once identified, into an appropriate motor response, then P300 latency is minimally affected. Similarly, if task instructions require the subject to respond prior to a complete evaluation of the stimulus, P300 latency shows only minor variation.

The relative magnitude of the S-R compatibility effects on RT and P300 reported above have been largely confirmed in a subsequent study by Magliero, Bashore, Coles, and Donchin (in preparation). In this study, large increases in RT (125-160 msec) were observed when incompatible RT responses were required. However, P300 latency showed only a small (14 msec) increase. These investigators varied the amount of noise that was introduced into the noise matrix and were able to demonstrate that RT, as well as P300 latency, increased monotonically with increases in the noise. This study confirms our hypothesis that P300 latency is primarily affected by stimulus evaluation processes. Apparently inconsistent with our conclusions, however, are the findings of Ragot and Renault (1981). In their study, subjects were required to respond to one colored light with the left hand and another colored light with the right hand. Two lights (one of each color) were paired on each side of a central fixation point so that each hand would respond to a light on the same side (compatible response) and opposite side (incompatible response). With hands in their normal positions, the incompatible response was associated with an increase of 28 msec in RT and 20 msec in P300 latency. Thus the relative increases in RT and P300 were comparable. The authors asserted that these data may indicate that response compatibility does affect P300 latency. The data, however, are subject to alternate interpretations. At issue is whether their manipulation of response compatibility also affected stimulus evaluation processes. As these authors note, the spatial position of the colored lights might have interfered with stimulus processing. Thus, what they label a compatibility effect may as well be due to the need to discriminate between the two spatial positions of the stimuli. Without additional experimental factors, it is not possible to determine whether their manipulation of S-R compatibility was independent of stimulus evaluation processes, effects upon which could account for the covariance between RT and P300. Ragot and Renault also introduced a manipulation in which the subjects responded with their hands held in a crossed position. This manipulation increased overall RT and also increased the magnitude of the spatial position effect, but had no effect upon P300 latency.

While our data specify information processes whose durations influence P300 latency, they do not specify the actual operations which P300 may reflect. To state that P300 is affected by the time required to evaluate the category membership of a stimulus does not say that P300 is a manifestation of that evaluation process. Rather, the P300 is in some way dependent upon the outcome of such a succession of processes. The speed-accuracy tradeoff data suggest that the 'P300 process' is not part of a linear path of processes leading to a response on a given trial. Rather, our view is that of a process which monitors the output of the stimulus evaluation process, perhaps to alter the subject's expectancies on subsequent trials. A theoretical discussion of P300 can be found in Donchin (1981).

Even though our data do not make strong theoretical statements concerning the actual process that is manifested by the appearance of P300, knowledge of what processes P300 latency depends or does not depend can be exploited in information processing experiments. In this sense P300 latency can be used as a metric, much the same way as reaction time. A recent example of the application of the strategy we advocate is provided by the experiment of Duncan-Johnson and Kopell (1981) who used the joint information provided by P300 and RT to determine that the Stroop interference effect had its primary effect upon response processes. Another example is provided by the work of Ford, Roth, Mohs, Hopkins, and Kopell (1979) in their analysis of RT and P300 latency variation in young and old subjects engaged in a Sternberg memory task. Increases in memory set size resulted in RT slopes which were steeper for the old than the young subjects. The P300 latency slopes were, however, similar for both sets of subjects leading the experimenters to conclude that some change in motor response, but not memory comparison time, differentiated the young and old subjects.

One finding of potential importance is the observation that even when P300 latency and RT covary, the change in RT for a given experimental manipulation is typically larger than the change in P300 latency. For example, adding 'noise' characters to the matrix display in the experiment reported above increased RT by 266 msec and increased P300 latency by 191 msec, only 75% of the RT effect. It is possible that the techniques used for estimating P300 latency on a single trial have a bias against long latency estimates, leading to artifactually shorter mean latency estimates. If, for example, P300s tended to be smaller in amplitude when later in latency, then latency measurement procedures might pick erroneous noise peaks (with a latency distribution centered in the search period) with greater frequency on those trials. If the artifact argument can be rejected, then our result indicates that P300 latency is sensitive to only some of the processes which are affected by the experimental variables. The more similar the slopes of P300 and RT, the more likely the effects of the experimental variables are restricted to processes upon which P300 latency is contingent. In the case of our stimulus discriminability manipulation, the interval between P300 and RT increased as a function of adding noise. This result suggests that stimulus discriminability directly or indirectly affects processes subsequent to those upon which the process manifested by P300 were contingent. The nature of these subsequent processes is unknown, but the behavioral evidence rules out response selection as a likely stage. It appears more likely that additional stimulus evaluation processes are affected by our stimulus discriminability manipulation, but that these processes are unnecessary for the elicitation of P300. This may indicate that the point at which the P300 is elicited depends upon when certain stimulus attributes become available, even though processing of other stimulus attributes is still ongoing.

Note: The research reported in this paper was performed at the Cognitive Psychophysiology Laboratory, University of Illinois, Champaign, Illinois, under Contract No. N00014-76-C-0002 from the Office of Naval Research with funds provided by the Defense Advanced Research Projects Agency.

Tutorials in ERP Research: Endogenous Components
A.W.K. Gaillard and W. Ritter (eds.)

13

SENSORY AND MOTOR ASPECTS OF THE CONTINGENT NEGATIVE VARIATION

John W. Rohrbaugh

University of Nebraska College of Medicine

Anthony W.K. Gaillard

Institute for Perception TNO

Evidence supporting a multi-component interpretation of the Contingent Negative Variation (CNV) is reviewed. Two major components are considered to form the CNV: an early component (herein called the "O wave"), and a terminal wave. It is argued that the O wave is a general response to salient or novel stimuli, which can be elicited in both CNV and non-CNV situations. The terminal wave is interpreted as being related to motor response processes, and is identified with the readiness potential component of the motor potential complex. Evidence for a "true CNV," having the traditionally ascribed anticipatory features, is considered to be amenable to interpretation in terms of the individual O wave and terminal CNV components.

INTRODUCTION

It is with some irony that this chapter on the Contingent Negative Variation (CNV) appears amidst a collection of papers dealing with cognitive aspects of the event-related potential. As the title suggests, it is our intent to argue that concerns with cognitive aspects of the CNV have been misplaced and overemphasized. We shall argue that the evidence in support of the mental priming, expectancy, association and attention properties traditionally ascribed to the CNV is scanty--that the appeal of these hypothesized properties is more a matter of conceptual allure than experimental demonstration. We shall argue further that the CNV waveform is created by the temporal juxtaposition or superimposition of several distinct waves, none of which is unique to the paired stimuli CNV situation, and none of which, therefore, is directly related to the cognitive processes thought to be tapped expressly by that situation.

A typical CNV is illustrated in Figure 1. The CNV depicted here is associated with the forewarned reaction time (RT) task--a task that has long been favored for the laboratory investigation of association, attention, preparation and expectancy. The discovery by Walter et al. (1964) of a slow EEG wave whose duration coextended with the foreperiod thus led to the embrace of the wave as an objective sign of these important psychological processes. The name given by Walter et al. to the wave, the "Contingent Negative Variation," describes its usual appearance as a broad negative wave that develops over the foreperiod between warning (S1) and imperative (S2) stimuli. Commensurate with a long-standing literature relating surface negativity to cortical activation, the negative polarity of the CNV was taken to implicate it in a process of "cortical 'priming' whereby responses to associated stimuli are economically accelerated and synchronized" (Walter et al., 1964, p. 383). The CNV since has been the focus of a large amount of research, which has buttressed its importance as a means for investigating mental faculties, but which has shown an eagerness to bestow the wave with somewhat more elaborate psychological properties. The CNV has been reported to be sensitive to a wide variety of task variables (including response, perceptual and cognitive variables) and sensitive also to an equally variegated assortment of subject variables (including age, sex, personality traits, neurological impairment, and psychiatric disposition).

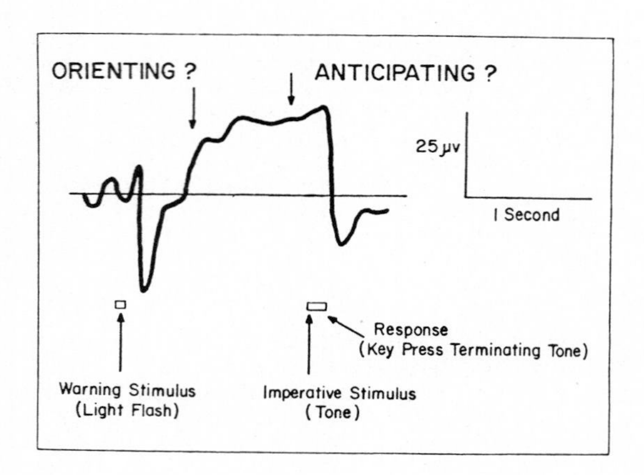

Figure 1. Prototypical CNV waveform and depiction of typical eliciting situation--S1 and S2 (separated by a short foreperiod), followed by a RT response. Negativity is indicated by an upward deflection in this and all subsequent figures. (From Callaway, 1975.)

Many of these relationships are, however, disconcertingly weak. Even the very basic association between the speed of RT and CNV amplitude has proved evanescent (see Rebert and Tecce, 1973). The abundance of positive, but weak, results, coupled with occasional difficulties in replication, has led to proposals that the CNV, or behavior, or both, are mediated by elusive intervening states or processes that are able to break down any close relationship between the two. Knott and Irwin (1973), for example, envisage a fluctuating background state of arousal upon which various CNV situations may be superimposed, but which may impose a "ceiling" upon CNV development. Similarly, McCarthy and Donchin (1978) picture a rolling mental "terrain" whose irregularities may yield vagaries in CNV amplitude despite a constant level of behavioral performance.

An alternative explanation proposes that the CNV comprises several separate waves, all having relationsips with experimental variables that are quite strong individually, but which are diluted in the aggregate. This notion is illustrated in Figure 1, where two processes have tentatively been ascribed to the CNV: orienting to the early phase, and anticipating to the late phase. The issues we shall address are implicit in this figure. Among these issues are the following: Must we indeed entertain two or more separate processes? Are these labels the appropriate labels? Do any of these processes depend specifically or uniquely upon the paired-stimuli situation for their elicitation? Are these processes different in admixture than in isolation?

We have chosen to use this figure to raise these issues as a way of emphasizing at the outset that the arguments we raise here are not new. This figure was published in 1975, and summarizes a variety of earlier evidence. The existence of multiple CNV waves was presaged by animal work that revealed regional differences in CNV morphology and amplitude (e.g Borda, 1970; Donchin et al., 1971) and was further attested to by human experiments that disclosed variations in the appearance of the CNV brought about by changes in the task, stimuli or recording site (Järvilehto and Frühstorfer, 1970; Näätänen and Gaillard, 1974; Otto et al., 1977; Syndulko and Lindsley, 1977). The arguments since have been made explicit by Loveless (see Loveless, 1979), Vaughan (1975) and others, as well as in our own work. The problem, of course, is that all possible arguments pertaining to the CNV have at some time or place been made (or so, at least, it often seems); what is needed is a winnowing among these alternative suggestions, rather than the formulation of novel interpretations. We are hopeful that the evidence summarized below will be helpful in this regard.

THE CNV AS MULTIPLE WAVES

The most persuasive evidence that the CNV comprises separate components comes from experiments in which the traditional, short foreperiod of 1 or 1.5 sec is lengthened to 3 sec or more, whereupon at least two distinct waves become apparent: one early and one late (Connor and Lang, 1969; Weerts and Lang, 1973; Loveless and Sanford, 1974a). Under these long-foreperiod conditions, the early peak is attained within the first sec or so after S1, and the second wave emerges shortly before S2. The biphasic nature of the CNV at long intervals is a ubiquitous effect that since has been confirmed in a large number of reports (see Gaillard, 1978, and Loveless, 1979, for reviews).

Figure 2 provides an illustration of this phenomenon with the CNV recorded in a forewarned RT task with a foreperiod of 4 sec. The CNV is shown in the bottom traces--for bilateral sites at C3 and C4 at the left, and for three sites along the midline at the right. The CNV was recorded under conditions in which the pitch of the tone S1 indicated to the subjects with which hand they were to respond to the visual S2.

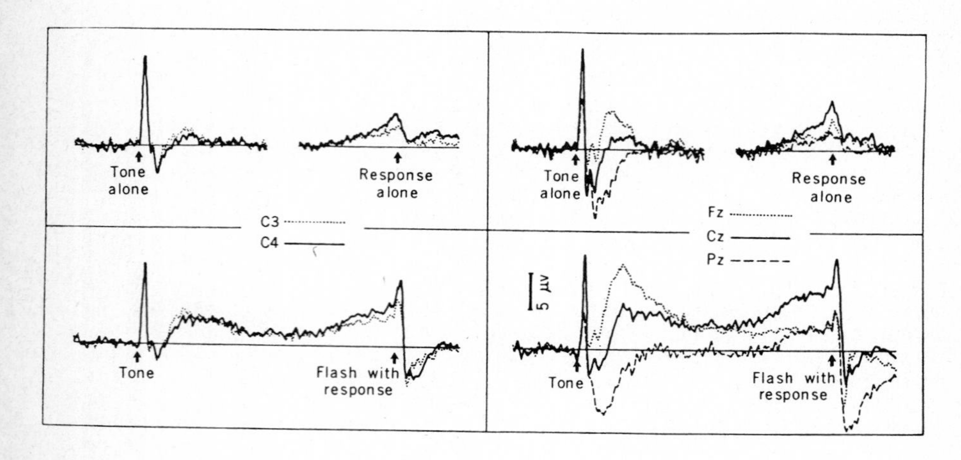

Figure 2. Event-related potentials elicited in a RT task with a 4 sec foreperiod (bottom panels), and individually by non-paired tones and uncued key presses (top panels). Records from midline frontal (Fz), central (Cz), and parietal (Pz) electrode sites are shown in the right panels, and for lateral left (C3) and right (C4) motor sites in the left panels. Responses were made with the left hand. (From Rohrbaugh et al., 1976.)

Under these conditions the CNV assumes a distinctly biphasic appearance. The late wave, we believe, is related to preparation for motor response. Evidence for this belief is presented in succeeding sections. In brief, we may note that it is larger preceding fast than slow RT responses (Harkins et al., 1976; Rohrbaugh et al., 1976; Grünewald et al., 1979a; Brunia and Vingerhoets, 1980), and that it is greatly attenuated when no motor response is required (Loveless, 1975; Gaillard, 1977; Kok, 1978; Lang et al., 1978; Perdok and Gaillard, 1979; Gaillard, 1980) or when attention is directed toward accurate rather than speedy response (Loveless and Sanford, 1974b; Gaillard and Perdok, 1980; Gaillard et al., 1980). Also, it shares many temporal and topographic features with the readiness potential component of the motor potential complex (first described by Kornhuber and Deecke, 1965; Gilden et al., 1966). This is apparent upon comparison of the CNV traces in the bottom of Figure 2 with the "Response Alone" traces in the upper panels, which were obtained from a condition in which subjects pressed the response key at their own pace. As shown in the right panels, the late CNV wave and readiness potential share a similar midline distribution, with the greatest amplitude at the vertex site. Moreover, as shown for C3 and C4 sites in the left panels, the late CNV wave shows the same lateral asymmetry as the readiness potential. The waves shown here are for left hand responses and show a preponderance contralateral to the hand of response; the asymmetries reversed correspondingly when right hand responses were given (see also Otto and Leifer, 1973; Syndulko and Lindsley, 1977; Brunia, 1980b; Gaillard, 1980; Gaillard et al., 1980).

The functions associated with the early CNV wave are less apparent. It seems, for the most part, to relate to the properties of S1. Since it peaks early in the foreperiod, and its peak latency or breadth is not altered by changes in the duration of the foreperiod (Loveless and Sanford, 1974a), it would seem not to be related to anticipation or expectancy for S2. The early CNV wave does, on the other hand, depend upon the modality (Gaillard, 1976), intensity (Loveless and Sanford, 1975; Connor and Lang, 1969), duration (Klorman and Bentsen, 1975) and task role (Kok, 1978; Gaillard and Perdok, 1979, 1980) of S1. Moreover, as shown for the "Tone Alone" responses in the upper panels of Figure 2, a similar negative wave, smaller but having the same topographic features, can be elicited even when no S2 or RT response follows--all of which have led us to conclude that the early CNV wave is elicited by S1 and is dependent upon its characteristics, but is not related specifically to the paired-stimuli CNV situation.

The simplest and most straightforward interpretation of the CNV, based upon this sort of evidence, proposes that the CNV is comprised solely of these separate early and late waves. The late wave can be identified with the readiness potential, whereas the early wave is a response to the S1. Either of these waves can be elicited individually, and neither is specific to the paired-stimuli CNV situation. The typical ramp-shaped CNV waveform at shorter foreperiods arises because the timing of stimuli and motor responses in the CNV situation is such as to yield fortuitously the overlap and summation of these separate waves. Whether or not there remains any vestige in this mixture of the "true CNV" (see Näätänen and Michie, 1979) is still to be proven. A variety of considerations, however, suggests that most CNV phenomena can be accounted for without recourse to a "true CNV" and that, if present, it is sure to be much smaller than has been customarily assumed.

THE EARLY CNV WAVE AS A GENERAL RESPONSE

Perhaps the most critical issue pertaining to the interpretation of the early CNV wave is whether or not its appearance depends in particular upon the paired-stimuli CNV situation. Considerable evidence indicates that it does not. An early characterization of the human auditory evoked potential includes a prominent late negative wave peaking at about 600 msec (Derbyshire and McCandless, 1964); subsequently, there have been occasional reports of protracted negative afterwaves elicited by single (unpaired) stimuli, which in some cases bear suggestive similarities to the early CNV wave (Haider et al., 1968; Symmes and Eisengardt, 1968; Cohen, 1969; Lang et al., 1975; N. Squires et al., 1975; Simons and Lang, 1976). Among the most thorough descriptions of such waves is that of Loveless (1976), who described a complex of broad negative waves following non-signal tone stimuli, and who also commented specifically on their importance in the interpretation of the CNV. Moreover, Loveless distinguished two separate slow negative components in the complex, designated N3 and N4. Both waves were reported to be largest over anterior regions, with a diminution in amplitude or positive values more posteriorly.

Waves similar to those described by Loveless (1976) are depicted in Figure 3, which illustrates also the multi-component nature of the wave complex. These data were obtained in response to single (unpaired) stimuli in a traditional "Oddball" paradigm. Subjects were asked to count infrequent tones at 1000 Hz (p=.25), intermingled with a series of frequent tones at 2000 Hz (p=.75). Several late waves, each responsive to stimulus probability, are apparent. The first peak is at about 350 msec, where it is positive at all three electrode sites. This wave can be identified with the P300 component commonly elicited under such conditions (Sutton et al., 1965). P300 is followed by additional components having substantial negative aspects. The first of these is negative anteriorly and positive posteriorly. The positive aspect of this intermediate component can be quite large, as it is here, and P300 seems often to be dwarfed by it or to be grafted upon it. This intermediate component is followed by yet another wave that accounts for the late negativity. The last wave is negative at all sites, but tends here (as in most situations) to be largest at or near central sites.

Although the individual components are not usually so distinct or pronounced as in Figure 3, a multi-component structure is routinely disclosed when Principal Components Analysis (PCA) is applied. This is true both in unpaired stimulus situations (Rohrbaugh et al., 1978, 1979) and in long-interval CNV situations (Plooij-van Gorsel, 1980; Sanquist et al., 1981; Lutzenberger et al., 1982.

The existence of multiple waves renders inappropriate or incomplete most of the names that have been applied to the complex. In part because of its dependence upon the intensity, significance or novelty of the eliciting stimulus, and its preponderance over frontal sites, the entire complex often is called the "O wave" (after Loveless and Sanford, 1974a) to

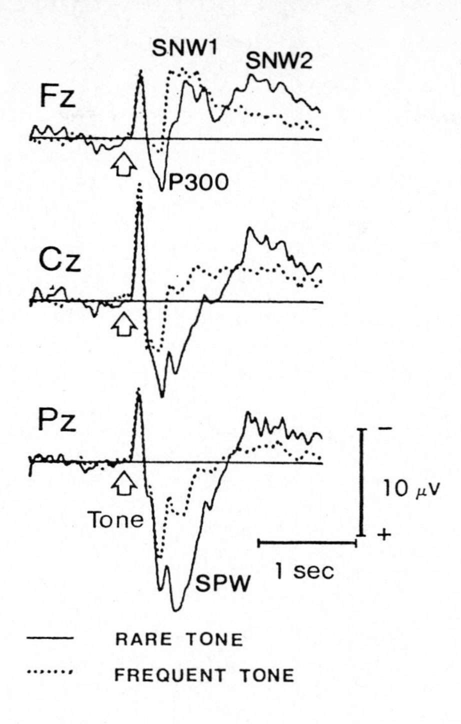

Figure 3. Event-related potentials elicited in an Oddball paradigm in which subjects were asked to count the rare (p=.25) tones. (J.W. Rohrbaugh, K. Syndulko, and D.B. Lindsley, unpublished data.)

denote an hypothesized link with orienting processes. Problems with this identification are that such a link has not been well established experimentally, and that neither the O wave nor the orienting reflex appear to be unitary processes. Also unsuitable are other common names for the complex; "slow negative wave," "negative afterwave," and "cortical negative wave" ignore the waves' positive aspect, "slow wave" pertains only to the earlier portion of the wave (as considered below) and "early CNV wave" does not recognize the likelihood that identical waves are recorded in non-CNV and CNV paradigms. We shall adopt here the "O wave" designation for the entire complex of waves, but more as a matter of convention than conviction. The extent to which these waves may be held to be manifestations of orienting activity is an important but largely unexplored problem. Since there is evidence that each of the components within the O wave is subject to independent variation, we have adopted the following descriptive terminology for the individual O wave components: "SNW1" for the first slow negative wave peaking over frontal areas at 500 to 700 msec, "SPW" for the slow positive wave over parietal areas at about the same (or slightly earlier) latency, and "SNW2" for the late, broadly distributed slow negative wave.

The precise number and identification of O wave components remains in question. PCA generally recognizes two components in the O wave, as described above. A possible interpretation is that the earlier of these two components (including SNW1 and SPW) may be identified with the "slow wave" component described by N. Squires et al. (1975). The latency and conditions of elicitation are appropriate, as is the general frontal-negative to posterior-positive gradient. Despite the preservation of this anterior to posterior gradient, however, the absolute amplitudes of SNW1 and SPW appear to vary independently. This raises some question as to whether SNW1 and SPW can be combined legitimately into a single component. Independent variation in the two aspects has been observed both as a function of task and stimulus modality, both in CNV and in unpaired stimulus situations. An example of task-related variation in unpaired stimulus situations is given in Figure 4. This figure serves also to indicate the magnitude and ubiquity of the O wave, as well as its sensitivity to a variety of task variables. These O waves have been obtained in response to tone stimuli in four tasks (described in detail in Rohrbaugh et al., 1978). The solid traces in each case depict records from a frontal (Fz) site, while the dotted traces are from a parietal (Pz) site. At one extreme, there is a large SNW1 but no SPW in the Count Stimuli condition, and at another extreme there is a substantial SPW but less corresponding SNW1 in the response to occasional omitted stimuli in the Omitted Stimuli condition (although a clear SNW2 is present in the latter condition). The Pitch Discrimination and Probability of Occurrence conditions yield SNW1s of comparable amplitude, but the SPW is nearly twice as large for the rare stimuli in the Probability of Occurrence condition. Examples in CNV situations of similar, seemingly independent, variation of negative and positive aspects may be found in Kok (1978), Gaillard and Perdok (1980), and Sanquist et al. (1981).

Both the positive and negative aspects of the early O wave appear to depend upon some degree of task relevance, or some inherently obtrusive property, for their elicitation (Loveless, 1976; Rohrbaugh et al., 1978, 1979). Beyond this, it appears as a general rule that the SPW is more sensitive to stimulus probability than is SNW1. Although on occasion an effect of probability on SNW1 can be seen (e.g. Rohrbaugh et al., 1978; see also N. Squires et al., 1975; Duncan-Johnson and Donchin, 1977; N. Squires et al., 1977), in many instances no effect is observed. The differential sensitivity of negative and positive aspects to stimulus probability is apparent in Figure 4, and in Figure 5. The data in Figure 5 were obtained in an experiment (Rohrbaugh, Syndulko, and Lindsley, unpublished data) in which subjects counted silently the number of target tones in an Oddball paradigm. The factors of target probability, and target/non-target designation, were varied parametrically. Records from midline electrodes are plotted here as a function of probability of occurrence and target status, and have been averaged over 9 subjects. The data were obtained in 4 separate runs of 100 stimuli each in which high (1500 Hz) and low (1000 Hz) tone bursts were presented in unpredictable order. Two runs each used probabilities of 0.2 and 0.8 for the high and low tones, respectively; in one run the subjects were asked to count the rare (high) target tones, and in the second run subjects counted the frequent (low) tones. Similarly, two runs employed probabilities of 0.4 and 0.6, with the rare tones designated targets in one run, and the frequent tones designated targets in the second run.

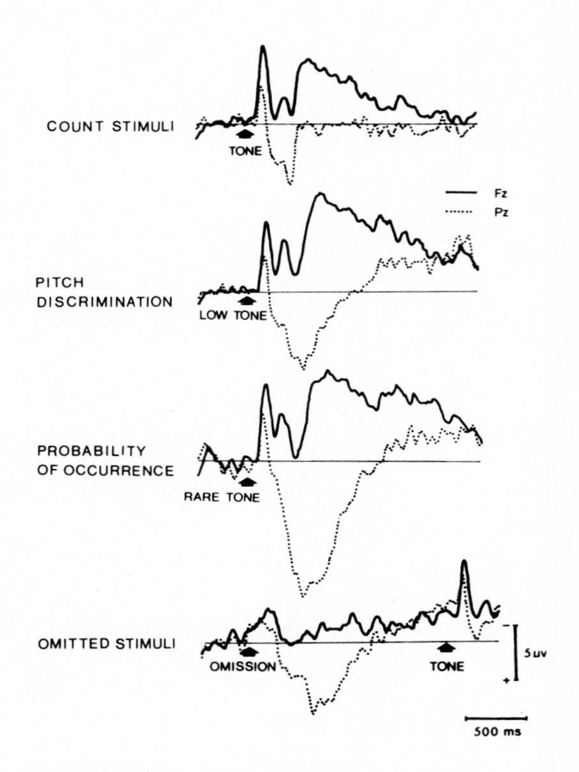

Figure 4. O waves obtained from frontal (Fz) and parietal (Pz) sites in four different acoustic tasks: A simple voiceless Counting task; a Pitch Discrimination task entailing the identification of high and low tones differing by 30 Hz; an Oddball paradigm entailing rare (p=.20) and frequent tones; an Omitted Stimulus paradigm in which occasional (p=.20) stimuli were omitted from the sequence. (From Rohrbaugh et al., 1978.)

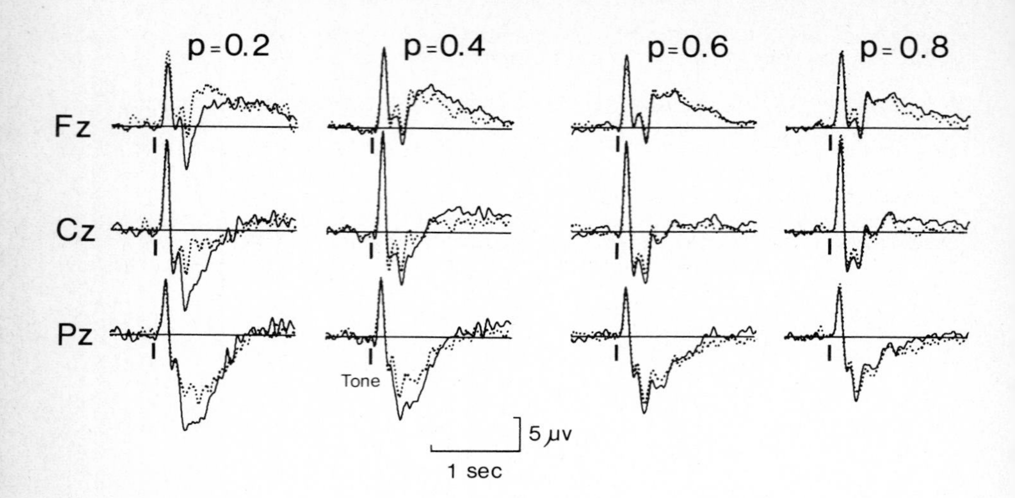

Figure 5. O waves obtained in Oddball paradigms for probabilities ranging from 0.2 to 0.8. Solid traces are from conditions in which stimuli were designated targets and were counted by the subjects; the dotted traces are from conditions in which the complementary stimuli were designated targets and were counted. (J.W. Rohrbaugh, K. Syndulko, and D.B. Lindsley, unpublished data.)

The potentials in all conditions show a P300 component (latency about 350 msec) that is positive at all electrode sites. This is followed by simultaneous frontal negativity (SNW1) and parietal positivity (SPW). SNW1 is relatively unaffected by the factors of stimulus probability and target designation. In contrast, the amplitude of P300 and the following SPW is a strong inverse function of probability, and, at low probabilities, is appreciably larger when elicited by target stimuli. SNW2 shows a small tendency to be largest for improbable targets in a manner similar to that for the preceding SPW (see also Figure 3).

FACTORS AFFECTING THE O WAVE: PARALLELS IN CNV AND NON-CNV SITUATIONS

Although the psychological determinants of the O wave components have not yet been precisely defined, it is known that the O wave is responsive to a wide spectrum of experimental variables. Examples of this are apparent in Figures 4 and 5, and are further elaborated below. This feature is important in two regards. One is that it suggests that the O wave warrants more extensive examination than it has received. The lateness of the O wave does not diminish in our minds the likelihood that it is a manifestation of important psychological phenomena: there are a variety of protracted storage, ruminative, response selection and adjustment processes whose temporal course may well follow pari passu that of the O wave.

A second respect in which the O wave variability is important, which is more relevant to the present context, has to do with the interpretation of the CNV: experimental variability in the O wave for unpaired stimuli parallels, and likely accounts for, comparable variability in the CNV.

Stimulus modality. One important source of variability is known to be stimulus modality. Using a simple stimulus counting task, Rohrbaugh et al. (1979) obtained results indicating that the O wave for acoustic stimuli is uniformly greater than that for visual stimuli, no matter what the intensity. This conclusion is in general accord with results from other investigators, which have indicated an overall tendency for the late activity following visual stimuli to be either small or predominantly positive (Roth et al., 1978; Kok and Looren deJong, 1980a, 1980b; Runchkin et al., 1980; Sanquist et al., 1981; Hömberg et al., 1981; Rösler, 1981). This difference is not always found (e.g. N. Squires et al., 1977), and seems to interact with task features (Rohrbaugh et al., 1979; Ritter et al., 1980), but is present in most instances. The intriguing correspondence between this O wave difference, and modality-related differences in the strength of orienting or activating functions, has been noted previously (Gaillard, 1976; Rohrbaugh et al., 1979.) Equally important is the existence of corresponding modality-related effects in the CNV. Despite earlier disclaimers to the contrary (e.g. Blowers et al., 1976), it now appears certain that the modality of S1 can, on occasion, have appreciable effects on CNV shape, distribution and amplitude (Gaillard, 1976; Gaillard and Näätänen, 1976; Simson et al., 1977b; Deecke et al., 1980; Ritter et al., 1980; Rohrbaugh et al., 1980; Sanquist et al., 1981). The effect parallels that seen for the O wave in response to unpaired stimuli, with smaller CNV amplitudes for visual than for acoustic warning stimuli.

Stimulus intensity. The effects of stimulus intensity have not been thoroughly examined, but they appear to be less pronounced than the effects associated with modality. One study (Rohrbaugh et al., 1979) failed to find any effect associated with intensity, either for visual or for acoustic stimuli. Elsewhere, however, we have observed small effects. An example is presented in Figure 6. These data were obtained in an experiment in which subjects counted silently a sequence of 20 tone or flash stimuli, superimposed on a continuous background stimulus. The intensities of target and background stimuli varied from condition to condition, but within each condition the intensities were kept constant. (The data were acquired under conditions similar to those described in Rohrbaugh et al., 1979.) The data plotted here are average amplitude values from the Fz electrode for the segment spanning 400 to 1000 msec poststimulus. Also, these data are averaged over subjects, but only the half of the subjects (n=4) who showed large O waves.

The effects of intensity were absent or considerably smaller in subjects whose O waves were small.

The plots in the top panel of Figure 6 are for O waves elicited by acoustic stimuli. In the left plot, the intensity of the target tone has been held constant at the intermediate value of 85dB SPL, while the intensity of the background white noise ranges from soft (65dB) to loud (85dB). The amplitude of the O wave is reduced when the background is loud. Conversely, when the background noise intensity is held constant at the intermediate value (75dB) and the intensity of the target tone is varied from soft (75dB) to loud (95dB), the O wave is largest for loud target tones. In contrast, the O waves for visual stimuli (bottom panel) hover around zero or slightly positive values, no matter what the intensity of the target flash or background field. (The flash stimuli here were produced by a Grass photostimulator set to intensities of 1, 4, or 16

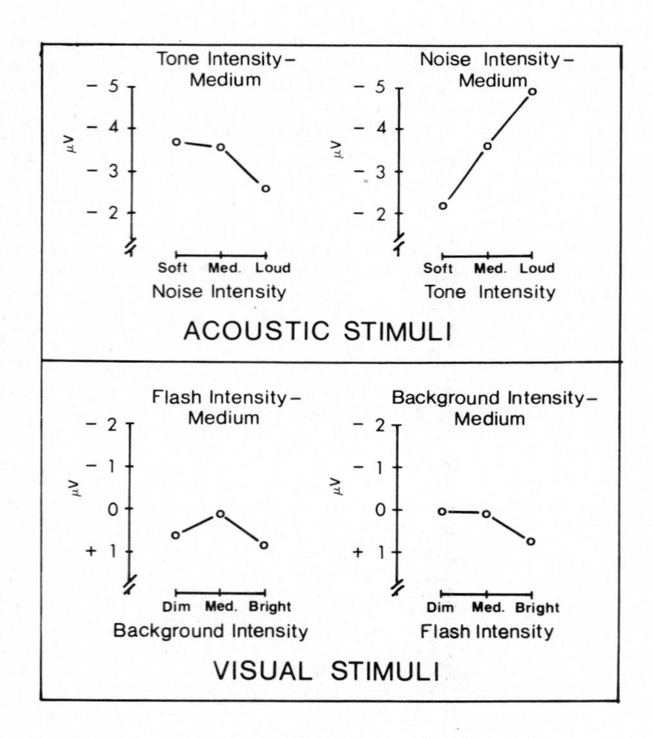

Figure 6. Average O wave amplitudes (during 400 to 1000 msec) in simple voiceless counting tasks involving acoustic (top panel) and visual (bottom panel) stimuli. The intensities of the target stimuli and constant background stimuli are varied. (J.W. Rohrbaugh, K. Syndulko, and D.B. Lindsley, unpublished data.)

[reduced by half through the optics of the display device], and the intensity of the background field was either 5, 20, or 80 ml.)

The effects of intensity revealed here for acoustically-elicited O waves, though small and apparently present only for those subjects whose O waves are large, are again paralleled by comparable findings for CNV amplitude. Although the conventional view is that the CNV is independent of such physical features as intensity, Loveless and Sanford (1975) note that support for this view comes from experiments in which the primary intent was "demonstrating that a well-developed CNV may be elicited by any perceptible stimulus" (p. 218), rather than systematically investigating the effects of intensity. Using white noise to demarcate the foreperiod, they observed systematic increases in early CNV amplitude for loud warning stimuli, particularly at short foreperiod durations (see also Ruchkin et al., 1981).

Stimulus duration. Yet another physical attribute of the eliciting stimulus that appears to affect O wave amplitude is stimulus duration. A number of suggestive parallels may be noted between the O wave and the "sustained potential" that accompanies prolonged stimulation. Sustained potentials in humans were first described in the early 1950's by Köhler and his co-workers (Köhler et al., 1952; Köhler and Wegener, 1955). The most thorough investigations have used auditory stimuli, although there are reports of sustained potentials in the visual modality as well (see below). For auditory stimulation, the onset of sustained potentials occurs within 100 to 200 msec (Picton et al., 1978a) and they attain their maximum over fronto-central regions (David et al., 1969; Keidel, 1971a; Picton et al., 1978a, 1978b). A number of stimulus attributes, in addition to duration, influence the amplitude and shape of the wave. Included are pitch, intensity and burst envelope. Picton et al. (1978a, 1978b) have observed also that the potentials are larger during attentive states, and they may be obtained readily during sleep (see also Hari et al., 1979a, 1979b).

Although less thoroughly investigated, sustained potentials have been reported for visual stimuli as well (Clynes, 1964; Rix and Korth, 1977; Lehmann et al., 1978; Korth and Rix, 1979; Järvilehto et al., 1978). The amplitude appears generally to be smaller than for acoustic stimuli (Järvilehto et al., 1978), commensurate with modality-related differences in O wave amplitude described above. The visual sustained potentials vary also with duration and intensity of the stimulus, and have been reported by one author (Keidel, 1971b, 1980) to be remarkably modality-specific in distribution, with a maximum amplitude over occipital regions (see Hari, 1980, for complementary findings in the auditory modality).

Upon inspection of the published examples of sustained potentials, they appear (in the auditory modality, at least) to attain an early maximum over frontal regions and then later to assume a more central representation. Aside from the respective breadth of the waves, this progression is not unlike that for the O wave--suggesting, perhaps, that the sustained potential comprises two negative waves. This possibility is supported by data from Järvilehto et al. (1978) and Hari et al., (1979a, 1979b), who found the early portion of the sustained potential to habituate more rapidly than the late. A likely possibility, then, is that the sustained potential is similar or identical to the O wave, but that one or both of the negative aspects may cascade for the duration of the stimulus. The possibility has obvious import for the interpretation of studies that have manipulated the duration of S1 (Klorman and Bentsen, 1975), as well as a large number of studies that have used sustained stimulation to demarcate the CNV foreperiod.

Lateral distribution. The O wave appears also to be distinctly asymmetrical in its lateral distribution. This feature is illustrated in Figure 7, where O waves are displayed for three tasks using tone burst stimuli (Rohrbaugh et al., 1981). In the Passive Listening condition, subjects listened to a sequence of tones (one tone per 15 sec), all at a standard frequency. In the Count condition, they were required to maintain a voiceless count of the number of tones presented, and in the Pitch Discrimination condition they were required to maintain a voiceless count of the standard tones when intermingled in sequence with other types of tones that differed in frequency by 30 Hz (difficult discrimination) or 400 Hz (easy discrimination). Waveforms are shown for bilateral frontal (F3, F4), central (C3, C4) and parietal (P3, P4) sites. The asymmetry is very prominent, particularly at frontal sites, and particularly in the Pitch Discrimination condition. The asymmetry appears to be confined to SNW2, and is essentially undiminished even at the end of the 3 sec epoch. These records are averaged over 9 subjects (one left handed), all of whom showed a right O wave preponderance in some degree. We (Rohrbaugh, Newlin, Varner, and Ellingson, in preparation) since have determined, in a larger sample of subjects, that occasional subjects show a reversal in this asymmetry, and that the direction and degree of asymmetry is weakly correlated (r=.40) with the strength and direction of the propensity to make lateral eye movements upon reflective thought (as measured with the procedure and test items described by Schwartz et al., 1975). Careful assessment of the magnitude and distribution of electro-oculographic artifact indicates that eye movements *per se* are not the source of this asymmetry.

Whether this asymmetry is amenable to reversal using some different type of task or stimulus is not known. The nature and distribution of the asymmetry is consistent with evidence pointing to a crucial role of right frontal areas in the perception of pitch and melody (Roland et al., 1981; Shapiro et al., 1981). A second possible interpretation is suggested by the speculations of Heilman and others (e.g. Heilman and Van Den Abell, 1979, 1980) that attention and orienting processes are principally functions of the right hemisphere. This interpretation is remarkably consistent with speculations linking the O wave to orienting processes. Whatever interpretation is eventually assigned to the O wave asymmetry, however, it is important to note the likelihood that asymmetries in the O wave per se may lead to corresponding lateral variation in the CNV. When seen in combination with motor-related asymmetries, a complex and subtle pattern of CNV asymmetries conceivably could emerge (c.f. McCallum, 1980). This is especially the case when the intended character of the warning stimulus itself is such as to engage differentially one or the other hemisphere (e.g. Rebert and Lowe, 1980).

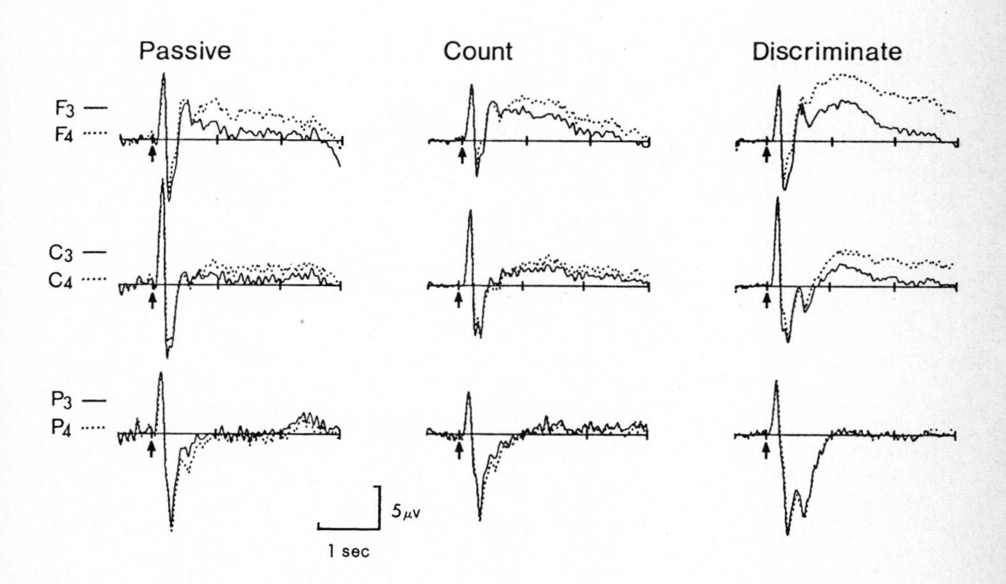

Figure 7. O waves obtained from bilateral frontal (F3, F4), central (C3, C4), and parietal (P3, P4) electrode sites in 3 conditions: A Passive Listening condition (left column), a simple voiceless Counting condition (center column), and a Pitch Discrimination condition in which the number of standard tones was counted when presented in irregular order with other tones differing in frequency by 30 or 400 Hz. (From Rohrbaugh et al., 1981.)

Stimulation rate. O wave amplitude has been shown to be a very strong function of stimulation rate (Rohrbaugh et al., 1979). Although we believe the O wave is frequently elicited and is likely to be present in a great many paradigms that are used to study the event-related potential, it is difficult to observe with the rapid rates of stimulation commonly used in the study of P300 and earlier components. Obversely, with long inter-trial intervals an O wave is obtained quite easily, even under conditions of passive observation. It is likely that this feature is one reason the O wave is so prominent in CNV paradigms, since the inter-trial intervals typically are long in CNV research so as to discourage the development of temporal expectancies from one trial to the next. Several reasons for the effect of stimulation rate may be entertained. The physiological or psychological mechanisms for O wave generation may have protracted recovery cycles, or perhaps the eliciting characteristics of novelty or salience are less pronounced for individual stimuli at rapid rates of stimulation (see Geer, 1966; Van Olst, 1971). Yet another possibility is that the O wave for individual stimuli in the run may accumulate or cascade in a manner similar to that proposed above for sustained potentials, so that an overall negative shift in baseline throughout the run is effected.

Response variables. A RT motor response seems to interfere with the development of the O wave. A representative example of this phenomenon is depicted in Figure 8, where the O wave has been obtained in two non-CNV tasks using auditory stimulation. These waveforms have been averaged over 6 subjects and are in response to short tone bursts at 1000 Hz. In the left column are data from two silent Counting conditions, one at the beginning of the experimental session and the second at the end. Prominent O waves are present, particularly at Fz, in both conditions. In contrast, the waves in the right column (from a RT condition) show only a small and atypically formed O wave. These waves were obtained in a condition in which subjects pressed a telegraph key non-discriminatively in response to the tones. A monetary incentive was provided for responses faster than the mean RT established in practice trials. We have observed similar attenuation in a variety of motor response tasks. It is not clear at present whether the O wave diminution represents an unfavorable compounding of the O wave with a separate post-response positive wave, or whether instead it reflects a response-related interruption or termination of the processes reflected in the O wave. Again, however, this feature may be relevant to the interpretation of the CNV, particularly a number of reports that CNV amplitude is diminished when the CNV trial is self-initiated, or when a motor response to S1 is required (McCallum and Papakostopoulos, 1972; Hamilton et al., 1973; Otto and Leiffer, 1973). In more typical CNV situations, this feature of the O wave also suggests a possible reason why the negativity following S2, though appreciable on occasion, is normally less dramatic than that following S1.

O wave parallels in CNV and non-CNV situations: An illustration. In preceding paragraphs, we have described the O wave as a complex of several distinct waves, subject to several important sources of variability related to physical and psychological characteristics of the stimulus. We also have taken note of representative instances in which O wave variability is mirrored in comparable CNV variability. A summary illustration of these points is provided in Figure 9, which

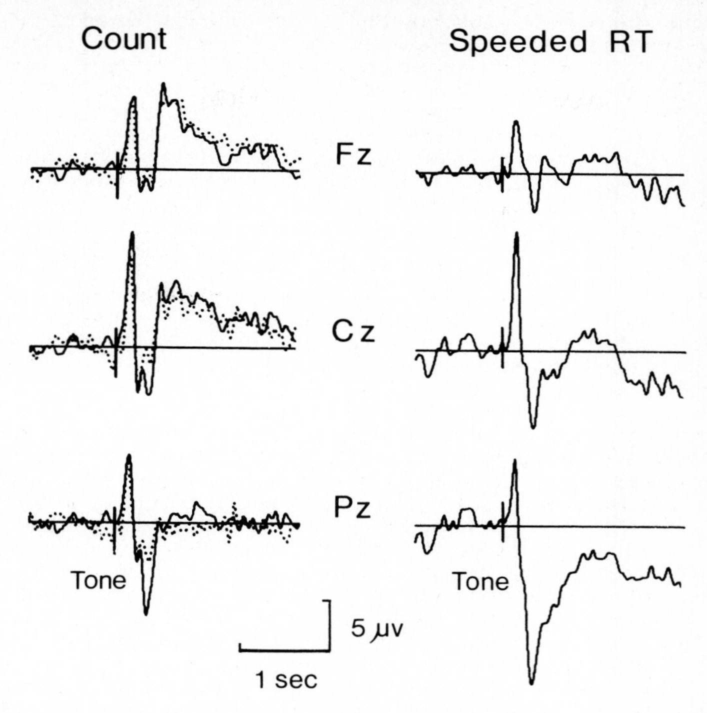

Figure 8. O waves obtained in simple voiceless counting tasks (left column) and a speeded RT task (right column). Records from 2 separate counting conditions are overlaid in the left column--one from the beginning of the experimental session (solid trace) and the second from the last condition of the session (dotted trace). (J.W. Rohrbaugh, K. Syndulko, and D.B. Lindsley, unpublished data).

displays CNVs from several different RT tasks with a foreperiod of 6 sec. These records are averaged over 5 subjects, with data from 3 midline sites on the left and data from lateral placements at C3 and C4 on the right. The S1 in all cases is a brief tone burst, S2 is a light flash, and the response is the depression of a telegraph key. The conditions are, respectively, simple RT, Go/No Go RT (where the shape of the S2 flash tells the subject whether or not to respond), Choice RT (where the shape of the S2 flash tells the subject with which hand to respond), and what is called here a Specific Warning condition. In the Specific Warning condition the subject responds on each trial with either the left or right hand; the pitch of the S1 tone tells him which.

The CNV is biphasic in appearance in all of these conditions. The terminal CNV wave is a centrally dominant wave that emerges in mid- foreperiod and attains its maximum amplitude near the end of the foreperiod. Evidence linking this terminal CNV wave to motor processes is summarized in succeeding sections. The early CNV wave is characterized by an early frontal peak and a broad negative wave that persists well into the foreperiod. This distribution and temporal progression is equivalent to that commonly found for the O wave in non-CNV situations, as described above. When S1 assumes the additional role of providing information about the eventual hand of response, as in the Specific Warning conditions, the amplitudes of both the negative and positive aspects of the early CNV wave are increased (see also Kok, 1978; McCarthy and Donchin, 1978; Gaillard and Perdok, 1980; Deecke et al., 1980; Sanquist et al., 1981). The increase in amplitude under these conditions matches corresponding increases seen when stimuli are made relevant, informative or rare in non-CNV situations (see Figure 4 above; Rohrbaugh et al., 1978).

Moreover, the factor structure for the early CNV wave, as determined by PCA, is equivalent to that usually seen for the O wave in non-CNV situations. As described above, that structure inevitably includes two separate O wave components--an early one showing a pronounced negative-to-positive gradient along the midline (including both SNW1 and SPW), and a second component that is broadly negative at all electrode sites (SNW2). An equivalent structure is present for the early CNV wave, as shown in Figure 10. The loadings for the first 4 PCA factors (of the covariance matrix with Varimax rotation) are displayed in the left column. Also plotted are the associated scores for each of the 4 factors at the midline (center column) and lateral electrode placements (right column). Among these factors (and others not depicted here), there are but three factors associated with slow activity during the foreperiod. One factor (Factor 1) is associated with the terminal CNV wave. Although there appears to be some variation in the magnitude of this factor across the experimental conditions (with the largest scores in the Choice RT condition), the effects were not consistent across subjects; neither the effect of condition nor its interaction with electrode site approaches statistical significance ($P > .10$). The remaining two foreperiod factors (Factors 3 and 4) are associated with late and early aspects, respectively, of the early CNV. The temporal courses of the loadings are remarkably similar to those seen for components of the O wave in non-CNV situations. Also similar are the distributions of the factor scores. For the second component of the early CNV (Factor 3) the distribution tends to be centrally dominant and to show a slight preponderance over the right hemisphere. A marginally significant effect ($p=.07$) is associated with condition, with the largest scores from conditions in which the information value of S1 is increased (the Specific Warning conditions). The earlier factor (Factor 4) shows the usual anterior-to-posterior gradient, with an increase in both SNW1 and SPW aspects under the specific warning conditions. The interaction between condition and electrode site is highly significant ($p < .001$).

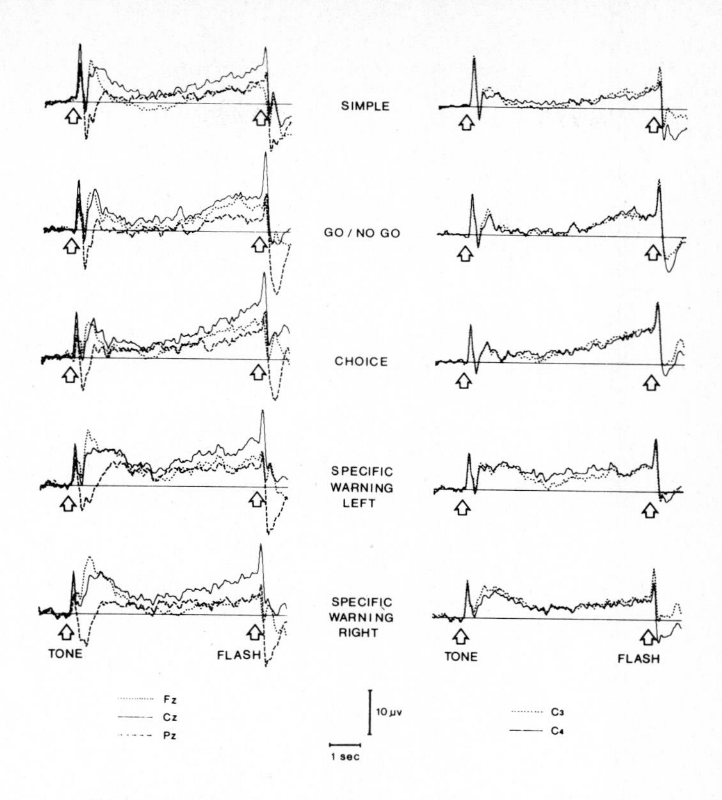

Figure 9. CNV waveforms obtained from 5 conditions with a foreperiod of 6 sec: a Simple RT task, a Go/No Go RT task, a Choice RT task, and a paradigm in which the pitch of the tone indicated whether the response was to be with the left hand (Specific Warning Left) or right hand (Specific Warning Right). All conditions except the Specific Warning Left required responses with the right hand. Records are from midline (Fz, Cz, Pz) sites (left column) and from bilateral motor area (C3, C4) sites (right column). (J.W. Rohrbaugh, K. Syndulko, and D.B. Lindsley, unpublished data.)

In sum, we believe the O wave is an important constituent in the CNV. The O wave can on occasion be quite large (15 to 20 µv in some subjects), and can persist in some instances for several seconds (see, for example, Figure 7 above). As such, the O wave provides an important determinant of CNV shape, amplitude and topographic representation. Its effect is especially pervasive at short foreperiods, wherein virtually any measurement at any latency throughout the foreperiod will include some measure of it. The variability seen in the O wave as a function of modality, task, intensity, laterality, inter-trial interval and response requirements applies to the CNV composite as well.

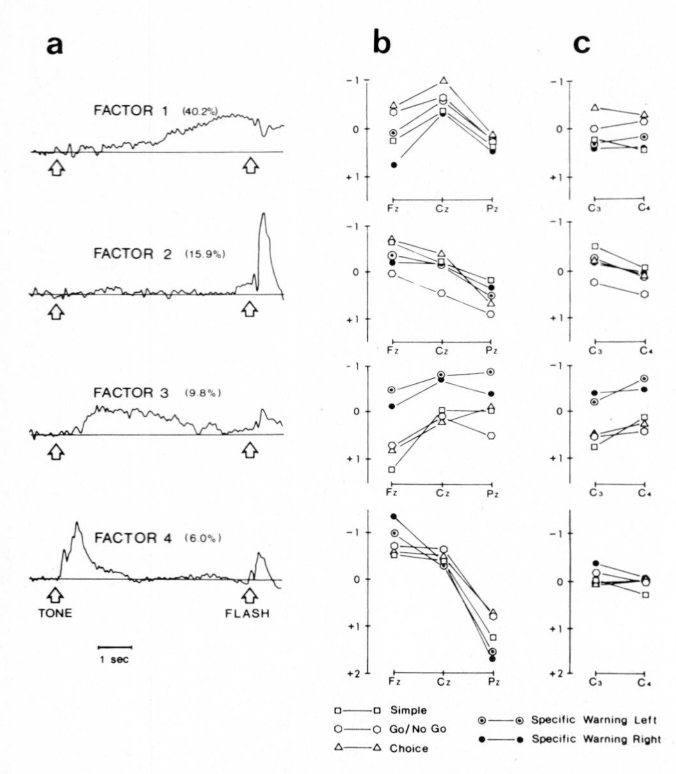

Figure 10. Results from PCA (of covariance matrix with Varimax rotation) of CNV records presented in Figure 9. Factor loadings are plotted for the time points (left column). Also plotted are the associated factor scores for midline records (center column) and bilateral motor area records (right column). (J.W. Rohrbaugh, K. Syndulko, and D.B. Lindsley, unpublished data).

Moreover, neither the elicitation of the O wave, nor the appearance of this variability as a function of the eliciting stimulus characteristics, depends in any strict sense upon a contingency between S1 and S2. As a minimum criterion, we would propose that if a wave is to be designated "contingent," an indispensible control condition is to demonstrate that the wave disappears when the contingency is removed. For the O waves obtained in response to single (nonpaired) stimuli and depicted in the above figures, this clearly is not the case.

There is an important precondition, however, that with but very few exceptions has not been satisfied. As a proviso for this demonstration it is necessary that some attempt be made to mimic the signal value associated with a warning stimulus, because it is this and not the contingency or association *per se* that is important. A variety of tasks that lend experimental significance to a stimulus, including discrimination, counting and detection tasks, are effective at this--at least the O waves bear strong similarities to those associated with warning stimuli.

As a corollary, it should be noted that the signal value of a warning stimulus might also be expected to vary as a function of the task associated with the imperative stimulus; it might well be, for example, that a warning stimulus has more salience or immediacy when a speeded RT is impending, or when a difficult judgment is upcoming, or when the foreperiod is very short, or when it signals choice as opposed to simple RT. The O wave thus might very well assume different appearances as a function of the imperative task features, but these changes reflect changes imparted by the task to the role of the warning stimulus, and not what is expected at the end of the foreperiod.

The differences here are subtle, but not trivial. Giving a stimulus a signal value by making it a warning stimulus causes that stimulus to elicit a prolonged O wave, but the effect can be recreated in varying degree by any other manipulation that lends salience to the eliciting stimulus. This conclusion is similar to that of Maltzman (1977) regarding the "first interval" conditioned electrodermal response (which is broadly analogous to the early CNV wave in long- interval paradigms). He concludes that the first interval response is an orienting response that is "generated by the participant's covert problem solving activity. It is not the response elicited by the CS signal as the result of the establishment of an association; it is a consequence of the discovered significance of the CS as a signal for the UCS." (p. 113).

THE TERMINAL CNV AS A MOTOR-RELATED POTENTIAL

Central to the interpretation of the terminal CNV is the problem of whether it represents an accompaniment of the motor response *per se*, or whether instead it is a manifestation of the abstract attentional or volitional determinants of the movement. The problem is similar to those faced elsewhere in, for example, the interpretation of heart rate responses (e.g. Obrist, 1976) or hippocampal electrical activity (e.g. Black, 1975). With respect to the latter issue, Vanderwolf and Robinson (1981) make the following observation: "Perception, attention, motivation, decision making, etc. occur during voluntary movement but thinking can also occur during behavioral immobility. ...It seems to follow that if a given brain electrical pattern is found to be consistently present during movement and consistently absent during the absence of movement, then it must be related to the movement in some way and not to those psychological processes which are presumed to occur both during movement and during immobility" (p. 505). Evidence pertaining to the terminal CNV seems clearly, if evaluated in view of this reasonable criterion, to relate it to preparation for movement rather than to some mental precursor. The relationship with movement is further attested to, as noted above, by the remarkable similarities between the terminal CNV and the readiness potential component of the motor potential complex. Relevant evidence has been reviewed by Gaillard (1978, 1980), who has culled the following arguments:

1. In studies using short foreperiods, the amplitude of the CNV is attenuated when no motor response to S2 is required (e.g. Järvilehto and Früstorfer, 1970; Peters et al., 1970; Donald, 1973; Näätänen et al., 1977; Syndulko and Lindsley, 1977). Irwin et al. (1966) have pointed out that the enhancement of the CNV amplitude by a subsequent motor response is of the same magnitude as the readiness potential that precedes any voluntary movement.

2. In studies using longer foreperiods, the terminal CNV is absent or greatly attenuated when no motor response to S2 is required (Loveless, 1975; Gaillard, 1977; Lang et al., 1978; Gaillard, 1980; Gaillard and Perdok, 1980).

3. The terminal CNV is affected by task variables such as foreperiod duration, foreperiod variability, and relative signal frequency, all of which, on the basis of choice RT studies, have been assumed to affect the level of motor preparation (Niemi and Näätänen, 1981). For example, Loveless and Sanford (1974a) compared a regular foreperiod condition with an irregular foreperiod condition and found that the O wave was present in both conditions, whereas the terminal CNV wave was observed only in the regular foreperiod condition. In general, these results run parallel to the behavioral effect; RT is shorter in the regular condition and increases with increasing foreperiod duration.

4. The amplitude of the terminal CNV is not increased preceding difficult discriminations as compared to easy ones, either in RT tasks or in signal detection tasks involving a delayed response. Also, no enhancement in amplitude is found preceding correct responses as compared to errors (Perdock and Gaillard, 1979).

5. In studies using foreperiods of 3 sec or longer, the terminal CNV is largely attenuated or even virtually absent in sensory tasks where no quick motor response is required (Perdock and Gaillard, 1979; Gaillard, 1980).

6. Larger CNV amplitudes are obtained when larger amounts of muscular effort are required for the response to S2 (Low and McSherry, 1968; Rebert et al., 1967).

7. The amplitude of the terminal CNV is enhanced under speed instructions. This result could also be taken as evidence for

a motor-preparation hypothesis. If the CNV were positively related to perceptual sensitivity, larger amplitudes would be expected under accuracy instructions, where, of course, fewer errors are made (Loveless and Sanford, 1974b; Gaillard et al., 1980; Gaillard and Perdok, 1980). An illustration of the relationship of the terminal CNV to speed instructions is presented in Figure 11 (from Gaillard, 1980), where vertex records (averaged over 10 subjects) from CNV tasks with a 4 sec foreperiod are presented. These records were obtained in three experimental conditions: a Speed condition in which the subject was instructed to respond as quickly as possible, even at the cost of some errors; an Accuracy condition in which the subject was instructed to produce quick RTs, but the importance of avoiding errors was emphasized; and a Detection condition in which the subject was instructed to delay his response by 1 sec following S2. Under all conditions S1 was a moderate tone and the subject was required to make a visual discrimination at S2. The final amplitude of the terminal CNV is nearly twice as large under Speed instructions as it is under Accuracy instructions, and delaying the response (the Detection condition) results in a dramatic diminution of the terminal CNV.

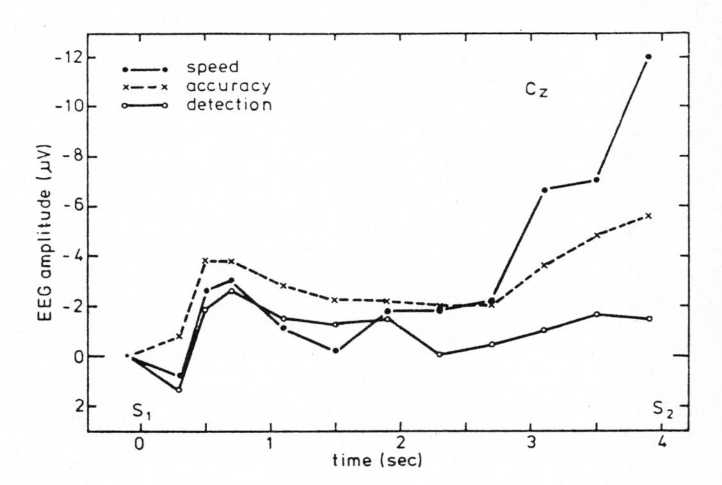

Figure 11. Potentials obtained from Cz in three conditions: Speed instructions (permitting some errors), Accuracy instructions (stressing error-free performance), Detection instructions (entailing a delayed motor response). S1 is a 70 dB tone, and S2 requires a visual discrimination. Data are averaged over 10 subjects. (From Gaillard, 1980.)

8. The amplitude of the terminal CNV varies with spontaneous trial-to-trial fluctuations in the speed of motor response, so that quick RTs are associated with larger terminal CNVs than

are slow RTs (Harkins et al., 1976; Rohrbaugh et al., 1976; Brunia and Vingerhoets, 1980).

9. The motor character of the terminal CNV is further supported by its topographical and morphological similarity to the readiness potential, as noted above with reference to Figure 2. Both are negative shifts most prominent over the motor cortex, which gradually develop beginning one or two sec before the execution of a motor response. Laterally, there is a tendency for both to be dominant over motor areas contralateral to the hand of response. The magnitude of the asymmetry for the terminal CNV is related to the speed of response, and is largest for subjects who show large asymmetries in the readiness potential (Rohrbaugh et al., 1976). When foot movements are required, the direction of asymmetry reverses so that the dominance is ipsilateral, both for the terminal CNV and for the readiness potential associated with spontaneous foot movements (Brunia, 1980a, 1980b; Brunia and Vingerhoets, 1980; Brunia and van den Bosch, 1981). The ipsilateral preponderance for foot movements is apparently related to the orientation of the contralateral projection area for the lower limb in the depth along the longitudinal fissure, which is such that its activity is more easily recorded over the opposite hemisphere (Brunia and Vingerhoets, 1980).

The identification of the CNV, or a portion of it, with the readiness potential is occasionally objected to on grounds that the two do not always share an identical topographic distribution. It is generally found that the CNV observed at short foreperiods includes a more prominent anterior representation than does the readiness potential (e.g. Tecce, 1972; McCallum, 1978). As noted elsewhere, however (Loveless, 1979; Gaillard, 1980), this difference can be attributed to the contribution of the frontal elements of the O wave, which combine with the centrally-dominant terminal CNV wave. In general, the anterior-posterior distribution of the CNV at short foreperiods will depend on the relative contribution of the individual components.

With respect to the lateral distribution, Donchin et al. (1978), deplore the "simple equation of late CNV negativity with the (readiness potential)" because it "ignores a wide range of established facts about both phenomena. It is reasonably well established that the (readiness potential) is asymmetrical, showing greater negativity over the hemisphere contralateral to the operative muscles. The general consensus with regard to the CNV is that it shows no consistent lateral asymmetry" (p. 369).

Our experience has been that the CNV lateralizes to the extent that the readiness potential itself lateralizes, and that the asymmetry of the readiness potential itself is not so universal a finding as the above quote would imply (c.f. Deecke et al., 1969, 1973, 1976). An example of a failure to observe laterality in the terminal CNV is apparent in Figure 9; subsequent examination determined that the readiness potential itself showed no appreciable laterality for the particular response used. Elsewhere, Donchin et al. have come to a similar conclusion. After having declared that "The CNV does not behave like a 'motor' potential" (Donchin et al., 1974) because it failed to show any response-related asymmetry, it subsequently was determined that the readiness potential itself also failed to lateralize for the specific thumb button press used (Donchin et al., 1977). Additionally, it must be noted that the most significant asymmetries in the pre-motion negativity are confined to the last 150 msec before the movement (Deecke et al., 1973), meaning that any normal RT latency would postpone the region of largest asymmetries until after S2, not during the CNV foreperiod.

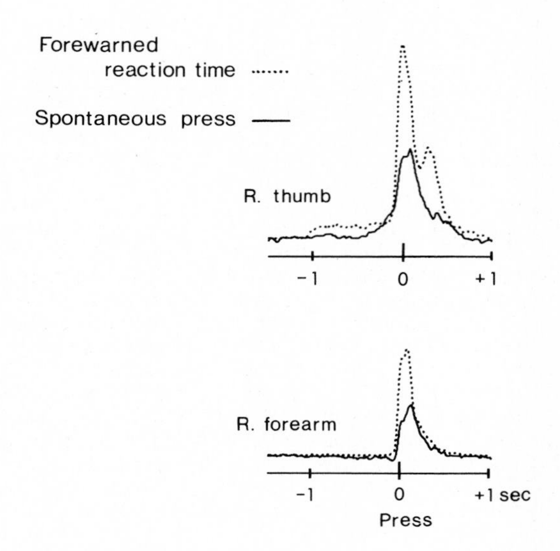

Figure 12. Integrated EMG from thumb and forearm muscles in a simple forewarned RT condition (dotted traces) and a Spontaneous Pressing condition (solid traces). (J.W. Rohrbaugh, J.L. Varner, and R.J. Ellingson, unpublished data.)

Another factor is that the movements themselves simply are not normally equivalent in uncued as opposed to RT situations, and so neither are the preceding negative waves likely to be equivalent. Movements in RT situations usually are made with greater speed, force and EMG activity, even in spite of stringent instructions to the contrary. An example of this difference is illustrated in Figure 12, where EMG data from two thumb press tasks are plotted (see also Gaillard, 1980). The top traces are from electrodes overlying the involved thumb abductor muscle, while those on the bottom are from electrodes overlying the forearm flexor muscles. The EMG has been rectified and integrated with a time constant of 0.5 sec. The solid records are from a Spontaneous Pressing condition (at an approximate rate of one press per 6 sec). The dotted records are from typical forewarned RT conditions, with a 1 sec foreperiod. The EMG activity associated with the forewarned RT conditions is substantially greater at both sites, despite frequent entreaties to the subjects that they standardize their movements.

Although we believe the terminal CNV is inextricably linked with motor preparation, it clearly is not the case that it determines the nature of the eventual movement with unremitting faithfulness. It is clear that movements can be executed reflexively or responsively--without the 1 or 2 sec preparation implicit in the readiness potential--and it is clear also that muscles are innervated by input from extrapyramidal as well as pyramidal sources. The readiness potential is likely to manifest only a portion of these control sources. It is not surprising, therefore, that the terminal CNV and response characteristics are subject to dissociation.

Examples of dissociation have been reported recently by Tecce et al. (1981a, 1981b) in distraction paradigms using a foreperiod of 1.5 sec. (Since the O wave can extend for several seconds, the terminal CNV measure at this foreperiod is not apt to be a very pure measure.) Distraction is found to decrease CNV amplitude, but the response-related EMG activity following S2 is greater than that associated with no-distraction conditions. Tecce et al. believe these data "call into serious question the motor reductionist view of late CNV development." Contrarily, we would note that, if the distraction procedures during the foreperiod are indeed as effective as purported, the comparisons range essentially between forewarned (no distraction) and non-warned (total distraction) responses. And under non-warned conditions (perhaps as a way of compensating for the lack of preparation) the EMG activity tends to be greater than for forewarned movements (Gaillard, 1980). The fact that movements can be executed successfully either with or without an unimpeded opportunity for preparation (and accompanying negativity) indicates only the obvious fact that a foreperiod is not an indispensible precursor to movement. An illustrative analogy is found in a report by Kutas and Donchin (1980), where EMG and readiness potential data are presented from spontaneous response conditions and in non-warned RT conditions (among others). Subjects were trained to give responses in both conditions with approximately the same force and EMG manifestations. In the spontaneous responding condition the responses were preceded by a readiness potential of 10 μv or more. In contrast, the non-warned responses were, of course, not preceded by any observable readiness potential. Are we to argue, from this example of dissociation between readiness potential and EMG, against a "motor reductionist" view of the readiness potential? In view of the strength of the evidence indicating the contrary, a more proper course may be to recognize that superficially comparable movements may be executed according to a variety of strategies, or, conversely, that superficially different movements may share some common preparatory element.

THE CNV AS NON-CONTINGENT WAVES: THE DEGREE OF GENERALITY

We have summarized evidence that indicates, in brief, that the CNV is composed of two major components (or clusters of components): the O wave and the readiness potential. Neither of these components, moreover, depends strictly upon the paired-stimulus CNV situation for its elicitation. In the following section we consider a number of objections that have been raised to this point of view, and sample also the range of CNV phenomena that can be accounted for with comfortable extensions of the model.

Foreperiod duration. Arguments about the composition of the CNV are based in part upon data obtained from situations in which the foreperiod is 3 or 4 sec. An immediate problem pertains to the extent to which conclusions from such conditions can be generalized to foreperiods of different length. At very long foreperiods, it is commonly found that the terminal CNV is attenuated in comparison to values at foreperiods of 3 or 4 sec (e.g. Loveless and Sanford, 1974a). This may be held most likely to reflect difficulties in precise time estimation of long intervals, and the consequent reluctance or inability on the subjects' part to engage in anticipatory motor behavior.

There remains the issue of the extent to which the model can be generalized to the customary, short CNV foreperiod of 1 or 1.5 sec. A variety of RT and conditioning effects appear to be optimum within these (or shorter) intervals, raising the possibility that qualitatively distinct brain potential manifestations emerge at these intervals as well (e.g. Frölich et al., 1980). At longer foreperiods, these distinct manifestations might be less prominent, and relegated proportionately to an indiscernible portion of the total waveform.

Several considerations indicate that such potentials, if present, are quite small and contribute little to the total waveform at any foreperiod. A number of authors (e.g. Irwin et al., 1966; Loveless, 1979) have noted that the amplitude and topographic representation of the composite CNV is of the same order as would be expected from the admixture of readiness potential and O wave components. It seems unlikely, moreover, that the readiness potential and O wave components (whose ubiquity has been noted above) would be supplanted or dwarfed by other components simply because the foreperiod is short.

We have addressed this issue experimentally in a method outlined in Figure 13 (from Rohrbaugh et al., 1980). Essentialy, CNV-like waveforms are constructed from separate O wave, readiness potential, and sensory related potentials, for comparison with actual CNVs from the same subjects. The left panels in Figure 13 depict records from an equivalent pairing of visual-auditory stimuli, while the right panels depict records from an auditory-visual pairing. Potentials (averaged over 8 subjects) are shown from 3 midline and 2 lateral sites. The top trace in each panel is a readiness potential obtained under conditions in which subjects pressed a telegraph key spontaneously with their right hands, with no external stimulation or strict enforcement of rate. The second trace is the event related potential (including the O wave) obtained for unpaired flashes

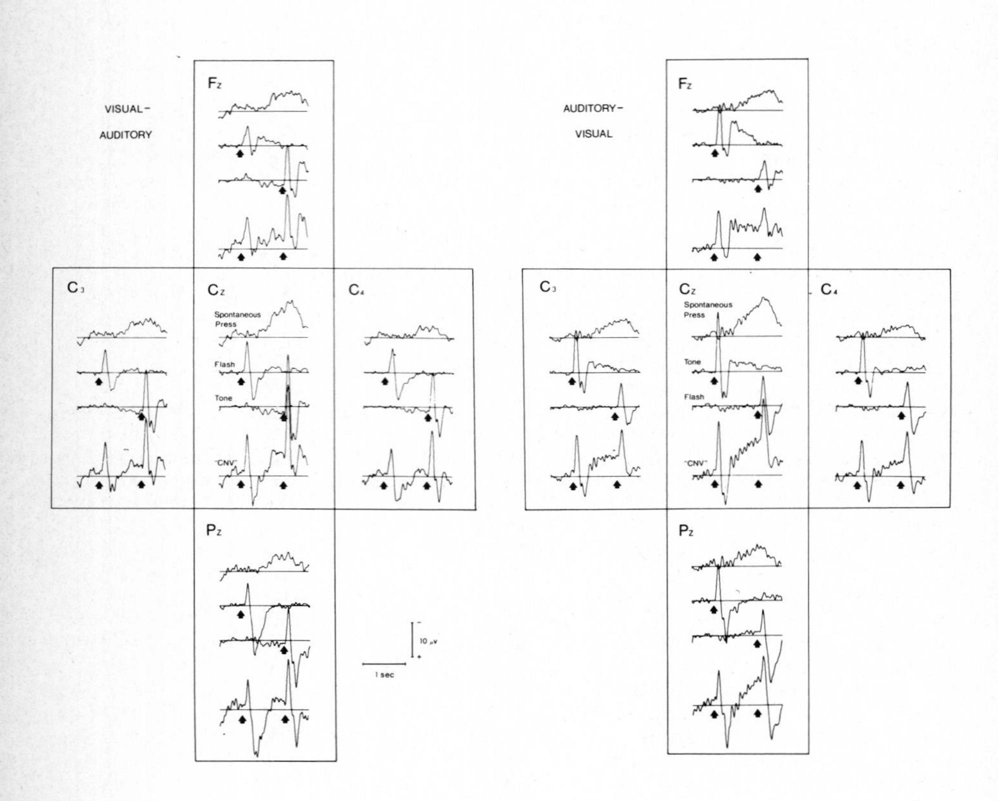

Figure 13. Potentials from 5 electrode sites for a visual-auditory sequence (left) and an auditory-visual sequence (right). Potentials are shown from electrodes overlying midline frontal (Fz), central (Cz), and parietal (Pz) sites, and bilateral left (C3) and right (C4) motor area sites. The bottom waveform is a synthesized waveform that has been constructed by adding separate potentials associated with uncued motor presses (top trace in each panel), and with individual stimuli whose records are staggered so as to simulate a foreperiod of 1 sec. (From Rohrbaugh et al., 1980.)

(left panels) or tones (right panels). The third traces are again potentials obtained from unpaired tones (left panels) or flashes (right panels), but the traces have been staggered with those above so as to simulate a foreperiod of 1 sec. The readiness potential is aligned so that the moment of response follows the second stimulus by delays equivalent to actual RT delays derived from subsequent CNV-RT conditions. Additionally, the readiness potential traces have been temporally smeared by deliberately introducing some averaging jitter equivalent to that injected by actual RT distributions. In the bottom traces, the separate waves have simply been added together, yielding waveforms typical of CNV waveforms.

The features described in previous sections are evident for the individual waves, and are preserved in the composite waveforms: the O waves are dominant frontally and are smaller in response to visual than auditory stimuli. The readiness potential is largest centrally, and shows a preponderance contralateral to the hand of response.

In acquiring these potentials, two steps have been taken to recreate the conditions representative of typical CNV situations. One is to introduce a simple counting task when O waves are obtained, in an attempt to mimic the signal value associated with a warning stimulus. The second step is to base the readines potential averages only upon those key presses that look like key presses made under actual CNV conditions. We observe that subjects typically make spontaneous key presses with less force, speed and EMG activity than when the presses are given in RT situations (see Figure 12). The trials included in the readiness potential averages here are selected so that the throw times of the key presses (i.e., the times taken from initial key displacement to full displacement) are similar to the throw times associated with actual RT responses.

The CNV-like waveforms so synthesized closely match actual CNV waveforms obtained from the same subjects under equivalent conditions, and a number of distinguishing characteristics are apparent for both actual and synthesized CNVs: a tendency to be largest at Cz (Cohen, 1969; Syndulko and Lindsley, 1977), the early rise and plateau at Fz (Järvilehto and Frühstorfer, 1970; Näätänen and Gaillard, 1974; Weinberg and Papakostopoulos, 1975; Syndulko and Lindsley, 1977), the late lateral predominance on the side contralateral to the hand of response (Otto and Leiffer, 1973; Syndulko and Lindsley, 1977; McCallum and Curry, 1980), and the tendency for CNVs with a visual S1 to be relatively smaller and later in their growth than for those with an auditory S1 (Gaillard and Näätänen, 1976).

The essential similarities among synthesized and actual CNVs are confirmed also by PCA. Separate PCAs of the covariance matrices associated with actual and synthesized CNVs disclose, in each case, two factors associated with slow activity during the foreperiod. The loadings for these factors are indicated by the heavy traces in Figure 14. The corresponding factors for actual and synthesized CNVs are remarkably similar in their appearance; one factor is associated with the O wave following the first stimulus (Synthesized CNV Factor 2 corresponding to Actual CNV Factor 3) and the other factor is associated with the readiness potential (Synthesized CNV Factor 3 corresponding to Actual CNV Factor 1). A single PCA including all data (both actual and synthesized CNVs) combined the factors that corresponded with one another in the separate analyses, yielding again only two factors associated with the slow activity during the foreperiod. Whereas analysis of variance of the associated factor scores showed significant variation as a function of electrode site and stimulus pairing order, no significant differences were found for the two foreperiod factors that permitted discrimination between actual and synthesized CNVs. A similar factor structure has been reported in a CNV situation by McCarthy and Donchin (1978), but it bears emphasizing that the factor structure on the left of Figure 14 has been obtained from an ensemble of individual components, none of which has anything to do with the paired-stimuli CNV situation. The major discrepancy between the factor structures here and those derived from long foreperiod situations (see Figure 10) is that the O wave is represented by only one, rather than two,

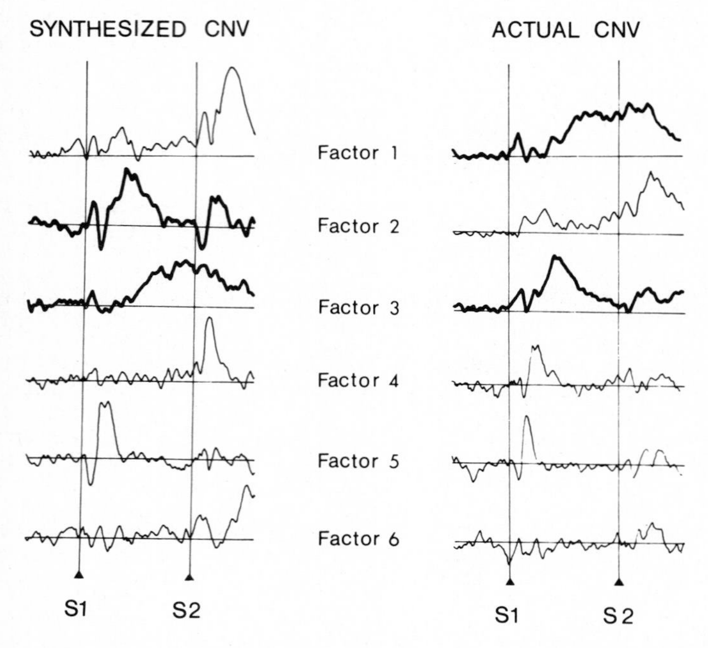

Figure 14. Factor loadings from PCA (of covariance matrix with Varimax rotation) of synthesized and actual CNV waveforms. The loadings associated with slow activity during the foreperiod are indicated by the heavy records. (After Rohrbaugh et al., 1980.)

factors. The reason for this discrepancy likely is that the late O wave factor (representing SNW2), given its central representation and late onset, is in position to be coalesced with the readiness potential factor when the foreperiod is 1 sec.

<u>Task elaborations: Judgment complexity, difficulty and accuracy.</u> An objection that has been raised to the above demonstration is that it pertains most clearly to the standard CNV paradigm having invariant S1 and S2, and a non-discriminative response. The extent to which the model may be generalized to more complicated or demanding situations remains untested. We do not see this as a significant problem. The effects upon the CNV associated with such manipulations are remarkably small or non-existent. Even the fundamental complication of requiring subjects to prepare for choice rather than simple RT seems not to yield consistent increases in CNV amplitude. One study (Donchin et al., 1975) has reported a statistically significant but small CNV increase for choice RT (although comparison is made difficult by the requirement that subjects fit both simple and choice RTs within a 350 to 500 msec window). Kutas and Donchin (1980) obtained a larger CNV preceding choice RT than simple RT, but only at a frontal site (suggesting a possible contribution of an O wave effect). The more usual finding is no difference (e.g. Donchin et al., 1972; Syndulko and Lindsley, 1977) or a marked decrement in CNV amplitude for choice RTs (e.g. Donchin et al., 1976; Poon et al., 1976; Kirst and Beatty, 1978). The decrement with choice RT may be interpreted within this context as reflecting a diversion from the motor character of the simple RT task (Gaillard and Perdok, 1980).

Similarly, at long foreperiods the terminal CNV appears not to be remarkably or consistently different when preparation is for choice as opposed to simple RT. On one occasion we have observed the terminal CNV wave to be somewhat larger (although not significantly so) preceding choice RT (see Figures 9 and 10 above). On other occasions essentially no difference is found (Plooij-van Gorsel, 1980), or the terminal wave is smaller preceding choice RT (Sanquist et al., 1981). The pattern of results in choice RT situations is consistent with proposals that motor programming involves principally the generation of abstract timing networks (e.g. Klapp, 1977), broad facilitatory substrates (e.g. Brunia, 1979), and postural compensation (e.g. Lee, 1980)--not solely commands to specific muscles. These generalized motor effects may, of course, be susceptible to distraction when attention is diverted to perceptual or judgmental aspects of the task, and they may on occasion be supplemented with specific, unilateral preparation as the task warrants or permits (see Gaillard et al., 1980).

Neither is there any consensus that the CNV is increased when judgment difficulty is increased, nor does the CNV appear reliably to be greater preceding accurate than faulty judgments. Hillyard, in his thoughtful review of the CNV literature (1973), concluded that "the degree to which the CNV predicts subsequent perceptual sensitivity is rather meager" (p. 165) and we know of no recent data to contraindicate this conclusion (see review by Gaillard, 1978). The more usual finding in recent reports is, as before, that the CNV is not related to difficulty or performance, or that the CNV is actually decreased as a function of difficulty or accuracy (Jenness, 1972; Paul and Sutton, 1972; Loveless and Sanford, 1974b; Warren, 1974; Roth et al., 1975, 1978; McCarthy and Donchin, 1978; Perdok and Gaillard, 1979; Gaillard and Perdok, 1980; Sanquist et al., 1981). The decrease when judgmental demands are great most likely indicates, again, that the motor aspects of the task have been compromised.

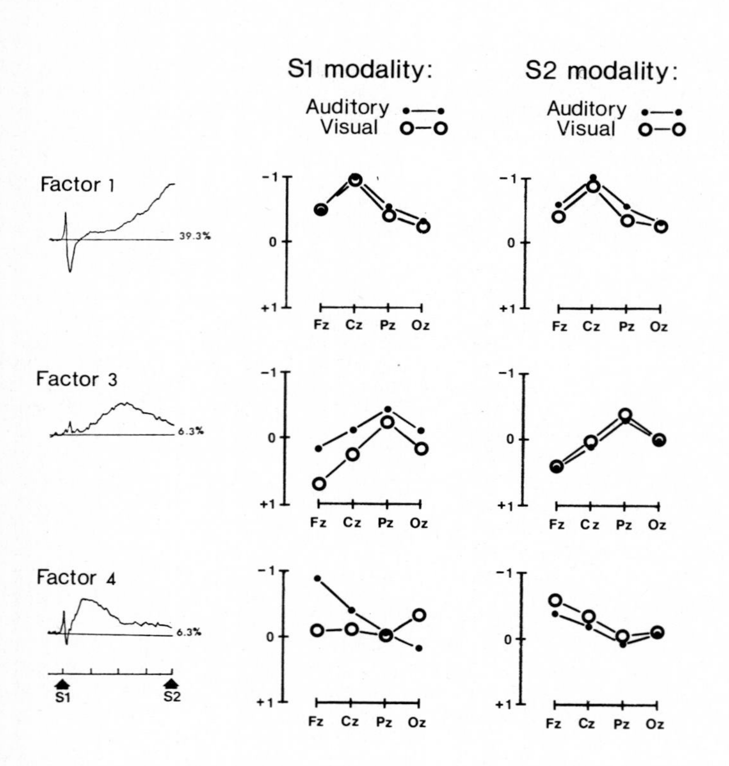

Figure 15. Results from PCA (of covariance matrix with Varimax rotation) of CNV records with a 4 sec foreperiod, in which modalities of S1 and S2 have been factorially crossed. The loadings for the 3 factors associated with slow activity during the foreperiod are plotted in the left column. In the center column are the associated factor scores, derived by collapsing across all variables except S1 modality. In the right column the scores have been collapsed across all variables except S2 modality. (Constructed from data presented in Sanquist et al., 1981.)

Those occasional instances in which a positive relationship between CNV amplitude and sensitivity has been observed (e.g. Hillyard et al., 1971) could be ascribed to properties of the O wave rather than anticipatory priming of the sensory and perceptual systems. An important feature of the orienting reflex (which, it is conjectured, the O wave represents) is the regulation of sensitivity. Performance might be expected to vary, therefore, along the temporal course of the O wave (Loveless, 1975). At any given foreperiod, a weak relationship might emerge from trial-to-trial fluctuations in attentiveness, since attentive states presumably would pose favorable conditions both for O wave generation in response to S1 and for ensuing performance (Gaillard, 1978).

Modality-specific anticipation. Equally uncompelling is the evidence in support of a modality-specific anticipatory factor in the CNV. In some instances purporting to demonstrate such a factor, the modality of S2 (and hypothesized modality of preparation) has been confounded with the modality of S1 (Poon et al., 1976; Simson et al., 1976, 1977; Syndulko and Lindsley, 1977). In these instances, the effects are equally likely to represent the previously described O wave variation as modality-specific anticipation. In other instances, the effects are extraordinarily small--no more than 1 or 2 μv--and border on the range of experimental error (Gaillard and Näätänen, 1976; Ritter et al., 1980; see Gaillard, 1978). A recent study by Sanquist et al. (1981), in which modalities of S1 and S2 were factorially crossed, and in which judgment difficulty and response requirements were parametrically varied, found no evidence in support of a modality-specific anticipatory factor. Data from Sanquist et al. (1981) are summarized in Figure 15. The data plotted here have been collapsed across difficulty and response variables so as to distill only those aspects bearing on modality influences on the CNV. (There were no significant interactions between modality and these variables.) The plots are based upon a PCA with Varimax rotation of the cross products matrix, but similar conclusions may be drawn from parallel analyses of the covariance matrix or the more usual amplitude measurements. As is the usual case at long foreperiods (see Figure 10 above), 3 factors are related to the foreperiod. Factor 1 is a terminal CNV factor, while Factors 3 and 4 represent the late and early O wave factors, respectively. In the middle column the magnitude of the factor scores has been plotted at the various midline electrode sites, with separate functions for auditory and visual S1s. The plots in the right column are a function of S2 modality. For the first factor, representing the terminal CNV, there are, as might be expected, no effects associated with S1 modality. More importantly, neither is there any suggestion of an effect associated with the modality of S2. In contrast, the two O wave factors are quite clearly related to S1 modality, with a significant main effect for Factor 3 and a significant interaction with electrode site for Factor 4. The interaction observed for Factor 4 is consistent with a pattern of activity (although not anticipatory activity) specific to the relevant sensory cortices.

CNV shape. As summarized above, factors of judgment complexity, difficulty, accuracy or modality seem to have little or negative effect on the CNV, and would seem to provide no test as to the adequacy of the two-component model of CNV formation. In marked contrast, a number of situations in which CNV variation is very great can be accounted for by the model quite well. We have noted in previous sections the ease whereby the effects of modality, intensity, response requirements, laterality and information content can be accommodated. Perhaps the most impressive source of CNV variability is the familiar and remarkable variation among individuals. The shape of the CNV, in particular, appears to be a promising feature in the characterization of psychiatric or neurological disposition (see Roth, 1977). The most common distinction is between so-called type A and type B, drawn by Tecce (1972) and others. Whereas type A has an abrupt rise early in the foreperiod, type B has a more ramp-shaped development that peaks at the time of S2. A more subtle classification system, distinguishing 4 types, has been proposed by Timsit-Berthier et al. (1973a) and shown in a large sample of patients to be related to psychiatric diagnosis. Yet another feature considered of importance is the prolongation of the CNV following S2, usually called the post-imperative negative variation (PINV; see review by Dongier et al., 1977).

Although the amount of variation across subjects is indeed remarkable, it does not exceed that which might be expected from separate variation of the separate constituents (Gaillard, 1978). As shown in previous figures, the O wave can assume a wide range of appearances as a function of experimental variables, and a comparable range of variation is typically seen over subjects. It is also not uncommon to observe a variety of amplitudes and shapes in the readiness potential for different tasks (e.g. Kutas and Donchin, 1974; Grünewald-Zuberbier and Grünewald, 1978; Taylor, 1978) or different individuals (Timsit-Berthier et al, 1973b; Karrer and Ivins, 1976; Deecke and Kornhuber, 1978; Warren et al., 1981). The range of variation demonstrated individually for the O wave and readiness potential would be sufficient in concert to encompass a prodigious array of CNV shapes, amplitudes and distributions. Accordingly, the type A shape would reflect the relative predominance of the early O wave, type B would reflect the predominance of the readiness potential, and the "dome shaped" CNV would represent a more balanced admixture. It is possible that the PINV might represent an unusually protracted O wave that persists beyond S2, or perhaps an O wave atypically elicited by S2 itself (Callaway, 1975).

Distraction. Another large and well documented CNV effect is that associated with distraction (e.g. Tecce et al., 1976; McCallum and Walter, 1968). The presence of extraneous stimuli or tasks during the foreperiod typically leads to a profound reduction in CNV amplitude. Again, these findings are compatible with the two-component model. The deleterious effects of competing mental demands upon the terminal CNV and associated motor performance have been described above (see Loveless and

Sanford, 1974b; Gaillard and Perdok, 1980). Although there appear to be no direct data bearing on this, it is extremely likely that the uncued readiness potential is susceptible to distraction as well. Certainly motor performance itself is; there is experimental evidence dating from the reports of Binet (1890) and Welch (1898) demonstrating that even so simple a motor act as maintaining the strength of a grip suffers in the presence of competing mental demands.

Data are available to indicate that the O wave also is sensitive to attentional factors. Some rudiment of task involvement appears necessary for its elicitation in the first place. It seems additionally to be important that attention be directed to the eliciting stimulus by making it relevant (Rohrbaugh et al., 1978).

IS THERE A TRUE CNV?

As we have considered in the previous section, the effects associated with foreperiod duration, task factors, modality, shape and distraction all are amenable to interpretation in terms of a two component model of the CNV. None of these factors can be considered to reveal unequivocally the existence of a "true CNV" in the composite waveform. In the following section, we consider what are perhaps the most frequently proposed examples of the "true CNV," namely those that can be obtained in the absence of an instructed motor response. In such cases, there is presumed to be no readiness potential contribution to the CNV waveform. Here, too, we detect little or no evidence in support of a "true CNV;" the effects are subject to interpretation in terms of the O wave, or the readiness potential, or some combination of the two.

CNVs with no instructed response: Contribution of the readiness potential. We must first note the difficulty of devising tasks that are devoid of motor activity. It is generally recognized, although only platitudinously honored, that detection, discrimination and judgment tasks are invariably accompanied by subtle or adventitious somatic activity. Sperry (1952) has suggested that, regardless of the manifest behavioral situation, the primary end product of the brain is motor output: "insofar as an organism perceives a given object, it is prepared to respond with reference to it." (p. 301). This activity is difficult to characterize comprehensively, and whether or not it is preceded by a specific readiness potential is not known.

In some situations, the extraneous motor activity may not be covert at all. It is not an uncommon practice, in particular, to tolerate EOG activity after the foreperiod--in some instances the problem apparently is so acute that records are not presented beyond S2. Yet, if these eye movements are large, are encouraged by scanning tasks or as a relief from the rigors of fixation during the foreperiod, and are time locked to S2, they may be considered as surely to be motor responses as are instructed key presses. Two recent studies, by Klorman and

Ryan (1980) and Simons et al. (1979), seem especially amenable to this interpretation. These studies report a departure from the usual finding of no appreciable terminal CNV in the absence of an instructed motor response. Both experiments shared several methodological similarities. One shared feature, which in view of the sustained potential characterized above may represent a complicating factor, is that both used a continuous tone to demarcate the foreperiod. Also, both used pictorial S2s that were laden with interest (nude females [Simons et al., 1979]) or affect (mutilation scenes [Klorman and Ryan, 1980]), and neither required a speeded or overt motor response. The terminal CNV wave observed under these conditions was somewhat smaller than when a RT response was required (Simons et al.), and had a more frontal representation (Klorman and Ryan).

A possible interpretation, which in view of the potential importance of these results merits examination, is that the terminal CNV in these cases reflects preparation for voluntary saccadic movements or other scanning-related movements. The projected images subtended a size of about 17 by 31 degrees visual angle (Klorman and Ryan), which would seem to be so large as to require some scanning movements in order to be apprehended. This is acknowledged by Klorman and Ryan, who reported that "Occular fixation during the slide was not required, because it would have interfered with viewing the slides" (p. 515). Precisely how the propensity for these movements might be distributed among the neutral, interesting or aversive viewing conditions is difficult to preordain. It is known that voluntary saccades of this sort are preceded by readiness potential-like negative shifts, whose amplitude and scalp distribution depend upon whether the saccades are spontaneously initiated or externally timed (Lehmann, 1971; Becker et al., 1972; Kurtzberg and Vaughan, 1977). Of particular relevance is one anecdotal report that the readiness potential is larger preceding gaze shifts to an interesting picture (again, a picture of a "beautiful girl") than to less engrossing pictures (Jung, 1969). It is also possible that changes in head orientation may accompany the ocular scanning.

<u>CNVs with no instructed response: Contribution of the O wave.</u> In most instances, however, we believe that CNVs in the absence of instructed motor responses can be explained without recourse to covert muscular activity. As described throughout this manuscript, a slow negative wave of substantial proportion can be recorded not only without a motor response, but without an S2, and we would propose that CNVs in response-free situations include the O wave as their sole constituent. We hardly can claim authorship for this proposal; Walter (1967) noted that CNVs could be obtained without a motor response, but noted further that CNVs could be obtained without an S2 as well, "provided that the subject attached some importance to the warning in the sense of 'making up his mind'" (p. 125).

Still, however, negative waves occurring in the absence of an instructed response are often considered anticipatory in nature, rather than an O wave to S1, and are pointed to on occasion as examples of the "true CNV." These reports frequently are difficult to evaluate. Often such waves are observed incidentally to some other purpose, and appropriate conditions are not available for assaying somatic activity or for comparing the waves with O waves obtained in non-contingent situations. A representative example is apparent in a report by Woods et al. (1980), wherein an elegant series of experiments examining temporal factors in decision making and P300 generation are described. The paradigm entails an interval of 3.3 sec bounded by warning and imperative flashes. At precisely demarcated positions within this interval, tone stimuli did, or did not, appear, and subjects were instructed to respond to the perceived number of tones with a verbal report following the terminating flash. The entire foreperiod is spanned by a broad negative wave, upon which the tone-evoked responses are interpolated.

The size (about 7 to 9 µv maximum), duration, and distribution of this wave are not disparate from values reported for the O wave in response to non-paired stimuli (c.f. Figure 7 above), which (in combination with even the slightest pre-verbalization readiness potential) could yield waveforms of the sort observed. The mid-interval tones might serve to reinforce or contribute additional negativity to this wave, as is suggested by the finding that the total accumulation of negativity was larger for intense, suprathreshold tones than for near threshold tones. (This result is opposite to what might be expected if the wave were related to perceptual or judgmental demands.) The finding that the wave persists even on trials in which no intervening tones were presented is not especially telling, since it is known that stimulus omissions in precisely timed situations are effective at eliciting their own O waves including a small, but nevertheless reliable, negative aspect (Rohrbaugh et al., 1978).

Perhaps the most systematic, and frequently cited, investigation of CNVs in the absence of an instructed motor response is that by Donchin et al. (1972). We (Rohrbaugh, Peters, Varner and Ellingson, unpublished data) have replicated this experiment as faithfully as our laboratory conditions permit, and have made the additional observation of actual O waves in response to non-paired stimuli. CNVs were obtained in several response and no-response conditions, each of which used a foreperiod of 1 sec. The S1 always was a click (produced by a 1 msec, 1 volt pulse applied to Grason-Stadler insert earphones), and S2 was equiprobably either a star or circle figure. The figures were luminous upon a dark background, and were displayed (using a shuttered projection tachistoscope) for a duration of 10 msec, with a size of 1 degree 42' visual angle. These stimuli were presented in four experimental conditions, given in counterbalanced order. Two conditions required motor responses, non-selectively in the "Both" condition, and selectively only to the star in the "Select" condition. The motor response entailed pressing a thumb button that was hand-held in a block of dental acrylic. The "Guess" and "Compute" conditions required no immediate overt response. In the "Guess" condition the subject was asked to predict the upcoming pattern, before the trial, while in the "Compute" condition he was given the base number of 545 and asked to keep a running computation, adding 7 for a star, and subtracting 7 for a circle.

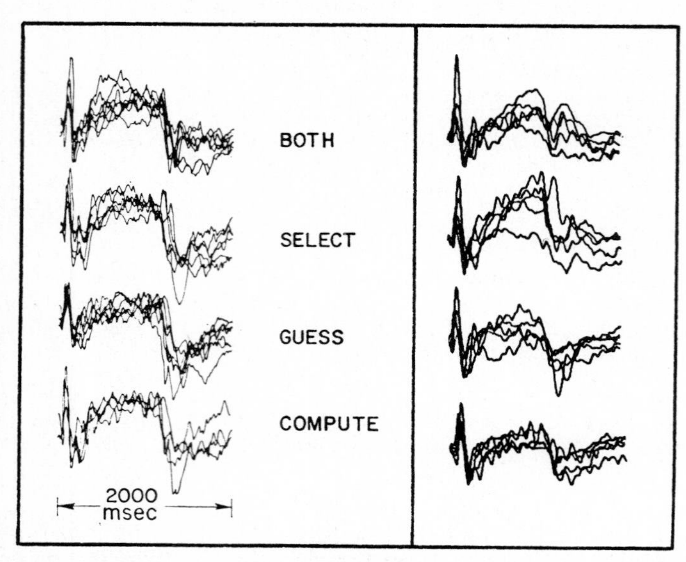

Figure 16. CNV records obtained from 4 CNV conditions by Donchin et al., 1972 (left panel) and a replication (right panel). (J.W. Rohrbaugh, J.F. Peters, J.L. Varner, and R.J. Ellingson, unpublished data.) Overlaid traces are from individual subjects.

CNV records obtained under these conditions are plotted in Figure 16. The records on the left are reproduced from the original Donchin et al. (1972) article, and depict overlaid traces of the vertex records from 7 subjects. On the right are the records we obtained from 5 subjects under the same conditions. The agreement between the two sets of records is quite good. Certainly, it is evident that some sort of CNV is obtained in no-response conditions (the "Guess" and "Compute" conditions) in both experiments.

In Figure 17 the records we obtained have been plotted in a different format. Baselines have been fitted through the 500 msec prestimulus interval, an amplitude calibration mark has been added, and the records have been averaged over the 5 subjects. Records from the two motor response conditions are overlaid in the left column, for 5 midline electrode sites. The CNVs in the "Select" condition are somewhat larger than those in the "Both" condition; this difference appeared in 2 of the 5 subjects (2 showed a small difference in the opposite direction) and did not approach statistical significance. In the center column, records from the two no-response conditions have been similarly overlaid. Two major differences are apparent: the CNVs in no-response conditions are substantially smaller than those in response conditions, and their representation is more frontal. These differences replicate those that are commonly reported (Järvilehto and Früstorfer, 1970; Peters et al., 1970; Donald, 1973; Näätänen et al., 1977; Syndulko and Lindsley, 1977; Herning et al., 1979; McCallum and Curry, 1981). The amplitude and frontal distribution of the CNVs in the no-response conditions also are suggestively compatible with those described above for the O wave. Data in the right column permit direct comparison with actual O waves obtained from the same subjects in a separate session under conditions involving no S2 or response. The waves shown are for two selected conditions, one a click intensity discrimination condition (in which subjects were asked to count voicelessly the number of occasional [p=.10] 10 dB weaker clicks)

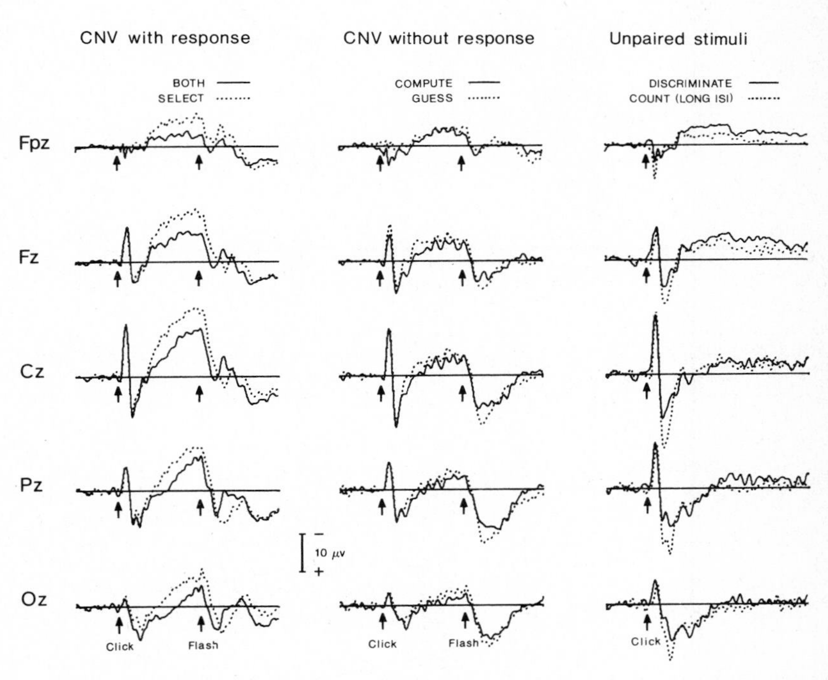

Figure 17. CNV records obtained for RT response conditions (left column) and no-response conditions (center column), plotted for 5 midline electrode sites. Records in the right column are O waves from two non-paired stimulus tasks, entailing an intensity discrimination (solid line) and a simple voiceless counting task (dotted line). (J.W. Rohrbaugh, J.F. Peters, J.L. Varner, and R.J. Ellingson, unpublished data.)

and a simple voiceless counting condition in which the inter-stimulus interval (ISI) averages 16 sec. The center and right columns compare favorably with one another (the major discrepancy being a greater posterior SPW in the O waves from the non-paired conditions), supporting again the likelihood that the CNV with no response is in essence an O wave, of the sort observed for non-paired stimuli.

CNVs with no stimuli: A "true CNV" in the readiness potential? As a final issue, one that again concerns the relationship between the CNV and motor response aspects, we should like to consider the possibility that the "true CNV" is embodied in a symmetrical component of the readiness potential. It is often found that the amplitude, but not the degree of asymmetry, of the readiness potential is related to force, excursion, precision and other movement qualities. From these findings, the existence of a separate symmetrical component has been inferred, and it is frequently suggested that this symmetrical component may represent a CNV-like wave having the various non-specific preparatory features traditionally ascribed.

Several objections may be raised to such a proposal, all of which stem from what may be seen as a tendency to inject an excessive measure of complication into the spontaneous movement paradigm. The CNV has fled from a paradigm involving two stimuli but no response, to another paradigm involving a response but no stimuli. This is a very large shift in paradigm, one that demands an extraordinarily protean conception of the CNV. The spontaneous movement paradigm is an unusually straightforward experimental paradigm, and we question the usefulness of introducing the CNV concept, with all the surplus meaning that adheres to it, to this situation. At some point, the criterion of what makes good sense operationally must be asserted. The paired-stimuli and spontaneous movement paradigms share in common no experimental operations at all, other than the facts that electrodes are attached and the subject is believed to have something on his or her mind during the testing session. But there are no operations in the spontaneous movement paradigm for assessing or assuring what it is that the subject has on his mind. He may indeed (as has been suggested) be mobilizing his mental resources, shifting his attention to the impending movement, or conjuring up imaginary warning and imperative stimuli. From what little our experimental operations are able to tell us about these mental processes, however, it is equally likely that his thoughts are wholly extraneous.

A few things about the spontaneous movement paradigm are known with certainty--that the subject is making a movement, and that the EEG potentials recorded while he is making that movement are related faithfully and lawfully to the mechanical and EMG aspects of that movement. The interests of simplicity discourage passing beyond these simple and obvious facts when seeking an explanation. And certainly there are several means whereby symmetrical waves related to the movement *per se* could arise. It is known that focal generators may lead to very broadly distributed scalp potentials, because of the vagaries of volume conduction and extra-cerebral current paths. More significantly, it is also known that movements often are accompanied by diffuse bilateral substrates of adventitious muscular activity and compensatory postural adjustments (e.g. Lee, 1981), that some movements (particularly those of the limbs and axial musculature) have an element of ipsilateral control (see Iverson, 1981), that a major element of motor programming is for coordinated spatio-temporal patterns, rather than specific muscles (e.g. Klapp, 1977), and that midline supplementary motor areas are engaged by many movements (e.g. Deecke and Kornhuber, 1978; Orgogozo et al., 1979). Any or all of these factors might reasonably contribute to a symmetrical readiness potential component and, we should think, might more fruitfully be explored than hypothetical mental factors.

CONCLUSIONS

In sum, we have argued that, in the situations to which we are accustomed, the CNV may be construed as including principally or solely the O wave and readiness potential. This is not to exclude dogmatically the possibility that other waves may on occasion be present. It is known that slow waves neurally are multifaceted in their origins (e.g. McCallum et al., 1973; Rebert, 1976). Existence is also known of a variety of spontaneous (e.g. Girton et al., 1973; Bauer and Nirnberber, 1981) and conditioned slow responses (e.g. Elbert et al., 1980) that cannot be identified readily with either the O wave or readiness potential. Our arguments are simply that in the vast majority of CNV situations, including those responsible for the origin of the CNV concept in the first place, the two-component model provides an adequate explanation. Neither do we wish to discredit the liklihood (or, indeed, the certainty) that perceptual and cognitive acts are preceded by preparatory neuronal processes (e.g. Roland, 1981); we question only the strength of the evidence linking the CNV to such processes. Nor do we wish to deny the possible existence of a "true CNV"; we conclude only that it is not routinely a contributing factor in the CNV waveform. If it does exist, it is considerably smaller than is usually proposed. If it is to be made accessible for study, paradigms different from those now available must be improvised, and greater care must be taken to isolate it from other waves with which it presumably is confluent. If its role is to be clarified, it will come not from the routine appeal to principles of neurophysiological activation, but from the establishment of strong functional relationships with experimental variables--relationships that have not been forthcoming. In the meanwhile, the two-component model would seem to account very well for most CNV phenomena. As a general strategy, it may be profitable to explore the O wave and readiness potential individually, and exhaust all explanations for CNV phenomena in terms of their joint variability before turning to more abstruse explanations.

We have collected and reiterated these arguments for several reasons. One is in response to the frequency with which the concept of the CNV as a unitary entity continues to be called upon, despite overwhelming evidence to the contrary. We have used the term "CNV" reluctantly--to identify the traditional conditions of observation. However, its use is problematic. As argued above, the wave is a composite of several waves, none of which is "contingent" in the usual sense. Neither should the importance of the positive aspect (SPW) be overlooked.

A second, and more important reason, is in hopes of fostering study of the individual components: unencumbered, in the case of the O wave, by the paradigmatic restrictions of the paired-stimuli situation, and free, in the case of the readiness potential, from concerns that it is supplanted if recorded in timed situations. The potential range of variation for the O wave is largely unexplored, as are its behavioral concomitants. Similarly, little is known about the possible range of variation for the readiness potential, whose study has been confined to situations involving stereotyped, spontaneous movements rather than the wealth of the movement repertory.

ACKNOWLEDGMENTS

The first author is supported by NIH Grant R01 MH35742-01. Some of the data described herein were collected under Office of Naval Research contract N00014-77-C-0325. The helpful comments of Nancy Squires, Walter Ritter, Jon Peters, and Robert Ellingson are gratefully acknowledged.

Tutorials in ERP Research: Endogenous Components
A.W.K. Gaillard and W. Ritter (eds.)

14

CEREBRAL POTENTIALS DURING VOLUNTARY RAMP MOVEMENTS IN AIMING TASKS

Gerhard Grünewald and Erika Grünewald-Zuberbier

Institut für Hirnforschung der Universität, D-4000 Düsseldorf, and Neurologische Universitätsklinik mit Abteilung für Neurophysiologie, D-7800 Freiburg i.Br., Federal Republic of Germany

INTRODUCTION

In recent neurophysiological studies of motor performance two classes of voluntary movements are often distinguished: a 'ballistic' and 'ramp' type of movement (e.g. Kornhuber (1971, 1974), DeLong and Strick (1974), Marsden, Merton, Morton, Hallett, Adam and Rushton (1977), Phillips and Porter (1977), Desmedt and Godeaux (1978)). Ballistic movements are brief (about 200 msec and less), fast, and thought to be 'preprogrammed', that is launched without peripheral guidance. Ramp movements are slow, smooth, carried out in more than 500 msec, and are highly responsive to control by peripheral sensory feedback.

Scalp recorded cerebral potentials related to self-paced voluntary movements in man have been studied mainly in ballistic movements, especially in brisk isotonic or isometric contractions of limb muscles (e.g. Kornhuber and Deecke (1965), Gilden, Vaughan and Costa (1966), Vaughan, Costa and Ritter (1968), Deecke, Scheid and Kornhuber (1969), Gerbrandt, Goff and Smith (1973), Deecke, Grözinger and Kornhuber (1976), Kutas and Donchin (1977, 1980), Goff, Allison and Vaughan (1978), Shibasaki, Barrett, Halliday and Halliday (1980a,b), Becker and Kristeva (1980), Brunia and Vingerhoets (1981), Brunia and van den Bosch (in press)). Characteristic examples are demonstrated in Figure 1. The first and second rows show EMG-triggered individual averages of the left and right precentral recordings for left and right unilateral rapid flexions of the index finger. The most obvious features of the potentials are: a gradually increasing negative shift over both hemispheres beginning 1 sec or more prior to EMG onset, which subsides rapidly with a complex positive deflection after the initiation of movement. As can be seen in the bipolar left vs. right recordings of the bottom row of Figure 1, the gradient of the slow negativity increases asymmetrically some 400 msec prior to EMG. The bipolar recording shows a negative deflection preceding right-sided movement and a positive deflection preceding left-sided movement. Thus, the negativity becomes larger contralateral to the responding side. This lateralization of the potential further increases shortly before the EMG onset. (1)

Subdural recordings have verified the cortical origin of these

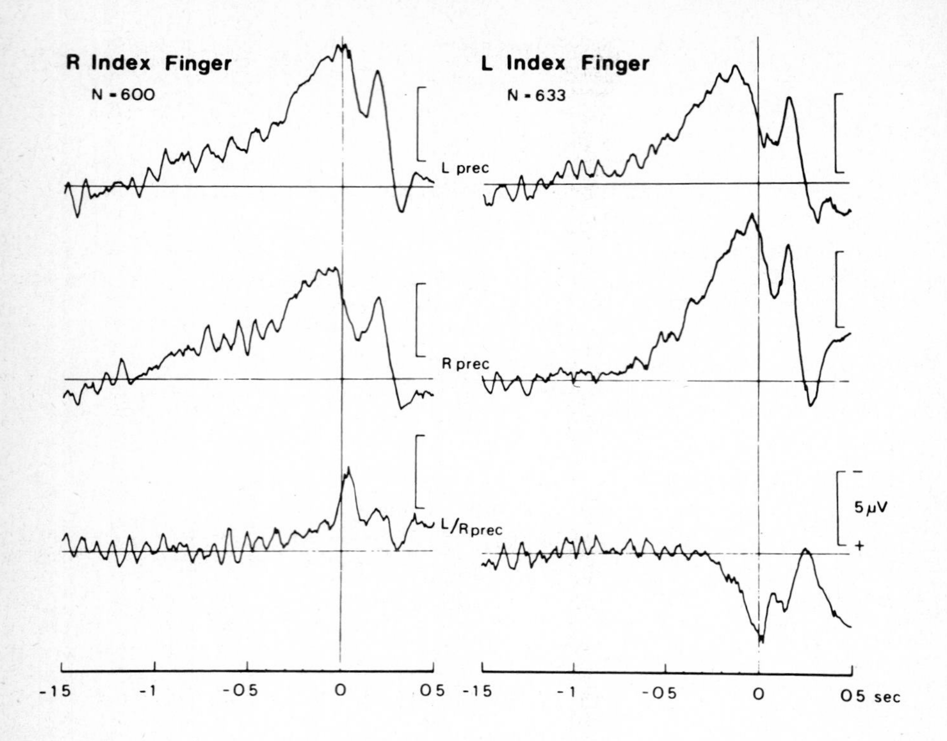

Figure 1. Examples of scalp-recorded cerebral potentials related to self-paced voluntary ballistic movements (Deecke et al. (1976)). Comparison of individual averages of left and right unilateral rapid flexions of the index finger. Bipolar recording (left vs. right precentral) in the bottom trace.

movement-related potentials in man (e.g. Papakostopoulos, Cooper and Crow (1975), Groll-Knapp, Ganglberger and Haider (1977), McCallum (1978), Haider, Groll-Knapp and Ganglberger (1981), and in monkey (e.g. Arezzo and Vaughan (1975, 1980), Hashimoto, Gemba and Sasaki (1979), Gemba, Hashimoto and Sasaki (1979), Pieper, Goldring, Jenny and McMahon (1980)).

The bilaterally symmetrical component of the premovement negative potential shift shows a wide distribution extending from the frontal to the parietal region with a maximum at the vertex (see Deecke and Kornhuber (1978)). It has been interpreted as indicative of a general preparatory state (Kornhuber and Deecke (1965) called this component the 'Bereitschaftspotential' (BP); see also Deecke (1978a), Näätänen and Michie (1979)). The late lateralized increase of the slow negativity is maximal over the precentral region and seems to be related more specifically to motor preparation (Shibasaki et al. (1980a), Kutas and Donchin (1980)). The increase in lateralization immediately before EMG onset might reflect some aspect of the cortico-spinal outflow (Deecke et al. (1976), Arezzo and Vaughan (1980)). The potentials after the onset of movement have usually been interpreted as reafferent activity evoked by the movement (Kornhuber and Deecke (1965), Deecke et al. (1976), Deecke (1978b)); they may, however, also include internal feedback from structures within the central nervous system (Vaughan, Gross and Bossom (1970), Arezzo and Vaughan (1975, 1980), Papakostopoulos (1978a)).

We have extended the investigation of movement-related potentials to ramp movements by using aiming tasks. Subjects had to perform smooth isotonic or isometric contractions in order to reach a specified amplitude of displacement or a specified level of force under predictable conditions. In the following, the main findings of these experiments are described.

POSITIONING MOVEMENTS

Figure 2A shows one of our positioning paradigms. (For a detailed description see Grünewald-Zuberbier, Grünewald, Runge, Netz and Hömberg (1981).) The index finger was attached to a lever coupled to the axle of a torque motor which produced a constant torque load of 23 Ncm. While continuously fixating an area of 0.5 degree diameter on a screen the subject had to produce self-paced flexions at the metacarpo-phalangeal joint of the index finger in order to reach precisely a target zone between 25 and 27 degrees of angular displacement, and to return to the starting position after each movement. In the starting position the lever rested against a backstop, so that there was no initial load and the agonist was in a relaxed state.

With a constant delay of 1.6 sec after the return movement of each trial a visual information feedback (usually referred to as 'knowledge of results'; see Bilodeau (1966)) was given, indicating a hit, an 'undershoot' error or an 'overshoot' error. The information feedback consisted of a 100 msec light signal from one of three diodes horizontally arranged in the fixation area. According to the direction of movement, for a right side action the left diode served as overshoot indicator and the right one as undershoot indicator, and vice versa.

Recordings were made of: (a) angular position and the force applied to the lever indicated by transducers (potentiometer and strain gages, respectively), (b) EMG through surface electrodes over the first dorsal interosseus muscle, which acts as a synergist to the long flexor muscles of the index finger, in some cases also of its antagonist, extensor digitorum longus muscle, (c) vertical EOG to control for eye movements, and (d) EEG from bilateral precentral and post-central positions, referred to a common reference consisting of the linked mastoids, using long time constants (5 sec).

In the single trial records of Figure 2B five vertical lines

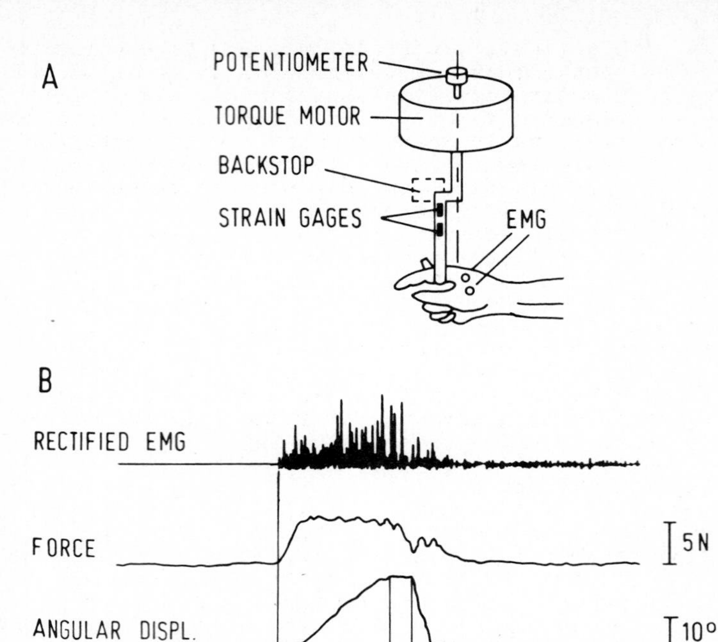

Figure 2.

A: Arrangement for one of our positioning experiments. The index finger is attached to a lever fixed on the shaft of a torque motor producing a constant load. Subjects had to perform self-paced flexions at the metacarpo-phalangeal joint in order to reach a specified angular displacement, and to return to the starting position after each movement. In the starting position the lever rested against a backstop. Angular position of the finger was measured by a potentiometer. Two strain gages on either side of the lever provided a measure of the subject's force output. EMG was recorded from the first dorsal interosseus muscle.

B: Single trial records of the positioning movement with the right index finger. The reference points for the averaging procedures were: 1 = onset of activity in the rectified EMG, 2 = point of maximal angular displacement, 3 = begin of the return movement, 4 = end of the return movement (closure of a switch when the starting position is reached), and 5 = onset of a visual information feedback projected in the fixation area and indicating a hit, an overshoot, or an undershoot error. These five points divide the trial epoch into six segments: A = the premovement-interval, B = the positioning-movement-interval, C = the stop-on-target-interval, D = the return-movement-interval, E = the constant feedback delay, and F = the post-feedback-interval.

represent the events which are taken as the reference points for the averaging procedures. Line 1 marks the onset of the full wave rectified flexor EMG activity, line 2 shows when the lever reaches the maximal angular displacement, lines 3 and 4 correspond to the lever beginning the return movement and reaching the starting position, respectively, and line 5 indicates the onset of the visual information feedback. By these five lines the trial is divided into six segments: the premovement-interval (A), the positioning-movement-interval (B), the stop-on-target-interval (C), the return-movement-interval (D), the constant delay of the information feedback (E), and the post-feedback-interval (F).

Figure 3 demonstrates individual averages of 35 trials of one right-handed subject, who was instructed to carry out the positioning movement at different speeds: rapidly (ballistic) (left column) and smoothly (ramp) (right column). Mean positioning movement times were 178 msec (SD 51) and 852 msec (SD 74), respectively. In the upper part of the Figure, for both the ballistic and the ramp movements, two separate averages are shown which are triggered (a) from the onset of flexor EMG activity, and (b) from the point of maximal angular displacement, in order to portray the potentials accompanying the initiation and the termination of the positioning movement. Averaging in the positioning-movement-interval was limited to the shortest interval length of all trials to be averaged. In the lower part of the Figure, average waveforms of the whole trial period (except part of interval E and interval F) are demonstrated. To reveal the average waveform in the action intervals B, C, and D, the intraindividual variability of their durations was eliminated by transforming each of them in the single trial into a standard interval of arbitrary but constant length before averaging (Grünewald, Grünewald-Zuberbier, Hömberg and Netz (1979)). That means, the potentials of these intervals were not averaged over the same time points, but over the corresponding interval positions. Thereby the average potentials of the standard intervals were adjusted to the mean durations of the intervals B, C, and D.

The EMG patterns shown in Figure 3 present the typical features as described in the literature for ballistic and for ramp movements (e.g. Marsden et al. (1977), Desmedt and Godeaux (1978), Desmedt (1981), Hallett and Marsden (1981)):

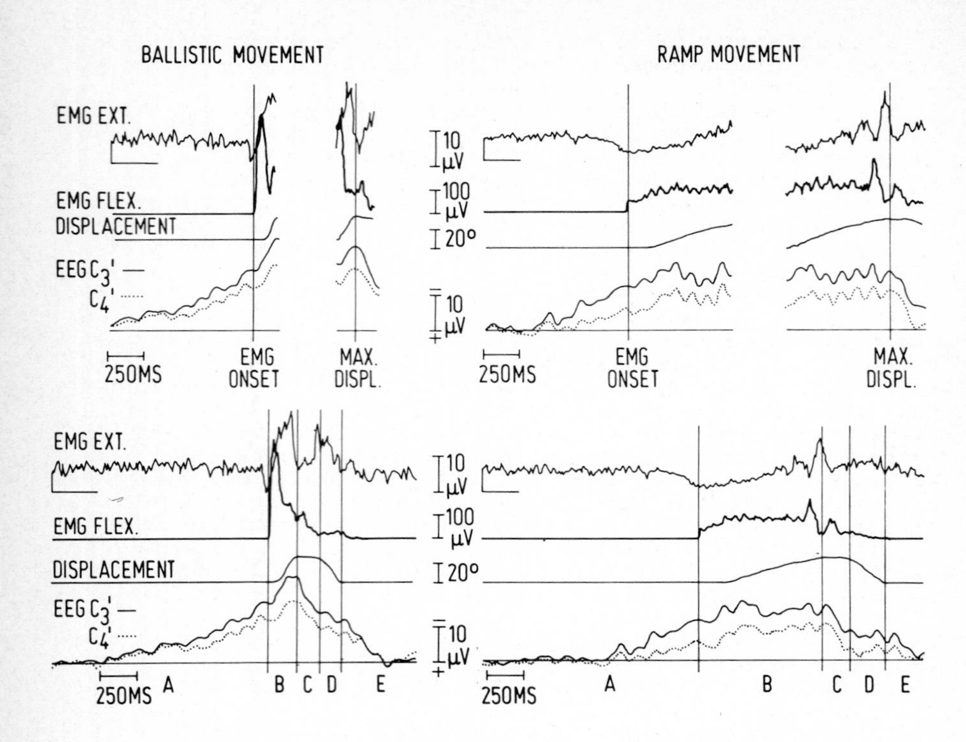

Figure 3. Positioning movements of the right index finger performed under ballistic and under ramp conditions. Individual averages (based on 35 trials) of rectified flexor EMG and extensor EMG (the leading edge represents the total voltage), angular displacement, and precentral left (C3') and right (C4') EEG (1 cm anterior to C3 and C4 in the 10-20-system).

Upper part: For both the ballistic and the ramp movements two separate averages are shown, triggered from the onset of flexor EMG activity and from the point of maximal angular displacement, respectively.

Lower part: Average waveform of the whole trial period (except part of interval E and interval F; see Figure 2B). Before averaging, each of the three action intervals B, C, and D was transformed per single trial into a standard interval in order to account for intraindividual variability in duration. The average potentials of the standard intervals were then adjusted to the individual mean durations of B, C, and D.

the former are characterized by alternating bursts of activity in agonist and antagonist muscles, the latter are characterized by a smooth onset of agonist contraction which continues throughout the movement up to its peak. (2)

In the precentral EEG averages of the ballistic movement (Fig. 3), the lateralized negative potential shift prior to EMG onset further increases up to a maximum after the agonist EMG peak, and is followed by a positive deflection after the target position has been reached. A negative potential peak following the EMG of self-paced ballistic movements has been described by various authors (Bates (1951), Gerbrandt et al. (1973), Wilke and Lansing (1973), Papakostopoulos et al. (1975), Gerbrandt (1978), Papakostopoulos (1978b, 1980), Shibasaki et al. (1980a)). The waveform of the ramp positioning movement resembles that in the ballistic movement considering the fact that the interval between initiation and termination of the movement is largely extended in the ramp mode. Here, too, the premovement negativity continues until the maximum displacement has been attained, and is followed by a positive deflection thereafter. A similar observation was already made by Vaughan et al. (1970). Using epidural recordings from the precentral motor cortex of monkeys trained to close a switch with a wrist extension, these authors demonstrated a sustained negative potential shift during prolonged contractions. This negative potential shift was largest contralateral to the movement. The positive deflection did not appear until near the end of the contraction.

The main difference between the EEG potentials accompanying ballistic and ramp movements (Fig. 3) consists in the amplitudes of the pre- and during-movement negativities, which are higher in ballistic movements. This observation is in agreement with results of Grünewald, Grünewald-Zuberbier, Netz, Hömberg and Sander (1979a). (3) The reason for the amplitudes being higher before ballistic than before ramp movements may be an association between BP amplitude and the rate of force change developed. Likewise, the amplitudes of the negative potential prior to ballistic isometric presses requiring a larger rate of force change are higher than those those with a low force output (Kutas and Donchin (1974, 1977), Becker and Kristeva (1980); see also Kristeva and Kornhuber (1980), and Hashimoto, Gemba and Sasaki (1980)). As stated by Goff et al. (1978): 'it is clear...that the largest...motor potentials are obtained with forceful contractions which are executed as rapidly as possible'.

Figure 4 demonstrates grand averages of the smoothly performed positioning movement over 7 right-handed subjects (verified by Oldfield's Inventory (1970)), as a function of responding side, cerebral region (pre- and postcentral) and hemisphere. The grand averages of the time-normalized intervals B, C, and D (see above) were adjusted to the group mean durations of these intervals. The negative potential shift preceding the EMG onset further increases briefly and then persists with a slow decrement until the maximum displacement has been reached. Approximately at this point a positive deflection appears. In the return-movement-interval D a second smaller negative wave occurs which is followed by a marked positive deflection

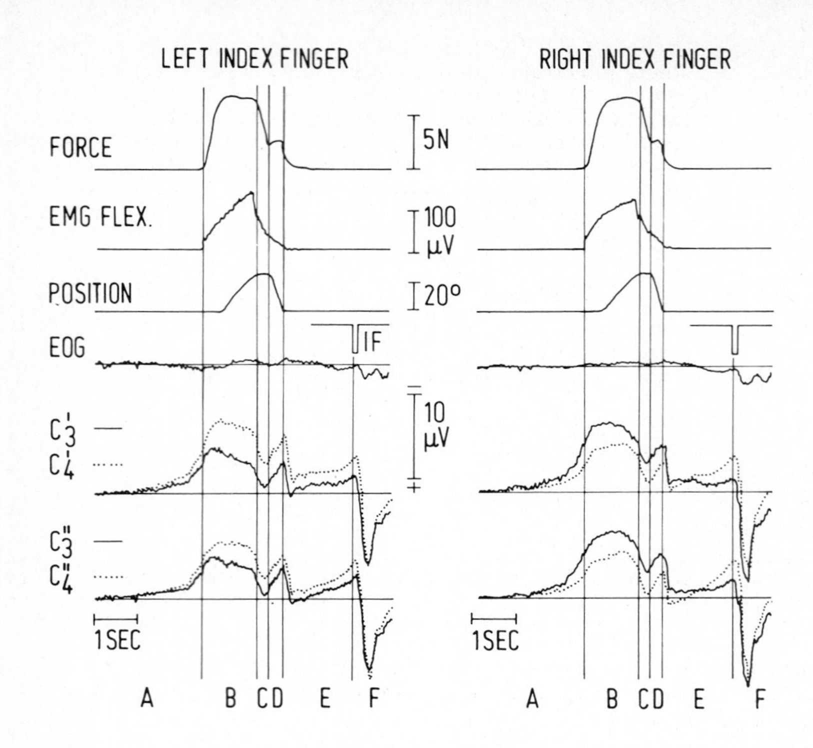

Figure 4. Ramp positioning movements of the left and right index finger. Grand averages across seven right-handed subjects. Recordings of force, rectified EMG, position, vertical EOG, and slow potential shifts in left and right precentral (C3', C4') and postcentral (C3'', C4'') EEG (1cm anterior and 2 cm posterior to C3, C4 positions). IF = information feedback. For definition of trial intervals A - F see Figure 2B. The group averages of the time-normalized intervals B, C, and D were adjusted to the group mean durations of these intervals.

(including a somatosensory evoked potential component) after reaching the backstop. Throughout the whole motor task the slow negative potential shift is consistently more negative over the hemisphere contralateral to the responding limb. This asymmetry begins 700 msec and more before the EMG onset (but it was statistically not present over parietal regions, as demonstrated in other experiments; see Grünewald-Zuberbier et al. (1981)).

In the constant interval E preceding the visual information feedback (IF) a stimulus-related negativity develops, the topography of which is different from that of the movement-

Table Ia.
Mean voltage (and S.D.) in µV of movement- and external feedback-related potentials in a self-paced positioning task. Group means of seven subjects performing with the right and with the left index finger (see Figure 4).

Criterion measurements:
1 = mean amplitude of the 200 msec interval preceding EMG onset,
2 = mean amplitude of the positioning-movement-interval,
3 = mean amplitude of the 200 msec interval prior to information feedback.
The baseline for these measurements was the mean amplitude of the initial 500 msec of the trial epoch.

	Left index finger			Right index finger		
Criterion	1	2	3	1	2	3
Electrode						
Prec.L C3'	-2.7 (2.1)	-4.2 (3.2)	-1.5 (2.9)	-4.6 (2.3)	-7.4 (3.5)	-1.3 (2.7)
Prec.R C4'	-4.5 (2.8)	-7.7 (4.1)	-3.4 (2.8)	-2.6 (1.7)	-5.0 (2.4)	-3.6 (3.3)
Postc.L C3''	-2.6 (2.2)	-4.0 (2.8)	-1.9 (3.5)	-4.4 (2.5)	-7.2 (3.3)	-1.8 (2.6)
Postc.R C4''	-3.6 (2.4)	-6.1 (3.2)	-3.4 (3.0)	-2.3 (1.9)	-4.7 (2.8)	-3.7 (3.3)

Table Ib.
F(1;6)-values of ANOVAs performed with responding side (left-right) and hemisphere (left-right) as factors.

	Precentral			Postcentral		
Criterion	1	2	3	1	2	3
Resp. Side A	0.0	0.0	0.0	0.1	0.4	0.0
Hemisphere B	0.1	2.6	25.8[b]	3.6	0.6	14.9[a]
A x B	53.9[c]	50.0[c]	0.9	61.3[c]	47.0[c]	0.9

a: $.005 < p < .01$ b: $.001 < p < .005$ c: $p < .001$

For the criteria 1 and 2, there are no significant overall main effects of responding side and hemisphere, but the interaction of both is significant, reflecting the contralateral preponderance of the movement-related negativities. For criterion 3 responding side and hemisphere do not interact, but there is a significant overall main effect of the hemisphere, reflecting the right-hemispheric preponderance of the feedback-related negativity.

related potentials. In the lateral distribution it is always larger over the right hemisphere, independently of the hand used. (4) The sequence of event-related potentials of the positioning task is terminated in F by the feedback-evoked potential complex.

All the potential features described are consistent over all the seven subjects investigated (see Grünewald-Zuberbier et al. (1981)). A statistical analysis of the data is presented in Tables Ia and Ib. Essentially the same results were found with different subjects in other positioning experiments requiring flexions at the wrist (Grünewald-Zuberbier and Grünewald (1978), Grünewald-Zuberbier, Grünewald and Jung (1978), Grünewald et al. (1979b), Grünewald-Zuberbier et al. (1981)). Taylor (1978) observed similar continued negativities after movement onset during the execution of a skilled serial positioning task (see also Coupland, Taylor and Koopman (1980)).

ISOMETRIC RAMP CONTRACTIONS

In order to answer the question, whether a movement of the joints is a necessary condition for the appearance of the potentials described in voluntary ramp contractions, the following experiment was conducted (Figure 5). (For a detailed

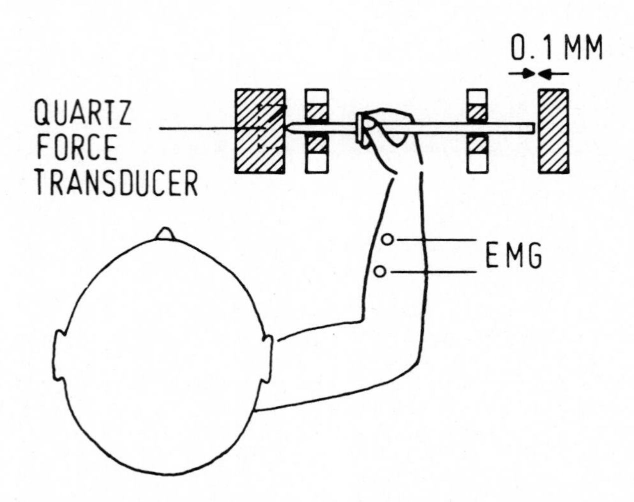

Figure 5. Schematic diagram of the experimental arrangement in the isometric ramp task. Subjects were instructed to produce self-paced isometric force ramps up to a specified force level by pressing a rod against a quartz force transducer with a wrist flexion, and to relax immediately thereafter. Each pressing began and ended with a minimal rod displacement which triggered a microswitch to indicate the beginning and the end of the action interval. EMG was recorded from the carpi radialis muscle.

description see Grünewald-Zuberbier, Grünewald, Schuhmacher and Wehler (1980a).) The subject was seated in a darkened room fixating a lighted circle of 0.5 degree diameter. He held a rod between thumb and index finger, his forearm being positioned on an armrest, and he was instructed to produce self-paced isometric force ramps up to a specified force level of 5 N by pressing the rod against a quartz force transducer, and to relax immediately thereafter. Each pressing began and ended with a minimal rod displacement which triggered a micro-switch to indicate the beginning and the end of the action interval. Visual feedback superimposed upon the fixation area was given only during the practice-trial-blocks preceding each of three 20-test-trial-blocks. The durations of the force ramp in this experiment commonly exceeded 700 msec.

Recordings were made of: (a) the exerted force, (b) the EMG of the flexor carpi radialis muscle, (c) the vertical EOG, (d) the EEG from bilateral pairs of electrodes over precentral, post-central, and parietal positions, all referred to linked mastoids, and (e) the nasal respiration to control for respiratory potentials (Grözinger, Kornhuber and Kriebel (1977)).

Grand averages of these recordings across six right-handed subjects who had to perform the task with their right and with their left hand are demonstrated in Figure 6. The four segments A, B, D, and E of the trial epoch correspond to those in Figure 4. It can be seen that the waveform and distribution of the contraction-related EEG potentials are essentially the same as were registered in the isotonic positioning task described above. Consequently, the appearance of the negative potentials during the aiming interval is independent of the type of muscular contraction, be it a change of limb position or an isometric change of force.

HOLD CONTRACTIONS

The voluntary motor tasks in question are characterized by slow phasic motor activity exerted to reach a specified goal. The question arises whether sustained voluntary contractions per se are sufficient for the appearance of the prolonged negative potentials described. Some light is shed on this problem by the investigation of a non-aiming, non-phasic type of sustained voluntary contraction, i.e. a simple hold contraction (see Grünewald-Zuberbier et al. (1980a)).

Under the same experimental conditions as in the experiment described before, subjects had to produce an unspecified isometric force level (between 3 and 10 N) and to maintain it for some seconds without any precision requirements. They were instructed to initiate and terminate contractions rapidly and to relax as completely as possible between presses. Thus, this voluntary motor task required a ramp-and-hold-contraction, the holding period being characterized by non-aiming, tonic motor activity.

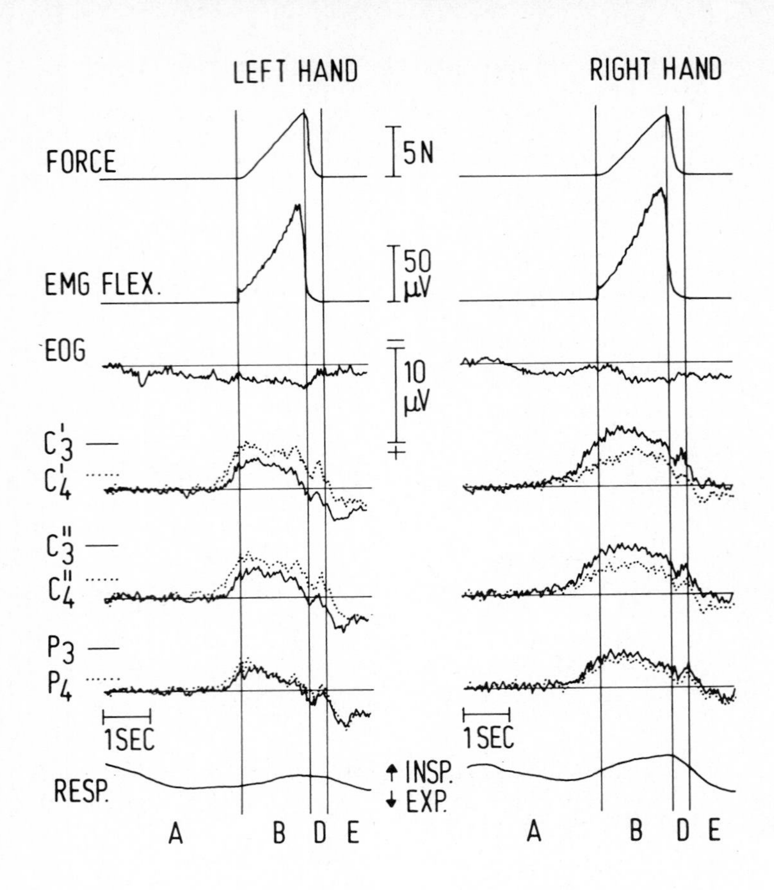

Figure 6. Isometric ramp contractions of the left and right hand (Grünewald-Zuberbier et al. (1980a)). Grand averages across six right-handed subjects. Recordings of force, rectified EMG, vertical EOG, nasal respiration, and slow potential shifts in the left and right precentral (C3', C4'), postcentral (C3'', C4''), and parietal (P3, P4) EEG. The A = pre-contraction-, B = ramp-contraction-, D = relaxation-, and E = after-relaxation-intervals correspond to the equally labelled segments in the ramp positioning task (Fig. 4). The equivalent for the stop-on-target-interval C was too short for being defined separately. The group averages of the time-normalized intervals B and D were adjusted to the group mean durations of these intervals.

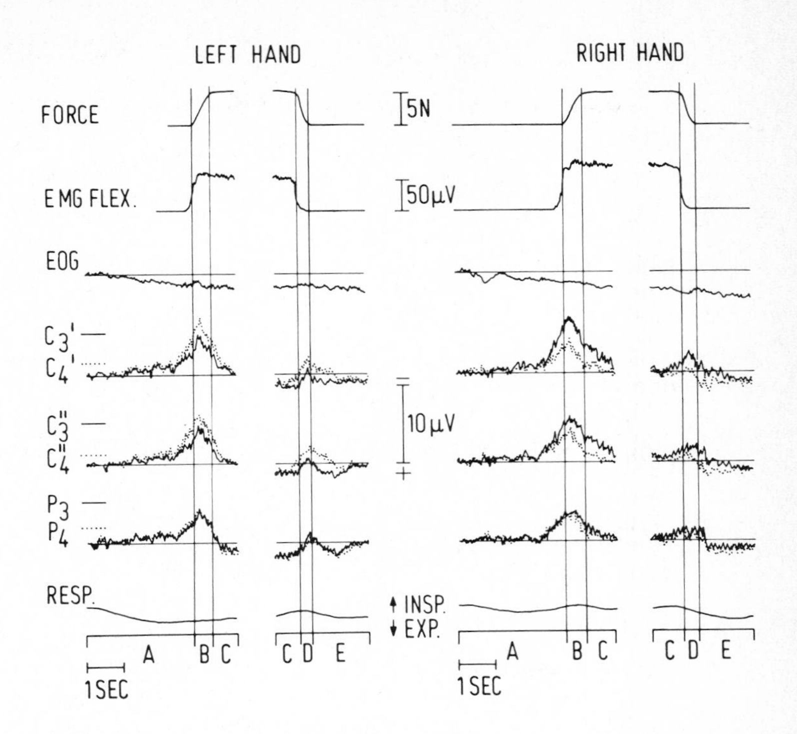

Figure 7. Fast-ramp- and hold-contractions of the left and right hand (Grünewald-Zuberbier et al. (1980a)). Subjects had to produce an unspecified force level and to maintain this level for some seconds without any precision requirements. Grand averages across six right-handed subjects of the same recordings as in Fig. 6. A = precontraction-, B = fast-ramp-contraction-, D = relaxation-, and E = after-relaxation-intervals. The holding-interval C is represented by two averages over the shortest holding phase of all individual trials: (a) time-locked to the beginning, and (b) time-locked to the end of the stable force period.

Figure 7 demonstrates grand averages (N = 6) of the potentials recorded under this condition, separately for right-hand and left-hand responses. The holding interval C is represented by two averages (a) time-locked to the beginning, and (b) time-locked to the end of the stable force period; (group mean durations of the complete holding period were 1986 msec (SD 570) for the right, and 1948 msec (SD 514) for the left hand). After the fast ramp contraction B the negative potentials at all electrode positions decrease; the waveform eventually tends to become positive, as can be seen with the left-hand response in Figure 7. The contralateral preponderance of the potentials also diminishes, and becomes insignificant approximately 600 msec after the beginning of the hold interval. The asymmetry reoccurs shortly before the relaxation interval D. A replication of the right-hand condition of the experiment with another group of seven subjects has fully confirmed these results (publication in preparation). Here also, a tendency of the potentials to become positive during the holding period could be observed. The results are consistent with the findings of Otto, Benignus, Ryan and Leifer (1977) and Otto and Benignus (1978) who described a prolonged positive shift over centroparietal regions during sustained isometric isotonic contractions (pressing a button or hand dynamometer for approximately 1 or 3 seconds). No consistent hemispheric differences of the potentials and no direct relations to the force level were found.

The data suggest that a mere sustained voluntary contraction is not sufficient for the appearance of the prolonged negative potentials described. Probably, goal-directed phasic motor activity is the essential factor for the potentials to develop.

Figure 8 comprehends in a simplified scheme the main features of the EEG potentials over the motor cortex and of the peripheral recordings accompanying the four classes of voluntary contractions described in this paper.

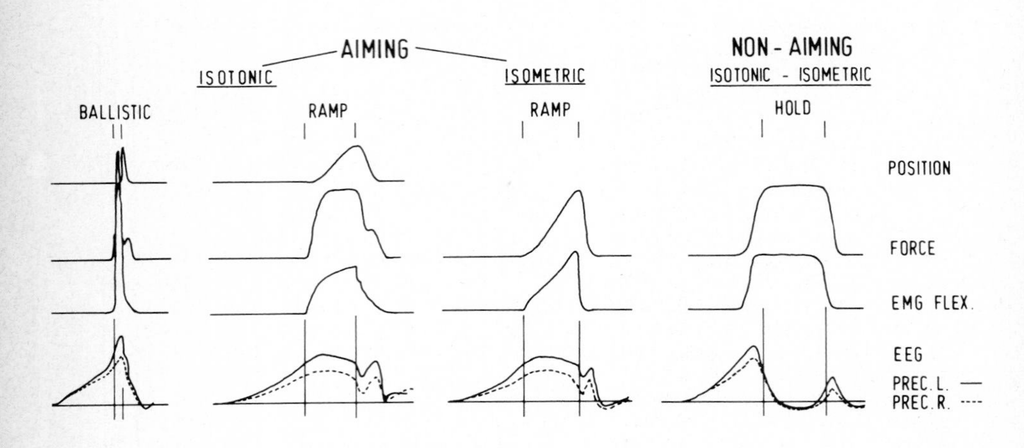

Figure 8. Main features of the EEG potentials over the motor cortex accompanying four types of voluntary contraction described in this paper: (a) Isotonic ballistic and (b) isotonic ramp contractions in order to reach a specified amplitude of displacement, (c) isometric ramp contractions in order to reach a specified level of force, and (d) hold contraction: maintenance of an unspecified force level for some seconds without precision requirements. Right-sided performance.

DISCUSSION

What is the functional significance of the prolonged negative potential shift during the execution of voluntary ramp contractions?

Obviously, the pre- and postcentrally lateralized part of this negativity (most apparent in bipolar left vs. right recordings) is a manifestation of neural activity related to the execution of the motor response. The association of the lateralized negative potential shift with phasic, but not with tonic motor activity has a parallel in movement-related changes of the pre-central EEG beta rhythm. Jasper and Penfield (1949) found that the beta rhythm from the contralateral hand area of the exposed precentral gyrus was blocked during the initiation and termination of voluntary contractions (clenching the fist and relaxing the grip), but not during sustained contractions (keeping the fist clenched). Furthermore, Roland, Larsen, Lassen and Skinhøj (1980) found that the regional cerebral blood flow in the contralateral motor cortex shows about 50 % lesser increase during sustained isometric finger contractions than during single finger flexions, in spite of greater muscular work in the former task. Our results are consistent with the suggestion of these authors that the motor cortex is involved chiefly in phasic voluntary contractions under normal conditions.

The question is, whether the lateralized (motor-specific) component represents mainly input from peripheral afferents evoked by the motor action, or is instead predominantly a manifestation of a centrally organized neural activity generated independently of peripheral feedback. As already mentioned in the introduction, the potentials after EMG onset in a ballistic movement have generally been attributed to peripheral reafference, mainly because it was claimed that the same components occurred after passive movements (Kornhuber and Deecke (1965), Deecke et al. (1976), Deecke (1978a)). However, Papakostopoulos et al. (1975) found that the potentials following passive finger displacement 'diminished, changed in waveform or disappeared completely when the subject himself was moving his finger' (see also Shibasaki et al. (1980b), and the evidence reviewed by Hazeman (1978), that sensory input is inhibited during movement). As a reliable potential after EMG onset of self-paced ballistic movements, these authors and Papakostopoulos (1978b, 1980) observed a negative peak with maximum amplitude over the contralateral precentral cortex. They refer to this transient negativity as the 'motor cortex potential', and consider it as the first signal of response-reafferent activity.

In parallel to this assumption it may be possible that the lateralized negative component observed during voluntary ramp contractions includes a sustained sensory evoked potential (see Keidel (1971a), Hillyard, Picton and Regan (1978), Näätänen and Michie (1979)) in response to the continuation of the somatosensory reafference. If so, the effective input cannot possibly arise from cutaneous receptors in the digital skin alone, for the sustained lateralized negative wave was still present after deep local anesthesia (2 x 1.5 ml 2% Carbostesin) of the index finger (unpublished observations in one subject). Furthermore, the lateralized component disappears during a voluntary isometric hold contraction with muscle spindles continuously discharging (Vallbo (1974)), which finding suggests that the negative potential is not related to these afferents.

From the available evidence it cannot yet be concluded to what extent peripheral reafference as opposed to centrally-generated motor commands contribute to the movement-related negative potential shift.

The positive deflection commonly designated as P2 (Vaughan et al. (1968)), the onset of which is associated with the termination, not with the initiation of the movement (Vaughan et al. (1970), Taylor (1978), our results), has been found to be resistant to upper limb deafferentiation by rhizotomy (Vaughan et al.(1970)). Therefore, these latter authors and Arezzo and Vaughan (1975, 1980) suggested that it is an index of internal feedback.

If one subtracts the asymmetric part of the potential during the voluntary ramp contraction there remains residual bilateral negativity. There is sound evidence to consider this residual as a separate component of the observed negativity, for it shows a differential response to experimental conditions (see Hillyard (1973)). Like the bilateral component of the BP (e.g. McAdam and Seales (1969), Becker, Höhne, Iwase and Kornhuber (1972), Papakostopoulos (1978b), Taylor (1978)) it was found to be sensitive to attention demands of the task (Grünewald et al. (1979b)), and to the subject's ability to concentrate (Grünewald-Zuberbier et al. (1980b), Grünewald and Grünewald-Zuberbier (1981)): increased amplitudes result from engagement with a more demanding task and are also reliably observed in children with higher abilities to concentrate. None of these factors has an influence on the lateralized component of both the BP and the potential during the voluntary ramp contraction. These findings suggest that the bilateral component of both potentials is associated with the degree of the motivational and attentional involvement of the subject in the task.

The psychological state with which these potentials are associated is characterized by the intention of the subject to initiate a planned action and to reach a desired goal (see Newell (1978), O'Connor (1981)), which features are lacking in the postural state of the hold contraction. Therefore, both the BP and the potential during the execution of the aiming movement have been called 'intention waves' (Jung (1980), Eccles (1980)). (5) The bilateral components of both potentials possibly are the expression of the same intentional engagement in the preparatory and execution phases of the goal-directed action. It was suggested that these potentials reflect the degree of activation in non-specific subcortico-cortical 'activation systems' (e.g. Näätänen and Michie (1979), Haider et al. (1981)).

Notes

This work was supported by the Deutsche Forschungsgemeinschaft (SFB 70) and by the Landesamt für Forschung, Nordrhein-Westfalen. We wish to thank C.H.M. Brunia and A. Gaillard for their helpful comments on an earlier version of this report. Address requests to: G. Grünewald, Institut für Hirnforschung der Universität, Universitätsstr. 1, 4000-Düsseldorf, West Germany.

(1) Premovement lateralized negative potential shifts have been demonstrated also for cued-response conditions, i.e. in reaction time tasks with and without a warning stimulus (e.g. Rohrbaugh, Syndulko and Lindsley (1976), Syndulko and Lindsley (1977), Gaillard (1978), Kutas and Donchin (1980)).

(2) The ballistic element immediately preceding the termination of the ramp movement in Figure 3 is not typical, it was observed only in this subject.

(3) Becker, Iwase, Jürgens and Kornhuber (1976), on the contrary, reported slow movements to be preceded by BPs of higher amplitudes than were ballistic movements. (Paradoxically, the Figures given in their paper seem to illustrate just the opposite relationship.) The reason for this discrepancy is not yet clear; it is even more surprising since the two studies employ rather similar tasks and similar velocity ranges. However, both studies agree that the BP begins earlier in slow movements than in rapid movements. The Grünewald et al. study especially showed an earlier onset of the BP lateralization before the EMG onset of slow movements.

(4) This right-hemispheric preponderance is also independent of stimulus modality, and it does not appear when the same stimuli are without any meaningfulness for a manipulospatial action, and have to be merely discriminated (Grünewald, Grünewald-Zuberbier, Hömberg and Schuhmacher (in press)).

(5) Also the contingent negative variation (CNV) (Walter, Cooper, Aldridge, McCallum and Winter (1964); for a review see Donchin, Ritter and McCallum (1978)), which occurs in the interval between a warning stimulus and an imperative stimulus, is considered as an 'intention wave' by these authors. Supplementing a statement from Donchin, Gerbrandt, Leifer and Tucker (1972), we could say: the general case of situations in which the three potentials (BP, CNV, and the potential during the execution of an aiming movement) can be elicited seems to consist of situations in which two events (external or internal) delimit a time interval, the termination of which marks start or completion of an action.

Tutorials in ERP Research: Endogenous Components
A.W.K. Gaillard and W. Ritter (eds.)

15

COGNITIVE COMPONENTS OF THE EVENT-RELATED BRAIN POTENTIAL: CHANGES ASSOCIATED WITH DEVELOPMENT

Eric Courchesne

Neurosciences Department, University of California at San Diego
La Jolla, California, 92093
and
Speech, Hearing and Neurosensory Center, Children's Hospital
San Diego, California, 92123

This report addresses some aspects of the relationship between normal development and cognitive components of event-related brain potentials (ERPs). It includes a description of the ERP components which have been associated with cognitive processes in normal infants, children and adolescents and a description of the way that these components might be changing with development. Also presented are some of the suggestions that researchers have made about the cognitive correlates of these components.

A major goal in the field of developmental human neurophysiology is to relate changes in ERP components to changes in brain structure and function associated with cognitive development. Although this line of research certainly has great potential, it is still in its infancy. More than eight cognitive components have been reported in human developmental research. The neural sources of cognitive components are unknown, although suggestive evidence has recently emerged that one cognitive component, P3b, may result from activity in medial temporal lobes (Halgren et al., 1980).[1] The functional significance of each of the cognitive components reported is only known in a general fashion.

Designs for ERP experiments have usually not been derived from the literature on cognitive development, although interpretations of ERP results occasionally have been (Courchesne, 1978a; Friedman et al., in press; Courchesne et al., 1981). Nonetheless, issues in cognitive development revolving around, for example, attention, learning, memory and language (Appleton et al., 1974; Pick et al., 1975) seem especially amenable to ERP techniques.

It can be anticipated, therefore, that the study of ERPs in cognitive development will prove important in resolving issues that are intractable to solution by other methods and in influencing the unfolding of new ideas. This is so because ERPs are a direct manifestation of the activity of the central nervous system; they do not depend upon a behavioral response system; different ERP components are probably associated with different cognitive processes; and the sequence and timing of components could provide clues to the order and timing of different stages of cognitive processing. These advantages may be illustrated by several examples.

First, behavioral studies of attention and memory in infants are often hampered by limited repertoires of infants, but ERP studies are not. For instance, we showed that the ERPs of 6 month olds may provide clues about recognition memory; the relative differences in speed of processing familiar and discrepant information; and the tendency of substantial neural activity to be elicited even by the repeated presentation of the same information (Courchesne et al., 1981). Furthermore, since cognitive components are not dependent upon the capacity of behavioral response systems, ERPs could help ascertain whether performance changes are due to cognitive development or to motor coordination development.

Second, since different components may represent different neurophysiological activity involved in cognitive processing, examination of age-related differences in these components could provide a rewarding means of mapping developmental trends and transitions. For example, the ERP components elicited by novel, "never-seen-before" pictures are quite different in children and adults (Courchesne, 1977, 1978a). The transition from the childhood pattern of components to the adulthood pattern is gradual and is not complete until late adolescence and early adulthood. This indicates that different neurophysiological and cognitive systems are involved, and that a stage model of the development of visual novel information processing with abrupt transitions between stages may not be as useful as other models. Indeed, the stage model seems even less adequate in the face of ERP data from analogous experiments with novel, never-heard-before sounds (Courchesne and Galambos, in preparation). These sounds elicited one component which changed very little across the ages of 4 to 44 years and another one which gradually decreased in amplitude across these ages. More detailed descriptions of these and other patterns of ERP change during development will be presented in later paragraphs.

Third, ERPs could help to ascertain the order and timing of cognitive processes. In studies of visual and auditory ERPs in children, adolescents and adults, we have recorded P3b and SW components (Courchesne, 1978a; Courchesne and Galambos, in preparation; see Literature Review below). Each component has been associated with a different stage of cognitive processing (e.g., Ruchkin et al., 1980). Our evidence indicates that, with development from childhood to adulthood, the cognitive processes reflected by P3b and SW occur sooner after a stimulus, but the order of these processes and the timing between them remain constant.

The use of ERP techniques can also aid the study of disorders of cognitive development. This is exemplified by studies of hyperactive children (Prichep et al., 1976; Loiselle et al., 1980), dyslexic children (e.g., Otto et al., 1976; Shelburne, 1978), and adolescents with autism (Novick et al., 1980; Courchesne et al., in preparation). Loiselle et al. (1980) compared hyperactive and normal children in several tasks. ERPs were recorded in an auditory selective attention task and behavioral responses were recorded in a dichotic listening task and in a vigilance task. Loiselle et al. found that hyperactive children deviated from normal when required to selectively attend to particular items in continuous streams of information. Hyperactive children did not deviate from normal when required to attend to short, discrete pairs of sounds. As another example, we recorded ERPs from adolescents with autism (Courchesne et al., in preparation). Our ERP data does not support previous hypotheses that these individuals are hypersensitive to novel information or misperceive novel information as non-novel and insignificant. Rather, our ERP data suggests that the defect may be in associating and integrating novel information with past experience.

In addition, ERPs may provide a means for making the distinction between delayed and abnormal cognitive development. Satterfield and Braley (1977) studied 6 to 7 and 10 to 12 year old normal and hyperactive children in an auditory paradigm. They concluded that the ERPs of hyperactive children provided evidence of abnormal rather than delayed development. In our laboratory, we have recently completed ERP experiments comparing a variety of visual and auditory cognitive ERP components between normal individuals (4 to 44 years of age), individuals with autism (13 to 21 years of age), and individuals with Down's syndrome (13 to 21 years of age) (Courchesne, Kilman, Lincoln, and Galambos). Individuals with autism and with Down's syndrome have ERPs which differ from each other and from age-matched and younger normal children. The ERP data is consistent with disordered cognitive development in autism and Down's syndrome, not with delayed development.

Despite the potential value of ERPs in the study of cognitive development, research on cognitive components in infants, children and adolescents has not been extensive. For example: (1) There have been only four detailed cross-sectional developmental studies (Courchesne, 1978a; Friedman et al., 1981, in press; Courchesne and Galambos, in preparation). (2) There has been only one longitudinal study published (Kurtzberg et al., 1979); this study followed one group of 4 to 6 year olds for two years. (3) Little has been reported on auditory ERP components (Prichep et al., 1976; Goodin et al., 1978a; Loiselle et al., 1980; Friedman et al., in press; Courchesne and Galambos, in preparation). (4) Few studies have incorporated direct child-adult comparisons (Shelburne, 1972, 1973; Courchesne, 1977, 1978a; Goodin et al., 1978a; Courchesne and Galambos, in preparation). (5) No study has compared the ERP patterns from different modalities. (6) No study has systematically varied parameters known to affect certain cognitive components (such as _a priori_ probability effects on P3b amplitude). (7) Few studies have attempted to explore the possibility of age-related changes in the locus of brain activity associated with cognition (Courchesne, 1977, 1978a; Friedman et al., 1981, in press; Kurtzberg et al., 1979; Courchesne and Galambos, in preparation). Most studies have used too few electrode sites to allow this possibility to be explored.

METHODOLOGICAL CONSIDERATIONS

In evaluating reports on cognitive components and development, several points about design and analysis should be kept in mind. Analyses should take into account the overlapping, multiple nature of cognitive components and the possibility that the amplitudes and latencies of components may change with development. Components are best characterized by the use of several criteria, including amplitude distribution across the scalp, latency, and responsiveness to experimental manipulations.

The use of multiple electrode sites helps to identify components with characteristic scalp amplitude distributions, such as the P3b or SW components. It also helps to distinguish between components with overlapping latency ranges. Finally, it helps to assess possible differences between components from different modalities which appear to be similar in other ways, such as in latency. A liberal display of ERPs from individual subjects and grand averaged ERPs from all subjects from several electrode sites would be desirable in publications of original research.

Latency criteria for assessing components should take into account both the multiple nature of components and the possibility that latency may vary with age and task. For example, in young children, a latency window of 200 to 500 msec following a stimulus may be too early and narrow to detect the P3b component. Such a window ignores the likelihood of longer latencies for this component in this age group and may simply result in the measurement of some activity other than P3b. Klorman et al. (1979) used such a window for P3b in a visual discrimination paradigm with normal children and reported latencies of 310 msec; yet in analogous visual paradigms P3b latencies have been reported to be at least 400 msec in adults and adolescents (e.g., Courchesne et al., 1977; Courchesne, 1978a; Friedman et al., 1978a, 1981; Kutas and Donchin, 1978) and at least 500 msec in children (Courchesne, 1978a). It seems likely, therefore, that their latency window excluded P3b to some extent, and instead included a 310 msec component which might be analogous to Friedman et al.'s (1978) enigmatic P350.

On the other hand, latency windows which are too broad may include more than just the component sought. Courchesne (1977, 1978a) defined P3b, in part, as the highest peak in the latency window from the peak of P2 (ca 240 msec) to 1200 msec. This window undoubtedly included P3b, but it could have included SW as well.

Analysis techniques other than designating latency windows must also be used with equal care. For example, principal components analysis may give erroneous output when ERP components shift significantly in latency.

When assessing components, it is helpful to consider the effects of experimental manipulations on component amplitude and latency. Appropriate designs should be employed to reduce the diversity of subject options (Sutton, 1969). Consideration of manipulation effects seems especially important when attempts are made to categorize components in infants and children according to terminology used with adult ERPs. Such attempts should be justified. Just because a positive peak in an infant has a latency of 300 msec is not by itself sufficient evidence that it is equivalent to the adult P3b.

Designs and analyses should recognize the need to distinguish between ERP components and extra-neural potentials, such as eye movements. For instance, in order to distinguish ERP components recorded over frontal cortex from eye movement and blink potentials, several electrodes should be placed around the eyes and over frontal cortex. These electrodes should be referenced to the same electrode as other scalp electrodes. Potentials detected by electrodes from frontally located sites may very well be from eye movement or blink, but they may also be from the central nervous system. Indiscriminate rejection of all trials with deflections in such electrode derivations runs the risk of rejecting a class of trials which evoke substantial frontal neural activity.

LITERATURE REVIEW

Infants

Few studies have reported on ERP components which might be related to cognitive processes in infants. Schulman-Galambos and Galambos (1978) recorded visual ERPs from 7-48 week old infants. Electrodes were placed at C3 and C4 and below one eye. Stimuli lasted 2 sec and consisted of either slides of cartoon characters or people, defocused slides, flashes of light on the face of one of the experimenters playing peek-a-boo, or flashes of light on a doll held by one of the experimenters. In response to these events, they recorded ERPs characterized by negative waves 500-650 msec in latency and sometimes also by positive waves 1000 msec in latency. They concluded that these waves were related to cognition. However, this conclusion must be tempered, since defocused slides produced ERPs with latencies and amplitudes like the ERPs produced by focused slides, the doll and the experimenter.

Courchesne et al. (1981) recorded visual ERPs from 6 month old infants. ERPs were evoked by tachistoscopically presented photographs of two human faces which occurred once every 1800-2800 msec. For each infant, one face was presented frequently (p=.88) and the other infrequently (p=.12); which face was chosen to be the infrequent face was counterbalanced across subjects. Both types of events elicited a 700 msec, negative Nc component and a 1360 msec, positive Pc component. Nc and Pc were largest over frontal electrode sites. Infrequently presented faces elicited larger and longer latency Nc components than frequently presented faces. We suggested that this was evidence that our infants "were able to remember the frequently presented face from trial to trial and to discriminate it from the discrepant face". Pc components did not show such amplitude or latency differences. P3b components typically found in such paradigms in children and adults were not detected.

On the basis of this study and previous data (Courchesne, 1978a), we suggested that Nc components might be related to the perception of attention-getting events or, alternately, that Nc amplitude might be inversely related to the "clarity, precision, or stability of the memory trace for an event" (Courchesne et al., 1981). However, these ideas must be considered conjectures because of the lack of extensive data on Nc components in infants.

Hofmann and Salapatek (1981) reported an attempt to determine if the P3b component so prominent in children and adults to infrequent stimuli could also be detected in the 300 to 600 msec range in 3 month olds. In their report, Figures 2, 3, 5 and 6 show ERPs at Pz and Cz to infrequent visual changes and frequently presented stimuli. In these figures, there is an absence of any P3b-like component consistently occurring across subjects in the 600 msec latency zone. It is not too surprising that P3b is not detectable in these figures. P3b latency is ca 500-700 msec even in 4-5 year old children and would be expected to be at least this latency in infants.

Future attempts to discern P3b in infants would be aided by using at least a 1000 msec time base; by using at least 1500 msec interstimulus intervals; by using frontal as well as parietal electrodes; by using stimuli which are short-lasting so that sustained sensory potentials are not evoked; and by using an artifact rejection routine which takes into account the likelihood that large cognitive components will be elicited from frontal areas (see last paragraph of Methodological Considerations section).

Children and Adolescents

Studies of children have also found Nc and Pc components or components very similar in waveshape to them (Symmes and Eisengart, 1971; Courchesne, 1977, 1978a; Neville, 1977). Symmes and Eisengart (1971) and Neville (1977) presented slides to children (5-11 years and 9-13 years, respectively). In the Symmes and Eisengart study, slides were of cartoons and simple objects; in the Neville study, they were 32 line drawings from the Peabody Picture Vocabulary Test. In both studies, the children were expected to remember the slides and were tested either after each slide (Neville) or after 16 slides (Symmes and Eisengart). Broad negative components ca 300-500 msec in latency were evoked by the slides in both studies; these components seem similar to Nc. They were 20 to 40 uV at Fz and Cz in the Symmes and Eisengart study and ca 11 uV at T5 and T6 in the Neville study.

Symmes and Eisengart suggested that this broad negativity reflects the perception of meaningful stimuli since their negative (Nc-like) components were large in response to the cartoons and objects, but small to defocused slides. Neville reported that the negative (Nc-like) components were larger and earlier over the right (T6) than over the left hemisphere in normal children, but symmetrical in non-signing deaf children. She suggested that this negativity reflected hemispheric specialization of the processing of non-verbal spatial information. Neville also reported a broad positive component ca 970 msec which seems similar to Pc. This component was not measured independently of the 400 msec negative component and, so, its behavior is unclear; her tables suggest little asymmetry effects in either normal or deaf children.

In contrast, several other studies of cognitive ERP components in children and adolescents have employed relatively simple discrimination tasks with a limited and usually explicit set of events (e.g., a square and a triangle or a 1000 c/sec tone and a 2000 c/sec tone) (Shelburne, 1973; Prichep et al., 1976; Friedman et al., 1978, 1981, in press; Goodin et al., 1978a; Grünewald et al., 1978; Karrer et al., 1979; Loiselle et al., 1980; Hömberg et al., 1981). Each of the tasks used was similar to those used in experiments on the adult P3b component. Typically, subjects were requested to pay special attention to a designated "target" event or category. These target events evoked central or parietal positive components ca 300-600 msec in latency which appear to be the same as the adult P3b.

In accord with the conclusions of some of the older reports on P3b in adults, Shelburne (1972) and Goodin et al. (1978a) suggested that the P3b in children reflects decision-making processes. In the Shelburne study, P3b occurred only at decision points. In the Goodin et al. report, the three children studied (6, 8 and 9 years old) had P3b latencies (ca 400 msec) which were longer than latencies in young adults and, therefore, concluded that decision processes in children must take longer than in young adults.

In a study using 11 to 13 year olds, Hömberg et al. (1981) examined the effect on P3b amplitude of stimuli signifying different amounts of monetary reward. Their findings indicate that P3b is influenced by factors also reported to influence the adult P3b-- for example, the subjective importance of an event and the subjective probability of an event category

(Friedman et al., 1975a; Courchesne et al., 1977, 1978; Courchesne, 1978b; Donchin, 1979). The central conclusion to be drawn from the several studies of P3b in older children and adolescents is that it is influenced by factors similar to those reported in studies of P3b in adults.

Although these studies have shown that several cognitive components may be recorded in people younger than 18 years of age, for various reasons most were not designed to assess changes in cognitive components associated with development. For instance, such ERP changes cannot be assessed adequately when only one group of children is studied, whether the group is defined narrowly (e.g., 11 year olds) or broadly (e.g., 5 to 12 year olds). Also, since we do not know what variability to expect in the cognitive components of children, developmental changes cannot be adequately assessed when few subjects are studied.

Several studies have been designed to assess ERP changes associated with development (Courchesne, 1978a; Kurtzberg et al., 1979; Friedman et al., 1981, in press; Courchesne and Galambos, in preparation).

Kurtzberg et al. (1979) have published the only longitudinal study of cognitive components. They recorded visual ERPs in a simple visual discrimination task from 5 to 8 year olds. Each subject was tested three times at one year intervals. A positive component was found over lateral scalp sites--central, parietal and temporal--in response to target events but not to non-target events. Across the three recording sessions, its mean latency changed from 400 msec to 325 msec and its amplitude became more prominent over right parietal sites. Mean reacton time (RT) to target events changed in a similar fashion decreasing from 340 msec to 280 msec. Kurtzberg et al. suggested that the ERP and RT changes reflect developmental changes of parietal association areas involved in non-verbal information processing.

These longitudinal developmental results give a hint of one of the important contributions that ERP studies could make to the study of cognitive development. In their study, the RT changes alone could only have suggested quantitative changes in the speed of information processing. However, in combination with the ERP data showing concomittant changes in the location of neural activity, the RT-ERP changes suggest that changes in the speed of processing may be attributable, in part, to functional alterations in some neural areas active in processing.

Kurtzberg et al.'s 325 msec positive component is similar in latency and scalp amplitude distribution to a P350 component reported in visual tasks in children (Friedman et al., 1978; 1981) and adults (Kutas and Hillyard, 1980b). Because visual P3b components in children, adolescents and adults are usually reported to have much longer latencies than 325 msec (Shelburne, 1973; Courchesne, 1977, 1978a; Friedman et al., 1978, 1981; Kutas and Donchin, 1978; Homberg et al., 1981), this 325 msec positivity does not seem likely to be P3b.

Friedman et al. (1981, in press) studied six groups between 12 and 17 years of age. Electrode sites in both reports were Fz, Cz, Pz, and Oz.

In one study, Friedman et al. (in press) had 70 subjects detect either a pitch change or a missing stimulus in a train of 1000 Hz tone pips. Amplitude changes were found across ages in N250 and the frontal negative counterpart of SW. Friedman et al. noted that Nc has also been reported to decrease with age (Courchesne, 1978a; see below) and raised the possibility that N250 and Nc may be similar components. No latency changes were found across ages in P3b, the posterior positive SW, and a frontal N250 component.

In another study, Friedman et al. (1981) used a continuous performance paradigm with visual stimuli. Thirty subjects pressed a button to the numeral 8 in random sequences of numerals 2 to 19. Four components beyond P240 (or P2) were identified: P350, P450, P550 and SW. No effect of age was found in these components. P450 was interpreted to be P3b and P550 to be the P4 component of Stuss and Picton (1978).

In essence, then, Friedman et al. (1981, in press) found that the latencies, scalp distributions, and amplitudes of auditory and visual P3b and posterior positive SW components in adolescents were not much different from those reported in adults. The most notable age-related change was the decrease in frontal negativity (N250 and SW) in the auditory paradigm.

Changes in Auditory and Visual ERPs during Development

We have completed a series of studies of changes in auditory and visual ERPs associated with development from childhood to adulthood (Courchesne, 1977, 1978a, 1979; Courchesne and Galambos, in preparation). The following age groups were studied: 4-5, 6-8, 10-13, 14-18, and 23-44 years. Analogous auditory and visual paradigms were used. Each paradigm incorporated two general types of events: (a) bizarre, "unrecognizable" or "novel" events not seen or heard before by the subject and (b) precisely specified events (i.e., the letters A and B and the sounds me and you). In the auditory paradigm, targets were the sound you and backgrounds were the sound me; in the visual paradigm targets were the letter A and backgrounds B. Subjects pressed a button as rapidly as possible to the presentation of target events (p=0.12) randomly interposed in sequences of background events (p=0.76). Also interposed in these sequences were the bizarre novel events (p=0.12).

ERPs were recorded from electrodes below the right eye (LoE), above the right eye (UpE), at the left outer canthus, at Fz, at Cz, and at Pz (see Figure 1). The reference electrode site was the right mastoid. The visual ERP data presented below is the combined data from one of our recently completed studies and a nearly identical visual study already published (Courchesne 1977, 1978a, 1979).

Target ERPs. Figure 1 shows the auditory and visual ERPs to targets and backgrounds in subjects of five age groups. Several results may be noted. In visual and auditory modalities in each age group, P3b and SW components were maximal in amplitude at Pz; were larger to infrequently-presented stimuli than to frequently-presented stimuli; and were about 200 msec apart. From this data, it appears that the processing stages reflected by P3b and SW are present across the age span from childhood to adulthood.

Differences were also found, however, between children and adults in P3b and SW. In both modalities, these two components decreased in latency with age in concert with RTs and reached adult values by adolescence

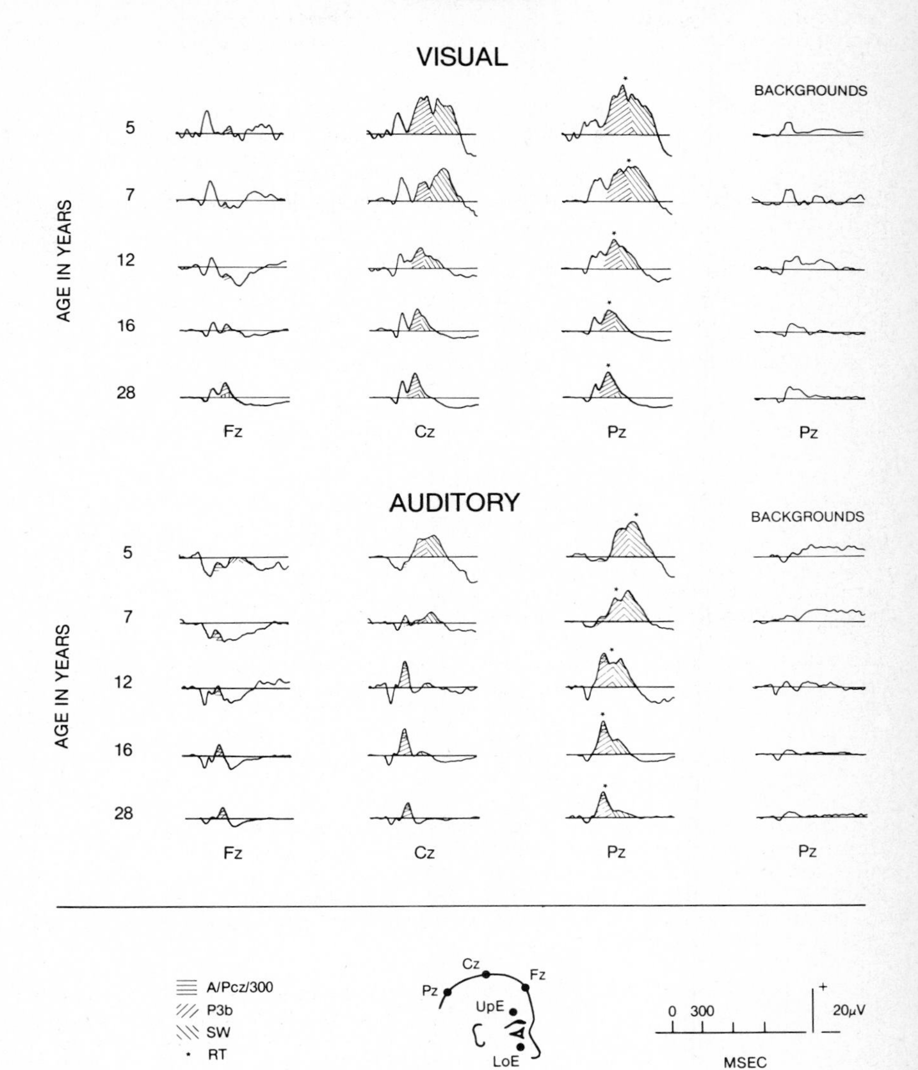

Figure 1

ERP responses to visual and auditory target events and background events. ERPs to targets at Fz, Cz and Pz and ERPs to backgrounds at Pz. Records show ERPs from 5 age groups. Each trace is a grand average of 5 or more subjects.

(Figure 1). SW decreased substantially in amplitude with age. Lastly, Nc and Pc amplitudes decreased with age. These changes could indicate that the processes reflected by P3b and SW increase in speed and efficiency with age. Whether these changes might also imply some qualitative changes as well, remains to be determined.

Novel ERPs. The ERPs to visual novels changed substantially with age in component composition, amplitude, and latency (upper panel of Figure 2). There were striking differences between the visual ERP waveforms of children and adults. The waveform transition from childhood to adulthood was gradual and not complete in adolescence. In children, waveforms were characterized by large, frontal Nc (ca 30 uV and 410 msec), frontal Pc (ca 30 uV and 900 msec), and posterior P3b components (ca 28 uV and 700 msec). The amplitudes of these components decreased with age and a 400 msec, fronto-central P3 component began to emerge through Nc in pre-adolescence and adolescence. This fronto-central P3 nearly replaced (or perhaps, obscured) Nc by adulthood. These data suggest that children and adults employ different neural systems to evaluate novel visual information.

The ERPs to auditory novels also changed with age (lower panel of Figure 2). Three components decreased in amplitude with age: Nc, Pc, and a component termed A/Ncz/800 (A=auditory; N=negative in polarity; cz=maximal in amplitude at Cz; 800=800 msec in latency). Low amplitude P3b and SW components at Pz appeared to decrease in latency. By pre-adolescence P3b fell within the latency range of another component: A/Pcz/300 (A=auditory; P=positive in polarity; cz=maximal in amplitude at Cz; 300=300 msec in latency). Thus, in pre-adolescence, adolescence and adulthood, novels elicited a positivity at Pz in the 250 to 360 msec range which is probably composed of two distinct components, A/Pcz/300 and P3b.

The A/Pcz/300 component to auditory novels is a striking exception to most of the other cognitive components mentioned so far. It changed only very little from young childhood to adulthood (from 323 to 278 msec and, at Cz, from ca 11 to 13 uV).

The overlapping of A/Pcz/300 and P3b in adults but not in young children, raises an important point about the value of developmental ERP research. In adults, overlapping components are easily disentangled only if they respond differently to experimental manipulations (e.g., to attentional demands, etc.), or if they have substantially different scalp amplitude distributions. If they do not differ in these ways, then examining developmental ERP data would provide the best means of identifying and studying components.

Unlike the similarity between visual and auditory ERPs to targets, the visual and auditory ERPs to novels were different from each other at all ages. Auditory novels elicited A/Pcz/300 and A/Ncz/800 components, but visual novels did not. Conversely, visual novels elicited the fronto-central P3 component, but auditory novels did not. Although Nc and Pc waves were identified in both modalities, they differed between modalities in their scalp amplitude distributions: auditory Nc and Pc waves were more frontal than visual.

Figure 2 shows that ERP components may be found at electrode sites around the eyes. These components are distinguished in a variety of ways from electro-retinographic, eye blink and eye movement potentials; details

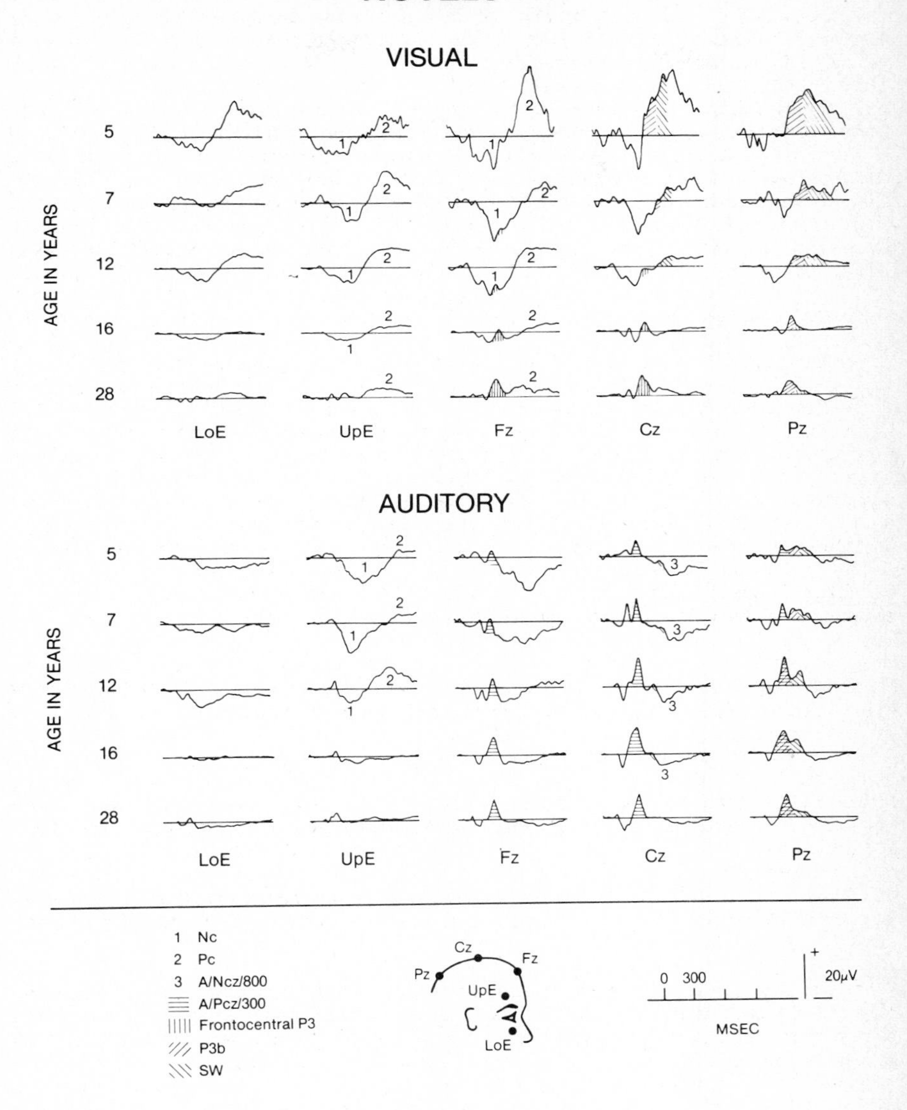

Figure 2

ERP responses to visual and auditory novel events. ERPs at LoE, UpE, Fz, Cz and Pz. Records show ERPs from 5 age groups. Each trace is a grand average of 5 or more subjects.

about these distinctions may be found in Courchesne (1977).

SW and Frontal Negativity. It has been suggested that the parietal positive SW is negative frontally (N.K. Squires et al., 1975; K.C. Squires et al., 1977). This suggestion implies that the larger the parietal positivity, the larger the frontal negativity. We did not find this relationship in our data. Figure 1 shows that in 5 year olds, targets evoke a large parietal positive SW and a corresponding small frontal positivity (i.e., volume conducted parietal SW), not the expected large negativity. As parietal SW gets smaller with age, frontal Nc-like negativities appear. It could be reasoned from this data that SW is not negative frontally. The frontal negativity at the SW latency may be a different component. Further investigations of SW and frontal negativities might benefit from using young children as subjects.

Summary of a Longitudinal ERP Study

Recently, we retested the majority of children and pre-adolescents who participated in our initial study of visual ERP changes associated with development (Courchesne, 1977, 1978a). The original 6-8 year olds were now 10-13 years old and the original 10-13 year olds were now 14-18 years old. The results of this longitudinal study confirm many of the findings in our initial cross-sectional study. For example, (1) Nc elicited by novel events decreased in amplitude with age; (2) in teenagers, the fronto-central P3 component evoked by novels nearly replaced Nc; and (3) the parietal P3b to targets decreased in latency with age.

Synopsis: Changes in Auditory and Visual ERPs during Development

Several points can be made from these studies:

(A) ERP components which may be associated with cognitive processes are different in children and adults; the exception is A/Pcz/300.

(B) These differences may be either quantitative or qualitative: for example, 1) P3b and SW latencies are longer in children than in adults; 2) the large Nc, Pc, and A/Ncz/800 waves in children are small in adults; and 3) the visual fronto-central P3 seen in adults is not seen in young children.

(C) The transition from childhood ERP waveforms to adulthood waveforms is gradual. This does not mean that the rate of change is necessarily constant.

(D) Components reach "maturity" at different ages. For example, A/Pcz/300 reaches maturity during childhood (perhaps before ?); P3b during pre-adolescence and adolescence; and the visual fronto-central P3 during young adulthood.

(E) Some of these cognitive components are elicited by stimuli in only one modality (e.g., A/Pcz/300 and the visual fronto-central P3) and some by stimuli in visual and auditory modalities (P3b, SW, Nc, and Pc). The latter components may have very similar characteristics in each modality (e.g., P3b) or somewhat different ones (e.g., Pc).

(F) At all ages, different types of information elicit different ERP patterns. For instance, in children, auditory targets elicit large P3b and SW components, while auditory novels elicit small P3b and SW components, but large A/Pcz/300 and A/Ncz/800 components.

SOME CONJECTURES

The Nc Component

Nc and components which appear to be very similar have been hypothesized to be associated with several cognitive processes. Symmes and Eisengart (1971) concluded that the negative (Nc-like) component obtained in their study reflected the perception of meaningful stimuli. However, this suggestion does not seem compatible with the evidence that these types of negative components have been recorded in 6 month old infants (Courchesne et al., 1981) and in response to bizarre, complex, nonsensical visual and auditory patterns, i.e., novels (Courchesne, 1977, 1978a; Courchesne and Galambos, in preparation).

In infants and children, Nc and Nc-like components seem to be elicited by a variety of attention-getting events such as pictures of cartoons or novel visual and auditory events. If Nc is elicited by the perception of attention-getting events, then, for example, Nc could be used to study developmental changes in the range of events and contexts that are attention-getting.

The Pc Component

The Pc component has been related to several processes, including: (1) the refining of existent categories or the generation of new ones based on new information and (2) the categorizing of events on the basis of their physical, concrete properties (Courchesne, 1978a, 1979). The second suggestion was based partly upon the following: In 6 month old infants, categorization abilities are not yet developed; in young childhood, the ability to organize or categorize information on the basis of physical, concrete properties develops (e.g., Inhelder and Piaget, 1964; Bruner, 1966; Rosch et al., 1976); and in later childhood and pre-adolescence, the ability to categorize on the basis of symbolic, linguistic processes begins, reaching maturity in adults. Coincidentally, Pc does not appear to be fully developed in 6 month olds; it is prominent during childhood; and it diminishes during pre-adolescence and adolescence and is small in adults. However, since very little information exists on the Pc component, these conjectures should be considered an invitation to further research.

The A/Pcz/300 Component

The A/Pcz/300 component is unlike any of the other cognitive components reported in developmental ERP studies. It changes little in latency, amplitude, and scalp amplitude distribution from 4 years of age to adulthood. It seems specific to the auditory modality and is largest in amplitude to previously unexperienced acoustical events (novels). It also has a rather narrow waveshape, lasting only 100 to 140 msec (lower panel of Figure 2). These characteristics are suggestive of a component reflecting automatic detection of biologically significant acoustical deviations.

Such a detector would be especially adaptive in the auditory modality since auditory events tend to be very ephemeral and timing is crucial.

The auditory P3a component reported in the adult ERP (e.g., Roth, 1973; N.K. Squires et al., 1975; Snyder and Hillyard, 1976) is similar in some respects to A/Pcz/300. As with A/Pcz/300, the adult P3a is elicited by acoustical deviations occurring infrequently in a sequence of identical sounds. It has a latency range of 220-300 msec, which overlaps with the latency range we found for A/Pcz/300 (250-360 msec). However, there are also some differences. For example, A/Pcz/300 is rather high in amplitude, while P3a generally is not. This difference may be attributable to the fact that our novels were more deviant than the deviations used in studies of P3a. Also, A/Pcz/300 and P3a have different scalp amplitude distributions: while A/Pcz/300 is largest in amplitude at parieto-central scalp sites, the P3a is largest at centro-frontal sites. These distribution differences indicate that A/Pcz/300 and P3a might be two different components. Even so, insufficient data exists to draw a definitive conclusion.

Might there be a visual analog of A/Pcz/300 ? Several visual components have been reported to be larger to infrequently presented physical deviations in a train of stimuli: for example, the P2 component (Courchesne et al., 1977, 1978; Kutas and Hillyard, 1980b), the P350 component (Friedman et al., 1978, 1981; Kurtzberg et al., 1979; Kutas and Hillyard, 1980), and the visual fronto-central P3 component (Courchesne et al., 1975).

The visual P2 has a shorter latency than A/Pcz/300. So, because visual components are typically longer in latency than their auditory analogs, the visual P2 is unlikely to be an analog of A/Pcz/300.

Like A/Pcz/300, the visual fronto-central P3 is elicited by novel information. If it were an A/Pcz/300 analog, it should occur in young children. Is it possible that it does, but is not seen because it is swamped by huge Nc waves? Although our evidence does not clearly support this possibility, it does not definitively rule it out either. The shading in the upper panel of Figure 2, intimates the possibility that a trace of the fronto-central P3 may be detectable in the grand averaged waveforms of 7 year olds. However, it is so small in our data that in individual subjects, it cannot be distinguished from noise. This stands in contrast to the robust aspect of A/Pcz/300 in 5 and 7 year olds.

The P350 seems to be a better candidate as an analog of A/Pcz/300. Its latency is slightly longer than A/Pcz/300, as would be predicted. P350 resembles A/Pcz/300 in several ways. It is largest at parieto-central sites. It occurs after P2 but before P3b. Its latency in childhood (ca 325 to 400 msec) is apparently little different from its latency in adolescence and adulthood (ca 350 msec; Friedman et al., 1978; Kutas and Hillyard, 1980b), if the 325-400 msec positivity in children reported by Kurtzberg et al. (1979) is a P350 as suggested above (Liturature Review section). Lastly, it might be overlapped by P3b in adult waveforms, as suggested by Friedman (in Kurtzberg et al., in press). Nonetheless, the functional significance of P350 is obscure; this precludes any conclusion about its status as an analog of A/Pcz/300.

The Visual Fronto-central P3 Component

The visual fronto-central P3 is curious. It emerges very late in development. It also appears to be elicited under a very special set of circumstances: bizarre, never-seen-before visual patterns must occur infrequently in a monotonous sequence of stimuli. After three or four presentations of such patterns, the amplitude of this component at Fz is only about 60% of the amplitude to the very first presentation and after more than nine presentations, the amplitude is less than 40% of the amplitude to the first presentation (Courchesne et al., 1975; Courchesne, 1978b). While the visual fronto-central P3 decreases after several presentations, P3b amplitude at Pz increases.

In reference to these amplitude changes, we proposed that the visual fronto-central P3 occurs when new concepts or modification of old ones are required in order to handle bizarre, novel information (Courchesne, 1978b). We also proposed that P3b occurs when available concepts are sufficient. Thus, initially, such novel information should evoke the fronto-central P3. After several presentations of such novel information, a concept should form. At this point, the fronto-central P3 should decrease while P3b should increase. This hypothesis about the fronto-central P3 is reminiscent of one of our hypotheses about Pc. Might Pc and the fronto-central P3 be reflecting developmentally different modes of creating new concepts or modifying old ones?

The P3b and SW Components

A variety of concepts have been ascribed to P3b. A sample of them includes information delivery (Sutton et al., 1967); selective recognition of target information in an attended stream of information (Hillyard et al., 1973; Hillyard and Picton, 1979); the evaluation of contextual hypotheses concerning the sequential structure of a train of events (K.C. Squires et al., 1976); event categorization (Friedman et al., 1975; Courchesne, 1978a, b); the delivery of subjectively relevant information (Campbell et al., 1979); and Shiffrin and Schneider's (1977) formulation of controlled information processing in unfamiliar tasks (Woods et al., 1980). Ruchkin et al. (1980) suggested that SW components are elicited by stimuli which substantially increase processing demands as a consequence of being very surprising or unexpected. For reviews about these waves and their possible associated processes, see chapters by Ruchkin and Sutton, Roth, and Rösler in this volume. The following will only briefly highlight some of the conclusions which can be drawn by considering these waves in a developmental framework.

The RTs and the latencies of P3b and SW to targets decrease with age and reach adult latencies by adolescence (Courchesne, 1977, 1978a; Friedman et al., 1981, in press; Courchesne and Galambos, in preparation). These changes show that the speed of information processing increases with age. Although the latencies of P3b and SW decrease with age, the time separating them remains constant at about 200 msec. Accordingly, P3b may reflect a rate limiting step in information processing and SW may be dependent upon the output from the P3b "processor". Viewing P3b and SW as sequential is consistent with Ruchkin et al.'s (1980) formulation wherein P3b represents an earlier cognitive stage of stimulus evaluation.

We found that SW amplitude is greater than P3b in response to targets in children, but smaller than P3b in adults. These findings are consistent with Ruchkin et al.'s (1980) idea about SW, since the level of unexpectedness or surprise about a target in 4 year olds should be greater than in adults. However, the novel events should have been much more unexpected and surprising than targets and yet, in children, they elicited much smaller SW components than did targets. How might this discrepancy be resolved?

It seems most parsimonious to suggest the obvious: Different processing problems are handled in different ways by the brain. Some of these ways are reflected by different ERP components. Moreover, age is a factor in these differences.

The following exemplifies this suggestion. Some evidence indicates that P3b amplitude deceases as the ease of information processing decreases. Such decreases in the ease of processing may result from greater difficulty in detecting a stimulus (Ruchkin et al., 1980); the need to form new concepts or modify old ones to handle novel events (Courchesne, 1978b); or semantic incongruity (Kutas and Hillyard, 1980). These three causes for decreases in the ease of processing seem to be associated with different components: namely, SW, the fronto-central P3 (in adults), and N400, respectively. As another example, decreases in the ease of processing caused by visual novelty are associated with different ERP patterns in children and adults.

In conclusion, it appears that complex changes occur in ERP components associated with cognitve development. Many cognitive components have been identified and others will no doubt be discovered. Researchers are just beginning to understand which cognitive processes are associated with these components. If it turns out that different ERP components and wave patterns are associated with different stages of cognitive development, developmental ERP research should become indispensible in the field of normal and abnormal cognitive development.

Acknowledgements

I wish to thank Walter Ritter, Marta Kutas, Alan J. Lincoln, and Rachel Y. Courchesne for their valuable comments on this chapter. The writing of this chapter was supported by NICHHD grant 5-R01-HD11154 awarded to R. Galambos and by Speech, Hearing and Neurosensory Center, a division of Children's Hospital and Health Center, San Diego, California.

Send reprint requests to: Dr. Eric Courchesne, Speech, Hearing and Neurosensory Center, Children's Hospital and Health Center, 8001 Frost Street, San Diego, 92123 (USA). Tel. (714) 292-3499.

Footnote

1. Throughout this article the term "P3b", coined by N.K. Squires et al. (1975), will be used to refer to a component variously termed "P3", "P300" and "LPC".

Tutorials in ERP Research: Endogenous Components
A.W.K. Gaillard and W. Ritter (eds.)

16

EVENT RELATED POTENTIALS AND LANGUAGE PROCESSES

Dennis L. Molfese

Department of Psychology and School of Medicine
Southern Illinois University
Carbondale, Illinois
U.S.A.

INTRODUCTION

Although nearly one hundred scientific papers purporting to deal with evoked potential correlates of language processes have been published over the last three decades, little specific information is currently known concerning the electrophysiological correlates of language. This is especially true in the areas of syntax, semantics, pragmatics, and sentence processing. While phonology has received more systematic attention than the other divisions of language, even there large tracts remain virtually untouched. Such poor progress in spite of the relatively large number of Event Related Potential (ERP) studies can be attributed to (1) the general lack of systematic research except in only a few studies; (2) a simplified view of language coupled with the absence of an overall linguistically oriented approach to guide stimulus selection and control as well as the experimental design; (3) the more prevalent interest in ERP demonstrations of laterality--a situation that may have distracted investigators from identifying specific ERP differences to language related stimuli; (4) the use of few electrode sites, with the majority of studies using two sites; (5) the use of inappropriate analysis procedures or the absence altogether of objective statistical techniques; (6) conclusions that over-generalize from the absence of effects or which ignore effects not in keeping with the hypotheses being advanced.

In spite of the problems outlined above, there are reports from a variety of laboratories studying language processes which indicate that ERP techniques may offer new insights into the study of neurolinguistics. It is to these studies that we now turn. This chapter will focus on research findings which reflect the ability of the ERP to differentiate between language materials and to distinguish language related materials from other types of stimuli. The controversy over the ability of the ERP to reflect hemisphere differences is covered elsewhere (see Rugg, this volume) and will not be specifically addressed here.

This chapter will outline research in four areas of language study: phonology, contextual effects (pragmatics), semantics, and sentence processing.

PHONOLOGY

Phonology, the study of the system of sound patterns that occur in a language, enjoys perhaps a special place in current linguistic thinking. The major phonological theories (Jakobson, 1939; Jakobson, Fant, & Halle, 1957; Chomsky & Halle, 1969) have been in place for a relatively long period of time. In addition, speech and hearing scientists over a forty year period have identified a number of the acoustic cues carried in the speech signal which convey differences in meaning to the listener (Liberman, Cooper, Shankweiler & Studdert-Kennedy, 1967). Computer speech synthesis programs have also been developed (Mattingly, 1967) which can be used to construct speech stimuli in such a way that virtually every acoustic dimension of the signal can be systematically controlled and manipulated. Such theoretical and technological work offer the electrophysiologist not only a conceptual framework for the development of experiments, but, in addition, valuable information concerning stimulus construction, manipulation, and control.

A. ERP differences to speech versus nonspeech stimuli:

Unfortunately, much of the early ERP research into aspects of language phonology did not make use of these advantages. Early research attempted to determine whether ERP studies could detect differences between language-relevant speech sounds and various types of nonspeech sounds (Cohn, 1971; Morrell & Salamy, 1971; Molfese, 1972; Molfese, Freeman, & Palermo, 1975; Barnet, de Sotillo, & Campos, 1974; Neville, 1974; Neville, 1980; Hillyard & Woods, 1979; Friedman, Simson, Ritter, & Rapin, 1975; Galambos, Benson, Smith, Shulman-Galambos, & Osier, 1975).

In the earliest of these studies, Cohn (1971) recorded auditory evoked responses (AER) to clicks and consonant-vowel-consonant (CVC) words from electrodes placed over the left and right side of the head 2 cm anterior to the external meatus and a second pair 2 cm from the midsagital line in the vertical coronal plane. For the word stimuli, 17 adults produced larger left hemisphere (LH) AERs for components between a negative peak (N30-N50) and a later positive peak (P125). A large positive component with a peak latency of 14 ms was reported over the right hemisphere (RH) leads in all 37 adults in response to the clicks. However, the latency of these AER components did not match those generally reported for auditory stimuli (Morrell & Salamy, 1971). In a related study, Morrell and Salamy (1971) recorded AERs to five natural speech nonsense syllables at frontal (F_7, F_8), rolandic (C_5, C_6) and temporal-parietal electrode locations of seven adults who reported the stimuli after a delay. The N90 component appeared largest over the left hemisphere temporal-parietal leads. A number of methodological problems, however, temper these results (Grabow, Aronson, Rose, and Greene, 1980). Grabow et al. in a modified replication of Morrell and Salamy, failed to find comparable effects. They attributed this in part to possible incorrect baseline measures made by Morrell and Salamy. However, their attempt to replicate may have failed because of increased response variability due perhaps to differences in stimulus rise times and durations.

In the first study which attempted to identify changes in AER responses over different developmental periods to speech and nonspeech stimuli, Molfese and his colleagues (Molfese, 1972; Molfese, Freeman, & Palermo, 1975) recorded AERs from T_3 and T_4 of 10 infants one week to 10 months, 11 children 4 to 11 years old, and 10 adults 23 to 29 years of age. The natural speech stimuli consisted of CV syllables /ba, dæ/, and monosyllable words, /bɔi, dɔg/. A piano chord and white noise burst served as the nonspeech stimuli. Stimulus duration and intensities were matched across stimulus sets. Amplitude measures were made for the N100-P160 components of adults, N143-P214 components of children, and, for infants, between the largest negative (N454) and positive peaks (P528). A ratio of left hemisphere peak to peak amplitudes to left hemisphere plus right hemisphere amplitudes was calculated, (L/L+R). In general, larger left hemisphere amplitudes were found for the speech stimuli in all groups of subjects while larger right than left hemisphere amplitudes were noted for the nonspeech material. The lateral differences were noted as more consistent and marked for the nonspeech materials. Such results indicate that from infancy onward AER measures can distinguish speech from non speech stimuli.

A later study by Barnet, de Sotillo and Campos (1974) reported similar effects in infants. They recorded AERs from normal and malnourished infants under one year of age in response to click stimuli as well as to speech stimuli (the infants' names). Only the normal infants showed larger amplitude left hemisphere than right hemisphere AERs to speech stimuli. Both groups of infants showed greater AERs to the nonspeech (click) stimuli in the right hemisphere. These findings are in agreement with those of Molfese, et al., (1975). More recently, Tanguay, Taub, Doubleday, & Clarkson (1977) attempted to "replicate" the adult results reported by Molfese, et al., (1975) but failed to find hemisphere effects. However, these two studies differed in at least 9 features. Tanguay et al. measured different portions of the AER and placed their electrodes at different sites than Molfese. They measured peak amplitude differences using N(176-204 ms) to P(209-236) instead of the N100-P160 measure used by Molfese. Tanguay also placed electrodes at C_3, C_4, W_1 (midway between T_5 and C_3) and W_2 (midway between T_6 and C_4) rather than at T_3 and T_4. In addition, Tanguay used a monoaural mode of presentation rather than the binaural procedure used by Molfese. As both Neville (1974) and Haaland (1974) note, language-related hemisphere effects are not found when a monaural mode of presentation was used. However, both Neville and Haaland report consistent hemisphere effects during binaural presentations. Such differences between binaural and monaural presentation have been noted in the behavioral literature and generally result in the absence of hemisphere differences in studies employing monaural procedures while noting consistent hemisphere effects in binaural studies.

The Tanguary et al. study also differed in filter settings of amplifiers, stimulus order, types of stimuli used, fixed versus varied interstimulus interval, length of the interstimulus internal, and task requirements. Given such differences between the two studies, it is not surprising that Tanguay did not replicate the earlier adult work by Molfese. Perhaps the only point that is surprising is that Tanguay et al. considered their study a replication.

A number of investigators have suggested that subjects should be involved in a behavioral task while the ERPs were recorded (Neville, 1974; Hillyard & Woods, 1979). This, they argue, would provide a means for controlling the subject's attention and involvement during testing. In addition, the behavioral data could also be used to interpret findings from the ERP recordings. Neville presented speech (digits) and nonspeech (clicks) sounds through earphones to 10 adults and recorded from electrodes placed at the central (C_5, C_6) sites over each hemisphere. Pairs of speech stimuli were presented in groups of three to each ear during the dichotic and monaural task. In a different condition, clicks were presented to one ear at a time during a series of two second periods. Subjects reported the stimuli heard after each series. Recall accuracy of stimuli presented to the right ear (and, consequently, thought to be processed primarily by the left hemisphere) were more accurate than for the left ear (right hemisphere). This finding is in agreement with earlier behavioral studies (Kimura, 1967). No ear effects were noted for the nonspeech materials. Neville reported that peak latencies for the N_1, P_2, and N_2 components were shorter in the left hemisphere than in the right hemisphere for the speech stimuli. She also noted that larger latency differences occurred over the left hemisphere electrode site in response to verbal and nonverbal stimuli than over the right hemisphere. As in the case of Tanguay, et al., no hemisphere differences were noted when a monaural mode of presentation was used.

Neville (1980) later recorded auditory ERPs from P_3, P_4, C_3, C_4 and one suborbital lead (to measure EOG) from twelve adults during a series of tasks. In Task 1 subjects were asked to recall dichotically presented words. Subjects correctly recalled 34% of the words presented to the left ear and 60% presented to the right ear. Amplitude measures of the late occurring baseline to N1 component showed a hemisphere effect only over the parietal leads. No behavioral or electrophysiological effects were found for dichotic presentation of melodies in Task 2. In Task 3 subjects were asked to recall two different four-letter words presented simultaneous to the two visual fields. Nine of 12 subjects correctly reported more words presented to the right than the left visual field. Amplitude asymmetries in the visual ERPs, measured from baseline to the N1 peak and from the N1 peak to the P2 peak were reported only over parietal leads with the left hemisphere lead showing the largest amplitude. No such asymmetries were found when the words were presented out of focus (Task 4). Neville went on to note that "when subjects perform demanding tasks which are designed to produce behavioral asymmetries, the concurrently recorded ERPs show evidence of functional asymmetries" (p 309).

Using such criteria as increased task demands, other investigators have reported ERP differences to speech and nonspeech stimuli. Hillyard and Woods (1979) recorded ERPs during the processing of high information content stimuli and reported some success in discriminating ERPs to words in a poem passage from ERPs elicited by tones of varying frequencies. Through the use of more natural language stimuli Hillyard and Woods suggested that more robust language related effects could be found. Subjects listened to monotone speech stimuli presented at approximately 80 words per minute as part of a coherent, rhymed poem. Subjects were instructed to listen closely to the content and informed that they would be required to answer difficult questions later. The words ranged in duration from 300 msec for monosyllable words to 600 msec for polysyllable words. The 400 Hz, 1000 Hz and 4000 Hz tones were presented at approximately the same rate. In a methodological improvement over many of the earlier speech-nonspeech studies, the speech and tone stimuli sequences were matched in range of loudness, interstimulus interval, and rise/decay times. Electrodes were placed on 12 adults at the vertex, at a left scalp region over Wernicke's area, and its right scalp homologue (no more specific electrode coordinates were given). The authors noted that the monosyllable stimuli elicited a 34% larger N1 component as well as a larger (18%) "sustained (negative) potential" from 100 to 500 msec relative to the tone stimuli. As Molfese, et al., and Neville had noted, Hillyard and Woods also reported that the N1 effect (peak latency ca. 100 msec) produced its largest effect over the left hemisphere electrode site although no such differences were noted for the tone stimuli.

Friedman, Simson, Ritter, and Rapin (1975), in a paper designed to assess hemisphere differences rather than speech-nonspeech differences, challenged earlier findings of laterality differences. They recorded AERs from electrodes placed over P_2 and midway between P_2 and the mastoid area on both sides of the head in response to 5 natural speech words and 5 nonspeech human sounds. In Condition 1 subjects were simply instructed to listen to the sounds. In Condition 2, subjects were instructed to lift a finger as soon as possible to indicate the occurrence of a signal stimulus (out of five possible stimuli). AERs were averaged across the different signal words, and across these same stimuli when they did not serve as signal items; the third average was based on AERs elicited to these items during Condition 1. A number of effects unique to the LH electrode site were noted. Results indicated that LH amplitude for the N100 component was larger for the signal than for the non signal stimuli. The LH latency of this component was also longer for words than for sounds (an effect that was opposite to that reported earlier by Neville, 1974, although such a difference could relate to task differences between the two studies). Signal words resulted in the largest N100 components obtained over the LH for all eight subjects. The amplitude of the P300 component showed a similar effect for the same subjects. Friedman, et al., did find differences in ERP responses elicited by speech and nonspeech stimuli that again appeared to center around the N100 component. Such differences appeared in all of their subjects. A related study by Galambos, Benson, Smith, Shulman-Galambos, and Osier (1975), involved the recording of AERs to two natural speech syllables and tones from eight adults with electrodes placed at C_Z, midway between T_3 - P_5 and midway between T_4 - T_6. Subjects heard one list of speech syllables in which one occurred frequently and the second syllable infrequently and then a second list in which 2 tones were presented in a similar fashion. The latency for both the N100 and the P300 components differed for the speech and tone stimuli. Galambos, et al., note that the "latency shift may mean that different processes occur in the identification of speech and non speech stimuli" (p. 281). In addition, the amplitude of the P300 component for the speech stimuli in contrast to the nonspeech stimuli was also noted to be "reasonably common for targets, principally in the P_3 region, with somewhat larger areas of difference on the left" (p. 281).

In a study which attempted to assess ERP differences to natural speech stimuli and tones, Lawson and Gaillard (1981a) recorded AERs and reaction time (RT) to ten different consonants preceeding the vowel /E/ plus a 1kHz tone. Electrodes were placed at F_z, C_z, P_z, T_3 and T_4 and AERs recorded to blocks of each stimulus. Analyses from six subjects generally indicated that RT and the N1(111-) latencies distinguished one set of stimuli (tone, /pe, te, ke/) from a second group (/we, je, me, ne, ve, fe/). N1-P2 peak amplitudes grouped the tone and /pe, te, ke, we/ responses together and distinguished them from the /fe, se/ stimuli. Lawson and Gaillard conclude that these results were not affected by vowel onset, and suggest that the distinctions found were due to acoustic rather than phonetic differences since responses to the tone stimuli were grouped with the stop consonants /p, t, k/. Differences in stimulus rise times (Ruhm & Jansen, 1969), loudness and duration (Skinner & Jones, 1968; Onishi & Davis, 1968) could have contributed to these effects.

In virtually all of the studies described thus far, the large negative ERP deflection centered around 100 msec post stimulus onset appeared to be the component most often studied and one which appeared to reflect differences between speech and nonspeech stimuli (Morrell & Salamy, 1972; Molfese, et al., 1975; Neville, 1974, 1980; Hillyard & Woods, 1979; Friedman, et al., 1975; and Galambos, et al., 1975). However, as noted at the beginning of this section, given the many features which characterize the differences between the speech and nonspeech materials employed, it is impossible to specify exactly what factors produce these effects.

One obvious difficulty with much of the research reviewed above concerns the failure to clearly restrict and define the differences between the language-related and language-unrelated stimuli. Speech sounds of the language carry specific acoustical characteristics in terms of frequency, formant structure and duration, number of formants, frequency transitions, rise and decay times, as well as multiple parallel and serial frequency modulations. Since such factors in all the studies reviewed above were not matched in terms of the nonspeech stimuli employed, it is impossible to determine if the earlier effects were due solely to the linguistic characteristics of the speech stimuli versus the nonspeech stimuli or to the multiple acoustic differences which separate these two classes of stimuli.

B. ERPs to phonetic stimuli:

1. Task effects. In the second group of studies to be reviewed here, researchers were able to restrict the stimulus characteristics such that they could go beyond the issue of speech versus nonspeech to address questions concerning how ERP techniques could be used to study specific phonological dimensions (Wood, Goff, & Day, 1971; Wood, 1975; Dorman 1974; Molfese, 1978A; Molfese, 1980A; Molfese, 1978B; Molfese & Hess, 1978; Molfese & Molfese, 1979A; Molfese & Molfese, 1979B; Molfese & Molfese 1980; Molfese, 1980B; Molfese & Erwin, 1981).

Wood, Goff, and Day (1971) conducted the first AER study of specific linguistic contrasts. They recorded AERs from temporal (T_3, T_4) and central (C_3, C_4) electrode sites over the two hemispheres while subjects were involved in either a phonetic or an acoustic identification task. In the phonetic task, subjects were asked to indicate whether a /b/ or a /d/ initial syllable occurred. These consonants differed only in formant 2 (F_2) and formant 3 transitions (F_3) while the fundamental frequency (F_0) remained unchanged. Such a task should require linguistic processing since individuals would have to make use of their knowledge of language sounds in order to successfully complete the task. In the non-linguistic acoustic task, subjects were to indicate which syllables had high (140 Hz) or low (104 Hz) F_0. Reaction time (RT) measures and the AERs elicited to the /ba/ syllable with F_0 = 104 Hz were recorded during both tasks. No differences in RT were found. In the analysis of the pre RT response periods of the AER, Wilcoxin sign tests indicated no differences in AER responses to the /ba/ syllable recorded during the phonetic and acoustic tasks over the right hemisphere electrode leads although AER responses differed between the two tasks over left hemisphere leads. Wood, et al., interpreted these results as indicating that only the left hemisphere is involved in phonetic distinctions. In a replication and extension of this work, Wood (1975) reported comparable findings with the consonants /b, g/. However, recent work by Grabow, Aronson, Offord, Rose, and Greene, (1980) has called these findings into question. They recorded AERs over T_3, T_4, C_3, C_4, C_5, and C_6 while subjects were involved in phonetic and acoustic tasks identical to those used by Wood, et al., (1971). No phonetic-acoustic effects were noted. The T_3 responses to all stimuli were more attenuated than T_4 responses. This latter effect, Grabow, et al., note, is in agreement with waveforms presented by Wood (1975). While the findings of Grabow, et al., differed from Wood's, so too did their methodology. The Grabow, et al., study did not involve a RT task. If task factors are critical to demonstrating reliable ERP-language effects as Neville (1974) and Hillyard and Woods (1979) have argued, then this factor may have contributed significantly to the failure to replicate. Grabow, et al., also digitized the AERs at a different rate (8 msec intervals vs. the differential sampling rate for early middle, and late components used by Wood et al.). Such differences as well as differences in determining baseline values for the AERs (Wood, personal communication) may have contributed to discrepancies in findings between the two studies.

Other researches have attempted to determine whether AERs could differentially distinguish between stimuli varying in different numbers of features (Lawson & Gaillard, 1982b). A number of feature theories advanced over the last three decades have postulated that since the identification of speech sounds involves the detection of features; sounds which differ in more than one feature should be discriminated more readily than stimuli which differ in only a single feature (Jakobson, Fant, & Halle, 1963; Chomsky & Halle, 1968). Lawson and Gaillard (1982b) reported limited support for this notion in a study which indicated that the number of phonetic features available for discrimination influenced the latency and amplitude of AERs. Either a tone or the syllable /ke/ were presented as infrequent targets in six conditions in which the number of phonetic and acoustic features varied between the target and the nontarget stimuli. The N_2 amplitude recorded from C_z to the speech target was found to be smallest for the condition in which target and nontargets differed in a single feature (place of articulation). In addition, the N_2 peak latency was shortest for the three feature distinction condition, although it did not differ between the two and three feature distinction conditions.

The series of studies by Wood attempted to identify a phonological mode of processing. In order to accomplish this, one stimulus was held constant while other stimuli and the task (as defined by instructions) were allowed to vary. ERPs to the stimulus that was common to the two tasks were then compared. Lawson and Gaillard sought to determine whether the number of abstract features might differentially affect AER responses. Other researchers have sought to isolate components of the ERP which might reflect the specific acoustic as well as the phonological characteristics of speech sounds. Research in this area has centered around two important cues for consonant perception: voicing contrasts and place of articulation contrasts.

2. Voicing contrasts. Voicing contrast or voice onset time (VOT) reflects the temporal relationship between laryngeal pulsing (e.g., vocal chord vibration) and consonant release (e.g., the separation of lips to release a burst of air from the vocal tract during the production of bilabial stop consonants such as /b, p/). Investigators report that adult listeners discriminate changes in VOT only to the extent that they assign unique labels to these sounds (Liberman, Cooper, Shankweiler, & Studdert-Kennedy, 1967). Listeners fail to discriminate between bilabial stop consonants with VOT values of 0 and +20 msec and identify both stimuli as /ba/. Stimuli with +40 and +60 msec VOT are identified as /pa/. While subjects

are unable to discriminate between 0 and +20 and between +40 and +60 msec stimuli, they do discriminate between and assign different labels to stimuli with VOT values of +20 and +40 msec. These stimuli are from different phoneme categories (/b/ vs /p/). The 20 msec difference in VOT between speech syllables is only detected when the VOT stimuli are from different phoneme categories. Consequently, changes in VOT appear to be categorical. Such findings are consistently reported with adults in identification and discrimination studies (Lisker & Abramsom, 1964, 1970; Liberman, Delattre & Cooper, 1958) as well as with young infants in studies which employed high amplitude non-nutritive sucking procedures (Eimas, Siqueland, Jusczyk, & Vigorito, 1971).

In the earliest ERP study of voicing contrasts, Dorman (1974) used an ERP habituation paradigm. AERs were recorded from the C_Z lead from 50 adults, ten of whom were assigned to one of five groups. Computer synthesized speech sounds with 250 msec durations were presented at a fixed ISI (1750 msec). Group 1 listened to a series of 20 msec stimuli and then the series was changed to the 0 msec set. On the second day, this group heard first the 20 msec stimuli and then the 40 msec stimuli. Group 2 heard the same stimuli as group 1 but the stimulus order was reversed. Group 3 listened to 20 practice trials of the within category stimulus (0 msec) and the standard (20 msec). A within category shift was then started immediately after pretraining. Group 4 heard stimuli from one category while Group 5 listened to randomly ordered series of stimuli from both within and between phoneme categories. AERs were averaged for the different groups for the different stimuli. The N1-P2 peak to peak amplitude (N1 range = 75-125 msec; P2 range = 175-225 msec) was then measured. The only amplitude effects noted were for AERs elicited by the shift stimuli when they came from a different category than the habituating stimuli. Dorman interpreted this effect as demonstrating that ERPs could reflect the categorical-like effects of voicing cues. Although Wood and Dorman manipulated only within speech acoustic elements, both noted that these manipulations influenced the N100-P200 complex of the AER. These were the same components as noted earlier which appeared sensitive to the more general speech-nonspeech distinctions (Morrell & Salamy, 1972; Molfese et al., 1975; Nevell, 1974, 1980; Hillyard & Woods, 1979; Friedman et al., 1975; Galambos et al., 1975).

Using a different procedure to study VOT, Molfese (1978a) recorded AERs from the left (T_3) and right (T_4) temporal regions of 16 adults during a phoneme identification task. Subjects were presented with randomly ordered series of synthesized consonant-vowel (CV) syllables that began with bilabial stop consonants varying in VOT values of 0, +20, +40, and +60 msec. Adults were instructed to press one button after each stimulus presentation if they heard a /b/ and a second button if they heard a /p/. Adults identified the CV with VOT values of 0 and +20 msec as /ba/ approximately 95% and 93% respectively, while the CV with VOT times of +40 and +60 msec were identified respectively 95% and 98% of the time as /pa/. AERs to each stimulus were recorded during the identification task. Subsequent analyses involving principal component analysis with varimax rotation identified four factors that accounted for 91% of the total variance. On the basis of an analysis of variance of the factor scores, two AER components recorded from the T_4 site, and which influenced the N110 and the P350 components, respectively, (peak latencies = 110,355 msec) varied systematically as a function of the phoneme category of the evoking stimulus. Stimuli with VOT values of 0 and +20 msec elicited a different AER waveform from the right hemisphere site than did the +40 and +60 msec stimuli. No differences in the AER waveforms were found between the VOT values within a phoneme category (i.e., no differences were found between the 0 and +20 msec responses or between the +40 and +60 msec responses). These AER patterns of responding were comparable to the behavioral responses given by these subjects during the testing session. Components of the left hemisphere AER differentiated between 0 and +60 msec stimuli and differentiated the 0 and +60 msec stimuli from +20 and +40 msec stimuli.

Molfese (1980a) conducted a subsequent study to determine whether the laterality effects noted with the VOT stimuli were elicited by only speech stimuli or whether similar electrophysiological effects could be found for both speech and nonspeech materials. This would allow some conclusions regarding similarities in the mechanisms which underlie perception of those different materials with similar temporal lags. Since the 0 and +20 msec VOT stimuli were found to produce differential right hemisphere electrode effects than the +40 and +60 msec VOT stimuli, similar patterns of responding should be elicited by nonspeech stimuli with comparable temporal lags. Molfese used four tone-onset-time (TOT) stimuli in which the two-tone stimuli differed from each other in the onset time of the lower tone in relation to the higher tone: (1) The lower tone began at the same time as the higher tone for the 0 msec stimulus; (2) the lower tone lagged 20 msec behind the higher tone for the +20 msec stimulus; (3) for the +40 msec stimulus, the lower tone was delayed 40 msec after the onset of the higher tone; (4) while for the +60 msec stimulus this delay was increased to 60 msec. Both tones ended simultaneously and overall stimulus duration was 230 msec for all stimuli.

AERs from four scalp electrode sites over each hemisphere were recorded from 16 college age females in response to a randomly ordered series, with a varied ISI, of TOT stimuli. The analysis procedures used in Molfese (1978b) were also employed here. Nine factors accounting for approximately 81% of the total variance were isolated. Analyses of the factor scores indicated that one portion of the AER, which influenced the P330 component common to all four electrode sites over the right hemisphere, categorically discriminated the 0 msec and +20 msec TOT stimuli from the +40 msec and +60 msec TOT stimuli. No such changes were noted to occur over the left hemisphere electrode sites at the same latency. The left hemisphere electrical activity at this latency did distinguish between TOT stimuli from within a category (e.g., it discriminated 0 from + 20 msec and +40 msec from +60 msec). A second AER component with a peak latency of 210 msec reflected the detection of phoneme category-like boundaries over both the left and right parietal regions (P_3, P_4). A third component which affected the N110-P190 amplitude indicated that categorical-like discrimination of these temporal cues occurred over both hemispheres at the parietal and central electrode sites. AERs to temporal information, then, were characterized by both bilateral component changes in the first half of the ERP which were localized in or near the temporal and parietal regions of both hemispheres as well as an additional but later occurring component present at all right hemisphere leads.

Similar effects have also been found with four-year-old children in a study involving velar stop consonants (/k,g/) (Molfese & Hess, 1978). They recorded AERs from the left (T_3) and right (T_4) temporal regions of twelve children (mean age = 4 years, 5 months) in response to randomly arranged series of synthesized consonant vowel syllables which varied in VOT for the intitial consonant (0, +20, +40, +60 msec). Following analysis procedures already described, they found as in the case of adults one late occurring AER component (peak latency = 444 msec) from the right hemisphere electrode site that varied systematically as a function of phoneme category, but that could not distinguish between VOT values within a phoneme category. A second, earlier occurring AER component complex (peak latencies of 198 for the negative peak and 342 msec for the positive peak) also discriminated between VOT values along phoneme boundaries. However, unlike that reported by Molfese (1978A) for adults, this complex was present in recording sites over both hemispheres.

This work was later extended to include newborn and infant populations (Molfese and Molfese, 1979a). In one experiment, the four consonant-vowel speech syllables used by Molfese (1978A) were presented to 16 infants 2 to 5 months old (mean: 3 months, 25 days). AERs to each stimulus were recorded from scalp electrodes at T_3 and T_4 locations. These results showed that a positive component of the cortical AER (peak latency = 920 msec) from the right hemisphere site discriminated between VOT values from different phoneme categories. A second negative component (peak latency 528 msec) of the AER which responded in a similar fashion was present over both hemispheres. In a separate experiment (Molfese and Molfese, 1979A), 16 newborn infants under 48 hours of age were tested in an attempt to determine the developmental onset of VOT discrimination as reflected in AERs. The same consonant-vowel speech stimuli and recording sites described above were used in the newborn study. No evidence of any phoneme categorical-like VOT effect similar to that found with older infants, children and adults was found. These findings were interpreted to suggest that the ability to discriminate VOT stimuli along phoneme boundaries develops after birth.

Several general findings have emerged from this series of studies: (1) perception of the VOT cue as reflected in the ERP indicates the involvement of several cortical regions, some of which are in the right hemisphere and some of which are common to both hemispheres; (2) ERP components elicited by temporal stimuli change with development; and (3) ERP components discriminate between stimuli from different phoneme categories.

3. Place of articulation contrasts. Several studies have been undertaken to identify the electrocortical correlates of acoustic and phonemic cues which are important to the perception of consonant place of articulation information (Molfese, 1978B; Molfese, 1980B; Molfese & Molfese, 1979B; Molfese & Molfese, 1980). In general, these studies indicated that multiple electrode regions (which include lateralized and bilateral activity) reflect activity related to the perception of cues such as second formant (F_2) transition and formant bandwidth. These findings agree with recent behavioral studies which utilized dichotic temporal processing procedures (Cutting, 1974). Cutting found that stimuli with speech formant structure or which contained an initial transition element were better discriminated by the right ear. Since the right ear is thought to have the majority of its pathways projecting to the left hemisphere, Cutting reasoned that such findings reflected differences in the processing capacities of the two cerebral hemispheres. Molfese (1978B) attempted to isolate and localize the neuroelectrical correlates of these cues by presenting a series of computer generated three formant consonant-vowel syllables in which the stop consonants varied in place of articulation (/b, g/). The /b/ initial and /g/ initial syllables were identical in all aspects except that the second formant transition rose for the /b/ initial syllable and fell in

frequency for the /g/ syllable. Other stimulus features varied included formant structure (nonspeech-like formants composed of sine-waves 1 Hz in bandwidth or by speech-like formants with bandwidths of 60, 90 and 120 Hz for formants 1, 2, and 3, respectively), and phonemic versus nonphonemic transitions (the direction of the frequency changes for formant 1 and formant 3 were either rising so as to produce a phonetic transition in the sense that it could characterize human speech patterns, or these transitions were falling and therefore occurred in a manner not found in an initial position in human speech patterns). Using the principal components analysis to isolate major features of the AERs recorded from the T_3 and T_4 electrode sites of 10 adults, Molfese reported six major AER components which accounted for 97% of the total variance. Analysis of variance on the factor scores for these factors identified two positive AER components (peak latencies = 170 msec and 430 msec) unique to the left hemisphere electrode site which varied systematically in response to changes in F_2 transitions. In a replication and extension of this work, Molfese (1980b) also found that electrical activity recorded over the left hemisphere discriminated consonant place of articulation information. In this study 20 adults were presented with a series of consonant-vowel syllables which varied in the initial consonant, /b,g/, and the final vowel, /i, æ ,ɔ /. AERs were recorded from 3 scalp locations over each hemisphere (T_3, T_4, T_5, T_6, P_3, and P_4). Utilizing the analysis outlined above, the third positive peak (P_3) with its major peak latency at 460 msec was found to reflect the ability of only the left hemisphere to differentiate between the consonants /b/ and /g/, independent of the following vowel. A second and earlier occurring positive AER component (peak latency = 170 msec) reflected a similar discrimination by electrodes over both hemispheres.

The findings from this last study are important in terms of their implications for the problem of perceptual constancy. Until quite recently (Stevens & Blumstein, 1978) acoustic scientists were unable to isolate a set of acoustic properties that are invariant for a particular consonant place of articulation. Although such invariance exists for vowels, acoustic cues for consonants change as a function of subsequent sounds. Consequently, speech scientists long assumed that consonant and vowel information was processed together as a unit. This electrophysiological study by Molfese (1980B) represents the first direct indication that the brain may in fact respond to consonant sound configurations independent of vowel contexts.

Two studies by Molfese and Molfese (1979B, 1980) attempted to determine at what point in development infants are able to differentially respond to such place of articulation contrasts. In the first study AERs were recorded from T_3 and T_4 electrode sites from 16 newborn infants in response to two stop consonants which differed only in F_2 transition and formant bandwidth. As with adults, one late negative AER component (peak latency = 630 msec) found only over the left hemisphere site differentiated between consonants. A second negative orthogonal AER component (peak latency = 192 msec) was detected by electrodes over both the left and right electrode sites and also distinguished between the stop consonants.

The second study (Molfese & Molfese, 1980) sought to determine whether left hemisphere preocesses are present in the responses of preterm infants and whether left hemisphere mechanisms in infants are sensitive to phonetic and nonphonetic transitions. Eleven preterm infants (mean conceptional age: 35.9 weeks) were tested after birth. AERs to stimuli identical to those used by Molfese (1978B) were recorded again from the T_3 and T_4

electrode sites. As was found with the full term newborns reported above, a left hemisphere process was identified in the AERs of the preterm infants which distinguished between transition cues in stimuli with speech formant structure. An additional AER component recorded over the left hemisphere differentiated between only the nonphonetic stop consonants (/b/ and /g/). This finding is similar to that reported by Molfese (1978) with adults except that the left hemisphere process for adults was sensitive to both phonetic and nonphonetic stimuli.

There appears to be some basic differences in the organization and localization of brain activity measured by electrophysiological techniques in response to the temporal information contained in VOT and TOT stimuli and to place of articulation contrasts. Although both speech relevant contrasts elicit simultaneous and identical discrimination responses from both hemispheres (bilateral processes), they differ in important respects. Voicing contrasts elicit an additional distinction made by right hemisphere activity while place of articulation contrasts evoke an additional left hemisphere response.

4. Vowel contrasts. Vowel perception seems to produce quite different distribution of ERP responses than the voicing and place stimuli described above. Molfese and Erwin (1981) recorded AERs from T_3, T_4, T_5, T_6, P_3, and P_4 electrode sites of 20 adults in response to the three synthetic vowels /i, æ, ɔ / and three nonspeech stimuli matched to the vowel stimuli in duration, rise time, peak intensity, and mean formant frequencies but differing in formant bandwidth. Nine factors accounting for 80% of the total variance were identified with PCA. Analyses of variance in the factor scores identified several factors which differentiated between vowel sounds. No single electrode site detected differences between all three vowels. Rather, anterior temporal electrode sites (T_3, T_4) discriminated /i/ from / æ / and parietal sites (P_3, P_4) discriminated /i/ from / æ / and /i/ from / ɔ /. As in the majority of behavioral studies, no hemisphere effects were found to interact with vowel identification. These findings were interpreted to indicate that a number of discrete mechanisms located over different regions of both hemispheres are involved in the processing of different vowel sounds. The data clearly do not support a single localized region as responsible for vowel detection.

In general, findings from studies in which phonological variables are manipulated identify changes in the N100-P200 complex as a result of these manipulations. While the effect of task factors remain unclear at present (Grabow et al., 1980), there appear to be a number of effects that appear reliable. Independent investigators have found that AERs reflect phoneme category differences when VOT is manipulated (Dorman, 1974; Molfese, 1978a, 1979; Molfese & Hess, 1978; Molfese & Molfese, 1979a). Moreover, in studies employing the PCA-ANOVA sequence, two components of the AER in adults are usually found to vary as a function of VOT: the N100-P200 complex recorded from electrodes over both sides of the head, and a second, later occurring component influencing P350. Similar effects appear in young infants sometime after 2 months of age and continue in this fashion into adulthood. When nonspeech stimuli are used which imitate the temporal delays of the VOT stimuli, somewhat similar effects are noted (Molfese, 1980). When a second cue, place of articulation, was manipulated, again bilateral changes in the N100-P200 complex were noted at all ages from preterm and newborn infants (Molfese & Molfese, 1979, 1980) to adults (Molfese, 1978, 1980), while left hemisphere lateralized responses which also varied systematically as a function of the consonant, occurred later in time during the P450 portion of the AER, a point different in time than found in the VOT studies.

CONTEXTUAL EFFECTS

The psycholinguistic literature in recent years has noted increasingly the importance of pragmatics, the use of context in determining an individual's interpretation of linguistic events (Bates, 1976). It is here that one can also see significant levels of success in the use of ERP procedures to study language. A number of studies conducted over the last decade suggest that ERPs to language related stimuli change as a function of stimulus context (Teyler, Roemer, Harrison, & Thompson, 1973; Matsumiya, Tagliasco, Lombroso, & Goodglass, 1972; Brown, Marsh, & Smith, 1973, 1976, 1979; Brown, Lehman, & Marsh, 1980).

The study by Teyler, et al., and subsequent work by Brown and his associates marks the advent and refinement of ERP experiments in which verbal information was used to provide a disambiguiating context for subsequent ambiguous material. Teyler, et al., (1973) recorded AERs from C_3 and C_4 of 10 college women while they listened to noun or verb phrases through earphones. Subjects were exposed to three phases. This included (1) disambiguous stimulus--verbal response: the subject hears the phrase "a rock" or "to rock" and then is instructed to think of the meaning while the AER to a click is recorded 1 to 5 seconds after the phrase and then the subject reports the meaning after an additional delay of 600 to 1500 msec; (2) ambiguous stimulus--verbal response: the word "rock" is presented and the subject responds to the click by saying the meaning; (3) ambiguous stimulus--nonverbal response: this is identical to phase 2 except the subject does not say the word after the click but is instructed to think of the meaning and then 10 to 45 seconds later report it. This later condition was to permit evaluation of possible motor involvement in phases 1 and 2. Effects reported included larger N100-P160 amplitudes for the left hemisphere electrode site for both nouns and verbs across all phases. There was an overall shorter component latency for verbs (181 ms) versus nouns (186 msec) across both hemisphere electrode sites as well as a lateralized left hemisphere latency difference for verbs (173 ms) versus nouns (188ms). Although a confounding of syntactical function (noun/verb) with work meaning (hard stone/movement) limits the ability to draw firm conclusions in this one study concerning the specific source of these effects, the use of the same stimulus (the click) to elicit ERPs across all conditions controls for a variety of acoustic factors that have been allowed to vary across many studies. Differences in ERP characteristics elicited by the evokving stimulus must necessarily reflect internal factors rather than the stimulus characteristics which are identical across conditions.

Brown, Marsh, and Smith (1973), in a related study, recorded AERs from F_7 and F_8 leads and positions 3 cm posterior and dorsal from the T_3 site on the left side of the head and from T_4 on the right of five adults in response to sentences presented through earphones. An experimental stimulus set consisted of three-word sentences in which the final word ("fire" or "duck") was the target word for ERP analysis. The control set contained these two words as the first word in a sentence. Subjects, for example, heard "Sit by the Fire" and "Ready, aim, Fire" as the experimental sentences in which the preceding words set up different interpretations for the word "fire". The control sentences, "Fire is hot" and "Fire the gun", in which the word "fire" occurred first in the sentence had no preceeding context to establish specific meaning. Subjects heard the sentences in a random order and were instructed to "visualize" the meanings of the sentences. No behavioral response was made. Correlational analyses indicated differences in AER waveform characteristics over the left hemisphere electrode sites for the different contexts. It should be noted that this effect is similar to that reported by Teyler, et al., (1973). Again, as in the Teyler, et al., work syntactic and semantic factors are confounded.

In a follow-up study, Brown, et al., (1976) recorded AERs from 15 adults from the same electrode sites as in their previous study while they listened over earphones to the sentence "It was /led/" Subjects heard four blocks of 60 repetitions, each in which the noun and verb meanings were referred to equally often. Before the presentation of the sentence, subjects were instructed before each block of trials to perceive the stimulus word as a verb (as in the case, "the horse was led"), or as a noun (e.g., "the metal was lead"). One hundred ERPs were averaged for the stimulus word in each context (64 points/500 msec). The ERPs were normalized and compared by a Pearson product-moment correlation. These correlations were then z-transformed for subsequent statistical analysis. This analysis found that the lowest correlations occurred in the left anterior lead (F_7), a result comparable to that reported by Brown, et al., (1973). A step-wise discriminant analysis for each electrode site to identify specific components of the ERP differences indicated that the later portions of the left hemisphere anterior (at 390 and 500 msec) and posterior (258 and 305 msec) electrical activity discriminated best the noun and verb meanings of /led/. A subsequent reanalysis of these data using a principal components analysis procedure and analysis of variance techniques (Brown, Marsh, & Smith, 1979) indicated a high degree of comparability in findings across the different analysis techniques, althought the PCA procedures provided additional information not apparent from the discriminant analysis. Seven factors accounting for 75.4% of the total variance were isolated. Of these, a larger N150 component occurred for nouns than for verbs over the left hemisphere leads. A second component with peak latencies beginning at 230 msec indicated greater positivity for noun than verb occurences at posterior electrode sites. A third component with peak latencies at 370 msec reflected the occurrence of a large component at this latency, though of opposite polarity, at left hemisphere anterior and right hemisphere posterior sites for noun interpretations.

A subsequent study by Brown, Lehmann, & Marsh (1980) attempted to study the effects of context across languages. In this group of experiments, seven native English speakers heard "the boatman rose" and "a pretty rose." A second group of native Swiss-German speakers heard in German, "e shöni chlini flüüge" (a pretty little fly) and "en Vogel chunnt z' flüüge" ("a bird comes flying"). An additional group of seven Swiss-German speakers heard only the second sentence as a filter-degraded speech phrase that had been previously judged to be unintelligible speech. The first two groups were told to think of the meaning while the last group was instructed before each block of the degraded signal to imagine that it was one or the other of the test phrases used for group 2. ERP responses for all three groups were averaged from the onset of the ambiguous word ("rose" "flüüge"). Twelve electrodes were placed at 5 cm intervals in a 3 x 4 transverse array centered around the vertex with the four corner electrodes at positions comparable to the three earlier studies of Brown, et al., (1973, 1979, 1979). Responses were digited and averaged for 640 ms after stimulus onset. Results based on analysis of locations of scalp field maxima and minima as well as PCA/ANOVA for each group identified as anterior-posterior distribution for early and middle ERP components. These components were more positive anteriorly for nouns and more negative posteriorly than verbs. These data agree with previous results reported by Brown et al. (1979) with the sentence "It was /led/". Given the use of probes and homophones for the evolving stimulus across these sets of studies, it is clear that physical differences in the evolving stimuli were not responsible for these effects. Indeed, similar effects were found across the Teyler et al. work, the three experiments of Brown, et al., (1980) and the fourth reported by Brown, et al., (1979) with very different acoustical propeties across stimulus words and their carrier phrases. The Teyler et al. and the Brown et al. studies consistently report ERP differences to noun versus verb stimuli. These effects again appear to center on the N100-P200 complex of the waveforms and on later portions between 260 and 500 msec after stimulus onset. In addition, as in the studies reported by Molfese and his colleagues, both bilaterally recorded response and lateralized responses were noted to differentiate between stimulus classes. Such consistencies across different acoustic patterns, languages, and lexical meanings suggests that "the effect must relate to some general aspect of the ... the noun/verb syntactic difference ..." (page 350). Even though syntactic functions of the words were confounded with word meanings in any one study, the consistent pattern of responses across studies and different stimuli strongly supports Brown's interpretation.

SEMANTICS

Semantics, or the study of language meaning, has been an area of study which has long provoked puzzled, and confounded philosophers, linguists, psycholinguists and psychologists. Today, no universally accepted model of semantics is held within any discipline although many views are available (see Chafe, 1970; Steinberg & Jakobovitz, 1971; Osgood, Suci, & Tannenbaum, 1957; Palermo, 1978). Scientists using electrophysiological techniques to study this area with few exceptions (for example, the work by Chapman and his coworkers) have proceeded without the use or benefit of any but a basic functional model.

The studies to be outlined here can be divided into two general groups: (1) studies which attempt to demonstrate whether words are discriminated from non-word letter groups or speech sounds (Bucksbaum & Fedio, 1969; 1970; Matsumiya, Tagliasco, Lombroso, & Goodglass, 1972; Molfese, 1979; Shelburne, 1972; 1973), (2) synonym-antonym distinctions (Thatcher, 1976), and (3) studies concerned with connotative or emotional meaning (Begleiter & Platz, 1969; Chapman, Bragdon, Chapman, & McCrary, 1977; Chapman, McCrary, Chapman, & Bragdon, 1978; Chapman, 1979; Chapman, McCrary, Chapman, & Martin, 1980).

A. Words vs. non-words

A series of studies by Bucksbaum and Fedio (1969, 1970) attempted to determine whether verbal and nonverbal materials would elicit different visual evoked potentials (VERs) from electrodes placed over the two hemispheres. In the first study, 10 college students with electrodes placed over O_1 and O_2 passively viewed three groups of stimuli: (1) high frequency, familiar words including nouns, verbs, and articles; (2) random dot patterns to simulate a three-letter word; (3) designs of random dots. Eight VERs were averaged for each condition and product-moment correlations calculated between all VERs for each condition. Correlation coefficients were then transformed to Fisher z-scores to normalize the distribution. A discrimination index was then calculated based on the differences between mean z-scores to similar and dissimilar stimuli. The latency of the positive peak (190 - 280 msec) was shorter for the words than either the dot or design patterns (based on 4 of 6 one-tail t tests). In addition, the "word" VERs were discriminated best from the "design" VERs using the discrimination index computed from the left hemisphere electrode site. However, when the index was calculated across successive time intervals, better discrimination appeared at more points over the right hemisphere lead. No between hemisphere effects were noted for the "word" stimuli. Unfortunately, the use of multiple one-tail t-tests limit some of the conclusions drawn by the authors since it increased the likelihood of Type 1 errors.

In a follow up to this study, Bucksbaum and Fedio (1970) recorded VERs from 16 college students who passively viewed three letter words and nonsense visual patterns composed of three random dot patterns presented for 50 msec to the left or right visual field while subjects fixated on a dot in the center of a CRT display. Analyses were patterned after those used in the first study. Both the left and the right occipital electrode sites were found to discriminate the word from the dot patterns. The authors conclude that since p-levels were larger for the left hemisphere electrode site, that the left hemisphere had some advantage in processing speech over the right hemisphere. However, magnitude of p-levels within the range in which the null hypothesis is rejected cannot legitimately be compared. Consequently, their conclusion appears inappropriate.

As noted earlier, Neville (1974, 1980) and Hillyard & Woods (1979) speculated that language related ERP effects could be demonstrated more reliably if subjects are required to process the information presented. The work by Matsumiya, et al., appears to reinforce this conclusion. Matsumiya, et al., (1972) recorded bipolarly from electrodes at P_3 and W_1 (the midpoint of a triangle made by the 10-20 positions P_3, T_3, T_5 over the left side of the head, and P_4 and W_2 (comparable to W_1 on the right side). Nine adult subjects participated in 4 different tasks. In Condition 1 (undiscriminated word task) subjects were instructed to indicate the number of words that were intermixed with nonspeech sounds. During Condition 2 (undiscriminated sounds) subjects tallied the number of noises intermixed with the words of Condition 1. During the "discriminated sound" task of Condition 3 only mechanical sounds were presented and subjects were to indicate the number of different sounds. Condition 4 (meaningful speech) required subjects only to listen to ten sentences (150 words) delivered in spaced speech. The AERs elicited by stimuli within a condition were averaged together over a 250 msec period after stimulus onset. Peak-to-peak amplitudes were then calculated as an R-value using a ratio of left hemisphere amplitudes to left hemisphere plus right hemisphere amplitudes for a W-wave (the summed amplitudes of P60 - N90 - P140 - N180). Large R-values (greater than .5) were interpreted as indicating more left hemisphere involvement. R-values were largest for 8 of 9 subjects in Condition 4. Condition 3 produced larger R values than Condition 2 and no differences were noted between conditions 1 and 2. Matsumiya, et al., interpreted these findings to indicate that "the significance of the auditory stimuli may be more relevant to the occurrence of the interhemispheric asymmetry in AERs than is the mere use of verbal versus nonverbal materials" (p 792). Task activity alone did not predict the effects here since Condition 4 only required subjects to listen passively while all other conditions required some overt activity. Matsumiya, et al., did not rule out the possiblity that their effects might have been due in part to speech sounds effects (citing Wood, et al., 1971). Nevertheless, the thrust of their conclusions suggest that the more meaningful the material, the larger the amplitude of the P60-N180 complex over the LH electrode sites.

The work by Shelburne marked an early attempt to assess specific differences in ERP responses to word and nonword orthographic stimuli. Shelburne (1972) recorded VERs from C_z, P_3, P_4 (referred to linked ears) while eight adults viewed consonant-vowel-consonant trigrams flashed for less than 10 msec one letter at a time at one second intervals. Subjects viewed four runs, each consisting of 50 CVC words and 50 CVC nonsense syllables. The first two letters of the word and its matched nonsense syllable were the same. Only the last letter determined whether the syllable was a word or not. The subject was instructed to work a toggle switch to indicate whether they viewed a word (Condition 1). No response was required for the nonsense syllable. In Conditon 2 subjects were to indicate whether they viewed a word or a nonsense syllable. Results indicated that the third letter of the trigram elicited larger late positive components (beginning at 285 msec) than letters 1 or 2. In a second experiment in which the first letter determined whether the subject was to see a word or not, letter 1 amplitude was larger than that for letter 3. This amplitude difference also occurred earlier in the waveform (between 165 and 245 msec). Effects for both experiments were greatest at the C_Z lead. No differences between word and nonword stimuli were noted for any letter position. Shelburne interpreted his data as reflecting a decision process (word versus nonword). Comparable effects were reported for 20 children between 8 and 12 years of age (Shelburne, 1973).

In an auditory study related to Shelburne's work, Molfese (1979) employed a discrimination task while recording auditory ERPs from T_3 and T_4 leads to determine whether ERP measures could differentiate between meaningful and non-meaningful natural speech stimuli. Subjects listened to randomly ordered series of three letter consonant-vowel-consonant (CVC) words and nonsense syllables. All stimuli contained the same vowel, / æ / (pronounced as the vowel in the word, "b<u>a</u>t"). The articulation features for the initial and final consonants were matched in terms of voicing and place of articulation for both the word and nonword stimulus groups. In addition, stimuli were computer edited so as to be comparable in peak intensity, duration, and rise/decay times. Subjects pressed one of two buttons to indicate they heard a word or a nonsense syllable. Responses to the different keys were counterbalanced across subjects. The averaged AERs from two electrodes, 10 subjects and four word and four nonsense syllable stimuli were submitted to a PCA. Using the eigen = 1.0 criterion, ten factors accounting for 93.2% of the variance were isolated. The factor scores from each factor were submitted to an analysis of variance. Four AER components, P60, N250, P300, and N370 changed differentially as a function of whether the stimulus was a word or nonword. Molfese speculated that the P60 component reflected the processing of coarticulated speech cues occurring during the initial consonant from the final consonant position in the stimulus. Such effects have been previously noted in behavioral studies (Ali, Gallagher, Goldstein, & Daniloff, 1971; Daniloff & Moll, 1968).

The basis for the discrepancy between the findings of Molfese and Shelburne for word-nonword discriminations are not obvious. The two studies differed in the mode of stimulus presentation (auditory versus visual) as well as in the type of analyses used. Perhaps either or both of these factors contributed to these differences.

B. Synonyms and antonyms

Thatcher (1976) found differences in ERPs between word synonyms and antonyms. Each trial involved the presentation of a random dot display (control) which was followed by a word display (consisting of one of 36 words) which, in turn, was followed by another random dot display and then a second word display that occurred with equal frequency as either an antonym, a synonym, or as unrelated to the first word. Physically identical stimuli, then, served as synonyms, antonyms, and neutral words. Subjects moved a lever in one direction if the word was a synonym, another if an antonym and in both directions if it was neutral. A late positive component (LPC) in addition to the P_3 component was found at all sites for synonyms (peak latency 440 msec) and antonyms (peak latency 460 msec) but not for the same words when they were unrelated to the first words displayed. No differences were, however, noted between responses to synonyms and antonyms althought ERPs elicited by both classes of words were distinguished using t-tests as different from the ERPs elicited by the neutral words in all eight of the subjects.

C. Connotative meaning

In the studies of connotative meaning, Begleiter and Platz (1969) were the first to demonstrate that ERPs were sensitive to differences in the emotional meaning of words. Eighteen male adults viewed a series of stimuli consisting of taboo words ("fuck"), neutral words ("tile"), and blank fields which were presented for 10 msec at two to four second intervals while the visually evoked ERP was recorded from the O_2 lead (referred to linked ears). In one connotation subjects were instructed to say the word following a signal that occurred after the ERP was sampled. Subjects were asked to do nothing in another condition. Peak to peak amplitude measures were larger for the N100 - P160 component elicited by taboo words while no differences were noted between the neutral words and the flash. The P160 - N400 amplitude was also larger for taboo than for neutral words, which in turn produced a larger response than the flash. The response condition produced a shorter N100 latency than that for the no response condition. A larger N100-P160 amplitude was also noted across all stimuli for the response but not the no-response condition. Begleiter and Platz concluded that both the connotative meaning of the stimulus as well as the demand for a response affected the amplitude of various ERP components.

In a follow up of this work, Chapman, Bragdon, Chapman, and McCrary (1977) found ERP differences as a function of connotative meaning. They placed one electrode one-third of the distance from C_Z to P_Z (CP_Z) on twelve adults while they viewed a series of 220 words which characterized the semantic categories of Potency (P), Evaluation, (E), and Activity, (A) from Osgood's Semantic Differential.[1] Twenty words in each list were derived from each semantic class such that each 20 word set was high (+) or low (-) on one dimension (E, P, or A) while it was neutral on the other two dimensions. In this way the six sets of 20 words on each list came from the P+, P-, E+, E-, A+, and A- groups. The words were then visually presented in a random order for 17 msec on a CRT display. An asterisk occurred for .5 sec as a warning one second before each word. Subjects were instructed to say each word aloud at the end of a 2.5 sec interval following each stimulus. The ERPs were averaged over the 20 different words of each class and then templates for the E+ and E- words were developed from the averaged ERPs recorded from three subjects. This template was then applied to all 12 subjects (including the three used to develop it). The VERs were normalized using a z-transformation and then t-tests were calculated for the correlated means of each subject's response against the template. Eight of 12 subjects' responses were found to be significantly correlated in this fashion. When the ERPs were classified into E+ or E- classes on the basis of the magnitude of the "E" template scores (larger template scores were classified as E+ word stimuli while the smaller measures were classified as E- stimuli), 81% of the ERPs (out of 36 pairs) were correctly classified from the three subjects involved in the development of the scoring template while 69% correct classification was reached for the other nine subjects. When words from the Potency dimension were sorted in this manner, correct classifications of 65% and 71% were reached for the 3 template and the 9 nontemplate subjects. The Activity scale resulted in the lowest levels of classification with 72% and 51%, respectively. Chapman, et al., point out that the semantic distances along the E, P, and A dimensions of the semantic differential correlate with the better classifications of "E" words than "P" words and "P" words with respect to "A" words. The semantic space distances were longest for the "E" words, the "P" words were separated by an intermediate distance, and the "A" words were closest in space.

In a subsequent study in which a different analysis strategy was employed, Chapman, McCrary, Chapman, and Bragdon (1978) recorded the VERs of 10 adults elicited again in response to the different word meanings specified along the Osgood Semantic Differential dimensions. Electrode placements, word lists, the subject task and averaging procedures were comparable to those used earlier by Chapman, et al., (1977). However, in this study the 120 averaged ERPs (10 subjects, 6 dimensions, 2 word lists) were standardized by a z-transformation for each subject and submitted to a principal components analysis (PCA) which generated 12 factors to characterize 94% of the total variance. The component or factor scores from the PCA were used as input variables for a series of discriminant function analyses. The six semantic classes were then used as the criterion variables for the discriminant function analyses and then the classification functions were cross validated using jack knife classification, then the second word list, and, finally, these functions were applied to new data from another subject not included in the original analyses.

Six unidimensional discriminant functions were run on the ERPs for each of the three dimensions (E, P, A) and two word lists. List 1 ERPs were classified at 100% accuracy for each of the dimensions. When this was applied to the second list, accuracy declined to 80% (although still well above the chance level of 50%). Classification accuracy was comparable for the E and P words, but poorest for the A words with 40% accuracy for the "+" dimension words and 50% for the "-" dimension words. When the classification functions were applied to the data of a new subject whose responses had not been included in the original analyses, the classification of list 1 respones were at 100% and list 2 at 73%. When a single multidiscriminant function was applied to all dimensions at the same time (E+, E-, P+, P-,

A+, A-), classification accuracy for ERPs elicited by list 1 (the development list) was 56.7% (chance = 16.7%), jackknife classification 42%, and the accuracy of classifying ERPs elicited by a new list (list 2) was 40%. Chapman, et al., concluded from these data that "the internal representations of meaning can be assessed by analyzing brain responses" (page 203). Subsequently, Chapman (1979) attempted to determine whether different semantic expectancies on the part of the subject might produce different ERP effects. To accomplish this, Chapman required ten subjects to rate each visually presented word from +3 to -3 on 15 different rating scales: five for Evaluation, five for Potency, and five for Activity. Before each run of 20 words, subjects were given a scale on which to rate those words (e.g., nice-awful). A fixation asterisk was then presented for 5 seconds followed by a .5 second interval and then the 17 msec duration stimulus word. After 2.5 seconds, the subject gave on signal the rating for that word. Averages over 510 msec at 5 msec intervals were based on 20 presentations. The averages from all subjects were then submitted to a PCA with varimax rotation until an eigen = 1.0 criterion was reached. A stepwise discriminant function with jack-knife classification was then performed on the factor scores from the PCA. Two separate multidiscriminant analyses were run on the data elicited by the two word lists. Thus, all six classes of words (E+, E-, P+, P-, A+, A-) were included in each analysis. The discriminant analysis correctly classified 43.5% of the ERP data in terms of the six stimulus classes (as opposed to a chance level of 16.7%). The jackknife classification success rate dropped to 31%. When the discriminant functions developed with the first word list were used to classify the second word list, correct classification declined still further to 26.8%, although it continued to remain significantly above chance levels.

In order to assess the effects of semantic expectancies on the ERPs, a discriminant function was then conducted on the three dimensions (E, P, and A) in light of the subject's ratings. Classification accuracy was again well above the 33.3% chance level with 47.4% accuracy for list 1 (the list on which the discriminant functions were developed), and 43.9% using the jackknife procedure. Chapman concluded from this that the semantic effects held up again even thought the task required of the subjects was more difficult than that required earlier (Chapman, et al., 1978). The analyses succeeded in discriminating ERPs elicited by different stimuli and tasks at well above chance levels. In addition, the ERP components which distinguished between the word classes occurred later (at peak latencies of 332 and 311 msec for lists 1 and 2, respectively) than components reflecting task effects (189 msec).

A subsequent study by Chapman, et al., (1980) reported findings which closely matched those of Chapman (1979). It appears from work reviewed in this section that electrophysiological techniques can be used to successfully discriminate and to some extent identify connotative and devotative meanings of language stimuli as reflected in the ERP.
As in other areas of ERP-language research, ERP components throughout the waveform were identified as sensitive to semantic elements.

V. Sentence processing

Relatively few researchers have attempted to study the impact of syntactical constraints on the ERP. Work by Thatcher (1976) and Matsumiya, et al., (1972) as well as by Brown and his associates described in previous sections characterize some of the early attempts in this direction.

More recently, researchers have used ERP techniques to assess language activities during sentence processing (Friedman, Simson, Ritter, & Rapin, 1975; Kutas & Hillyard, 1980a, 1980b, 1980c; Papanicolaou, 1980; Erwin, 1981). In the first, Friedman, et al., (1975) recorded VERs from college-age adults with active electrodes placed at C_Z and half way between the P_Z and mastoid locations on each side of the head.

Subjects viewed first a slide with two centered dots and then at 2.4 second intervals six successive slides with one word each and then again a slide with two dots. The six word series depicted one of three sentences, "The wheel is on the axle," "The heel is on the shoe," and "The peel is on the orange." In Condition 1, the first grapheme of the second word in each series was missing. It was included in Condition 2. Subjects were instructed to report the second word during the 2.4 sec intersentence interval.

Results indicated a larger amplitude for the P300 component for the last word over both conditions. In addition, the baseline-P300 amplitude for the last word in Condition 1 was larger than that for condition 2 for both lateral electrode sites in comparison to the C_Z lead. The latency of the P300 component also varied as a function of condition with longer latencies for the word carrying information necessary for sentence understanding (word 6 in Condition 1 and word 2 in Condition 2). Several lateralized effects were also noted with a larger baseline N1 component amplitude reported for word 2 in Condition 2 and longer P300 latencies for words 2 and 6 in both conditions. This latter effect was also noted for the C_Z lead. A right hemisphere effect was noted only for Condition 2. Friedman, et al., suggested that their findings indicate that the P300 component acted as a "marker" to indicate syntactic closure.

Kutas and Hillyard (1980a, b, c) in a related series of studies, also noted special ERP effects when the final word in the sentence resolved or failed to resolve in an expected manner the information contained in the sentence. The procedures closely resembled those employed by Friedman, et al. Given the comparability of effects, only the Kutas and Hillyard (1980a) paper is reviewed here. Fourteen adult subjects viewed 160 different seven word sentences presented one word at a time. The 160 different seven word sentences consisted of 100 sentences ending with a semantically appropriate (congruent) seventh word, 20 sentences ending with a semantically inappropriate final word, 20 additional sentences with the appropriate final word presented in larger size type, and 20 sentences with an inappropriate word ending printed in the larger size type. Subjects were instructed to read each sentence silently and carefully in order to answer questions later at the end of the experiment. Electrodes were placed at F_Z, C_Z, P_Z, W_1, and W_2. W_1 and W_2 were defined as 30% of the interaural distance and 12.5% posterior of the vertex site. After all sentences were presented, subjects then viewed the word "Station" repeated 80 times in large size type and then 80 times in regular size type as part of a physical control condition. After this, subjects were then given a recognition/recall questionnaire in which they were asked to identify sentences seen previously on the basis of the first six words and then to write in the seventh word. Analyses of midline ERP responses based on area (relative to prestimulus baseline) identified a negative wave with a peak between 400 to 500 ms that occurred independent of type size for only the incongruous seventh word. This effect occurred at the central electrode sites. Based on sign tests and t-tests, the left hemisphere positive component area was found to be larger than the right hemisphere area for both the ERPs to the words 1 to 6 as well as to the last word. The physically deviant stimuli (large type) produced a larger positive area centered around P560 (300-600 ms), in particular for the C_Z and P_Z sites across the semantic conguent/incongruent dimension. No mention was made of lateral electrode effects. No P560 effect was reported for repeated presentations of "station" in large and small type. Instead, a larger amplitude P200 was noted for F_Z and C_Z for large size type stimuli. The authors concluded that such "late positivity associated with the infrequent physical deviations could not be attributed solely to the change in the physical characteristics of the elicitive word" (p 107). Measures of the ERP were sensitive to differences in sentences between expected (congruent) and unexpected (incongruent) endings.

Papanicolaou (1980) has been the only researcher thus far to identify changes in ERP patterns while subjects were involved in different language tasks. He recorded ERPs from frontal (F_7, F_8), temporal (T_3, T_4) and parietal (50% of the distance between T_3 - P_3 and T_4 - P_4) scalp locations of 8 adults. Each subject listened to a meaningless sequence of words presented at a rate of 200 words per minute. Four word passages were presented with normal intonation to simulate sentence inflections. At irregular intervals during each second paragraph, concrete nouns from the Rocsh (1975) category norms appeared. These words came from one of four semantic categories, contained one of four specific consonant sounds, and occurred at one of two loudness levels. A one Hz strobe light flash occurred throughout each passage. All subjects received the 4 different experimental conditions which included (1) a control condition in which subjects were instructed to attend only to the strobe light and to ignore the passages; (2) a semantic condition in which subjects were instructed to count the number of words from the same category; (3) a phonetic condition in which subjects counted the number of specific consonant sounds; and (4) an acoustic condition whereby subjects were to count the number of loud or soft words. Peak amplitude measures from P1 (mean latency = 92 msec) to N1 (mean latency = 133 msec) for each experimental condition were divided by the P1-N1 control condition amplitude to obtain a ratio of amplitude change. A decrease in electrical activity over the left frontal site was found for only the phonetic condition. He also noted that for both the phonetic and semantic tasks there was a marked decrease in left hemisphere amplitude ratios while right hemisphere amplitude ratios increased dramatically. Papanicolaou speculated that these results suggest that the VERs elicited during these tasks indicated an increase in LH processing which interferred in some manner with the VERs elicited by the strobe and resulted in smaller VERs here than in the control condition (see also Shucard, Shucard, & Thomas, 1977).

CONCLUDING REMARKS

As noted at the beginning of this chapter, a number of studies conducted over the past decade have attempted to study language-brain relationships using evolved potential procedures. While numerous methodological flaws and the absence of a theoretical framework have limited the contribution of this work, a number of general conclusions appear warranted. First, ERP measures indeed do appear sensitive to various language relevant stimulus manipulations including phonological cues such as voicing and place of articulation contrasts as noted by work from Dorman (1974) and Molfese and his colleagues. ERP measures are also sensitive to syntactic/denotative semantic manipulations as evidenced by work from Brown's laboratory (Brown et al., 1973, 1976, 1980). Connotative meaning manipulations also produce changes in the ERP waveform (Begleiter & Platz, 1969; Chapman et al. 1977, 1978, 1979, 1980). It is the impression of this writer that the most convincing data concerning the ability of ERP procedures to assess language related phenomenon come from laboratories which have undertaken a more systematic approach to study specific linguistically relevant dimensions.

Task effects have been shown to have a great impact on ERP-language studies. Work by Neville (1974, 1980), Kutas and Hillyard (1980a, 1980b, 1980c), Hillyard and Woods (1979), Papanicolaou (1980), and Matsumiya et al. (1972) all suggest that various task manipulations will influence the ability of the ERP to detect difficiences in language related stimuli.

Responses to language relevant manipulations are not restricted to any one single component or portion of the ERP. Studies which have undertaken more traditional amplitude-latency measures and have restricted their analyses to the classic components such as N100 and P300 generally report changes in these components as a function of stimulus and/or task manipulations (Neville, 1974, 1980; Matsumiya et al., 1972, Sorman, 1974; Galambos et al. 1975; Friedman et al. 1975) whereas studies which involve PCA, area measures, or slow wave analysis prodedures identify changes in both the tradiitonal component structures as well as in other ERP components (Kutas & Hillyard, 1980a, 1980b, 1980c) as evidenced in the series of studies by Brown et al., Chapman et al., Kutas & Hillyard, and Molfese and his associates.

It would appear that the time is indeed ripe to make significant inroads into our understanding of brain-language relationships using ERP techniques. To accomplish this, however, researchers must utilize the extensive ERP literature which covers the area of stimulus influences on the ERP waveform (e.g., rise/decay times, loudness levels, masking) within a linguistically sophisticated theoretical framework. In doing this, researchers could then eliminate many of the confounds which have plagued ERP-language research in the past. At the same time, they could make use of the extensive language-behavior research to aid in the design and construction of experiments. The use of a strong theoretical model would aid researchers in the development of a research program which could then evaluate in depth many of the neurolinquistic questions unanswered today.

FOOTNOTES

[1]Osgood's work on the psychophysics of semantic meaning has been well defined, objectively measured, and widely tested. Based on semantic differential measures on a large number of scales, Osgood noted that a large percentage of total variance in judgments of verbal meaning can be accounted for by three underlying dimensions: evaluation, potency, and activity (E, P, A). These are described as an evaluative factor (represented by scales such as good-bad, pleasant-unpleasant, and positive-negative); a potency factor (represented by scales such as strong-weak, heavy-light, and hard-soft), and an activity factor (represented by scales such as fast-slow, active-passive, and excitable-calm). Semantic meaning, then, may be specified by the location of a word on Osgood's E, P, and A dimensions.

Tutorials in ERP Research: Endogenous Components
A.W.K. Gaillard and W. Ritter (eds.)

17

THE RELATIONSHIP BETWEEN EVOKED POTENTIALS AND LATERAL ASYMMETRIES OF PROCESSING

M.D. Rugg

Psychological Laboratory, University of St. Andrews, U.K.

Following a summary of relevant methodological issues, the recent literature concerning EP studies of lateral asymmetries is reviewed. Sections of the review deal with auditory and visual EPs, with studies of late positive components and with 'probe' experiments. It is concluded that few results in this field can be regarded as secure, and that investigators should pay more attention to task variables when designing experiments intended to detect processing-related asymmetries in EPs.

INTRODUCTION

After over a decade of research effort the question of whether the well documented lateral asymmetries of processing occurring in the human brain have reliable scalp-recorded electrophysiological correlates is still open. The utility of establishing the existence of such correlates is clear, enabling the pursuit of neuropsychological questions in normal individuals which are not easily answered solely by employing behavioural techniques. In the past decade a number of reviewers have discussed the literature generated in this field in detail (eg. Butler and Glass, 1976; Donchin et al 1977ab; Friedman, 1978a; Marsh, 1978; Hillyard and Woods, 1979) and it is the principal intention of the present chapter to discuss recent studies not previously reviewed and to deal only selectively with earlier investigations. Moreover, the chapter is confined to a discussion of studies employing evoked potentials (EPs) and does not deal with investigations of the spontaneous EEG (see Beaumont, this volume) the CNV or Motor Potentials. Further, the emphasis of the review is on those studies which have studied the lateral distribution of EPs in experiments employing 'high level' meaningful material, rather than those investigating what Friedman (1978a) has termed the 'lower order' asymmetries sometimes associated with simple, unstructured stimuli.

METHODOLOGICAL CONSIDERATIONS

Many of the relevant methodological points regarding the design and conduct of experiments designed to detect task-dependent asymmetries in scalp-recorded data are covered in an excellent section in Donchin et al

(1977a). In combination with the suggestions of Donchin et al (1977c) for the conduct of EP studies in general, these provide good guidance for satisfactorily performing asymmetry experiments. A number of points, however, are worthy of repetition and expansion.

A. Choice of reference electrode

In asymmetry experiments, as with EP studies carried out in other fields, it is necessary that the determination of the locus of any differences in activity between channels can be made unambiguously; a common reference is therefore desirable. Ideally, the reference should be on a site unlikely to be affected by the variables manipulated in the experiment, and it therefore follows that asymmetrical reference sites should be avoided whenever possible as these may be differentially affected by manipulations intended to alter the lateral distribution of EPs. The use of mastoids or earlobes is undesirable in view of the fact that they may be within the potential field generated by EPs, particularly Visual EPs (Lehtonen and Koivekko, 1971). Perhaps the most appropriate site for many EP studies is an active midline site, e.g. mid-frontal, although this choice of site is of no use if it is intended to study late slow components, due to their widespread distribution. Under these circumstances a reference placed away from the head (e.g. the balanced non-cephalic reference of Stephenson and Gibbs, 1951) is perhaps the only really appropriate solution.

B. Establishment of task-related asymmetries

As noted by Donchin et al (1977ab) the demonstration merely of an asymmetry in an EP component or components as a result of some stimulus/task combination is not an acceptable basis for the inference that the EPs in question are indexing task-dependent asymmetrically mediated processing. It is necessary to demonstrate that the observed asymmetries are processing dependent, and to this end the use of a control condition, or a stimulus/task combination designed to produce asymmetries in the opposite direction, is mandatory as a within subject control. Such a design establishes that any asymmetries observed, assuming that they show themselves to be condition-specific, cannot be due in any simple way to variables such as underlying anatomical asymmetries, or asymmetric electrode placement. Clearly, the most powerful design is that which uses the same stimuli or class of stimuli in all conditions, varying only the task instructions given to the subjects. In such an experiment any task-related asymmetries can then be reliably ascribed to differences in cognitive variables.

C. Choice of task

A tremendous variety of cognitive tasks have been employed in research into EP correlates of asymmetries of processing. Many of these have been the invention of the workers carrying out a particular study, with little recognition of the vast behavioural literature concerning perceptual asymmetries related to tasks associated with auditory and visual stimuli. In as much as it is the intention of investigators to produce reliable task-related asymmetries of processing, it makes sense to choose tasks which have been well validated by behavioural or clinical techniques, and not to rely on the intuition that a task appears 'verbal' and can therefore be expected to engage the left hemisphere. By the same token, the monitoring of subjects' performance means that criteria of task engagement can be set and allows a check on the relative difficulty of the tasks employed. Clearly, it is desirable that task difficulty should not be confounded with other task variables as in such circumstances it may not be possible to ascribe task- related asymmetries in EPs solely to the different processing demands (in terms of laterality of processing) of the tasks employed.

D. Choice of analysis techniques

Relatively few investigators to date have reported EP asymmetry experiments employing one of the several multivariate procedures now available for EP analysis (Donchin and Heffley ,1978); they have for the most part been content to perform relatively simple peak measurements. In view of the unknown range of relationships which might exist between EPs and asymmetries of processing this would seem an unduly restrictive approach. Investigators in the area have also been content to assume that averaged EPs represent accurately the nature of their constituent single trials, and that observed differences in EP component amplitudes do not reflect systematic differences in latency jitter. As the theoretical interpretation of the asymmetry of amplitude in an EP component would be different depending on whether it was caused by hemisphere differences in latency jitter, or by 'genuine' amplitude differences, this would seem to be a question well worth pursuing. A number of techniques are now available for estimating the latency variability of components in single trials (e.g. the Woody Filter, Wastell,1977).

E. Subject variables

The desirablity of employing subject populations which are as homogeneous as possible with respect to cerebral lateralisation is obvious. It has been known for many years that handedness is an important factor in this regard, and more recent evidence suggests that familial sinistrality and sex may also be associated with differences in cerebral organisation (e.g. Hecaen et al, 1981; McGlone, 1981). It would therefore seem wise to avoid mixing these variables in asymmetry experiments.

STUDIES OF AUDITORY EPs (AEPs)

Quite a number of studies have searched for EP correlates of the lateralised activity assumed to occur during speech processing. Among the earliest to do so were Morrell and Salamy (1971) who elicited AEPs with nonsense words and reported that, in temporo-parietal leads, N1-P2 amplitude was of a greater amplitude from the left hemisphere than from the right. No non-linguistic control condition was employed, however, and as subjects were required to report each stimulus verbally there is no way of determining what aspect of the task produced the effect. In the same year Wood et al (1971) reported that AEPs recorded from temporal regions and elicited by CV stimuli showed asymmetries in the latency range 5Ø-2ØØ msec (left greater than right) when processed in a phoneme discrimination task, but not when associated with pitch discrimination. Subsequently, Wood (1975) reported a replication of this finding and Smith et al (1975) a failure to replicate.

The studies of Morrell and Salamy (1971), Wood et al (1971) and Wood (1975) are of particular interest as they have recently been the subjects of replication attempts by Grabow et al (198Øa,b). Grabow et al (198Øa) elicited AEPs with phonemic stimuli and tones, recording from homotopic frontal, central and parietal sites. No evidence of any lateralisation in the resulting AEPs was found, irrespective of whether or not subjects verbally responded to each of the stimuli (as in the Morrell and Salamy study). In the experiment of Grabow et al (1980b) the procedure of Wood et al (1971) was followed and the findings of the original investigators were not replicated. Indeed, Grabow et al reported a lower amplitude for an N1-P2 component recorded from the left temporal region, compared to that from a homolateral right hemisphere electrode, in both task conditions. This effect was absent from central leads and, assuming it is not an artefact of the proximity of the reference sites (linked mastoids) to the temporal electrodes, may be at variance with the commonly held assumption that the EPs associated with an asymmetrically processed stimulus will be larger in the hemisphere predominantly engaged in the processing.

The studies of Grabow et al (1980ab) add further weight to the doubt expressed by previous investigators as to whether AEPs to speech stimuli demonstrate reliable processing dependent asymmetries (e.g. Friedman et al 1975; Galambos et al 1975). However, some other recent studies suggest that such pessimism may be unjustified. These include the work of Molfese (this volume), who has conducted a series of experiments the results of which suggest that there are circumstances in which AEPs index lateral asymmetries in the processing of speech sounds.

Hillyard and Woods (1979) report an experiment which attempted to assess the lateral distribution of AEPs during natural speech processing. The linguistic stimuli consisted of words forming a coherent rhyming poem, AEPs being formed from a 500 msec epoch following the onset of each word. Recordings were taken from the vertex and left and right temporo-parietal areas. The control condition consisted of listening to a series of tones matched to the words for intensity and duration. The resulting AEPs were observed to be larger when elicited by the words, and to exhibit an asymmetry in the form of a larger negativity in the latency range 50-200 msec in the left hemisphere, this asymmetry being absent to the tone stimuli. This finding is of great interest, and suggests that the use of more naturalistic situations and high information loads may be an important component in the search for reliable speech processing related asymmetries in AEPs. However, judgement must be reserved until it is shown that the observed changes in asymmetry are related specifically to the variable of linguistic processing, and not simply to task engagement (subjects were required to listen and answer questions to the poems, and to listen passively to the tones). The experiment is thus in need of replication with a more appropriate control than that used in the original experiment.

In as much as a conclusion can be drawn at present from studies of AEP asymmetries related to the processing of auditory stimuli it would seem that the paradigms originally employed in this area by Morrell and Salamy (1971) and Wood et al (1971) do not produce reliable results. More promising perhaps is the development of these types of studies along the lines reported by Molfese (1978,1979,this volume), using carefully structured stimuli and a more powerful means of analysis (Principal Components Analysis) than that formerly employed. The approach described by Hillyard and Woods (1979) is also worthy of furthur investigation, and demonstrates the possible benefits of of using experimental paradigms which allow subjects to process coherent sets of stimuli in a relatively natural way.

STUDIES OF VISUAL EPs (VEPs)

In contrast to behavioural experiments investigating lateral asymmetries in the processing of visual stimuli, EP studies are not tied to the presentation of stimuli in the lateral visual fields. A number of studies have employed a divided visual field paradigm, however, and these will be reviewed separately from those in which midline presentation was employed.

A. Midline stimulation

Several investigations have studied the lateral distribution of VEPs elicited by the presentation of verbal and non- verbal stimuli across the visual midline. The first such study was carried out by Buchsbaum and Fedio (1969), who reported that VEPs elicited by words and random dot patterns were more dissimilar from the left hemisphere than the right. Whilst this study suffers from a considerable number of methodological defects (e.g. no task instructions to subjects, who had only to observe the stimuli passively; a large number of pairwise comparisons, giving rise to the possibility of type I error) it remains one of the few widely quoted as demonstrating VEP correlates of lateral asymmetries of processing. Several studies utilising midline visual stimulation and verbal or visuospatial tasks have failed to find evidence of task dependent asymmetries in VEP data (e.g. Shelburne, 1972,1973; Friedman et al, 1975; Rugg and Beaumont, 1978a,1979). All of these studies employed ostensibly verbal stimuli and tasks in at least one condition. For example, Shelburne's paradigm required subjects to perform a discrimination between words and non-words on the basis of a letter by letter presentation. As would be expected, the last letter, which determined whether or not the letter string formed a word, gave rise to a large late positive component, but no sign of any reliable asymmetries in any VEP component. Rugg and Beaumont (1978a) required subjects, in separate conditions, to discriminate between letters on the basis of whether the letter name contained an 'ee' sound, or between shapes on the basis of whether or not these were symmetrical. VEPs were recorded from lateral occipital placements referenced to the vertex. Whilst a P2-N2 component was observed to be larger in amplitude to the letter stimuli this effect was bilateral, and no task-dependent asymmetries were found. Subsequently, Rugg and Beaumont (1979) performed a similar experiment using lateral parietal placements referenced to linked mastoids, and investigated the large late positive component elicited by both 'target' and 'non-target' stimuli in these tasks. Once more, the VEPs differentiated the two experimental conditions, in terms both of the latencies and amplitudes of the P4ØØ components associated with the verbal and visuospatial stimuli. However, no indication of any task-dependent asymmetries was found for any of the components analysed.

Thatcher (e.g. Thatcher and April, 1976; Thatcher 1977ab) has reported a series of studies using what he has termed the 'background information probe' paradigm. This involves the presentation of neutral control stimuli (dot patterns) a meaningful stimulus (e.g. a three letter word), an unpredicatable number of control stimuli and finally another meaningful stimulus which must be matched with the previous one according to some criterion, e.g. semantic, phonological etc. The detailed presentation of the copious amounts of data presented in these papers makes the task of giving a brief description of Thatcher's results impossible. Of particular relevance in the present context, however, is that Thatcher has reported that hemisphere asymmetries in VEPs to the meaningful stimuli were observed in experiments in which the two stimuli were compared on some linguistic dimension. These asymmetries were most pronounced at parietal and temporal sites, and particularly prominent in the P400 component following the second meaningful stimulus.

A further study to have reported asymmetries in VEPs to midline stimuli is that of Chapman and McCrary (1979). In this experiment the task involved subjects viewing a series of flashes, of which two were blank, two illuminated letters and two illuminated numbers. The subjects' task was to compare either the numbers or the letters, in each case the other stimuli being irrelevant. The main across subjects comparison involved stepwise discriminant analysis, and the descriptors which distinguished best between control flashes and those illuminating the stimuli were characteristics of the VEPs from the left hemisphere. As no analysis of individual components was presented by these investigators, the locus of these effects in terms of the components giving rise to them is not known. Moreover, as no control task was utilised involving, say, visuospatial processing, it is not possible to conclude that this result reflects the specialisation of the left hemisphere for the particular tasks employed; it may be that the left hemisphere is simply more variable in its response to different stimuli than the right, irrespective of the nature of the stimuli and tasks.

Hink et al (1980) have recently reported a study of VEPs in a task involving the reading of Japanese words written in either kanji (ideographic) or hiragana (phonetic) scripts. The design of the experiment was conceived so as to control for the effects both of task difficulty and physical differences between the stimuli. The major finding was that the P1-N1 component of the VEPs, recorded from lateral parietal sites, demonstrated an asymmetry (left greater than right) for the hiragana script significantly greater than that associated with the kanji script. This finding is consonant with the neurological observation that the integrity of the left occipito-parietal area is necessary for the reading of hiragana but not kanji script (Sasanuma,1980), and may be a neurophysiological correlate of anatomical differences in the processing of the two types of material.

Rugg (in press) has conducted a study employing various types of letter string. The basis of this study was the suspicion that many previous VEP studies in this area did not take into account the fact that many aspects of tasks involving written language may be accomplished by either hemisphere, particularly if the material consists of concrete words (e.g. Marcel and Patterson, 1977). Two lexical decision tasks were employed, one requiring discriminations between randomly intermixed regular English nouns and pronounceable non-words, e.g. STONE vs STINE, the other requiring discriminations between pseudohomophones and non-words, e.g. STOAN vs STINE. It was argued that the former task could be accomplished bilaterally, both hemispheres having the competence to distinguish such words from non-words. In the case of the latter task however, which can only be accomplished by the use of a phonological code, much of the task-relevant processing of the letter strings should be lateralised, there being a range of evidence suggesting that the right hemisphere cannot employ phonological codes (Coltheart, 1980; Zaidel and Peters, 1981). VEPs were recorded from left and right inferior parietal sites and Pz, referenced to linked mastoids. They consisted of fixed-latency P100, N155 and P300 components, and a variable latency P670 component (see fig.1).

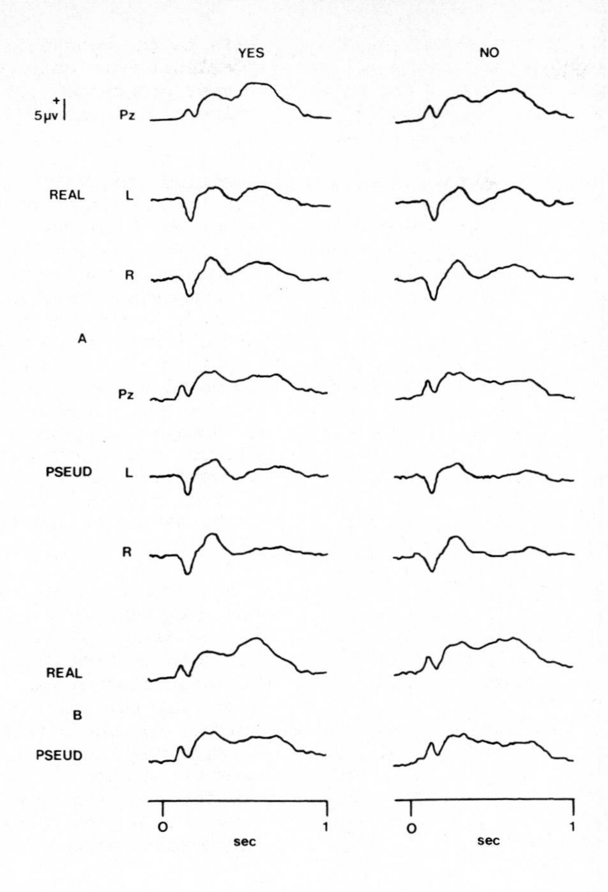

Figure 1. Grand average (8 subjects) VEPs from Rugg 1982 (in press). A. VEPs from Pz and left and right inferior parietal sites (L,R) for YES and NO lexical decisions between real words and non-words (REAL) and pseudohomophones and non-words (PSEUD). B. Grand average VEPs from the six subjects showing a clear P670 component in all conditions, Pz only.

This latter component's latency was partially associated with response time, being longer to NO responses and in the pseudohomophone task. Most pertinent to the present review is that none of these components showed task or stimulus dependent asymmetries in either amplitude or latency. A subsequent study (Rugg, in preparation) employed the same tasks in a more complex design, all three types of stimuli being present in the same list. Once again, no processing-related asymmetries were observed.

The results of these two studies could be interpreted as further evidence justifying the pessimism of other investigators who have failed to find task-related asymmetries in VEPs (e.g. Friedman et al ,1977). An alternative explanation is that the initial decoding of English letter strings, an automatic process involving parallel visual and phonological processing, gives rise to cerebral activity which is widespread over the hemispheres. Thus, there may be a greater likelihood of observing VEP asymmetries in verbal material when the task involves some significant amount of post- recognition verbal processing. Rugg's studies, summarised above, employed tasks requiring no such post-recognition processing and, according to the above argument, should therefore be unlikely to show task-dependent VEP asymmetries.

Support for this hypothesis is available from other VEP work. As noted previously, Thatcher(1977ab) has reported VEP asymmetries with verbal material in his 'background information probe paradigm'. These tasks require the storage and retrieval of information, operations which if carried out with verbal material would be expected to give rise to asymmetric cerebral activity. A final set of experiments to be reviewed in this section offers further support for the hypothesis. Kutas and Hillyard (198Øabc) have investigated VEPs in a paradigm in which the subject is confronted by a sequence of seven consecutively appearing words which make up a sentence. In a proportion of the sequences, the last word is either physically deviant, giving rise to a late positive complex, or semantically deviant, in which case a large N4ØØ component was observed in the absence of any prominent late positivity. Of particular relevance in the present context is the finding (reported in most detail in Kutas and Hillyard 1980a) that the VEPs to the first six words in the sequences exhibited a region (4ØØ-7ØØ msec post stimulus) of enhanced positivity from the left temporo-parietal area. This was most prominent in their dextral subjects with no familial sinistrality and was absent in VEPs to either the pre-sequence warning stimulus or a repetitively presented single word. These results are in accord with the view, expressed above, that task-related asymmetries are most likely to be observed when verbal material, following recognition, must be manipulated and stored, as occurs during the comprehension of sentences.

To conclude this section, it is argued that the rather mixed results so far obtained with VEP studies of language processes may be a consequence, in part, of a lack of appropriate tasks. Future work should take into account the fact that reading and related activities not only have multiple components but can often be accomplished by more than one means (e.g. 'visual' vs 'phonological' reading routes). The determination of which components are associated with asymmetries in concurrently recorded VEPs is a desirable goal for the future.

B. VEPs to stimuli presented in the lateral visual fields

Studies which have employed the presentation of stimuli in the lateral visual fields are of particular interest in that they employ one of the most popular behavioural paradigms in use at present for the investigation of lateral asymmetries of processing (see Beaumont, 1982 for reviews). A comprehensive review of VEP studies in this paradigm will be found in Rugg (1982). A number of studies have employed relatively unstructured stimuli with the intention of determining whether the resulting VEPs reflect the decussated structure of the retino-cortical pathways. Setting aside the special case of unilateral stimulation of the macular region with reversing checkerboard stimuli (which gives rise to a 'paradoxical' lateralisation of the P100 component of the checkerboard VEP; see Halliday et al, 1979) the evidence suggests that unilateral stimulation of the visual fields results in VEPs whose initial components are of a shorter latency and larger amplitude from the directly stimulated hemisphere (i.e. the hemisphere contralateral to the visual field of stimulation), an effect well illustrated in the data of Andreassi et al (1975). The VEPs studied by these workers were elicited by a small cross briefly exposed in one or other lateral visual field and the response required of the subjects was to count each occurrence of the stimulus. Ledlow et al (1978a) utilised as stimuli small squares, half of which were filled with a cross. Subjects made a go-nogo response on the basis of whether or not the squares were filled. The resulting VEP data were essentially similar to those of Andreassi et al (1975). In both studies recordings were made from lateral occipital regions referenced to linked mastoids.

The extent to which these visual field dependent asymmetries have any bearing on the well-documented asymmetries of responding associated with lateralised visual field stimulation has been questioned by Ledlow et al (1978a), who argued that this was unlikely in view of the stability of VEP asymmetries in the face of manipulations affecting the size of visual field-dependent behavioural asymmetries in reaction time. This objection is of doubtful validity as it can be argued that the variables affecting reaction time influence processes subsequent to the process of callosal transfer of information, it being this latter process that is reflected in the latency differences in VEP components recorded from homolateral sites. A more serious problem is that simple manual or vocal reaction times show a difference of up to 10 msec for responses initiated by the directly stimulated hemisphere compared to those initiated by the indirectly stimulated one, this difference presumably reflecting interhemispheric transmission time across the corpus callosum (Bashore, 1981; Milner and Lines, in press). However, the latency differences between the earliest homologous components in the data of Andreassi et al (1975) and Ledlow et al (1978a) are in the region of 18-19 msec, values inconsistent with the view that the interhemispheric transfer of the information necessary for the performance of behavioural tasks by an indirectly stimulated hemisphere is reflected by such latency differences.

A number of investigations have studied VEPs with unilaterally presented stimuli the processing of which might be expected asymmetrically to engage the cerebral hemispheres. The first such study was that of Buchsbaum and Fedio (1970), who utilised words and matched random dot patterns as stimuli and reported that the comparison of VEPs from directly and indirectly stimulated hemispheres showed them to be more dissimilar than VEPs recorded from each hemisphere when they were stimulated indirectly. In addition, it was reported that the VEPs to each type of stimulus also differed, this effect, as in the study of Buchsbaum and Fedio (1969), being greater in the VEPs from the left hemisphere. The same criticisms are applicable to this study as were raised to the previous one by these authors and the additional point made that the cross-correlational analysis technique employed allows no information as to which components contributed to the reported asymmetries.

Several studies employing lateralised meaningful material have been reported since that of Buchsbaum and Fedio (1970). Ledlow et al (1978b) and Rugg and Beaumont (1978b) both utilised lateralised letter stimuli requiring, in different conditions, verbal or spatial processing. In the experiment of Ledlow et al the stimuli consisted of letter pairs, presented either on the midline or in either visual field, with VEPs being recorded from lateral parietal sites. Subjects were required, on separate runs, to make either physical or name matches to the stimuli. Task dependent asymmetries in amplitude were superimposed upon visual field dependent amplitude asymmetries in several components. In particular, a late positive component was observed to be larger in the directly stimulated right hemisphere during the physical matching task with the opposite effect obtaining in the directly stimulated left hemisphere with respect to the name matches. The data pertaining to midline stimulus presentation is not reported. Rugg and Beaumont (1978b) utilised, in separate experiments, tasks involving the detection of letters containing the sound 'ee', or the detection of letters containing a right angle, with VEPs being recorded from occipital placements. A visual field by hemisphere interaction was observed in respect of the latencies of the initial VEP components and the amplitude of an N1-P2 component. When the subjects' task was verbal the change in amplitude in the N1-P2 component as a function of direct versus indirect stimulation was the same in both hemispheres (less amplitude in the indirectly stimulated hemisphere). In contrast, when the spatial detection task was employed no such change in amplitude was observed in the left hemisphere, sensitivity to the visual field of stimulation being confined to the right hemisphere. This effect was interpreted by the authors as reflecting differences in the asymmetries of processing associated with the two tasks.

Three further studies in this paradigm are of interest. Neville (1978) used as stimuli lateralised line drawings of common objects and required her subjects, deaf and hearing children, to view and identify them. A principal finding was that VEPs (from occipital placements) to these stimuli showed no visual field dependent asymmetries either in component amplitudes or latencies, and instead, a consistent right hemisphere advantage for these variables. In contrast, the deaf childrens' VEPs demonstrated a visual field by hemisphere interaction for the amplitudes of three prominent, late, components (greater amplitudes in the directly stimulated hemisphere) with no evidence of the right hemisphere advantage seen in the VEPs from the normal children.

Neville (1980) summarises a number of investigations of EP correlates of asymmetries of processing. Of particular interest are two experiments carried out on the same sample of subjects, one in which different four letter words were exposed simultaneously in each half-field, and another in which line drawings were unilaterally exposed. In the first experiment, N1 and N1-P2 amplitude was found to be greater at the left parietal electrode, this asymmetry not being present at central sites. The same components in VEPs to the line drawings were asymmetric over the central areas, having a greater amplitude over the right side, but were symmetrical over parietal sites. These results add weight to the argument (see below) that task-dependent VEP asymmetries are more likely to be observed with divided visual field presentation. They also indicate the desirability of recording from homolateral electrodes placed over more than one brain region in experiments of this type.

Gott et al (1977) reported an experiment in which intact individuals and commissurotomised patients served as subjects. Stimuli consisted of unilaterally or bilaterally presented shapes and words which had to be matched with a central test stimulus. The resulting VEPs were analysed in a rather uninformative fashion, and, in particular, the opportunity to assess the contribution of the forebrain commissures to the morphological similarity in VEPs which is observed between directly and indirectly stimulated hemispheres in normals is denied as Gott et al did not perform the relevant comparisons. Inspection of Gott et al's figure 4 intriguingly suggests, particularly in the case of later components, that the absence of forebrain commissures does not prevent this asymmetry from occurring.

The results of the relatively few studies carried out using as stimuli unilaterally presented 'meaningful' material give rise to more optimism about the sensitivity of VEPs to asymmetries of processing than do the those from studies employing midline stimulation. It would be interesting to perform an experiment which investigated lateral asymmetries in VEPs to lateralised and midline stimuli in the same subjects, with the same stimuli and tasks (whilst Ledlow et al 1978b employed lateralised and midline stimuli, they did not report the data from the midline conditions). The suspicion is raised by the brief review above that it is easier to obtain task dependent VEP asymmetries with lateralised rather than midline stimuli. If confirmed, this would cast doubt on the generality of the findings from behavioral experiments employing the divided visual field paradigm, the implication being that generalising from such experiments may overestimate the degree to which functional asymmetries of processing occur as a consequence of normal dichoptic viewing.

LATE COMPONENTS

Paradigms giving rise to prominent late positive components (i.e. the P3ØØ family; see Pritchard 1981 for a review) have been employed in a number of asymmetry experiments. Of those employing auditory (e.g. Friedman et al 1975; Galambos et al 1975) or visual (e.g. Thatcher 1976,1977ab; Rugg and Beaumont 1979;Rugg, in press; Ledlow et al 1978b) stimuli, only the studies of Thatcher and Ledlow et al have reported stimulus or task dependent asymmetries in a late component. That P3ØØ components are sensitive to manipulations of linguistic parameters was demonstrated in a series of experiments reported by Friedman et al (1977), who argued that such components might be a useful tool in the study of linguistic processing in spite of their apparent insensitivity to lateral asymmetries of processing. As noted previously the possibility exists that many investigations in this field have not used stimuli and tasks whose processing would, in the light of recent evidence, be thought to be unilaterally mediated. The degree to which P3ØØ and related components can be shown to reflect asymmetries of processing is an important question and bears on whether these components should be regarded as representing a diffuse, undifferentiated response to particular classes of events as suggested by, for example, Desmedt et al (1979).

Two studies employing somatosensory stimuli have a bearing on these issues. Desmedt and Robertson (1977) utilised a somatosensory detection task involving the stimulation of the fingers of one hand. They reported that although the middle latency negative components in the resulting SEPs (recorded from lateral central derivations) were lateralised predominantly to the hemisphere contralateral to the hand stimulated, the P400 component elicited by rare 'target' stimuli was bilaterally distributed. When an 'active touch' paradigm was employed (involving the palpation of a perspex edge to locate an irregularity) the resulting SEPs exhibited a P400 component lateralised to the right hemisphere irrespective of the hand stimulated. Desmedt and Robertson interepreted this result as reflecting the dominance of the right hemisphere in the performance of a high-level tactile task, a conclusion consonant with much clinical data. Although these authors considered the P400 elicited in this paradigm not to be equivalent to the 'classical' P300 component, unlike the P400 occurring in their detection task, it is arguable that, due to its relatively long latency, their 'active touch' P400 could be regarded as belonging to the P300 family, and perhaps be related to the sustained asymmetric positivity reported by Kutas and Hillyard (1980b, 1981).

Barrett et al (1980) also used a somatosensory detection task, and reported that the lateral distribution of the amplitude of the N150-P300 component in the resulting SEPs (recorded from lateral central sites) varied with the handedness of the subject and side of stimulation. N150-P300 amplitude was greatest over the hemisphere ipsilateral to the hand stimulated when this was the preferred hand, this result breaking down when the non-preferred hand was stimulated. Unfortunately, the experimental design used by these investigators, as they note themselves, makes it impossible to determine whether the observed asymmetries are the result of changes predominantly in the N150 or P300 component.

In the light of the studies of Kutas and Hillyard (1980b, 1981) Thatcher (e.g. Thatcher 1977a), Ledlow et al (1978b), Desmedt and Robertson (1977) and, possibly, Barrett et al (1981) it would seem premature to dismiss the possibility that the lateral distribution of late positive components may reflect asymmetries of processing. A more systematic investigation of this issue using stimulus/task combinations externally validated with respect to their being mediated unilaterally would appear highly desirable.

PROBE EXPERIMENTS

Thus far the majority of the experiments discussed have employed informational stimuli as a means of eliciting EPs. An alternative is to engage subjects in a task and elicit EPs with irrelevant, 'probe' stimuli, the assumption being that concurrent cognitive activity should influence the processing of such 'probes' and that if such activity is lateralised, then this should be reflected in the lateral distribution of the EPs to the probe stimuli. The first such study along these lines was performed by Galin and Ellis (1975). They utilised as their tasks the construction from memory of a Kohs Blocks figure (a putative right hemisphere task) and writing from memory (putative left hemisphere activity). VEPs were recorded using flash stimuli with an ISI of 3 secs, the derivations being lateral temporal and parietal sites referenced to the vertex. They reported that ratios (right hem./left hem.) of both EEG alpha power and VEP 'power' were significantly smaller in the blocks condition compared to the writing task. This effect was considered indicative of differences in hemisphere asymmetries of processing in the two tasks. This conclusion is unwarranted for two reasons. Firstly, in the absence of data pertaining to absolute levels hemisphere ratio data are uninterpretable, as bilateral changes in amplitude of the variable in question will give rise to apparent shifts in its lateral distribution (Donchin et al 1977b). Hence, it is not possible to determine from ratio data alone whether or not a 'true' shift in the lateral distribution of some EEG variable has occurred as a result of an experimental manipulation. Secondly, Galin and Ellis made no attempt to monitor fixation accurately or, more importantly, to control for differential motor activity between the two tasks. Noting these and a number of other points of criticism, Mayes and Beaumont (1977) attempted a replication of Galin and Ellis (1975), using motor and non-motor versions of the tasks. They found no evidence of any task-dependent asymmetries in their data. Unfortunately, EEG in this study was recorded bipolarly between occipital and parietal regions, leaving the study open to the criticism that lateralised effects which were relatively widespread over each hemisphere would not have been detected, due to common mode cancellation. Beaumont and Mayes (1978) reported a furthur experiment in the probe paradigm, using on this occasion a vertex common reference and recording from lateral central and parietal derivations. Imaginal 'spatial' (determining whether letters of an imagined sentence contained curves) and 'verbal' (deciding which letters of the same sentence contained an 'ee' sound) tasks were employed. Again, no lateral asymmetries related to the task variable were observed. However, this study is open to the criticism that the tasks employed were not validated and that subjects' performance was not monitored, making it impossible to ascertain degree of task involvement.

Rasmussen et al (1977) have also reported a probe experiment employing flash stimuli, using mental arithmetic as a task. Compared with a control condition requiring attention to the stimulus, performing mental arithmetic gave rise to a greater amplitude ratio (right/left) in an N15Ø-P22Ø component, implying, in the authors' view, that the left hemisphere gave rise to relatively smaller VEPs than the right in the task condition compared to the control. In view of the fact that no data regarding absolute amplitude levels were given, the same criticisms apply to this experiment as to that of Galin and Ellis (1975).

Recently, Papanicolaou (198Ø) used visual probe stimuli in an experiment involving the monitoring of auditory input. Subjects were required to attend to the probe stimuli, or to detect words of different semantic categories, detect consonants, or detect acoustically different items. VEPs were recorded from lateral frontal, temporal and temporo-parietal sites. While performance was at a similar level in each of the three tasks, those involving semantic and phonetic processing gave rise to VEPs with P1-N1 components which, compared to the control condition, were enhanced in the right hemisphere and reduced in the left, an effect found at all sites. Unfortunately Papanicolaou presented no representative waveforms, and no mention is made of any steps taken to eliminate trials containing artefacts from EOG and other sources of contamination. Notwithstanding these criticisms, this study is the best demonstration to date that the lateral distribution of VEPs to 'probes' may be influenced by a concurrent cognitive task.

A somewhat more complex paradigm has been employed by Shucard et al (1977) with auditory stimuli. The stimuli in this experiment consisted of pairs of tone pips separated by a 2 second interval. The tasks involved monitoring either musical passages or spoken text, AEPs to the tone pairs (which were embedded in the passages) being recorded from left and right temporal areas referenced to the vertex. The peak-to-peak amplitudes of the measurable components of the AEPs to each member of the tone pairs were greater from the putatively task engaged hemisphere, this effect being greater in the AEPs to the second of the tones. Subsequently Shucard et al (1981) reported that the direction of these asymmetries was reversed when a mastoid reference was employed, asymmetries being in the direction of lower AEP amplitudes from the task engaged hemisphere. It is arguable that, as far as is discernable from the original report of Shucard et al (1977), these studies are methodologically the most satisfactory of those conducted in the 'probe' paradigm. It is noteworthy that the task dependent asymmetries reported by these investigators were greater in the AEPs to the second of the tone pairs, an effect regarded as demonstrating differences in the 'habituation recovery cycle' between engaged and non-engaged hemispheres. As the 'true' effect on the AEPs to the tones in this experiment is a relative attenuation of those recorded from the temporal areas of the engaged hemisphere (showing as an apparant enhancement in the Shucard et al (1977) study because of the active reference used), this study is consistent with those reviewed above which have found similar asymmetries using visual probe stimuli.

On the basis of the studies reviewed the probe paradigm would appear to be worthy of furthur investigation, allowing the study of less 'unnatural' tasks than those usually associated with EP studies. It is also worth noting that if it is the case that concurrent cognitive activity reliably alters the lateral distribution of EP activity then the resolution of the debate over the extent to which 'neutral' visual or auditory stimuli give rise to asymmetrically distributed EP components (compare, for example, Harmony et al, (1973) and Rhodes et al (1975)) may have to take into account whether or not a systematic bias favouring a lateralised cognitive activity (e.g. covert verbalisation) was caused by the conditions of the studies in question.

CONCLUSIONS

What can be concluded from a survey of this kind ? Firstly, it would seem clear that there is still a long way to go before it can be stated that EPs give rise to reliable correlates of lateral asymmetries of processing. There are a number of possible reasons for this, a major one of which is that investigators have paid too little attention to the behavioural literature concerning stimuli and tasks most likely to be mediated asymmetrically. Combined with the large number of methodologically suspect studies in the field and investigators' propensity for employing different experimental paradigms, it is hardly surprising that a coherent body of results has not been reported.

Another reason for the confusion in the field is the conflict of aims on the part of many investigators. Experiments concerned with lateral asymmetries in EPs may be performed with the goal either of determining the extent to which EPs are sensitive to asymmetries of processing, or to obtain new information about the mediation of a particular task-related cognitive process. These two goals often appear to be combined in current EP research, although there would seem to be little point in pursuing the latter until the former is achieved. Thus, the finding of negative results may at present often be interpreted with equal validity either as evidence that EPs are not sensitive to asymmetries of processing or that the stimulus/task combination employed does not give rise to lateralised processing.

A further problem in this area is that because of the relative paucity of knowledge of the neurophysiological and cognitive concomitants of EPs it is rarely possible to make crucial predictions which test specific hypotheses; the field of enquiry is heavily data-driven. Because of this, those inferences which have been made from EP data have been on a post hoc basis and have contributed little to the resolution of the problems in the wider field of hemisphere specialisation of function.

It would, however, be premature to dismiss EPs as a means of providing information about cerebral asymmetries complementary to that provided by behavioural evidence. The number of well-conducted studies in the area is very small at present and a much more substantial body of evidence is required before a final assessment of their worth.

Acknowledgements

I should like to thank G.Barrett, A.Gaillard, R.Johnstone A.D.Milner and W.Ritter for their comments on earlier versions of this chapter.

Tutorials in ERP Research: Endogenous Components
A.W.K. Gaillard and W. Ritter (eds.)

18

THE EEG AND TASK PERFORMANCE: A TUTORIAL REVIEW

J. Graham Beaumont

Department of Psychology
University of Leicester,
U.K.

Methodological problems in studying the effects of cognitive task upon the on-going EEG are discussed, with particular reference to the investigation of cerebral lateralisation. Recording montages and parameters, the selection of cognitive tasks, subject variables and the presence of anatomical asymmetries are given particular attention. The current literature is then briefly reviewed. It is concluded that although there is a trend toward the finding of asymmetries, applying rigorous methodological criteria, no clear and reliable task related effects have been demonstrated. Finally, coherence analysis and its potential is discussed, with illustrations.

Striking developments have occurred in the field of human neuropsychology in the last two decades. Among the most important has been the creation of experimental human neuropsychology as an independent and active field of research, and one in which electrophysiological techniques have played an increasingly important role in the past ten years. While techniques have been developed which have enabled study of the organisation of the brain for higher mental functions in normal intact subjects, with a remarkable degree of success in analysing brain-behaviour relationships, the inference from studies of human performance to the organisation of the brain has been more or less indirect. These techniques, most notably dichotic listening and the divided visual field method, have relied upon the logic of presenting certain stimuli to subjects so that task performance may be studied as a function of the known lateral reception of such stimuli within the brain. This methodology, while partially validated by studies of brain damaged and commissurotomised or 'split brain' subjects, depends upon the design of very tightly controlled experiments in which differences in performance with lateralised stimuli can be attributed to lateral differences in the brain with respect to the processing of these stimuli. (Dimond and Beaumont, 1974; Kinsbourne 1978; Beaumont, 1982). Inferences about cerebral organisation may be drawn, but they are necessarily indirect and somewhat distant, and depend upon the difficult task of associating cognitive parameters with neurological variables.

The introduction of electrophysiological measures into this paradigm has served two purposes. Firstly, the cognitive paradigm has been seen to be a useful one for those interested in investigating the effects of psychological tasks upon the EEG. Secondly, by introducing electrophysiological parameters, a range of exciting possibilities has opened up. These centre around the opportunity to observe, for the first time, the correlation between cognitive and neural processes occurring together in real time. The importance of this opportunity is sometimes overlooked. While there is insufficient space here to discuss the philosophical issues known as the mind-body problem, physiological psychology has always been bedevilled by the difficulties of constructing models which relate psychological and physiological concepts (see Bunge, 1980). The study of electrophysiological changes associates with mental task performance provides the opportunity to observe the concurrent operation of the two classes of process and to resolve issues of the interrelationship between body and mind in a direct and empirical manner. For this reason studies of the effects of task performance on the EEG are of especial importance for psychology and the neurosciences, and deserve particular and careful consideration.

Electrophysiological studies of task performance fall into two relatively distinct areas: studies of average evoked responses and studies of the 'on-going' EEG. The AER studies are being dealt with elsewhere, and we shall therefore consider task effects upon the EEG. As the most active, and fruitful, region of study has been that of lateral cerebral differences, the discussion will concentrate on these studies, although the points to be made apply also in a general manner to other related studies.

There have been few extensive previous reviews of this area, although the literature up to 1974 was assessed by Butler and Glass (1976), and Marsh (1978) provides a more recent but selective review. The outstanding contribution has been that of Donchin, Kutas and McCarthy (1977), partially reproduced with minor additions as Donchin, McCarthy and Kutas (1977). Galin (1978) has also provided a valuable discussion of some methodological points.

The clearest point to be made by Donchin, Kutas and McCarthy was the inadequate design of many, if not all, the studies up to the date of their review. They made a number of methodological criticisms and recommendations, but there has not been clear evidence in the subsequent literature that their critique has been heeded. This review will therefore concentrate first upon methodological issues, some of which naturally reflect points made by Donchin, Kutas and McCarthy, in the light of which the current literature can subsequently be evaluated.

METHODOLOGICAL PROBLEMS

A. Montages

Some of the following discussion may seem unnecessarily elementary to those familiar with EEG techniques. Nevertheless, the studies of task effects have not been characterised by clear thinking in the selection of recording montages, and even these elementary points bear repetition.

The EEG trace is always a record of the potential difference between the activity detected at two electrodes. Either both electrodes are considered to be 'exploring' electrodes, as in bipolar or common reference recording, or else a relatively 'inactive' site is chosen for one of the two electrodes, monopolar or unipolar recording. If we are concerned to detect lateral differences in the activity of the two cerebral hemispheres, then bipolar recording with a pair of electrodes placed over each hemisphere is quite inappropriate. At best, if a difference is observed between the left and right hemisphere derivations, the locus of the effect cannot be deduced. At worst, this montage can fail to detect differences which are actually there. This is illustrated in Figure 1. Although there is greater amplitude at the two right hemisphere electrodes (R1, R2) than at the two left hemisphere sites (L1, L2), there is no difference between the left (L) and right (R) derivations.

Having selected either 'common reference' or monopolar recording, it is important to employ a common reference site. In monopolar studies which have used an ipsilateral reference such as the ipsilateral ear, even given careful matching of impedance at the reference sites, there is no guarantee that the reference will be truly inactive. Even the slight contamination of one reference site by the activity under study, and there is evidence that this may extend to the earlobes and mastoids (Stephenson and Gibbs, 1951; Lehtonen and Koivikko, 1971), may be sufficient to produce artifactual asymmetry. Figure 2 illustrates this problem. The appearance of slight activity at the right ear (RE) results in the right lead (R) being of lesser amplitude than the left lead (L).

The reference must therefore be common, and equidistant from the active electrodes if it is not to share more activity in common with one of the electrodes by its proximity. There still remains the problem of selecting a relatively indifferent site for the reference. The solution has been most commonly to select the vertex (Cz) as an appropriate site. Nevertheless, problems remain for any on-the-head site, which cannot be considered truly inactive. One of the most thorny, even for midline reference, can produce the situation which is represented in Figure 3. Here activity at the exploring electrodes (L1, R1) is of the same frequency and amplitude, but is out of phase. The activity at the reference contains some activity at the same frequency, but which is in phase with one of the two lateral electrodes. The result, a dramatic difference in amplitude between the left and right derivations, can clearly be seen.

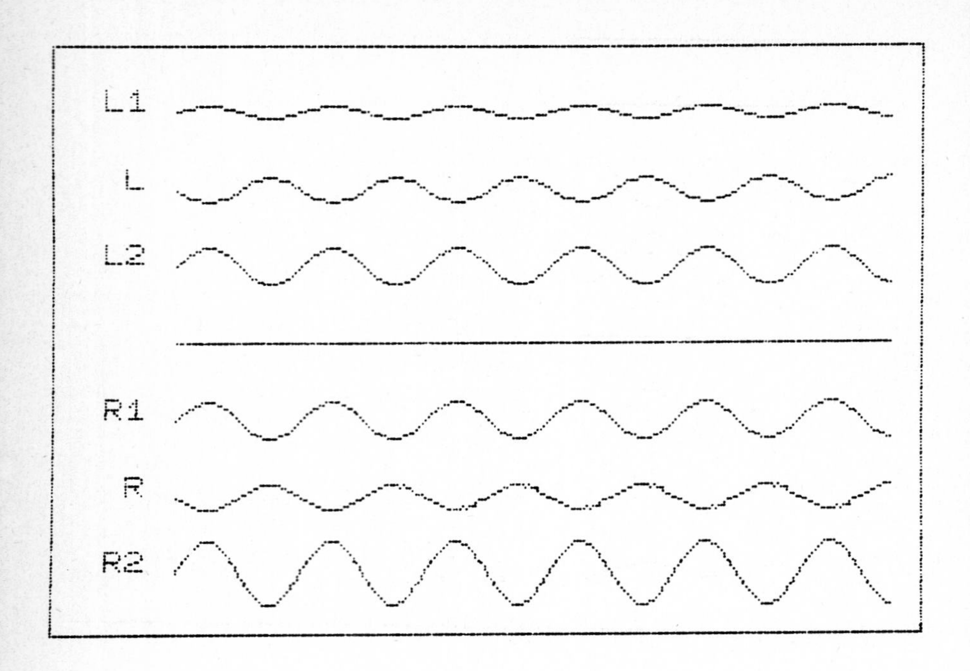

Figure 1
Bipolar recording may mask lateral asymmetry. The left hemisphere sites (L1,L2) and right hemisphere sites (R1,R2) yield identical recordings (L and R) despite greater power in the right hemisphere than the left.

A common alternative has been to place the reference as a pair of linked electrodes at the mastoid processes or upon the earlobes. The difficulty here is that unless the impedances of the two electrodes are very carefully matched, and are monitored during the course of the experiment, then one of the two electrodes will contribute more than the other to the linked average activity and the reference becomes asymmetric (Mowery and Bennett, 1957). As already indicated, as even mastoids or earlobes cannot be considered truly inactive sites, this is an important concern, especially as without carefully controlling the impedances, the variables operate entirely outside the control of the experimenter.

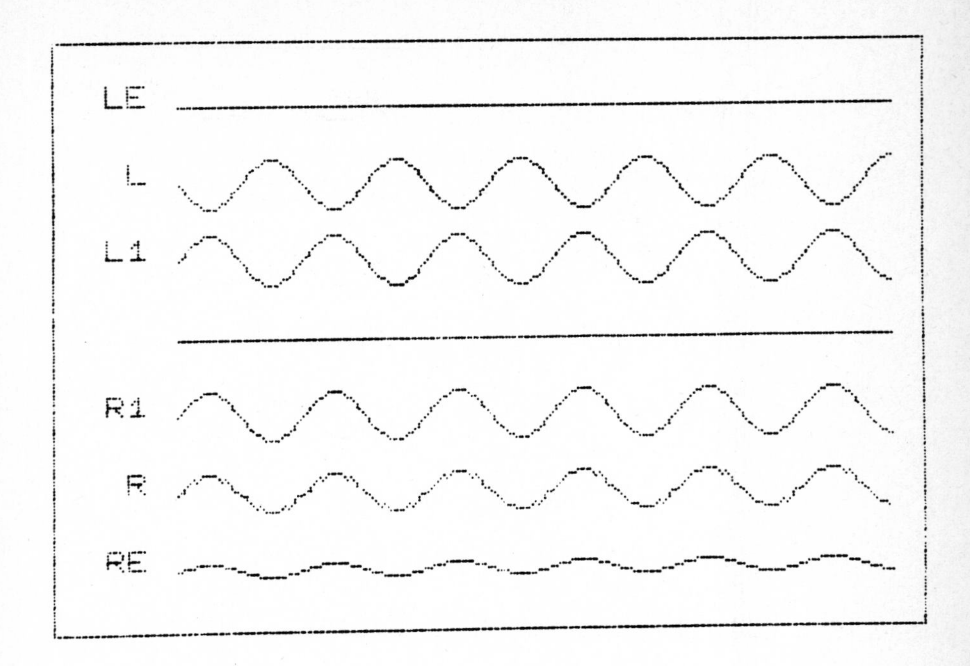

Figure 2
Effect of active ipsilateral reference. The left ear reference (LE) for the left hemisphere exploring electrode (L1) yields recording L. The right ear reference (RE) contains some activity in common with the right hemisphere electrode (R1) with the result that R is of less amplitude than L.

It would therefore appear that Cz may be an acceptable reference site, with the caveat that use of this site may present problems of interpretation of the lateral derivations. Fz may also be an acceptable reference site, if eye movements are carefully monitored, and the behavioural task does not demand significant task related deviations of gaze. However, there are distinct advantages to a midline off-the-scalp reference site, and both note and chin sites have occasionally been employed. Although these sites are more difficult to work with, they are more acceptable than the vertex or Fz location, and should be more widely employed. There would seem to be distinct advantages to an off-the-head reference and any study of coherence or phase relationships (see below) will only be interpretable if such a reference is used. Although I know of no study of task effects in on-going EEG employing such a reference, the balanced non-cephalic reference (Stephenson and Gibbs, 1971) deserves serious consideration.

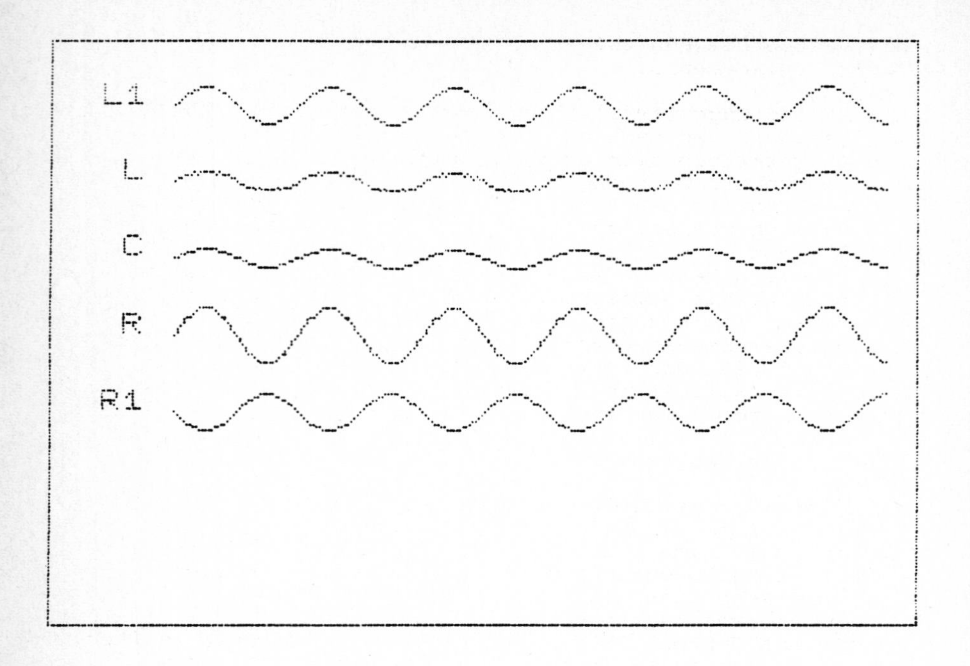

Figure 3
Contamination of common reference (C).
Activity at exploring electrodes (L1, R1)
is of same amplitude but out of phase.
Some activity in phase with L1 is present
at C. As a result there is a difference
in amplitude between the recordings L and R.

Is there evidence that choice of reference site can influence EEG asymmetries? At least three studies have compared different reference sites and have found that it does. Amochaev and Salamy (1979) found that the intrasubject stability of task effect with a variety of tasks was related to the site of the reference electrode. They found a task effect with bipolar derivations and an ipsilateral ear reference, but not when electrodes were referred to Cz. Davidson, Taylor, Saron and Snyder (1980) recorded activity at P3 and P4 referred to Cz, linked ears and the nose. They examined the 9 to 11 c/sec band under four task conditions in right handers with or without familial sinistrality. For those with familial sinistrality the asymmetry was to be observed with all reference sites and whether the data was expressed as raw power derivations or reduced to lateral ratio scores. However, for those without familial sinistrality, only ratio scores with the linked ear reference showed a task effect. There was therefore a difference between the effects observed with different reference placements.

Beaumont and Rugg (1979) also employed both Fz and linked mastoids under two task conditions. With coherence analysis, although reference site contributed a main effect, higher levels of coherence being found with the Fz reference, this factor did not interact with task or hemisphere effects. However, with respect to power analysis of the alpha band, there was not only a main effect of reference site, but also an interaction between reference site and the difference between electrodes within hemispheres. With the anterior temporal montage used in this study, there was also an interaction between reference site and task condition. All these results argue for great care in the selection of reference sites, and the further explicit investigation of the contribution of the reference site to observed lateral asymmetries.

B. Anatomical asymmetries

This problem is probably the most challenging for EEG studies of cerebral lateralisation at the present time. It has been known for some time that there are gross lateral differences in the surface topography of the cerebral hemispheres (LeMay and Geschwind, 1978). The most important of these for our concerns is the course of the lateral fissure (Rubens, Mahowald and Hutton, 1976). Firstly, the juncture of the transverse temporal sulcus with the temporal lobe shows a marked asymmetry between the left and right hemispheres. Secondly, there is a striking asymmetry in the posterior course of the lateral fissure, with a short horizontal ramus joining a sharply angulated posterior ascending ramus on the right, as shown in Figure 4.

The implication of this asymmetry is clear. Electrodes placed at symmetrical sites with reference to the 10-20 system will be recording different activity from the left and right cortices by virtue of being placed over different cortical sites, particularly if placed at T3 and T4, P3 and P4 or at some temporo-parietal placement defined with reference to these sites. Even placements at T5 and T6 might well be affected. It is perhaps not merely a coincidence that most of the studies which have reported asymmetries have recorded activity from such sites, and the effects of anatomical asymmetry may well be confounded with task effects. It is certainly the case that given the kind of cognitive task which is generally employed, these are the areas of the cortex in which we might expect to observe asymmetries in activity, and a solution has to be found to the problem of underlying cortical asymmetry.

This problem is a serious one, and we cannot continue to ignore it. There seem two possible approaches to a solution, although a solution has yet to be found. The first might involve detection of the lateral asymmetry in a given individual by some physical investigation, the use of computerised axial tomography being the obvious candidate. While this solution may be feasible,

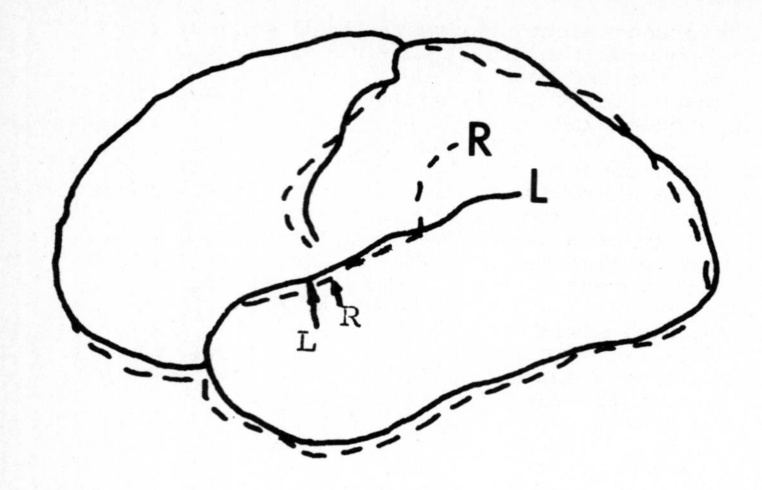

Figure 4
Composite tracing of course of central and lateral fissures showing lateral asymmetries between right (R) and left (L). Smaller letters indicate junction of lateral temporal lobe with transverse temporal sulcus. (Redrawn after Rubens, Mahowald and Hutton, 1976).

the cost and difficulty in relation to most experimental investigations rules out its use for the present. The second route to a solution might be the detection of asymmetries by electrophysiological means. It is not yet certain whether this will be possible, but preliminary investigations are proceeding in our own laboratory to establish such a method. If electrode placements were related to cortical features independently established from preliminary recordings, rather than at homologous sites, then some advance might well be made towards assessing the significance of underlying cortical asymmetries.

Two minor problems are associated with the problem of anatomical asymmetry. The first is the possibility of slight asymmetry of the medial plane. This seems a less significant problem than the asymmetry of the lateral fissure, but if even a small proportion of subjects had an asymmetrical medial plane, so that a midline reference electrode was in fact recording more activity from one of the hemispheres, then artifactual asymmetries in the EEG would easily be found in the combined group averages. Yet a further argument in favour of an off-the-head reference site.

I also have some concern about whether even electrodes placed according to the 10-20 system are really placed at symmetrical sites on the skull. The degree of error may not be sufficiently great for this to be a significant problem, but it is unlikely that the error is random, and particular care should be given to electrode placement.

C. Recording parameters

It has always been a surprise to me that so many studies have concentrated upon the alpha band. At a rough count, of 120 studies, only 8 have explicitly looked at bands of activity other than alpha. In an historical context, this is easily understood. The effect of mental effort upon the alpha rhythm has long been known, and by extension it is natural to move on to examine the effect of different types of mental activity upon alpha abundance. Nevertheless, in that the model assumes that mental effort produces a reduction in alpha, alpha being replaced by something else, it would seem more reasonable to ask what the 'something else' is, and to try to measure it. The strategy of observing alpha seems like trying to understand the organisation and production of a factory by sitting in nearby pubs and cafes and watching the off-duty workers come in and out. This may be an easier (and more pleasant) way to conduct the research, but there are probably more effective strategies which could be employed. Examination of reports which have investigated other bands of activity (Dolce and Decker, 1975; Dolce and Waldeier, 1974; Doyle, Ornstein and Galin, 1974; Ehrlichman and Wiener, 1980; Gevins, Zeitlin, Doyle, Yingling, Schaffer, Callaway and Yeager, 1979; Gevins, Zeitlin, Yingling, Doyle, Dedon, Schaffer, Roumasset and Yeager, 1979; Gevins, Zeitlin, Doyle, Schaffer and Callaway, 1979; Giannitrapani, 1975; Rugg and Venables, 1980; Shepherd and Gale, 1980; Spydell, Ford and Sheer, 1979) suggest at least that different effects are to be observed in different bands, and that investigation of activity other than alpha is to be recommended.

The selection of a method for quantifying the EEG is at present purely pragmatic. Most studies have selected power as the relevant variable, with a variety of methods being employed to estimate it. These range from integrators to devices which transduce the duration of activity at a certain frequency beyond a criterion amplitude, and to devices which rely purely upon amplitude. All these methods of course confound amplitude and duration, and are therefore possibly missing features of importance. Phase relationships are also not preserved in this data. Undoubtedly, methods based upon full spectral analysis, usually by the Fast Fourier Transform, are to be preferred. Integrated power measures at discrete frequency intervals can be obtained with such methods, and its variance among short epochs also extracted. Further analyses, including the calculation of coherence and phase relationships (see below), can also be derived, giving the method significant advantages.

Certain other measures, including 'cortical coupling' (Callaway and Harris, 1974), and the 'alpha control ratio' (Goodman, Beatty and Mulholland, 1980), have been employed, but do not seem superior to full spectral analysis. The examination of the peak frequency within frequency bands (Pfurtscheller, Maresh and Schuy, 1977) and of interhemispheric synchrony (Marsh 1978, Surwillo, 1971), deserve more attention.

The final point about parameters is an important one which has been emphasised before, but not fully heeded. This is the use of ratio scores. There are only two reasons for expressing between-hemisphere scores as ratios. The first is for clarity in data representation, and there is some small justification for this use. The second is to reduce between-subject variance in group data. The unfortunate result is that a score is derived which confounds changes due to the task in the two hemispheres with the between-hemisphere interaction. Overall changes at right and left hemisphere derivations are confounded with specific increases or decreases at the lateral derivations. A further problem in some studies is the statistical analysis of untransformed ratio scores which are not, in their raw form, suitable for parametric statistical analysis. The solution is simple, and that is to treat the lateral derivations as a separate factor in the research design and to work from the raw lateral scores. If between-subject variance is a problem, then within-subject normalisation of the data (Ornstein, Johnstone, Herron and Swencionis, 1980) may be carried out, and is certainly preferable to the calculation of ratio scores. The use of repeated measures ANOVA designs also helps with the problem of subject variance, although it does not entirely resolve it. For the reasons above, any study which employs only an analysis of ratio scores should be treated with great caution. At best the results may be difficult to interpret and at worst they may be misleading.

Some recent studies have performed both forms of analysis, and it has to be admitted that in a number the effects observed with analysis of the raw scores and the derived ratios are essentially similar (Ehrlichman and Wiener, 1979; Gevins et al. (1979a); Herron, Galin, Johnstone and Ornstein, 1979). In a further study, Bennett and Trinder (1977) performed an ANACOVA to take into account the general differences in basal scores, although this may not be an entirely satisfactory procedure because of the model of the relationship between basal and task scores which is assumed. Also, although Robbins and McAdam (1974) only analysed ratio scores, they present the mean results for their individual subjects. I have performed the ANOVA upon their tabulated means, and this analysis does support their analyses of the ratio scores. However, two studies which have reported both forms of analysis did find task effects with ratio scores which were not to be observed as an interaction between task and hemisphere upon analysis of the raw scores. Davidson et al. (1980) found an effect for their group with negative familial sinistrality, and a linked ears reference, for ratio scores which are absent in their raw score analysis. Given these problems with ratio scores, investigators should cease to use them, except as an aid to the display of data already analysed by a more satisfactory design.

D. Cognitive tasks

A factor often neglected by those interested in electrophysiological parameters is the selection of appropriate cognitive tasks. This is more hazardous than might at first appear. Differences between the cerebral hemispheres in the processing of cognitive tasks have been summarised in a variety of ways, but most often loosely in terms of 'verbal' and 'nonverbal' material or processes. It is clear (Beaumont, 1981) that this dichotomy, as the more sophisticated dichotomies, is not entirely valid. It is premature to form any clear model of what comprises the basis of observed asymmetries in hemisphere specialisation. These asymmetries result, in so far as we can be sure they reflect cerebral asymmetries in organisation at all, from a complex interaction of attentional variables, cognitive set and strategy, the nature of the stimulus material and response mode, and a host of subject variables. Despite the enormous body of research using the divided visual field techniques, for example, there are few really secure and reliable effects which can be predicted. Beyond the right visual field advantage for words, and the left visual field advantage for unfamiliar faces and some other perceptual features, there is little that the literature unequivocally supports. Tasks which are expected to engage the left or right hemisphere should therefore be selected with care. Only tasks which have previously been shown to yield reliable asymmetries should be employed, and preferably only in a context in which the performance asymmetry can be observed in parallel with the EEG recordings. This is rarely attempted, and too many unfounded assumptions have been made about the engagement effects of particular tasks.

An associated point is the effect of task difficulty. Three studies have examined task difficulty (Dumas and Morgan, 1975; Galin, Johnstone and Herron, 1978; McKee, Humphreys and McAdam, 1973), but only in the study of McKee et al. was a clear relationship of increased asymmetry with increasing task difficulty found. This may be due to there being a more complex relationship between difficulty and lateralised processes, or simply due to the inherent difficulty of quantifying task difficulty. The important point is that for a satisfactory demonstration of lateralised EEG parameters it is essential to be able to demonstrate a double dissociation between two parallel tasks, one which engages the right hemisphere and one which engages the left. Evidence of asymmetry based upon a single task is insufficient evidence in view of the many potential sources of artifact (see below), and evidence of a task-related shift in asymmetry in comparison with a rest condition is suspect as a result of the uncontrolled nature of 'rest' conditions and the confounding effects of task engagement. Independent asymmetries must therefore be shown to result from two tasks, and tasks which are carefully matched. This matching must be with respect to all variables except those that are hypothesised to underlie the relative specialisation of the cognitive processes involved. In particular they must be matched for difficulty, for stimulus parameters, for response mode, and for either the nature of the material or the cognitive processes involved. Such tasks are very difficult to find, even with regard to performance research in which the additional requirements of EEG recording can be ignored. Nevertheless, the employment of such tasks should be the goal of investigators in this field, and the tasks employed should be assessed against this standard in evaluating any report of lateral asymmetries. Beaumont and Rugg (1979) provide an example of at least the attempt to employ two validated and matched tasks, and the studies of Gale, Brown, Osborne and Smallbone (1978), Haaland and Weitz (1976), Rebert (1977a) and Tojo (1978) attempts to devise matched tasks.

Performance measures are also important to ensure task engagement as well as to provide data of task performance which may be correlated with other dependent variables. Care must be exercised in selecting response modes which do not generate EEG artifacts, but appropriate designs are possible, and even if covert activity is preferred, then a global assessment of performance may be even more important.

Finally, a note about 'rest' conditions, already mentioned. It is naive to assume that subjects instructed to 'just relax' do not engage in significant cognitive activity. They may, if anxious, continue to think about the experiment, revert to planning the essay they were about to write, or rehearse material from the lecture they have just attended. They may start to think about fixing their motorbike, wonder what colour they ought to paint the kitchen, or dwell on the beautiful features of the partner they danced with last night. In short, the situation is hopelessly uncontrolled from a cognitive viewpoint, so as to make the use of rest conditions as a 'baseline' in this area of research almost worthless. Particularly where subject variables, especially sex, are to be studied, rest conditions may be a powerful source of confounding.

E. Artifactual sources of asymmetry

There are a variety of artifactual sources of asymmetry, some of which have already been noted. In addition to anatomical asymmetries, asymmetrical application of the electrodes, and the problems of baseline conditions, there are at least five other sources of artifact which deserve mention.

The first relates to the consideration of baseline conditions and is the reported 'resting' asymmetry in the distribution of alpha. Several studies (among them: Butler and Glass, 1974; Giannitrapani, 1966; Marsh 1978; Smyk and Darwaj, 1972; Wieneke, Deinema, Spoelstra, Storm van Leeuwen and Versteeg, 1980) have demonstrated greater right abundance in resting alpha, although there is also the study of Cohen, Bravo-Fernandez, Hosek and Sances (1976) which fails to support this. Marsh (1978) also reports an asymmetry in the habituation of alpha blocking. Whether these effects are due to a true 'resting' asymmetry, or simply to a bias in the form of cognitive activity in which subjects engage at rest, they should be accounted for in assessing the significance of asymmetries observed under task performance. They are a further argument for the double dissociation research design. Rebert's (1977b) attempt to control for the problem by making stimulus presentation contingent upon interhemispheric alpha conditions is also worthy of note.

The second source of artifact is the lateral asymmetry of skull thickness (Leissner, Lindholm and Petersen, 1970), which may be attenuating the EEG from the left hemisphere. This may indeed explain the resting asymmetry.

Thirdly, there are response artifacts. If overt performance is required during the period from which EEG samples are recorded, artifacts related to the mode of response may well be present. Even if the response follows EEG sampling there may be preparatory processes which are more prominent in the hemisphere controlling the hand of response or the hemisphere generating speech output. A partial solution is to require covert activity, although this does not rule out covert performance. Monitoring of the EEG from the appropriate cortical areas may also be relevant, but once again, a better solution is the double dissociation paradigm with carefully matched responses in the parallel conditions.

The fourth type of artifact is allied to the last point. There may well be asymmetrical muscle activity over the scalp (Galin, 1978) which in its lower frequency range is capable of contaminating the faster ranges of beta. Care should be taken to avoid aliassing, and monitoring of EMG may be a wise precaution. We do not know the extent of such activity, or the degree of asymmetry present, but this is an important potential source of artifact.

Finally there is the presence of lateral eye movements (Ehrlichman and Weinberger, 1978). Lateral deviations of gaze are known to occur following the presentation of materials or tasks which engage the left or right hemisphere. That these may occur following the presentation of stimuli in EEG experiments in either central vision or the auditory modality, is very likely. They should be monitored and gaze direction either controlled, or trials discarded in which gaze deviation occurs.

F. Subject variables

Various subject variables have been studied in the course of previous studies, notably sex and handedness. Studies which have examined the effects of preferred handedness have included: Butler, Carter and Glass, 1977; Davidson, Schwartz, Pugash and Bromfield, 1976; Davidson et al., 1980 (in the form of familial sinistrality); Eberlin and Mulholland, 1976; Giannitrapani, 1966; O'Connor and Shaw, 1978. Sex has been a variable in the studies of Beaumont, Mayes and Rugg, 1978; Butler, Carter and

Glass, 1977; Davidson et al., 1976; Gale et al., 1978; Trotman and Hammond, 1979. While the evidence for the effects of these variables may not be entirely clear, it at least demonstrates that they are of sufficient importance that they should be explicitly controlled in any research design.

Other subject variables have been: musical training (Davidson and Schwartz, 1977; Hirshkowitz, Earle and Paley, 1978); age (McLeod and Peacock, 1977); field dependance (O'Connor and Shaw, 1978; Oltman, Semple and Goldstein, 1979); occupation (Dumas and Morgan, 1975; Ornstein and Galin, 1973); academic success (Wiet and Goldstein, 1979). Subject strategy is also a highly relevant variable (Davidson and Schwartz, 1976), and attempts should be made to control this variable by manipulation of the cognitive aspects of the task, rather than its being employed as a source of post hoc explanation of observed effects, as has often been the case.

Lastly, there is currently much concern in the main body of experimental neuropsychological literature over the variance between subjects which is masked by group data (Colbourn, 1978). There may well be strong asymmetries exhibited by only a minority of subjects which determine the effects which are extracted statistically. In line with trends in experimental psychology, the number of individual subjects exhibiting the effect should at least be reported, and preferably tabulation of subjects' scores should be presented. It may well be that more intensive investigation of single subjects, rather than group designs, would be rewarding.

THE CURRENT LITERATURE

Discussion of these methodological points at length has left me insufficient space to deal in detail with much of the current literature. This has been deliberate in that I think that recognition of the methodological problems is the most important aspect of this field at the present time. Also, while the review of Donchin et al. (1977a, b) is not critical with respect to particular studies, and is now a little dated, it forms an excellent general overview of the field. A more recent bibliography is also available (French, 1980).

It seems to me that a critical analysis of the literature now available would not reach a conclusion dramatically different from that of Donchin et al. (1977b). This was that there is 'adequate support for the assertion that the ratio of EEG power over hemispheres is sensitive to task variables ..... (but) These trends are far from conclusive' (p.373). One can still adopt this view. Taking the available studies as a whole, there is general support for the idea of lateral asymmetries dependent upon psychological task. However, if one were to adopt a much harder line, and to examine each study in the light of the fore-going methodological critique, then few studies would remain to provide us with evidence. In fact, I do not think that any one of the studies (including our own) can be regarded as entirely methodologically sound. Which approach one adopts is a matter of choice, but I must admit to being suspicious of the former. In the light of these considerations, and making as few allowances for methodological laxity as possible, it seems sensible to pose three questions of the literature.

A. Are there good studies which report an asymmetry with task?

Among the published studies, if we adopt not too strict a methodological criterion, and particularly if we accept data recorded with a vertex reference, there are a number which do appear to provide evidence of task-related asymmetries.

In the study by Davidson et al.(1980), as already mentioned, an asymmetry was observed with both vertex and nose reference, for alpha, in a separate factors analysis. This effect was however only for right handers with sinistrals in the family, the tasks were simply selected as likely to engage asymmetric mechanisms, and the question of the control of response factors is unclear. Ehrlichman and Wiener (1979) with a series of similarly unvalidated tasks, and a vertex reference, also found reliable between-hemisphere differences in integrated alpha, although their subsequent work (1980) points to the influence of covert verbalisation in mediating this asymmetry. Whether covert verbalisation may introduce artifactual asymmetry into the EEG is uncertain, but possible.

Rather arbitrarily selected tasks were also used by Moore (1979), again with a Cz reference. The tasks were counting 'the' or 'a' in an auditory passage, or listening to music, and were clearly not matched for difficulty. A task-by-hemisphere interaction was found, but resulted from a difference on the verbal, and presumably more difficult, task. This task also involved counting during the sampling epochs, and thus minor vocalisation. A similar experiment by Moore and Haynes (1980) using listening tasks which were intended to be better matched, did also find the task by hemisphere interaction, and did not require counting, although the verbal task may have encouraged rehearsal.

Rebert and Low (1978) employed a nose reference, and the tasks were again listening to material as a verbal task, while the spatial counterpart was performing an imaginary block rotation task. It has to be admitted that this is a rather odd task, but they did find a double dissociation of alpha power accompanying the tasks. A range of verbal and spatial tasks were used by Ornstein et al.(1980) with Cz as the reference, and subject normalised raw power measures, and they report task by hemisphere interaction at the central recording sites.

The studies of Grabow, Aronson, Greene and Offord (1979) are particularly interesting. They used a midjaw reference, and also used an assortment of tasks thought to engage right or left hemispheres. The interesting aspect of their report is that while they found little evidence of an asymmetry for direct between-hemisphere tasks, or lateral asymmetry adjusted for resting asymmetry, when the task results were adjusted for the effect of the previous task, a much clearer asymmetry emerged.

Lastly, although the subjects cannot be considered to have normal brains, the study of Kamp and Vliegenthart (1977), with frontal intracerebral electrodes, and presenting two classes of words, reports differential changes at the homologous recording sites.

B. Are there good studies which report no asymmetry with task?

This question is more difficult to answer, partly because of the logic of hypothesis testing, as well as of the influence of selective reporting. Nevertheless, some interesting failures to find an asymmetry have appeared.

The study of Ornstein et al. (1980) has already been cited, and while task effects were found at central electrode sites, they were not found at parietal sites. Similarly, Rebert and Mahoney (1978) using temporal and parietal recording with verbal and nonverbal target detection tasks, found no task effect in a full analysis of the raw power measures. Their reference was, however, linked mastoids.

Tucker (1976), using visuospatial tasks designed to be synthetic or analytic in cognitive demands, together with a vocabulary task, but with an undeclared reference, seems to have failed to observe a task related asymmetry in EEG power, despite the claims of the abstract, although interesting sex differences were found. Dolce and Waldeier (1974) and Dolce and Decker (1975) applied multivariate analysis to spectral parameters, and although they found interesting and complex task effects in beta, theta and delta, no task effect in alpha seems to have been present.

A study by McCarthy and Donchin (1978) employed ingenious task materials derived from studies with commissurotomy patients requiring structural or functional matches to figural stimuli. This is a good attempt at finding matched tasks, and is to some degree validated. They used frontal, central and parietal sites, but referred them to linked mastoids. Although an effect of match type was reported, which they attributed to relative task difficulty, no task dependent asymmetries in alpha were found.

Lastly, there are also studies which have found asymmetries, but not in the direction predicted. Rebert (1977a), whose subjects detected either words or dot patterns; found right hemisphere alpha to be enhanced by the nonverbal dot detection task. Similarly, Beaumont and Rugg (1979), although they found the predicted asymmetry in intrahemispheric coherence, found specific task related alpha enhancement among the power effects observed with a pair of matched and validated tasks. Either the effects of these studies are artifactual, or else, if we conclude that specific task related alpha power enhancement is to be observed, the model we must adopt of task related EEG power effects becomes considerably more complex, and the expected effects of task engagement more difficult to predict. The already widespread use of *post hoc* reasoning to explain task effects becomes in this situation even more dangerous.

C. Are there good studies which find effects but which suggest the operation of artifacts?

Many of these studies have already been cited during the discussion of methodological problems. In particular the effects of reference site have already been discussed, but the study of Amochaev and Salamy (1979) is particularly important among these. Similarly, the confounding effects of task difficulty have been included, with the study of McCarthy and Donchin (1978) as a notable example.

Haynes (1980) raises another important issue. He found an asymmetry, and a task difference, using the vertex reference, temporo-parietal sites and analysing raw scores. However, his tasks were to listen to sentences and either to imitate the sentence following its presentation, or to construct a novel sentence. EEG was recorded during presentation of the stimulus sentence. Left hemisphere activation was found only when imitation was demanded. Haynes discusses this in the light of possible motor programming mechanisms for speech. Other explanations are of course possible, including a complex relationship between task difficulty and the observed asymmetry. In the imitation condition, subjects may engage in more direct subvocal rehearsal of the stimulus material. However, if the motor programming hypothesis is accepted, then it is important to know if this is an aspect of the basic processes which underlie lateral spekcialisation (and curiously specific if it is - this would not be supported by the human performance literature), or results from an artifact of subvocal preparation, or other response related activity, which might suggest an alternative interpretation of at least some of the asymmetries reported.

The studies of Gevins with other workers (Gevins et al. 1979a, b, c) are particularly relevant here. They conducted a series of studies which attempted to control for stimulus characteristics, the effects of limb and eye movements, and performance related factors such as the subject's ability and engagement. They present a considerable amount of data, and it is only unfortunate that their reference site was either the ipsilateral ear or linked ears. Nevertheless, their conclusion was quite clearly that when intertask differences in stimulus parameters, efferent activities and other performance factors were controlled, then the asymmetries which they had otherwise observed, disappeared.

Having posed our three questions, and having drawn some samples from among the published studies, what are we to conclude? At best we should conclude that, although there is a trend towards the finding of task related asymmetries, the mechanisms which underlie these asymmetries are complex and poorly understood. Clear and reliable task related effects have not as yet been demonstrated. At worst, we should be very suspicious of any claim to have demonstrated such asymmetries, and conclude that artifacts may be responsible for the lateral differences which have been reported. Certainly, claims in this area should be cautious, and future investigations based upon much more rigorous and carefully controlled research designs.

COHERENCE ANALYSIS

I should like, before finally concluding, to mention one form of analysis which I think deserves more attention than it has received to date: coherence analysis. Like so many new developments in our techniques, it is not without its peculiar problems, but the form of analysis, or some development of it, carries the promise of enriching the models we can construct of task related effects in the EEG.

Power analysis, or variants of it, has always seemed a meagre way to inspect a system which is generally conceived as a complex network of interrelated functional parts. Indeed, one of the methodological problems of analysing power changes arises because we try to look at relative changes in this network through a series of discrete windows. One advance is to directly measure the interrelationship between the activity at different sites and to observe the change in this interrelationship with cognitive task involvement. Coherence analysis seems to be the best method which we have at present of doing this.

Coherence is the measure of the correlation of two signals as a function of frequency. Shaw and Ongley (1972) were among the first to see the value of this form of analysis, and pointed out that the presence of related regions of desynchronised activity, in the frequency domain, could be detected independently of the effects of the amplitude of any given component within that frequency domain. The important value of coherence analysis is its independence from the effects of both the power and the phase relationship of the two derivations being studied. Coherence is theoretically independent of power, and studies both in our laboratory as well as by Shaw (personal communication) have confirmed this with studies of human EEG. Coherence therefore circumvents some of the methodological problems associated with power analysis discussed above.

A second advantage of this form of analysis is that in order to construct coherence spectra with adequate degrees of freedom, the normal practice is to record a series of short (1 or 2 sec) periodograms, and to average across these epochs. An epoch of 1 sec gives 1 c/sec resolution in the spectrum, and of 2 sec a resolution of 0.5 c/sec. There are theoretical reasons why averaging over short epochs is superior to analysing a single long epoch. This form of recording is ideally suited to psychological tasks where the experimenter is interested in a brief period of cognitive activity which may be repeated over a series of task trials. Indeed, there is no reason why even shorter epochs should not be analysed, giving the opportunity for even finer resolution in the analysis of cognitive task components. How short the epoch may be before the method becomes unreliable seems not yet to have been established for nonstationary waveforms such as the EEG. Subsequent smoothing of the coherence spectrum can be carried out over adjacent frequencies, giving good statistical reliability from relatively small amounts of EEG data. Averaging can of course be carried out over the conventional bands of activity.

Given that spectral analysis seems the best way of analysing EEG data, even if one is interested only in power, then the opportunity to derive power, peak power frequency, coherence and cross-phase spectra (given a sufficiently high degree of coherence) from the Fast Fourier Transform coefficients is a further powerful argument for employing this form of analysis. Further discussions of the background and practical application of coherence analysis are to be found in Cooper, Osselton and Shaw (1980), Orr and Naitoh (1976), and Shaw (1980).

The one difficulty, and it is a serious one, has been clearly pointed out by Shaw (Shaw, Brooks, Colter and O'Connor, 1979). That is that the same change in coherence may result from (a) a change at the common reference site, (b) a change at one or both of the active electrodes, (c) a change at all three electrodes. This results from the essentially correlational nature of the method: for example, if one active site becomes desynchronised, then a larger proportion of the common signal variance will be contributed by the reference site. Activation and deactivation may also have the same effects. Use of a common reference exacerbates the difficulty, but cannot be avoided in this field of investigation. Possibly methods based upon the spectra from an array of electrodes may help to solve the problem, or methods which combine coherence with other measures. Despite this difficulty, the method seems to warrant further investigation.

Interestingly, the method seems to have been employed to examine lateralisation in psychiatric patients rather than in normal subjects, and recent reports of the examination of such patients are to be found in Flor-Henry, Koles, Howarth and Burton (1979), Shaw et al. (1979), and Weller and Montague (1979). With normals there are studies which show that coherence can be related to manual preference (Giannitrapani, 1975; O'Connor and Shaw, 1978) and to field dependence (O'Connor and Shaw, 1978). Orme-Johnson (1977) found coherence changes contingent upon Transcendental Meditation. Wienecke et al's (1980) normative data also, most usefully, includes information on coherence measures.

Task effects have also been more directly demonstrated. Busk and Galbraith (1975) examined visual-motor practice and found that more difficult tasks produced higher coherence levels, while practice led to a decrease in coherence, presumably through a reduction in task difficulty. A verbal learning task was studied by Rugg and Venables (1980) and interhemispheric coherence, in both alpha and theta, predicted relative levels of recall of high and low imagery words. Beaumont et al (1978) using two verbal and two spatial tasks, and recording from temporal and parietal sites, reported interhemispheric asymmetries, independent of power effects, which were related to both cognitive task and sex.

In an attempt to undertake a more tightly controlled study, Beaumont and Rugg (1979) used two parallel tasks known from clinical evidence to be associated with the anterior left and right temporal lobes, respectively. Two reference sites were also employed, Fz and linked mastoids, to assess the effect of reference on both power and intrahemispheric coherence measures. The power effects, which are summarised in Figure 5, demonstrated the difficulty of assessing task dependent changes independently of reference site. Task and hemisphere effects, although of

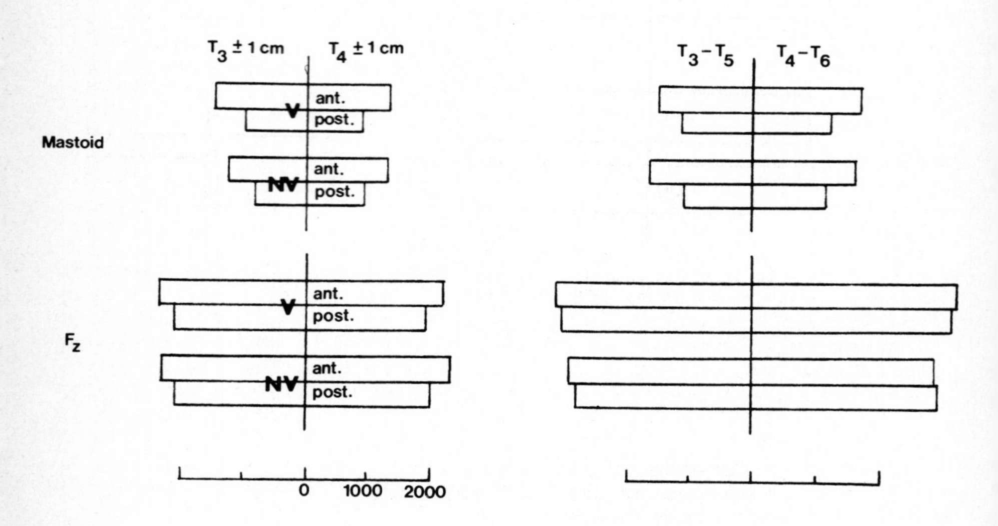

Figure 5
Alpha power at two montages of two pairs of homolateral sites referred to either linked mastoids or Fz, while subjects engaged in verbal (V) or nonverbal (NV) tasks. The power at anterior (ant.) and posterior (post.) electrodes in each montage is indicated (arbitrary units).

task specific alpha enhancement, were found in EEG power at the anterior temporal sites, but interacted with the reference site. The results of coherence analysis, Figure 6 revealed clear effects of both hemisphere and task, although no interaction between them, which were independent of reference site.

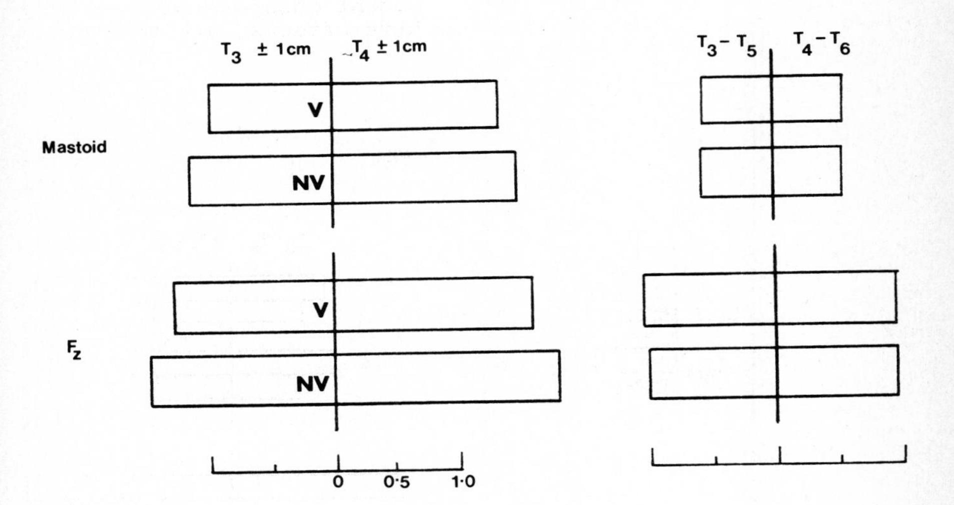

Figure 6
Coherence in the alpha band within hemispheres for two montages of homolateral electrode pairs referred to linked mastoids or Fz, while subjects engaged in verbal (V) or nonverbal (NV) tasks. (units are z-transformed coherence).

The results were highly specific to the anterior temporal electrodes, and did not extend to the posterior temporal lobe. These results, while still leaving many questions unanswered, at least argue for the value of considering coherence alongside power analysis.

Coherence analysis, or some similar method, may enable us to examine the complex interrelationships between different areas of the cortex as it engages in cognitive tasks, and with an accuracy of temporal resolution which has not hitherto been expected for studies of the on-going EEG.

CONCLUSION

It is clear that studies of the on-going EEG have not been as fruitful as studies of average evoked responses in providing evidence for models of brain-behaviour relationships. Despite the advantages of EEG recording, as against AER studies, of the use of more natural tasks, and of the smaller number of ad hoc decisions which must be made in data reduction, the failure to consider methodological issues clearly must make us cautious about the value of past studies in this area.

Specifically with reference to studies of cerebral lateralisation, while there is evidence for lateral asymmetries being observed in the EEG dependent upon psychological task, there are equally studies which have failed to find such asymmetries, and yet others which point to the variety of confounding and artifactual effects which may operate. It would not be proper to say that there is unequivocal evidence for hemisphere differences in the EEG associated with cognitive tasks. In view of the current reappraisal of lateral asymmetries in human performance research, and the more cautious approach advocated by some to this research, caution over the results of EEG studies seems not inappropriate.

Nevertheless, this area of research still seems well worth pursuing. Given a more rigorous approach to methodological problems, it is still possible that studies of the EEG will form an important bridge between psychological and physiological studies of human behaviour, and enable us to perform direct studies of the cortical activity associated with cognitive function, validating and extending current models of neuropsychological function.

Acknowledgement

I should like to acknowledge the contribution of valuable discussions with Dr. Mick Rugg, Dr. John Shaw and Chris French. Deficiencies in this work are however entirely my own.

BIBLIOGRAPHY

To reduce the size of this compound list of references (ca 800), it was necessary to introduce the following abbreviations:

EEG j. for Electroencephalography and Clinical Neurophysiology.
Bio.Psy. for Biological Psychology.
Psy.Phys. for Psychophysiology.
Desmedt, Vol. 1 for J.E. Desmedt (Ed.), Attention, Voluntary Contraction and Event-Related Potentials, Progress in Clinical Neurophysiology (Vol. 1). Basel: Karger, 1977.
Desmedt, Vol. 3 for J.E. Desmedt (Ed.), Language and Hemispheric Specialization in Man: Cerebral Event-Related Potentials, Progress in Clinical Neurophysiology (Vol. 3). Basel: Karger, 1977.
Desmedt, Vol. 6 for J.E. Desmedt (Ed.), Cognitive Components in Cerebral Event-Related Potentials and Selective Attention, Progress in Clinical Neurophysiology (Vol. 6). Basel: Karger, 1979.
Kimmel et al. for H.D. Kimmel, E.H. van Olst & J.F. Orlebeke (Eds.), The Orienting Reflex in Humans. Hillsdale, N.J.: Erlbaum, 1979.
Kornhuber & Deecke for H.H. Kornhuber & L. Deecke (Eds.), Motivation, Motor and Sensory Processes of the Brain: Electrical Potentials, Behaviour and Clinical Use, Progress in Brain Research (Vol. 54). Amsterdam: Elsevier, 1980.
Otto for D.A. Otto (Ed.) Multidisciplinary Perspectives in Event-Related Brain Potential Research, EPA-600/9-77-043. Washington, D.C.: U.S. Government Printing Office, 1978.

Adam, N. & Collins, G., I. Late components of the visual evoked potential to search in short-term memory. EEG j., 1978, 44, 147-156.

Adams, J. & Benson, D., Task-contingent enhancement of the auditory evoked response. EEG j., 1973, 35, 249-257.

Ali, L., Gallagher, T., Goldstein, J. & Daniloff, R., Perception of co-articulated nasality. Journal of the Acoustical Society of America, 1971, 49, 538-540.

Amochaev, A. & Salamy, A., Stability of EEG laterality effects. Psy.Phys., 1979, 16, 242-246.

Anderson, J.R., Cognitive Psychology and its Implications. San Francisco: Freeman, 1980.

Andreassi, J.L., Okamura, H. & Stern, M., Hemispheric asymmetries in the visual cortical evoked potential as a function of stimulus location. Psy.Phys., 1975, 12, 541-546.

Appleton, T., Clifton, R. & Goldberg, S., The development of behavioral competence in infancy. In: F.D. Horowitz (Ed.), Review of Child Development Research, Vol. 4. Chicago: The University of Chicago Press, 1974, pp. 101-186.

Arezzo, J., Legatt, A.D. & Vaughan, H.G. Jr., Topography and intracranial sources of somatosensory evoked potentials in the monkey. I. Early components. EEG j., 1979, 46, 155-172.

Arezzo, J., Pickoff, A. & Vaughan, H.G. Jr., The sources and intracerebral distribution of auditory evoked potentials in the alert rhesus monkey. Brain Research, 1975, 90, 57-73.

Arezzo, J. & Vaughan, H.G. Jr., Cortical potentials associated with voluntary movements in the monkey. Brain Research, 1975, 88, 99-104.

Arezzo, J.C. & Vaughan, H.G. Jr., Intracortical sources and surface topography of the motor potential and somatosensory evoked potential in the monkey. In: Kornhuber & Deecke, 1980, pp. 77-83.

Arezzo, J.C., Vaughan, H.G. Jr. & Legatt, A.D., Topography and intracranial sources of somatosensory evoked potentials in the monkey. II. Cortical components. EEG j., 1981, 51, 1-18.

Attneave, F., Some informational aspects of visual perception. Psychological Review, 1954, 61, 183-193.

Bagshaw, M.H., Kimble, D.P. & Pribram, K.H., The GSR of monkeys during orientation and habituation and after ablation of the amydgala, hippocampus, and inferotemporal cortex. Neuropsychologia, 1965, 3, 111-119.

Ball, M.J., Neurofibrillary tangles and the pathogenesis of dementia: A quantitative study. Neuropathology and Applied Neurology, 1976, 2, 395-400.

Ball, M.J., Topographic distribution of neurofibrillary tangles and granulovacuolar degeneration in hippocampal cortex of aging and demented patients. A quantitative study. Acta Neuropathologica, Berlin, 1979a, 42, 73-80.

Ball, M.J., Histotopography of cellular changes in Alzheimer's disease. In: K. Nandy (Ed.), Senile Dementia, Amsterdam: Elsevier North-Holland Biomedical Press, 1979b, pp. 89-104.

Ball, M.J. & Nuttal, K., Neurofibrillary tangles granulovacuolar degeneration, and neuron loss in Down's Syndrome: quantitative comparison with Alzheimer's disease. Annals of Neurology, 1980, 7, 462-465.

Ball, M.J. & Nuttal, K., Topography of neurofibrillary tangles and granulovacuoles in hippocampi of patients with Down's Syndrome; quantitative comparison with normal ageing and Alzheimer's disease. Neuropathology and Applied Neurobiology, 1981, 7, 13-20.

Banquet, J.P., Renault, B. & Lesèvre, N., Effect of task and stimulus probability on evoked potentials. Bio.Psy., 1981, 13, 203-214.

Barrett, G., Halliday, A.M. & Halliday, E., Evoked potential asymmetries related to handedness and side of stimulation. In: Kornhuber & Deecke, 1980, pp. 761-766.

Barry, R.J., A factor-analytic examination of the unitary OR concept. Bio.Psy., 1979, 8, 161-178.

Bashore, T.R., Vocal and manual time estimates of interhemispheric transmission time. Psychological Bulletin, 1981, 89, 352-368.

Bates, E., Language and Context. New York: Academic Press, 1976.

Bates, J.A.V., Electrical activity of the cortex accompanying movement. Journal of Physiology, 1951, 113, 240-257.

Bauer, H. & Nirnberger, G., Concept identification as a function of preceding negative or positive spontaneous shifts in slow brain potentials. Psy.Phys., 1981, 18, 466-469.

Beaumont, J.G. (Ed.), Divided Visual Field Studies of Cerebral Organisation. London: Academic Press, 1981.

Beaumont, J.G. & Mayes, A., Do task and sex differences influence the visual evoked potential? Psy.Phys., 1977, 14, 545-550.

Beaumont, J.G., Mayes, A.R. & Rugg, M.D., Asymmetry in EEG alpha coherence and power: effects of task and sex. EEG j., 1978, 45, 393-401.

Beaumont, J.G. & Rugg, M.D., The specificity of intrahemispheric EEG alpha coherence asymmetry related to psychological task. Bio.Psy., 1979, 9, 237-248.

Becker, W., Hoehne, O., Iwase, K. & Kornhuber, H.H., Bereitschaftspotential, Prämotorische Positivierung und andere Hirnpotentiale bei sakkadischen Augenbewegungen. Vision Research, 1972, 12, 421-436.

Becker, W., Iwase, K., Jürgens, R. & Kornhuber, H.H., Bereitschaftspotential preceding voluntary slow and rapid hand movements. In: W.C. McCallum and U.R. Knott (Eds.). The responsive brain. Bristol: Wright, 1976, pp. 99-102.

Becker, W. & Kristeva, R., Cerebral potentials prior to various force deployments. In: Kornhuber & Deecke, 1980, pp. 189-194.

Becker, D.E. & Shapiro, D., Direction attention toward stimuli affects the P300 but not the orienting response. Psy.Phys., 1980, 17, 385-389.

Begleiter, H. & Platz, H., Cortical evoked potentials to semantic stimuli. Psy.Phys., 1969, 6, 91-100.

Begleiter, H., Porjesz, B. & Tenner, M., Neuroradiological and neurophysiological evidence of brain deficits in chronic alcoholics. Acta Psychiatrica Scandinavia, Suppl. 286, 1980, 62, 3-14.

Bennett, J.E. & Trinder, J., Hemispheric laterality and cognitive style associated with transcendental meditation. Psy.Phys., 1977, 14, 293-296.

Berlyne, D.E., Craw, M.A., Salapatek, P.H. & Lewis, J.L., Novelty, complexity, incongruity, extrinsic motivation and the GSR. Journal of Experimental Psychology, 1963, 66, 560-567.

Bernstein, A.S., Electrodermal lability and the OR: Reply to O'Gorman and further exposition of the "significance hypothesis". Australian Journal of Psychology, 1973, 25, 147-154.

Bernstein, A.S., The orienting response as novelty and significance detector: Reply to O'Gorman. Psy.Phys., 1979, 16, 263-273.

Bernstein, A.S., The orienting response and stimulus significance: Further comments. Bio.Psy., 1981, 12, 171-185.

Bernstein, A.S. & Taylor, K.W., The interaction of stimulus information with potential stimulus significance in eliciting the skin conductance orienting response. In: Kimmel et al., 1979, pp. 499-519.

Bickford, R.G., Jacobson, J.L. & Cody, D.T.R., Nature of average evoked potentials to sound and other stimuli in man. Annals of the New York Academy of Sciences, 1964, 112, 204-223.

Biederman, I. & Kaplan, R., Stimulus discriminability and S-R compatibility: Evidence for independent effects on choice reaction time. Journal of Experimental Psychology, 1979, 86, 434-439.

Bilodeau, I., Information feedback. In: E.A. Bilodeau (Ed.), Acquisition of skill. New York: Academic Press, 1966, 255-296.

Binet, A., La Concurrence des états psychologiques. Revue Philosophique de la France et de l'etranger, 1980, 24, 138-155.

Black, A.H., Hippocampal electrical activity and behavior. In: R. Isaacson & K. Pribram (Eds.), The hippocampus. Vol. 2. Neurophysiology and behavior. New York: Plenum Press, 1975, pp. 129-167.

Blowers, G.H., Ongley, G.C. & Shaw, J.C., Implications of cross-modality stimulus permutations for the CNV. In: W.C. McCallum & J.R. Knott (Eds.), The responsive brain. Bristol: Wright, 1976, pp. 20-25.

Borda, R.P., The effect of altered drive states on the CNV in rhesus monkeys. EEG j., 1970, 29, 173-180.

Boucher, J. & Warrington, E.K., Memory deficits in early infantile autism: Some similarities to the amnestic syndrome. British Journal of Psychology, 1976, 67, 73-87.

Broadbent, D.E., Perception and Communication. London: Pergamon, 1958.

Broadbent, D.E., Stimulus set and response set: Two kinds of selective attention. In: D.I. Mostofsky (Ed.), Attention: Contemporary Theory and Analysis. New York: Appleton-Centry-Crofts, 1970, pp. 51-60.

Broadbent, D.E., Decision and Stress. New York: Academic Press, 1971.

Broadbent, D.E. & Gregory, R., Stimulus set and response set: The alternation of attention. Quarterly Journal of Experimental Psychology, 1964, 16, 309-317.

Broekhoven, L.H., Brooker, B.H., Czigler, M. & Donald, M.W., Maximum likelihood estimation of the accuracy of discrimination performance in the absence of an overt response to every stimulus. Psychometrika, in press.

Brooker, B.H., The development of selective attention in learining disabled and normal boys: An auditory evoked potential and behavioural analyses. Doctoral dissertation, Queen's University, Kingston, Ontario, Canada, 1980.

Brown, C.H., The relation of magnitude of galvanic skin responses and resistance levels to the rate of learing. Journal of Experimental Psychology, 1937, 20, 262-278.

Brown, W.S., Lehmann, D. & Marsh, J.T., Linguistic meaning related differences in evoked potential topography: English, Swiss-German, and Imagined. Brain and Language, 1980, 11, 340-353.

Brown, W.S., Marsh, J.T. & Smith, J.C., Contextual meaning effects on speech evoked potentials. Behavioral Biology, 1973, 9, 755-761.

Brown, W.S., Marsh, J.T. & Smith, J.C., Evoked potential waveform differences produced by the perception of different meanings of an ambiguous phrase. EEG j., 1976, 41, 113-123.

Brown, W.S., Marsh, J.T. & Smith, J.C., Principal component analysis of ERP differences related to the meaning of an ambiguous word. EEG j., 1979, 46, 709-714.

Bruner, J.S., On cognitive growth: I and II. In: J.S. Bruner, R.R. Olves & R.M. Greensfield (Eds.), Studies in Cognitive Growth. New York: Wiley, 1966.

Brunia, C.H.M., Some questions about the motor inhibition hypothesis. In: Kimmel et al., 1979, pp. 241-258.

Brunia, C.H.M., Motor preparation, recorded on the cortical and spinal level. In: G.E. Stelmach & J. Requin (Eds.), Tutorials in motor behavior. Amsterdam: North Holland, 1980a, pp. 399-419.

Brunia, C.H.M., What is wrong with legs in motor preparation? In: Kornhuber & Deecke, 1980b, pp. 232-236.

Brunia, C.H.M. & Bosch, W.E.J. van den, The influence of response side on the readiness potential prior to finger and foot movements. Sixth International Conference on Event Related Slow Potentials of the Brain, EPIC VI, Lake Forrest, 1981, in press.

Brunia, C.H.M. & Vingerhoets, A.J.J.M., CNV and EMG preceding a plantar flexion of the foot. Bio.Psy., 1980, 11, 181-191.

Brunia, C.H.M. & Vingerhoets, A.J.J.M., Oppposite hemisphere differences in movement-related potentials preceding foot and finger movements. Bio.Psy., 1981, 13, 261-269.

Buchsbaum, M. & Fedio, P., Visual information and evoked responses from the left and right hemisphere. EEG j., 1969, 26, 266-272.

Buchsbaum, M. & Fedio, P., Hemispheric differences in evoked potentials to verbal and nonverbal stimuli in the left and right visual fields. Physiology and Behavior, 1970, 5, 207-210.

Buchwald, J.S., Generators of auditory evoked potentials. In: E.J. Moore (Ed.), Handbook of Electrocochleography and Brainstem Electrical Responses. New York: Grune and Stratton, 1981, in press.

Burger, P.C. & Vogel, F.S., The development of pathological changes of Alzheimer's disease and senile dementia in patients with Down's Syndrome. American Journal of Pathology, 1973, 73, 457-476.

Bunge, M., The Mind-Body Problem. Oxford: Pergamon Press, 1980.

Busk, J. & Galbraith, G.C., EEG correlates of visual-motor practice in man. EEG j., 1975, 38, 415-422.

Butler, R.A., The effect of changes in stimulus frequency and intensity on habituation of the human vertex potential. Journal of the Acoustical Society of America, 1968, 44, 945-950.

Butler, S.R., Carter, J.C. & Glass, A., Sex differences in alpha EEG asymmetries and genetic factors influencing cerebral dominance. (abstract). EEG j., 1977, 43, 533-534.

Butler, S.R. & Glass, A., EEG correlates of cerebral dominance. In: A.H. Riesen & R.F. Thompson (Eds.), Advances in Psychobiology, (Vol. III). New York: Wiley, 1976, pp. 219-272.

Cala, L.A., Jones, B., Wiley, B. & Mastaglia, F.L., A computerized axial tomography (C.A.T.) study of alcohol induced cerebral atrophy --- in conjunction with other correlates. Acta Psychiatrica Scandinavia, Supp. 286, 1980, 62, 31-40.

Callaway, E., Schizophrenia and evoked potentials. In: H. Begleiter (Ed.). Evoked Brain Potentials and Behavior. New York: Plenum Press, 1979, pp. 517-524.

Callaway, E., Brain electrical potentials and individual differences. New York: Grune & Stratton, 1975.

Callaway, E. & Harris, P.R., Coupling between cortical potentials from different areas. Science, 1974, 183, 873-875.

Callaway, E., Tueting, P. & Koslow, S.H., Event-Related Brain Potentials in Man. New York: Academic Press, 1978.

Campbell, K.B., Courchesne, E., Picton, T.W. & Squires, K.C., Evoked potential correlates of human information processing. Bio.Psy., 1979, 8, 45-68.

Chafe, W., Meaning and the Structure of Language. Chicago: The University of Chicago Press, 1970.

Chapman, R.M., Evoked responses to relevant and irrelevant visual stimuli while problem solving. Proceedings of the 73rd Annual Convention of the American Psychological Association, 1965, pp. 177-178.

Chapman, R.M., Connotative meaning and averaged evoked potentials. In: H. Begleiter (Ed.), Evoked Brain Potentials and Behavior. New York: Plenum Press, 1979, pp. 171-196.

Chapman, R.M. & Bragdon, H.R., Evoked responses to numerical and non-numerical visual stimuli while problem solving. Nature, 1964, 203, 1155-1157.

Chapman, R.M., Bragdon, H.R., Chapman, J.A. & McCrary, J.W., Semantic meaning of words and average evoked potentials. In: Desmedt, Vol. 3, 1977, pp. 36-47.

Chapman, R.M. & McCrary, J.W., Hemispheric differences in evoked potentials to relevant and irrelevant visual stimuli. In: D. Lehmann and E. Callaway (Eds.), Human Evoked Potentials: Applications and Problems. New York: Plenum Press, 1979, pp. 55-68.

Chapman, R.M., McCrary, J.W., Chapman, J.A. & Bragdon, H.R., Brain responses related to semantic meaning. Brain and Language, 1978, 5, 195-205.

Chapman, R.M., McCrary, J.W., Chapman, J.A. & Martin, J.K., Behavioral and neural analyses of connotative meaning: Word classes and rating scales. Brain and Language, 1980, 11, 319-339.

Chesney, G.L. & Donchin, E., Predications, their confirmation and the P300 component. Proceedings of the 18th Annual Meeting, Society for Psychophysiological Research. Psy.Phys., 1979, 16, 174.

Chomsky, N. & Halle, M., The Sound Pattern of English. New York: Harper & Row, 1968.

Clynes, M., Unidirectional rate sensitivity as a biological function. Annals of the New York Academy of Sciences, 1961, 92, 946-969; 1962, 98, 806-845.

Clynes, M., Kohn, M. & Lifshitz, K., Dynamics and spatial behavior of light evoked potentials, their modification under hypnosis, and on-line correlation in relation to rhythmic components. Annals of the New York Academy of Sciences, 1964, 112, 468-508.

Cohen, J., Very slow brain potentials relating to expectancy: The CNV. In: E. Donchin & D.B. Lindsley (Eds.), Average evoked Potentials. Washington, D.C.: NASA SP-191, 1969, pp. 143-198.

Cohen, B.A., Bravo-Fernandez, E.J., Hosek, R.S. & Sances, A. Jr., Period analysis of the electroencephalogram: normative data. Psy.Phys., 1976, 13, 591-595.

Cohn, R., Differential cerebral processing of noise and verbal stimuli. Science, 1971, 172, 599-601.

Colbourn, C.J., Can laterality be measured? Neuropsychologia, 1978, 16, 283-289.

Coltheart, M., Deep dyslexia: a right hemisphere hypothesis. In: M. Coltheart, K. Patterson and J.C. Marshall (Eds.), Deep Dyslexia. London: Routledge and Kegan Paul, 1980.

Connor, W.H. & Lang, P.J., Cortical slow-wave and cardiac rate responses in stimulus orientation and reaction time conditions. Journal of Experimental Psychology, 1969, 82, 310-320.

Cooper, R., Osselton, J.W. & Shaw, J.C., EEG Technology (3rd ed.). London: Butterworths, 1980.

Corby, J.C. & Kopell, B.S., The effect of predictability on evoked response enhancement in intramodal selective attention. Psy.Phys., 1973, 10, 335-346.

Corteen, R.S., Skin conductance changes and word recall. British Journal of Psychology, 1969, 60, 81-84.

Coupland, S.G., Taylor, M.J. & Koopman, R.F., EEG landscapes: an application of computer cartography. Psy.Phys., 1980, 17, 413-417.

Courchesne, E., Event-related brain potentials: A comparison between children and adults. Science, 1977, 197, 589-592.

Courchesne, E., Neurophysiological correlates of cognitive development: changes in long-latency event-related potentials from childhood to adulthood. EEG j., 1978a, 45, 468-482.

Courchesne, E., Changes in P3 waves with event repetition: Long-term effects on scalp distribution and amplitude. EEG j., 1978b, 45, 754-766.

Courchesne, E., From infancy to adulthood: The neurophysiological correlates of cognition. In: Desmedt, Vol. 6, 1979, pp. 224-242.

Courchesne, E., Courchesne, R.Y. & Hillyard, S.A., The effect of stimulus deviation on P3 waves to easily recognized stimuli. Neuropsychologia, 1978, 16, 189-199.

Courchesne, E. & Galambos, R., From childhood to adulthood: Changes in cognitive components of the auditory event-related brain potential. In preparation.

Courchesne, E., Ganz, L. & Norcia, A.M., Event-related brain potentials to human faces in infants. Child Development, 1981, 52, 804-811.

Courchesne, E., Hillyard, S.A. & Courchesne, R.Y., P3 waves to the discrimination of targets in homogeneous and heterogeneous stimulus sequences. Psy.Phys., 1977, 14, 590-597.

Courchesne, E., Hillyard, S.A. & Galambos, R., Stimulus novelty, task relevance and the visual evoked potential in man. EEG j., 1975, 39, 131-143.

Courchesne, E., Kilman, B.A., Galambos, R. & Lincoln, A.J., Cognitive processing of novel information in autism: Event-related brain potentials. In preparation.

Craik, F.I. & Blankstein, K.R., Psychophysiology and human memory. In: P.H. Venables & M.J. Christie (Eds.), Research in Psychophysiology. New York: Wiley and Sons, 1975, pp. 389-417.

Crowder, R.G. & Morton, J., Precategorical acoustic storage (PAS). Perception and Psychophysics, 1969, 5, 365-373.

Dainer, K.B., Klorman, R., Salzman, L.F., Hess, D.W. & Davidson, P.W., Learning-disordered children's evoked potentials during sustained attention. Journal of Abnormal Child Psychology, 1981, 9, 79-94.

Daniloff, R. & Moll, K., Coarticulation of lip rounding. Journal of Speech and Hearing Research, 1968, 11, 707-721.

Daruna, J.H., The late positive component of event-related brain potentials and personality. Unpublished doctoral dissertation, University of Illinois at Chicago Circle, 1981.

David, E., Finkenzeller, P., Kallert, S. & Keidel, W.D., Akustischen Reizen zugeordnete Gleichspannungsänderunger am intakten Schädel des Menschen. Pflüegers Archiv, 1969, 309, 362-367.

Davidson, R.J. & Schwartz, G.E., Patterns of cerebral lateralisation during cardiac biofeedback versus the self-regulation of emotion: sex differences. Psy.Phys., 1976, 13, 62-68.

Davidson, R.J. & Schwartz, G.E., The influence of musical training on patterns of EEG asymmetry during musical and non-musical self-generation tasks. Psy.Phys., 1977, 14, 58-63.

Davidson, R.J., Schwartz, G.E., Pugash, C.E. & Bromfield, E., Sex differences in patterns of EEG asymmetry. Bio.Psy., 1976, 4, 119-138.

Davidson, R.J., Taylor, N., Saron, C. & Snyder, M., Individual differences and task effects in EEG measures of hemispheric activation: II: effects of familial sinistrality. Psy.Phys., 1980, 17, 312.

Davis, H., Principles of electric response audiometry. The Annals of Otology, Rhinology, and Laryngology, 1976, Suppl. 28, 85, No. 3, Part 3.

Deecke, L., Functional significance of cerebral potentials preceding voluntary movement. In: Otto, 1978, pp. 87-91.

Deecke, L., The role of the Bereitschaftspotential and potentials accompanying the execution of movement. In: Otto, 1978, pp. 640-641.

Deecke, L., Becker, W., Grözinger, B., Scheid, P. & Kornhuber, H., Human brain potentials preceding voluntary limb movements. In: W.C. McCallum & J.R. Knott (Eds.), Event-related slow potentials of the brain: Their relations to behavior. EEG j., 1973, Suppl. 33, 87-94.

Deecke, L., Grözinger, B. & Kornhuber, H., Voluntary finger movement in man: Cerebral potentials and theory. Biological Cybernetics, 1976, 23, 99-119.

Deecke, L. & Kornhuber, H.H., An electrical sign of participation of the mesial "supplementary" motor cortex in human voluntary finger movement. Brain Research, 1978, 159, 473-476.

Deecke, L., Niesser, A. & Ziller, B., Modality (visual and tactile) and stimulus predictability influence contingent negative variation and reaction time. In: Kornhuber & Deecke, 1980, pp. 301-308.

Deecke, L., Scheid, P. & Kornhuber, H.H., Distribution of readiness potential, pre-motion positivity, and motor potentials of the human cerebral cortex preceding voluntary finger movements. Experimental Brain Research, 1969, 7, 158-168.

De Long, R.G., A neuropsychologic interpretation of infantile autism. In: M. Rutter & E. Schopler (Eds.), Autism: A Reappraisal of Concepts and Treatment. New York: Plenum, 1978.

De Long, M.R. & Strick, P.L., Relation of basal ganglia, cerebellum, and motor cortex units to ramp and ballistic limb movements. Brain Research, 1974, 71, 327-335.

Delse, F.C., Marsh, G.R. & Thompson, L.W., CNV correlates of task difficulty and accuracy of pitch discrimination. Psy.Phys., 1972, 9, 53-62.

Derbyshire, A.J. & McCandless, G.A., Template for the EEG response to sound. Journal of Speech and Hearing Research, 1964, 7, 95-102.

Desmedt, J.E., The size principal of motoneuron recruitment in ballistic or ramp voluntary contractions in man. In: J.E. Desmedt (Ed.), Motor unit types, recruitment and plasticity in health and disease. Basel: Karger, 1981, pp. 97-136.

Desmedt, J.E. & Debecker, J., Wave form and neural mechanism of the decision P350 elicited without pre-stimulus CNV or readiness potential in random sequences of near-threshold auditory clicks and finger stimuli. EEG j., 1979, 47, 648-670.

Desmedt, J.E., Debecker, J. & Robertson, D., Serial perceptual processing and the neural basis of changes in event-related potential components and slow potential shifts. In: Desmedt, Vol. 6, 1979, pp. 53-79.

Desmedt, J.E. & Godeaux, E., Ballistic skilled movements: Load compensation and patterning of the motor commands. In: J.E. Desmedt (Ed.), Cerebral motor control in man: Long loop mechanisms. Basel: Karger, 1978,pp.21-55.

Desmedt, J.E. & Robertson, D., Search for right hemisphere asymmetries in event-related potentials to somatosensory cueing signals. In: Desmedt, Vol. 3, 1977, pp. 172-187.

Desmedt, J.E. & Robertson, D., Differential enhancement of early and late components of the cerebral somato-sensory evoked potentials during forced-paced cognitive tasks in man. Journal of Physiology, 1977, 271, 761-782.

De Swart, J.H. & Das-Smaal, E.A., Orienting reflex and uncertainty reduction in a concept-learning task. In: Kimmel et al., 1979, pp. 549-555.

Didday, R.L. & Arbib, M.A., Eye movements and visual perception: a "two visual systems" model. International Journal of Man-Machine Studies, 1975, 7, 547-569.

Dimond, S.J. & Beaumont, J.G., (Eds.), Hemisphere Function in the Human Brain. London: Elek Science, 1974.

Dolce, G. & Decker, H., Application of multivariate statistical methods in analysis of spectral values of the EEG. In: G. Dolce & H. Kunkel (Eds.), CEAN: Computerized EEG Analysis. Stuttgart: Fischer Verlag, 1975, pp. 157-171.

Dolce, G. & Waldeier, H., Spectral and multivariate analysis of EEG changes during mental activity in man. EEG j., 1974, 36, 577-584.

Donald, M.W., Parallel specific and nonspecific late components in the somatic evoked response. EEG j., 1972, 33, 450 (abstract).

Donald, M., Discussion. In: W.C. McCallum & J.R. Knott (Eds.), Event-related slow potentials of the brain: Their relations to behavior. EEG j., 1973, Suppl. 33, 241-242.

Donald, M.W., Topography of evoked potential amplitude fluctuations. In: W.C. McCallum & J.R. Knott (Eds.), The Responsive Brain. Bristol: Wright, 1976, pp. 10-14.

Donald, M.W., Limits on current theories of transient evoked potentials. In: Desmedt, Vol. 6. 1979, pp. 187-199.

Donald, M.W., Memeory, learning and event-related potentials. In: Kornhuber & Deecke, 1980, pp. 615-628.

Donald, M.W. & Little, R., The analysis of stimulus probability inside and outside the focus of attention, as reflected by the auditory N_1, and P_3 components. Canadian Journal of Psychology, 1981, 35, 175-187.

Donald, M.W. & Young, M.J., Habituation and rate decrements in the auditory vertex potential during selective listening. In: Kornhuber & Deecke, 1980, pp. 331-336.

Donald, M.W. & Young, M.J., The time course of selective neural tuning in auditory attention. Experimental Brain Research, 1982.

Donchin, E., Brain electrical correlates of pattern recognition. In: G.F. Inbar (Ed.), Signal Analaysis and Pattern Recognition in Electrical Engineering. New York: Wiley, 1975.

Donchin, E., Event-related brain potentials: A tool in the study of human information processing. In: H. Begleiter (Ed.), Evoked Brain Potentials and Behavior. New York: Plenum Press, 1979, pp. 13-88.

Donchin, E., Surprise! ..... Surprise? Psy.Phys., 1981, 18, 493-513.

Donchin, E., Callaway, E., Cooper, R., Desmedt, J.E., Goff, W.R., Hillyard, S.A. & Sutton, S., Publication criteria for studies of evoked potentials (EP) in man. In: Desmedt, Vol. 1, 1977, pp. 1-11.

Donchin, E., Gerbrandt, L.A., Leifer, L. & Tucker, L., Is the contingent negative variation contingent on a motor response? Psy.Phys., 1972, 9, 178-188.

Donchin, E. & Heffley, E., Multivariate analysis of event-related potential data: A tutorial review. In: Otto, 1978, pp. 555-572.

Donchin, E. & Heffley, E.F., The independence of the P300 and the CNV reviewed: A reply to Wastell. Bio.Psy., 1979, 9, 177-188.

Donchin, E. & Isreal, J.B., Event-related potentials and psychological theory. In: Kornhuber & Deecke, 1980, pp. 697-715.

Donchin, E., Johnson, R., Herning, R. & Kutas, M., Covariation of the magnitude of the CNV and P300 as a function of the subjects task. In: W.C. McCallum & J.R. Knott (Eds.), The Responsive Brain. Bristol: Wright, pp. 76-80.

Donchin, E., Kubovy, M., Kutas, M., Johnson, R. & Herning, R., Graded changes in evoked response (P300) amplitude as a function of cognitive activity. Perception and Psychophysics, 1973, 14, 319-324.

Donchin, E., Kutas, M. & Johnson, R. Jr., The CNV does not behave like a "motor" potential. EEG j., 1974, 37, 434 (abstract).

Donchin, E., Kutas, M. & McCarthy, G., Electrocortical indices of hemispheric utilization. In: S. Harnad, R.W. Doty, L. Goldstein, J. Jaynes & G. Krauthamer (Eds.), Lateralization in the Nervous System. New York: Academic Press, 1977a, pp. 339-384.

Donchin, E., McCarthy, G. & Kutas, M., Electroencephalographic investigations of hemispheric specialization. In: Desmedt, Vol. 3, 1977b, pp. 212-242.

Donchin, E., Otto, D., Gerbrandt, L.K. & Pribram, K.H., While a monkey waits: Electrocortical events recorded during the foreperiod of a reaction time study. EEG j., 1971, 31, 115-127.

Donchin, E., Ritter, W. & McCallum, W.C., Cognitive psychophysiology: The endogenous components of the ERP. In: E. Callaway, P. Teuting & S.H. Koslow (Eds.), Event Related Brain Potentials in Man. New York: Academic Press, 1978, pp. 349-411.

Donchin, E., Tueting, P., Ritter, W., Kutas, M. & Heffley, E., On the independence of the CNV and the P300 components of the human averaged evoked potential. EEG j., 1975, 38, 449-463.

Donders, F.C., On the speed of mental processes. In: W.G. Koster (Ed. and trans.), Attention and Performance II. Amsterdam: North-Holland, 1969, pp. 412-431.

Dongier, M., Dubrovsky, B. & Engelsmann, F., Event-related slow potentials in psychiatry. In: C. Shagass, S. Gershon & A.J. Friedhoff (Eds.), Psychopathology and Brain Dysfunction. New York: Raven Press, 1977, pp. 339-352.

Dorman, M.F., Auditory evoked potential correlates of speech sound discrimination. Perception & Psychophysics, 1974, 15, 215-220.

Doyle, J.C., Ornstein, R. & Galin, D., Lateral specialisation of cognitive mode: II EEG frequency analysis. Psy.Phys., 1974, 11, 567-578.

Dumas, R. & Morgan, A., EEG asymmetry as a function of occupation, task and task difficulty. Neuropsychologia, 1975, 13, 219-228.

Duncan-Johnson, C.C., P300 latency: a new metric of information processing. Psy.Phys., 1981, 18, 207-215.

Duncan-Johnson, C.C. & Donchin, E., On quantifying surprise: The variation of event-related potentials with subjective probability. Psy.Phys., 1977, 14, 456-467.

Duncan-Johnson, C.C. & Kopell, B.S., The Stroop-effect: Brain potentials localize the source of interference. Science, 1981, 214, 938-940.

Eason, R., Harter, M. & White, C., Effects of attention and arousal on visually evoked cortical potentials and reaction time in man. Physiology and Behavior, 1969, 4, 282-289.

Eberlin, P. & Mulholland, T., Bilateral differences in parietal-occipital EEG induced by contingent visual feedback. Psy.Phys., 1976, 13, 212-218.

Eccles, J.C., The human psyche. The Gifford lectures, University of Edinburgh 1978-1979. New York: Springer, 1980.

Ehrlichman, H. & Weinberger, A., Lateral eye movements and hemispheric asymmetries: a critical review. Psychological Bulletin, 1978, 85, 1080-1101.

Ehrlichman, H. & Wiener, M.S., Consistency of task-related EEG asymmetries. Psy.Phys., 1979, 16, 247-252.

Ehrlichman, H. & Wiener, M.S., EEG asymmetry during covert mental activity. Psy.Phys., 1980, 17, 228-235.

Eimas, P., Siqueland, E., Jusczyk, P. & Vigorito, J., Speech perception in infants. Science, 1971, 171, 303-306.

Elbert, T., Rockstroh, B., Lutzenberger, W. & Birbaumer, N., Biofeedback of slow cortical potentials I. EEG j., 1980, 48, 293-301.

Ellis, W.G., McCulloch, J.R. & Corley, C.L., Presenile dementia in Down's Syndrome --- ultrastructural identity with Alzheimer's disease. Neurology, 1974, 24, 101-106.

Erwin, R.J., Neuroelectrical correlates of ambiguous sentence processing. Unpublished doctoral dissertation. Southern Illinois University at Carbondale, 1981.

Fitzgerald, P.G. & Picton, T.W., Temporal and sequential probability in evoked potential studies. Canadian Journal of Psychology, 1981, 35, 188-200.

Flor-Henry, P., Koles, Z.J., Howart, B.G. & Burton, L., Neurophysiological studies of schizophrenia, mania and depression. In: J. Gruzelier and P. Flor-Henry (Eds.), Hemisphere Asymmetries of Function and Psychopathology. Amsterdam: Elsevier/North Holland, 1979, pp. 189-222.

Ford, J.M., Does P300 reflect template match/mismatch? In: Otto, 1978, pp. 181-188.

Ford, J.M. & Hillyard, S.A., Event related potentials (ERPs) to interruptions of a steady rhytm. Psy.Phys., 1981, 18, 322-330.

Ford, J.M., Roth, W.T. & Kopell, B.S., Auditory evoked potentials to unpredictable shifts in pitch. Psy.Phys., 1976, 13, 32-39.

Ford, J.M., Roth, W.T., Dirks, S.T. & Kopell, B.S., Evoked potential correlates of signal recognition between and within modalities. Science, 1973, 181, 465-466.

Ford, J.M., Roth, W.T. & Kopell, B.S., Attention effects on auditory evoked potentials to infrequent events. Bio.Psy., 1976a, 4, 65-77.

Ford, J.M., Roth, W.T. & Kopell, B.S., Auditory evoked potentials to unpredictable shifts in pitch. Psy.Phys., 1976b, 13, 32-39.

Ford, J.M., Roth, W.T., Mohs, R.C., Hopkins, W.F. III & Kopell, B.S., Event-related potentials recorded from young and old adults during a memory-retrieval task. EEG j., 1979a, 47, 450-459.

Ford, J.M., Hink, R.F., Hopkins, W.F., Roth, W.T., Pfefferbaum, A. & Kopell, B.S., Age effects on event-related potentials in a selective attention task. Journal of Gerontology, 1979b, 34, 388-395.

Fraisse, P. & Voillaume, Les repères du sujet dans la synchronisation et la pseudo-synchronisation. L'Année Psychologique, 1971, 71, 339-369.

Fredrikson, M. & Öhman, A., Heart rate and electrodermal orienting responses to visual stimuli differing in complexity. Scandinavian Journal of Psychology, 1979, 20, 37-42.

French, C.C., The use of electroencephalographic techniques in the study of lateralisation. Unpublished manuscript, Department of Psychology, University of Leicester, 1980.

Friedman, D., The late positive component and orienting behavior. In: Otto, 1978, pp. 178-180.

Friedman, D., Lateral asymmetry of evoked potentials and linguistic processing. In: Otto, 1978, pp. 258-260.

Friedman, D., The effects of reaction time quartile revisited. Poster presentation, 21st Annual Meeting of the Society for Psychophysiological Research, Washington, D.C., 1981.

Friedman, D., Brown, C., Cornblatt, B., Vaughan, Jr., H.G. & Erlenmeyer-Kimling, L., Changes in the late task-related brain potentials during adolescence. Proceedings of the Sixth International Conference on Event-Related Slow Potentials of the Brain, Lake Forest, Illinois (June, 1981), in press.

Friedman, D., Hakerem, G., Sutton, S. & Fleiss, J.L., Effect of stimulus uncertainty on the pupillary dilation response and the vertex evoked potential. EEG j., 1973, 34, 475-484.

Friedman, D., Ritter, W. & Simson, R., Analysis of nonsignal evoked cortical potentials in two kinds of vigilance tasks. In: Otto, 1978, pp. 194-197.

Friedman, D., Simson, R., Ritter, W. & Rapin, I., Cortical evoked potentials elicited by real speech words and human sounds. EEG j., 1975a, 38, 13-19.

Friedman, D., Simson, R., Ritter, W. & Rapin, I., The late positive component (P300) and information processing in sentences. EEG j., 1975b, 38, 255-262.

Friedman, D., Simson, R., Ritter, W. & Rapin, I., CNV and P300 experimental paradigms for the study of language. In: Desmedt, Vol. 3, 1977, pp. 205-211.

Friedman, D., Vaughan, Jr., H.G. & Erlenmeyer-Kimling, L., Stimulus and response related components of the late positive complex in visual discrimination tasks. EEG j., 1978, 45, 319-330.

Friedman, D., Vaughan, Jr., H.G. & Erlenmeyer-Kimling, L., Multiple late positive potentials in two visual discrimination tasks. Psy.Phys., 1981, 18, 635-649.

Frölich, W.D., Floru, R., Glanzmann, P., Juris, M., von Knoblauch zu Hatzbach, L. & Nist, W., The temporal development of early and late CNV in a simple discrimination paradigm: The effects of motor preparation and average reaction time. In: Kornhuber & Deecke, 1980, pp. 730-735.

Frowein, H.W. & Sanders, A.F., Effects of visual stimulus degeneration, S-R compatibility, and foreperiod duration on choice reaction time and movementtime. Bulletin of the Psychonomic Science, 1978, 12, 106-108.

Fruhstorfer, H., Habituation and dishabituation of the human vertex response. EEG j., 1971, 30, 306-312.
Fruhstorfer, H., Soveri, P. & Järvilehto, T., Short-term habituation of the auditory evoked response in man. EEG j., 1970, 28, 153-161.
Furedy, J.J. & Poulos, C.X., Short-interval classical SCR conditioning and the stimulus-sequence-change-elicited OR: The case of the empirical red herring. Psy.Phys., 1977, 14, 351-359.
Gaillard, A.W.K., Effects of warning-signal modality on the contingent negative variation (CNV). Bio.Psy., 1976, 4, 139-154.
Gaillard, A.W.K., The late CNV wave: Preparation versus expectancy. Psy.Phys., 1977, 14, 563-568.
Gaillard, A.W.K., Slow brain potentials preceding task performance. Doctoral disstertation, Institute for Perception TNO, Soesterberg, The Netherlands, 1978.
Gaillard, A.W.K., Cortical correlates of motor preparation. In: R.S. Nickerson (Ed.), Attention and Performance VIII. Hillsdale, N.J.: Lawrence erlbaum, 1980, pp. 75-91.
Gaillard, A.W.K. & Näätänen, R., Modality effects on the contingent negative variation in a simple reaction-time task. In. W.C. McCallum & J.R. Knott (Eds.), The Responsive Brain. Bristol: Wright, 1976, pp. 40-45.
Gaillard, A.W.K. & Perdok, J., Slow cortical and heart rate correlates of discrimination performance. Acta Psychologica, 1979, 43, 185-198.
Gaillard, A.W.K. & Perdok, J., Slow brain potentials in the CNV-paradigm. Acta Psychologica, 1980, 44, 147-163.
Gaillard, A.W.K., Perdok, J. & Varey, C.A., Motor preparation at a cortical and at a peripheral level. In: Kornhuber & Deecke, 1980, pp. 214-218.
Galaburda, A. & Sanides, F., Cytoarchitechtonic organization of the human auditory cortex. Journal of Comparative Neurology, 1980, 190, 597-610.
Galambos, R., Benson, P., Smith, T., Schulman-Galambos, C. & Osier, H., On hemispheric differences in evoked potentials to speech stimuli. EEG j., 1975, 39, 279-283.
Gale, A., Brown, A., Osborne, K. & Smallbone, A., Further evidence of sex differences in brain organisation. Bio.Psy., 1978, 6, 203-208.
Galin, D., Methodological problems and opportunities in EEG studies of lateral specialization. In: Symposium on Neurological Bases of Language Disorders in Children; Methods and Directions for Research. National Institute of Neurological and Communicative Disorders and Stroke, 1978.
Galin, D. & Ellis, R.R., Asymmetry in evoked potentials as an index of lateralised cognitive processes: relation to EEG alpha asymmetry. Neuropsychologia, 1975, 13, 45-50.
Galin, D., Johnstone, J. & Herron, J., Effects of task difficulty on EEG measures of cerebral engagement. Neuropsychologia, 1978, 16, 461-472.
Geer, J.H., Effect of interstimulus intervals and rest-period length upon habitation of the orienting response. Journal of Experimental Psychology, 1966, 72, 617-619.
Gemba, H., Hashimoto, S. & Sasaki, K., Slow potentials preceding self-paced hand movements in the parietal cortex of monkeys. Neuroscience letters, 1979, 15, 87-92.
Gerbrandt, L.K., Goff, W.R. & Smith, D.B., Distribution of the human average movement potential. EEG j., 1973, 34, 461-474.
Gerbrandt, L.K., Methodological criteria for the validation of movement-related potentials. In: Otto, 1978, pp. 97-104.
Gevins, A.S., Zeitlin, G.M., Doyle, J.C., Yingling, C.D., Schaffer, R.E., Callaway, E. & Yeager, C.L., Electroencephalographic correlates of higher cortical functions. Science, 1979a, 203, 665-668.

Gevins, A.S., Zeitlin, G.M., Yingling, C.D., Doyle, J.C., Dedon, M.F., Schaffer, R.E., Roumasset, J.T. & Yeager, C.L., EEG patterns during 'cognitive' tasks. I: Methodology and analysis of complex behaviors. EEG j., 1979b, 47, 693-703.

Gevins, A.S., Zeitlin, G.M., Doyle, J.C., Schaffer, R.E. & Gallaway, E., EEG patterns during cognitive tasks. II: Analysis of controlled tasks. EEG j., 1979c, 47, 704-710.

Giannitrapani, D., Attenuation of the EEG in left and right preferrant subjects. Proceedings of the American Psychological Association, 1966, pp. 127-128.

Giannitrapani, D., Spectral analysis of the EEG. In: G. Dolce & H. Kunkel (Eds.), CEAN: Computerised EEG Analysis. Stuttgart: Fischer Verlag, 1975, pp. 384-402.

Gilden, L., Vaughan, H.G. Jr. & Costa, L.D., Summated human EEG potentials with voluntary movements. EEG j., 1966, 20, 433-438.

Girton, D.G., Benson, K.L. & Kamiya, J., Observation of very slow potential oscillations in human scalp recordings. EEG j., 1973, 35, 561-568.

Glaser, E.M. & Ruchkin, D.S., Principles of Neurobiological Signal Analysis. New York: Academic Press, 1976.

Goff, E.R., Allison, T. & Vaughan, H.G., Jr., The functional neuro-anatomy of event-related potentials. In: E. Callaway, E.P. Tueting & S.H. Koslow (Eds.), Event-related brain protentials in man. New York: Academic Press, 1978, pp. 1-79.

Goff, W.R., Matsumiya, Y., Allison, T. & Goff, G.D., Cross-modality comparisons of average evoked potentials. In: E. Donchin & D.E. Lindsley (Eds.), Average Evoked Potentials. Washington, D.C.: NASA SP-191, 1969, pp. 95-141.

Goff, G.D., Matsumiya, Y., Allison, T. & Goff, W.R., The scalp topography of human somatosensory and auditory evoked potentials. EEG j., 1977, 42, 57-76.

Gomer, F.E., Spicuzza, R.J. & O'Donnel, R.D., Evoked potential correlates of visual item recognition during memeory-scanning tasks. Physiological Psychology, 1976, 4, 61-65.

Goodin, D., Squires, K.C., Henderson, B.H. & Starr, A., Age-related variation in evoked potentials to auditory stimuli in normal human subjects. EEG j., 1978a, 44, 447-458.

Goodin, D., Squires, K.C. & Starr, A., Long latency event-related components of the auditory evoked potential in dementia. Brain, 1978b, 101, 635-648.

Goodin, D.S., Squires, K.C., Henderson, B.H. & Starr, A., An early event-related cortical potential. Psy.Phys., 1978c, 15, 360-365.

Goodman, D.M., Beatty, J. & Mulholland, T.B., Detection of cerebral lateralization of function using EEG alpha-contingent visual stimulation. EEG j., 1980, 48, 418-431.

Gott, P.S., Rossiter, V.S., Galbraith, G.C. & Saul, R.E., Visual evoked response correlates of cerebral specialization after human commissurotomy. Bio.Psy., 1977, 5, 245-255.

Grabow, J.D., Aronson, A.E., Greene, K.L. & Offord, K.P., A comparison of EEG activity in the left and right cerebral hemispheres by power-spectrum analysis during language and non-language tasks. EEG j., 1979, 47,460-472.

Grabow, J.D., Aronson, A.E., Offord, K.P., Rose, D.E. & Greene, K.L., Hemispheric potentials evoked by speech sounds during discrimination tasks. EEG j., 1980a, 49, 48-58.

Grabow, J.D., Aronson, A.E., Rose, D.E. & Greene, K.L., Summated potentials evoked by speech sounds for determining cerebral dominance for language. EEG j., 1980b, 49, 38-47.

Graham, F.K., Distinguishing among orienting, defense, and startle reflexes. In: Kimmel et al., 1979, pp. 137-167.

Greenberg, H.J. & Graham, J.T., Electroencephalographic changes during learning of speech and nonspeech stimuli. Journal of Verbal Learning and Verbal Behavior, 1970, 9, 274-281.

Greene, R.L., Dengerink, H.A. & Staples, S.L., To what does the terminal orienting response respond? Psy.Phys., 1974, 11, 639-646.

Grings, W.W., Verbal-perceptual factors in the conditioning of autonomic responses. In: W.F. Prokasy (Ed.), Classical Conditioning: A Symposium. New York: Appleton-Century-Crofts, 1965, pp. 71-89.

Grings, W.W., Anticipatory and preparatory electrodermal behavior in paired stimulation situations. Psy.Phys., 1969, 5, 597-611.

Grings, W.W. & Dawson, M.E., Complex variables in conditions. In: W.F. Prokasy & D.C. Raskin (Eds.), Electrodermal Activity in Psychological Research. New York: Academic Press, 1973, pp. 203-254.

Grings, W.W. & Sukoneck, H.I., Prediction probability as a determiner of anticipatory and preparatory electrodermal behavior. Journal of Experimental Psychology, 1971, 91, 310-317.

Groll-Knapp, E., Ganglberger, J.A. & Haider, M., Voluntary movement-related slow potentials in cortex and thalamus in man. In: Desmedt, Vol. 1, 1977, pp. 164-173.

Grossberg, S., Do all neural models really look alike? A comment on Anderson, Silverstein, Ritz and Jones. Psychological Review, 1979, 85, 592-597.

Groves, P.M., De Marco, R. & Thompson, R.F., Habituation and sensitization of spinal interneuron activity in the actue spinal cat. Brain Research, 1969, 14, 521-525.

Groves, P.M. & Thompson, R.F., Habituation: a dual process theory. Psychological Review, 1970, 77, 419-450.

Grözinger, B., Kornhuber, H.H. & Kriebel, J., Human cerebral potentials preceding speech production, phonation, and movements of the mouth and tongue, with reference to respiratory and extracerebral potentials. In: Desmedt, Vol. 3, 1977, pp. 87-103.

Grünewald, G., Grünewald-Zuberbier, E. & Netz, J., Late components of average evoked potentials in children with different abilities to concentrate. EEG j., 1978, 44, 617-625.

Grünewald, G. & Grünewald-Zuberbier, E., Ereignisbezogene EEG-Potentiale bei Kindern mit unterschiedlicher Konzentrations-fähigkeit. In: W. Jahnke (Ed.), Beiträge zur Methodik in der differentiellen, diagnostischen und klinischen Psychologie. Festschrift zum 60. Geburtstag von G.A. Lienert. Meisenheim: Hain-Druck, 1981, pp. 116-127.

Grünewald, G., Grünewald-Zuberbier, Hömberg, V. & Netz, J., Cerebral potentials during smooth goal-directed hand movements in right-handed and left-handed subjects. Pflügers Archiv, 1979b, 381, 39-46.

Grünewald, G., Grünewald-Zuberbier, Hömberg, V. & Schuhmacher, H., Hemispheric asymmetry of feedback-related slow negative potential shifts in a positioning movement task. Sixth international conference on event related slow potentials of the brain, EPIC IV, Lake Forest, 1981, in press.

Grünewald, G., Grünewald-Zuberbier, E., Netz, J., Hömberg, V. & Sander, G., Relationships between the late component of the contingent negative variation and the Bereitschaftspotential. EEG j., 1979a, 46, 538-545.

Grünewald-Zuberbier, E. & Grünewald, G., Goal-directed movement potentials of human cerebral cortex. Experimental Brain Research, 1978, 33, 135-138.

Grünewald-Zuberbier, E., Grünewald, G., Hömberg, V. & Schuhmacher, H., Two components of slow negative potential shifts during smooth goal-directed hand movements. In: Kornhuber & Deecke, 1980b, pp. 755-760.

Grünewald-Zuberbier, E., Grünewald, G. & Jung, R., Slow potentials of the human precentral and parietal cortex during goal-directed movements (Zielbewegungspotentiale). Journal of Physiology, London, 1978, 284, 181-182.

Grünewald-Zuberbier, E., Grünewald, G., Schuhmacher, H. & Wehler, A., Scalp recorded slow potential shifts during isometric ramp and hold contractions in human subjects. Pflügers Archiv, 1980a, 389, 55-60.

Haaland, K., The effect of dichotic, monaural, and diotic verbal stimuli on auditory evoked responses. Neuropsychologia, 1974, 12, 339-345.

Haaland, K.Y. & Weitz, R.T., Interhemispheric EEG activity in normal and aphasic adults. Perceptual and Motor Skills, 1976, 42, 827-833.

Haider, M., Groll-Knapp, E. & Ganglberger, J.A., Event-related slow (DC) potentials in the human brain. Review of Physiology Biochemistry and Pharmacology, 1981, 88, 125-197.

Haider, M., Groll, E. & Studynka, G., Orienterungs und Bereitschaftspotentiale bei unerwarten Reizen. Experimental Brain Research, 1968, 5, 45-59.

Halgren, E., Squires, N.K., Wilson, C.L., Rohrbaugh, J.W., Babb, T.L. & Crandall, P.H., Endogenous potentials generated in the human hippocampal-formation and amygdala by infrequent events. Science, 1980, 210, 803-805.

Hallett, M. & Marsden, C.D., Physiology and pathophysiology of the ballistic movement pattern. In: J.E. Desmedt (Ed.), Motor unit types, recruitment and plasticity in health and disease. Basel: Karger, 1981, pp. 331-346.

Halliday, A.M., Barret, G., Blumhardt, L.D. & Kriss, A., The macular and paramacular subcomponents of the pattern evoked response. In: D. Lehmann and E. Callaway (Eds.), Human Evoked Potentials: Applications and Problems. New York: Plenum Press, 1979.

Hamilton, C.E., Peters, J.F. & Knott, J.R., Task initiation and amplitude of the contingent negative variation (CNV). EEG j., 1973, 34, 587-592.

Hannes, M., Sutton, S. & Zubin, J., Reaction time: Stimulus uncertainty with response uncertainty. The Journal of General Psychology, 1968, 78, 165-181.

Hansen, J.C. & Hillyard, S.A., Endogenous brain potentials associated with selective auditory attention. EEG j., 1980, 49, 277-290.

Hari, R., Sensory evoked sustained potentials in man. Doctoral Dissertation, University of Helsinki, 1980.

Hari, R., Sams, M. & Järvilehto, T., Auditory evoked transient and sustained potentials in the human EEG: I. Effects of expectation of stimuli. Psychiatry Research, 1979a, 1, 297-306.

Hari, R., Sams, M. & Järvilehto, T., Auditory evoked transient and sustained potentials in the human EEG: II. Effects of small doses of ethanol. Psychiatry Research, 1979b, 1, 307-312.

Harkins, S.W., Thompson, L.W., Moss, S.F. & Nowlin, J.B., Relationship between central and autonomic nervous system activity: Correlates of psychomotor performance in elderly men. Experimental Aging Research, 1976, 2, 409-423.

Harmony, T., Ricardo, J., Otero, G., Fernandez, G., Llorente, S. & Valdes, P., Symmetry of the visual evoked potential in normal subjects. EEG j., 1973, 35, 237-240.

Harter, M.R. & Guido, W., Attention to pattern orientation: Negative cortical potentials, reaction time and the selection process. EEG j., 1980, 49, 461-475.

Harter, M.R. & Previc, F.H., Size-specific information channels and selective attention: Visual evoked potential and behavioural measures. EEG j., 1978, 45, 628-640.

Harter, M.R. & Salmon, L.E., Intramodality selective attention and evoked cortical potentials to randomly presented patterns. EEG j., 1971, 32, 605-613.

Harvey, N. & Treisman, A.M., Switching attention between the ears to monitor tones. Perception and Psychophysics, 1973, 14, 51-59.

Hashimoto, S., Gemba, H. & Sasaki, K., Analysis of slow cortical potentials preceding self-paced hand movements in the monkey. Experimental Neurology, 1979, 65, 218-229.

Hashimoto, S., Gemba, H. & Sasaki, K., Premovement slow cortical potentials and required muscle force in self-paced hand movements in the monkey. Brain Research, 1980, 197, 415-423.

Hauser, S.L., DeLong, R.G. & Rosman, N.P., Pneumographic findings in the infantile autism syndrome: a correlation with temporal lobe disease. Brain, 1975, 98, 667-688.

Haynes, W.O., Task effect and EEG alpha asymmetry: an analysis of linguistic processing in two response modes. Cortex, 1980, 16, 95-102.

Hazemann, P., Effects of movement on sensory output. In: Otto, 1978, pp. 105-106.

Hebb, D.O., Organization of Behaviour. New York: John Wiley & Sons, 1949.

Hecaen, H. & Albert, M.L., Human Neuropsychology. New York: John Wiley & Sons, 1978.

Hecaen, H., De Agostini, M. & Mouzen-Moutes, A., Cerebral organization in left-handers. Brain and language, 1981, 12, 261-284.

Heilman, K.M. & Van Den Abell, T., Right hemsipheric dominance for mediating cerebral activation. Neuropsychologia, 1979, 17, 315-321.

Heilman, K.M. & Van Den Abell, T., Right hemispheric dominance for attention: The mechanism underlying hemispheric asymmetries of inattention (neglect). Neurology, 1980, 30, 327-330.

Herning, R.I., Jones, R.T. & Peltzman, D.J., Changes in human event related potentials with prolonged delta-9-tetrahydrocannabinol (THC) use. EEG j., 1979, 47, 556-570.

Herron, J., Galin, D., Johnstone, J. & Ornstein, R.E., Cerebral specialisation, writing posture and motor control of writing in left-handers. Science, 1979, 205, 1285-1289.

Hillyard, S.A., The CNV and human behavior. In: W.C. McCallum & J.R. Knott (Eds.), Event-related slow potentials of the brain: Their relations to behavior. EEG j., 1973, Suppl. 33, 161-171.

Hillyard, S.A., Selective auditory attention and early event-related potentials: A rejoinder. Canadian Journal of Psychology, 1981, 35, 159-174.

Hillyard, S.A., Hink, R.F., Schwent, V.L. & Picton, T.W., Electrical signs of selective attention in the human brain. Science, 1973, 182, 177-180.

Hillyard, S.A. & Picton, T.W., Event related brain potentials and selective information processing in man. In: Desmedt, Vol. 6, 1979, pp. 1-52.

Hillyard, S.A., Picton, T.W. & Regan, D., Sensation, perception and attention: Analysis using ERPs. In: E. Callaway, P. Tueting & S.H. Koslow (Eds.), Event-related Brain Potentials in Man. New York: Academic Press, 1978, pp. 223-321.

Hillyard, S.A., Squires, K.C., Bauer, J.W. & Lindsey, P.N., Evoked potential correlates of auditory signal detection. Science, 1971, 172, 1357-1360.

Hillyard, S.A. & Wood, D.L., Electrophysiological analysis of human brain function. In: M.S. Gazzaniga (Ed.), Handbook of Behavioral Neurobiology, Vol. 2. New York: Plenum Press, 1979, 345-378.

Hink, R.F., Kaga, K. & Suzuki, J., An evoked potential correlate of reading ideographic and phonetic Japanese scripts. Neuropsychologia, 1980, 18, 455-464.

Hink, R.F., Hillyard, S.A. & Benson, P.J., Event-related brain potentials and selective attention to acoustic and phonetic cues. Bio.Psy., 1978, 6, 1-16.

Hink, R.F., Van Voorhis, S.T., Hillyard, S.A. & Smith, T.S., The diversion of attention and the human auditory evoked potential. Neuropsychologia, 1977, 15, 597-605.

Hirshkowitz, M., Earle, J. & Paley, B., EEG alpha asymmetry in musicians and non-musicians: a study of hemispheric specialisation. Neuropsychologia, 1978, 16, 125-128.

Hofmann, M.J. & Salapatek, P., Young infants' event-related potentials (ERPs) to familiar and unfamiliar visual and auditory events in a recognition memory task. EEG j., 1981, 52, 405-417.

Hömberg, V., Grünewald, G. & Grünewald-Zuberbier, E., The variation of P300 amplitude in a money-winning paradigm in children. Psy.Phys., 1981, 18, 258-262.

Inhelder, B. & Piaget, J., The Early Growth of Logic in the Child. New York: Norton, 1964.

Irwin, D.A., Knott, J.R., McAdam, D.W. & Rebert, C.S., Motivational determinants of the "contingent negative variation". EEG j., 1966, 21, 538-543.

Isreal, J.B., Chesney, G.L., Wickens, C.D. & Donchin, E., P300 and tracking difficulty: Evidence for multiple resources in dual-task performance. Psy.Phys., 1980a, 17, 259-273.

Isreal, J.B., Wickens, C.D., Chesney, G.L. & Donchin, E., The event-related brain potential as an index of display-monitoring workload. Human factors, 1980b, 22, 212-224.

Iverson, S.D., Motor Control. British Medical Bulletin, 1981, 37, 147-152.

Jakobson, R., Fant, G. & Halle, M., Preliminaries to Speech Analysis. Cambridge: Massachusetts Institute of Technology Press, 1963.

Järvilehto, T. & Frühstorfer, H., Differentiation between slow cortical potentials associated with motor and mental acts in man. Experimental Brain Research, 1970, 11, 309-317.

Järvilehto, T., Hari, R. & Sams, M., Effect of stimulus repetition on negative sustained potentials elicited by auditory and visual stimuli in the human EEG. Bio.Psy., 1978, 7, 1-12.

Jasper, H. & Penfield, W., Electrocorticograms in man: Effect of voluntary movement upon the electrical activity of the precentral gyrus. Archiv Psychiatrie und Zeitschrift für Neurologie, 1949, 183, 163-174.

Jenness, D., Auditory evoked response differentiation with discrimination learning in humans. Journal of Comparative and Physiological Psychology, 1972, 80, 75-90.

John, E.R., Functional Neuroscience: Vol. II. Neurometrics: Clinical Application of Quantitative Electrophysiology. Hillsdale, N.J.: Erlbaum, 1977.

John, E.R., Ruchkin, D.S. & Vidal, J.J., Measurement of event-related potentials. In: E. Callaway, P. Tueting & S.H. Koslow (Eds.), Event-Related Brain Potentials in Man. New York: Academic Press, 1978, pp. 93-138.

Johnson, R., Electrophysiological manifestations of decision making in a changing environment. Doctoral dissertation, University of Illinois, 1979.

Johnson, R. Jr., P300 amplitude and probabilistic judgments. In: Kornhuber & Deecke, 1980, pp. 723-729.

Johnson, R. Jr., P300: A model of the variables controlling its amplitude. In: J. Cohen, R. Karrer & P. Tueting (Eds.), Annals of the New York Academy of Sciences, In press.

Johnson, R. Jr. & Donchin, E., On how P300 amplitude varies with the utility of the eliciting stimuli. EEG j., 1978, 44, 424-437.

Johnson, R. Jr. & Donchin, E., P300 and stimulus categorization: Two plus one is not so different from one plus one. Psy.Phys., 1980, 17, 167-178.

Johnson, R. Jr. & Donchin, E., Sequential expectancies and decision making in a changing environment: An electrophysiological approach. Psy.Phys., 1982, 19, 183-200.

Johnston, V.S. & Chesney, G., Electrophysiological correlates of meaning. Science, 1974, 186, 944-946.

Johnston, V.S. & Holcomb, P.J., Probability learning and the P3 component of the visual evoked potential in man. Psy.Phys., 1980, 17, 396-400.

Jung, R., Discussion of W.C. McCallum, The contingent negative variation as a cortical sign of attention in man. In: C.R. Evans & T.B. Mulholland (Eds.), Attention in Neurophysiology. London: Buttersworths, 1969, pp. 59-60.

Jung, R., Perception and action. In: J. Szentágothai, M. Palkovits & J. Hámori (Eds.), Regulatory functions of the CNS. Motion and organization principles. Adv. Physiol. Sci., New York: Pergamon Press, 1981, 1, pp. 17-36.

Just, M.A. & Carpenter, P.A., Eye fixations and cognitive processes. Cognitive Psychology, 1976, 8, 441-480.

Kahneman, D., Attention and Effort. Englewood Cliffs, N.J.: Prentice-Hall,1973.

Kamp, A. & Vliegenthart, W.E., Spectral changes of the EEG in human frontal cortex related to stimulus-response tasks (abstract). EEG j., 1977, 43, 566.

Kandel, E.R., Cellular basis of behavior; an introduction to behavioral neurobiology. San Francisco: Freeman, 1976.

Karis, D., Bashore, T., Fabiani, M. & Donchin, E., P300 and memory. Psy.Phys., 1981, 19, 328.

Karlin, L., Cognition, preparation and sensory-evoked potentials. Psychological Bulletin, 1970, 73, 122-136.

Karlin, L. & Martz, M.J. Jr., Response probability and sensory evoked potentials. In: S. Kornblum (Ed.), Attention and Performance IV. New York: Academic Press, 1973, pp. 175-184.

Karrer, R. & Ivins, J., Event-related slow potentials in mental retardates. In: W.C. McCallum and J.R. Knott (Eds.), The Responsive Brain. Bristol: John Wright, 1976, pp. 154-157.

Karrer, R., McDonough, B., Warren, C. & Cone, R., CNV during memeory retrieval by adolescent and pre-adolescent children. Paper presented to the Nineteenth Annual Meeting of the Society for Psychophysiological Research. Cincinnati, Ohio, 1979.

Keidel, W.D., D.C. --- potentials in the auditory evoked response in man. Acta Otolaryngologia, 1971a, 71, 242-248.

Keidel, W.D., What do we know about the human cortical evoked potential after all? Ohren- Nasen- und Kehlkopfheilkunde, 1971b, 198, 9-37.

Keidel, W.D., Late cortical-evoked response in clinical use. Audiology, 1980, 19, 16-35.

Kerkhof, G.A., Event-related potentials and auditory signal detection: Their diurnal variation for morning-type and evening-type subjects. Psy.Phys., in press.

Kimmel, H.D., van Olst, E.H. & Orlebeke, J.F. (Eds.), The orienting reflex in humans. Hillsdale, New Jersey: Erlbaum, 1979.

Kimmel, H.D., Piroch, J. & Ray, R.L., Monotony and uncertainty in the habituation of the orienting reflex. In: Kimmel et al., 1979, pp. 425-442.

Kimura, D., Functional asymmetry of the brain in dichotic listening. Cortex, 1967, 3, 163-178.

Kinsbourne, M., Asymmetrical Function of the Brain. Cambridge University Press, 1978.

Kirst, S. & Beatty, J., Processing strategy and distraction affect reaction time and the CNV. Bulletin of the Psychonomic Society, 1978, 12, 71-73.

Klapp, S.T., Response programming, as assessed by reaction time, does not establish commands for particular muscles. Journal of Motor Behavior, 1977, 9, 301-312.

Klee, M. & Rall, W., Computed potentials of cortically arranged populations of neurons. Journal of Neurophysiology, 1977, 40, 647-666.

Klinke, R., Frühstorfer, H. & Finkenzeller, P., Evoked responses as a function of external and stored information. EEG j., 1968, 25, 119-122.

Klorman, R. & Bentsen, E., Effects of warning-signal duration on the early and late components of the contingent negative variation. Bio. Psy., 1975, 3, 263-275.

Klorman, R. & Ryan, R.M., Heart rate, contingent negative variation, and evoked potentials during anticipation of affective stimulation. Psy.Phys., 1980, 17, 513-523.

Klorman, R., Salzman, L.F., Pass, H.L. Borgstedt, A.D. & Dainer, K.B., Effects of methylphenidate on hyperactive children's evoked responses during passive and active attention. Psy.Phys., 1979, 16, 23-29.

Knott, J.R. & Irwin, D.A., Anxiety, stress and the contingent negative variation. EEG j., 1968, 24, 286-287.

Kohlenberg, R.J., Instructions to ignore a stimulus and the GSR. Psychonomic Science, 1970, 19, 220-221.

Köhler, W., Held, R. & O'Connell, D., An investigation of cortical currents. Proceedings of the American Philosophical Society, 1952, 96, 290-330.

Köhler, W. & Wegener, J., Currents of the human auditory cortex. Journal of Cellular and Comparative Physiology, 1955, 45, Suppl. 1, 25-54.

Kok, A., The effect of warning stimulus novelty on the P300 and components of the contingent negative variation. Bio.Psy., 1978, 6, 219-233.

Kok, A. & Looren de Jong, H., Components of the event-related potential following degraded and undegraded visual stimuli. Bio.Psy., 1980a, 11, 117-133.

Kok, A. & Looren de Jong, H., The effect of repetition of infrequent familiar and unfamiliar visual patterns on components of the event-related brain potential. Bio.Psy., 1980b, 10, 167-188.

Kornhuber, H.H., Motor functions of cerebellum and basal ganglia: The cerebellocortical saccadic (ballistic) clock, the cerebellonuclear hold regulator and the basal ganglia ramp (voluntary speed smooth movement) generator. Kybernetik, 1971, 8, 157-162.

Kornhuber, H.H., Cerebral cortex, cerebellum, and basal ganglia: An introduction to their motor functions. In: F.O. Schmitt & F.G. Warden (Eds.), The Neurosciences. Third study program. Cambridge, Mass: MIT Press, 1974, pp. 267-280.

Kornhuber, H.H. & Deecke, L., Hirnpotentialänderungen bei Willkürbewegungen und passiven Bewegungen des Menschen: Bereitschaftspotential und reafferente Potentiale. Pflügers Archiv für die Gesamte Physiologie des Menschen und der Tiere, 1965, 284, 1-17.

Korth, M. & Rix, R., Der Einflüss der Vigilanz auf die Gleichspannungsantwort im visuell evozierten Potential des Menschen. Albrecht von Graefes Archiv für Ophthalmologie. Vereinigt mit Archiv für Augenheilkunde, 1979, 210, 141-150.

Kristeva, R. & Kornhuber, H.H., Cerebral potentials related to the smallest human finger movement. In: Kornhuber & Deecke, 1980, pp. 177-182.
Kuno, Y., The significance of sweating in man. Lancet, 1930, 1, 912-915.
Kurtzberg, D. & Vaughan, H.G. Jr., Electrophysiological observations on the visuomotor system and visual neurosensorium. In: J.E. Desmedt (Ed.), Visual Evoked Potentials in Man: New Developments. Oxford: Clarendon Press, 1977, pp. 314-331.
Kurtzberg, D. & Vaughan, H.G. Jr., Topographic analysis of human cortical potentials preceding self-paced and visually triggered saccades. Brain Research, in press.
Kurtzberg, D., Vaughan, H.G. Jr., Courchesne, E., Friedman, D., Harter, M.R. & Putnam, L.E., Developmental aspects of event-related potentials. In: J. Cohen, R. Karrer & P. Tueting (Eds.), Annals of the New York Academy of Sciences, 1982, in press.
Kurtzberg, D., Vaughan, H.G. Jr. & Kreuzer, J., Task-related cortical potentials in children. In: Desmedt, Vol. 6, 1979, pp. 216-223.
Kutas, M. & Donchin, E., Studies of squeezing: Handedness, responding hand, response force, and asymmetry of readiness potential. Science, 1974, 186, 545-548.
Kutas, M. & Donchin, E., The effect of handedness, of responding hand, and of response force on the contralateral dominance of the readiness potential. In: Desmedt, Vol. 1, 1977, pp. 189-210.
Kutas, M. & Donchin, E., Variations in the latency of P300 as a function of variations in semantic categorizations. In: Otto, 1978, pp. 198-201.
Kutas, M. & Donchin, E., Preparation to respond as manifested by movement-related brain potentials. Brain Research, 1980, 202, 95-115.
Kutas, M. & Hillyard, S.A., Event-related brain potentials to semantically inappropriate and surprisingly large words. Bio.Psy., 1980a, 11, 99-116.
Kutas, M. & Hillyard, S.A., Reading between the lines: Event-Related brain potentials during natural sentence processing. Brain and Language, 1980b, 11, 354-373.
Kutas, M. & Hillyard, S., Reading senseless sentences: Brain potentials reflect semantic incongruity. Science, 1980c, 207, 203-205.
Kutas, M., McCarthy, G. & Donchin, E., Augmenting mental chronometry: the P300 as a measure of stimulus evaluation time. Science, 1977, 197, 792-795.
Lang, P.J., Gatchel, R.J. & Simons, R.F., Electro-cortical and cardiac rate correlates of psychophysical judgment. Psy.Phys., 1975, 12, 649-655.
Lang, P.J., Ohman, A. & Simons, R.F., The psychophysiology of anticipation. In: J. Requin (Ed.), Attention and Performance VII. New York: Erlbaum, 1978, pp. 469-485.
Lawson, E.A. & Gaillard, A.W.K., Evoked potentials to consonant-vowel syllables. Acta Psychologica, 1981a, 49, 17-25.
Lawson, E.A. & Gaillard, A.W.K., Mismatch negativity in a phonetic discrimination task. Bio.Psy., 1981, 13, 281-288.
Lee, W.A., Anticipatory control of postural and task muscles during rapid arm flexion. Journal of Motor Behavior, 1980, 12, 185-196.
Leeuwenberg, E.L., Structural information of Visual patterns. Paris & The Hague: Mouton, 1967.
Ledlow, A., Swanson, J.M. & Kinsbourne, M., Differences in reaction times and average evoked potentials as a function of direct and indirect neural pathways. Annals of Neurology, 1978a, 3, 525-530.

Ledlow, A., Swanson, J.M. & Kinsbourne, M., Reaction times and evoked potentials as indicators of hemispheric differences for laterally presented name and physical matches. Journal of Experimental Psychology; Human Perception and Performance, 1978b, 4, 440-454.

Legatt, A., Short-latency auditory evoked potentials in the monkey. Unpublished doctoral dissertation. Albert Einstein College of Medicine,1981.

Lehmann, D., EEG, evoked potentials, and eye and image movements. In: P. Bach-y-Rita, C.C. Collins & J.E. Hyde (Eds.), The Control of Eye Movements. New York: Academic Press, 1971, pp. 149-174.

Lehmann, D., Skrandies, W. & Lindenmaier, C., Sustained cortical potentials evoked in humans in binocularly correlated, uncorrelated and disparate dynamic random-dot stimuli. Neuroscience Letters, 1978, 10, 129-134.

Lehtonen, J.B. & Koivikko, M.J., The use of a non-cephalic reference electrode in recording cerebral evoked potentials in man. EEG j., 1971, 31, 154-156.

Leissner, P., Lindholm, L.E. & Petersen, I., Alpha amplitude dependence on skull thickness as measured by ultrasound technique. EEG j., 1970, 29, 392-399.

LeMay, M. & Geschwind, N., Asymmetries of the human cerebral hemispheres. In: A. Caramazza & E.B. Zurif (Eds.), Language Acquisition and Language Breakdown: Parallels and Divergencies. Baltimore: Johns Hopkins University Press, 1978, pp. 311-328.

Levitt, R.A., Sutton, S. & Zubin, J., Evoked potential correlates of information processing in psychiatric patients. Psychological Medicine, 1973, 3, 487-494.

Liberman, A., Cooper, F., Shankweile, D. & Studdert-Kennedy, M., Perception of the speech code. Psychological Review, 1974, 74, 431-461.

Liberman, A., Delattre, P. & Cooper, F., Some cues for the distinction between voiced and voiceless stops in initial position. Language and Speech, 1958, 1, 153-167.

Lisker, L. & Abramson, A., A cross-language study of voicing in initial stops: Acoustical Measurements. Word, 1964, 20, 384-422.

Litzelman, D.K., Thompson, L.W., Mechalewski, H.J., Patterson, J.V. & Bowman, T.E., Visual event-related potentials and depression in the elderly. Neurobiology of Aging, 1980, 1, 111-118.

Loiselle, D.L., Stamm, J.S., Maitinsky, S. & White, S.C., Evoked potential and behavioral signs of attentive dysfunctions in hyperactive boys. Psy.Phys., 1980, 17, 193-201.

Lorente de No, R., A study of Nerve physiology. Part 2, Vol. 132. New York: Rockefeller Institute, 1947.

Loveless, N.E., The effect of warning interval on signal detection and event-related slow potentials of the brain. Perception and Psychophysics, 1975, 17, 565-570.

Loveless, N.E., Distribution of responses to non-signal stimuli. In: W.C. McCallum & J.R. Knott (Eds.), The Responsive Brain. Bristol: Wright, 1976, pp. 26-29.

Loveless, N.E., Event-related slow potentials of the brain as expressions of orienting function. In: Kimmel et al., 1979, pp. 77-100.

Loveless, N. & Sanford, A.J., Effects of age on the contingent negative variation and preparatory set in a reaction-time task. Journal of Gerontology, 1974a, 29, 52-63.

Loveless, N.E. & Sanford, A.J., Slow potential correlates of preparatory set. Bio.Psy., 1974b, 1, 303-314.

Loveless, N. & Sanford, A.J., The impact of warning signal intensity on reaction time and components of the contingent negative variation. Bio.Psy., 1975, 2, 217-226.

Low, M.D. & McSherry, J.W., Further observations of psychological factors involved in CNV genesis. EEG j., 1968, 25, 203-207.

Lukas, J.H., Human auditory attention: The olivo-cochlear bundle may function as a peripheral filter. Psy.Phys., 1980, 17, 444-452.

Lutzenberger, W., Elbert, T., Rockstroh, B. & Birbaumer, N., Principal component analysis of slow brain potentials during six second anticipation intervals. Bio.Psy., 1981, 13, 271-279.

Lutzenberger, W., Schandry, R. & Birbaumer, N., Habituation of the components of the AEP to stimuli of different intensities. In: Kimmel et al., 1979, pp. 123-128.

MacCorquodale, K. & Meehl, P.E., On a distinction between hypothetical constructs and intervening variables. Psychological Review, 1948, 55, 95-107.

MacKay, D.M., Cerebral organization and the conscious control of action. In: J. Eccles (Ed.), Brain and Conscious Experience. New York: Springer Verlag, 1966, pp. 422-445.

MacLean, V., Öhman, A. & Lader, M., Effects of attention, activation and stimulus regularity on short-term "habituation" of the averaged evoked response. Bio.Psy., 1975, 3, 57-69.

Mackworth, N.H. & Otto, D.A., Habituation of the visual orienting response in young children. Perception and Psychophysics, 1970, 7, 173-180.

Maltzman, I., Orienting in classical conditioning and generalization of the galvanic skin response to words: An overview. Journal of Experimental Psychology; General, 1977, 106, 106-119.

Maltzman, I., Orienting reflexes and classical conditioning in humans. In: Kimmel et al., 1979, 323-351.

Maltzman, I., Orienting reflexes and significance: A reply to O'Gorman. Psy.Phys., 1979a, 16, 274-282.

Mäntysalo, S., Alho, K. & Näätänen, R., Inter-stimulus-interval and N2 deflection of the evoked potential. In preparation.

Marcel, A.T. & Patterson, K., Word recognition and production: reciprocity in clinical and normal studies. In: J. Requin (Ed.), Attention and Performance VII. Hillsdale, N.J.: Erlbaum, 1977, pp. 209-226.

Marsden, C.D., Merton, P.A., Morton, H.B., Hallett, M., Adam, J. & Rushton, D.N., Disorders of movement in cerebellar disease in man. In: F.C. Rose (Ed.), The physiological aspect of clinical neurology. Oxford: Blackwells, 1977, pp. 179-199.

Marsh, G.R., Asymmetry of electrophysiological phenomena and its relation to behavior in humans. In: M. Kinsbourne (Ed.), Asymmetrical Function of the Brain. Cambridge: Cambridge University Press, 1978, pp. 292-317.

Marsh, G.R. & Thompson, L.W., Effects of age on the contingent negative variation in a pitch discrimination task. Journal of Gerontology, 1973, 28, 56-62.

Matsumiya, Y., Tagliasco, V., Lombroso, C. & Goodglass, H., Auditory evoked response: Meaningfulness of stimuli and interhemispheric asymmetry. Science, 1972, 175, 790-792.

Mayes, A. & Beaumont, G., Does visual evoked potential asymmetry index cognitive activity? Neuropsychologia, 1977, 15, 249-256.

McAdam, D.W. & Seales, D.M., Bereitschaftspotential enhancement with increased level of motivation. EEG j., 1969, 27, 73-75.

McCallum, W.C., Relationships between Bereitschaftspotential and contingent negative variation. In: Otto, 1978, pp. 124-130.

McCallum, W.C., Brain slow potential changes elicited by missing stimuli and by externally paced voluntary responses. Bio.Psy., 1980, 11, 7-19.

McCallum, W.C. & Curry, S.H., The form and distribution of auditory evoked potentials and CNVs when stimuli and responses are lateralized. In: Kornhuber & Deecke, 1980, pp. 767-775.

McCallum, W.C. & Curry, S.H., Late slow wave components of auditory evoked potentials: Their cognitive significance and interaction. EEG j., 1981, 51, 123-137.

McCallum, W.C. & Papakostopoulos, D., The effects of sustained motor activity on the contingent negative variation. EEG j., 1972, 33, 446 (abstract).

McCallum, W.C., Papakostopoulos, D., Gombi, R., Winter, A.L., Cooper, R. & Griffith, H.B., Event related slow potential changes in human brain stem. Nature, 1973, 242, 465-467.

McCallum, W.C. & Walter, W.G., The effects of attention and distraction on the contingent negative variation in normal and neurotic subjects. EEG j., 1968, 25, 319-329.

McCarthy, G., The relationship of P300 and reaction time: An additive factors study. Unpublished dissertation, University of Illinois, Champaign, 1980.

McCarthy, G. & Donchin, E., The effects of temporal and event uncertainty in determining the waveforms of the auditory event related potential (ERP). Psy.Phys., 1976, 13, 581-590.

McCarthy, G. & Donchin, E., Brain potentials associated with structural and functional visual matching. Neuropsychologia, 1978, 16, 571-585.

McCarthy, G. & Donchin, E., A metric for thought: A comparison of P300 latency and reaction time. Science, 1981, 211, 77-79.

McCarthy, G., Kutas, M. & Donchin, E., Detecting errors with P300 latency. Proceedings of the 18th Annual Meeting, Society for Psychophysiological Research. Psy.Phys., 1979, 16, 175.

McClelland, J.L., On the time relations of mental processes: An examination of systems of processes in cascade. Psychological Review,1979,86,287-330.

McGlone, J., Sex differences in human brain asymmetry: a critical survey. Behavioural and Brain Sciences, 1980, 3, 215-263.

McKee, G., Humphrey, B. & McAdam, D.W., Scaled lateralization of alpha activity during linguistic and musical tasks. Psy.Phys., 1973, 10, 441-443.

McLeod, S.S. & Peacock, L.J., Task-related EEG asymmetry: effects of age and ability. Psy.Phys., 1977, 14, 308-311.

Megela, A.L. & Teyler, T.J., Habituation and the human evoked potential. Journal of Comparative and Physiological Psychology, 1979, 93, 1154-1170.

Miller, J., Discrete vs. continuous stage models of human information processing: In search of partial output. Journal of Experimental Psychology, Human Perception and Performance, in press.

Mitzdorf, U. & Singer, W., Excitatory synaptic ensemble properties in the visual cortex of the macaque monkey: A current source density analysis of electrically evoked potentials. Journal of Comparative Neurology, 1979, 187, 71-84.

Molfese, D.L., Electrophysiological correlates of categorical speech perception in adults. Brain and Language, 1978, 5, 25-35.

Molfese, D.L., Left and right hemisperic involvement in speech perception: Electrophysiological correlates. Perception and Psychophysics, 1978, 23, 237-243.

Molfese, D.L., Cortical involvement in the semantic processing of coarticulated speech cues. Brain and Language, 1979, 7, 86-100.

Molfese, D.L., The phoneme and the engram: Electrophysiological evidence for the acoustic invariant in stop consonants. Brain and Lanaguage, 1980, 9, 372-376.

Molfese, D.L., Hemispheric specialization for temporal information: Implications for the processing of voicing cues during speech perception. Brain and Language, 1980, 11, 285-299.
Molfese, D.L. & Erwin, R.J., Intrahemispheric differentiation of vowels: Principal component analysis of auditory evoked responses to computer synthesized vowel sounds. Brain and Language, 1981, 13, 333-344.
Molfese, D.L., Freeman, R.B. Jr. & Palermo, D.S., The ontogeny of lateralization for speech and nonspeech stimuli. Brain and Language, 1975, 2, 356-368.
Molfese, D.L. & Hess, T.M., Speech perception in nursery school age children's sex and hemisphere differences. Journal of Experimental Child Psychology, 1978, 26, 71-84.
Molfese, D.L. & Molfese, V.J., Hemisphere and stimulus differences as reflected in the cortical responses of newborn infants to speech stimuli. Developmental Psychology, 1979, 15, 505-511.
Molfese, D.L. & Molfese, V.J., Cortical responses of preterm infants to phonetic and nonphonetic speech stimuli. Developmental Psychology, 1980, 16, 574-581.
Moore, W.H. Jr., Alpha hemisperic asymmetry of males and females on verbal and nonverbal tasks: some preliminary results. Cortex, 1979, 15, 321-326.
Moore, W.H. Jr. & Haynes, W.O., A study of alpha hemispheric asymmetry for verbal and nonverbal stimuli in males and females. Brain and Language, 1980, 9, 338-349.
Moray, N., Attention: Selective Processes in Vision and Hearing. New York: Academic Press, 1969.
Morrell, L.K. & Salamy, J.G., Hemispheric asymmetry of electrocortical response to speech stimuli. Science, 1971, 174, 164-166.
Mowery, G.L. & Bennett, A.E., Some technical notes on monopolar and bipolar recordings. EEG j., 1957, 9, 337.
Näätänen, R., Evoked potential, EEG, and slow potential correlates of selective attention. Acta Psychologica, 1970, 33, 178-192.
Näätänen, R., Selective attention and evoked potentials in humans - a critical review. Bio.Psy., 1975, 2, 237-307.
Näätänen, R., Orienting and evoked potentials. In: Kimmel et al., 1979, pp. 61-75.
Näätänen, R., The N2 component of the evoked potential: a scalp reflection of neuronal mismatch of orienting theory? In: J. Strelau, F. Farley & A. Gale (Eds.), Biological Foundations of Personality and Behaviour. Washington: Hemispere Press (in press a.).
Näätänen, R., Processing negativity: an evoked-potential reflection of selective attention. Psychological Bulletin, (in press b).
Näätänen, R. & Gaillard, A.W.K., The relationships between certain CNV and evoked potential measures within and between vertex, frontal and temporal derivations. Bio.Psy., 1974, 2, 95-112.
Näätänen, R., Gaillard, A.W.K. & Mäntysalo, S., S_2 probability and CNV. Activitas Nervosa Superior (Praha), 1977, 19, 142-144.
Näätänen, R., Gaillard, A.W.K. & Mäntysalo, S., Early selective-attention effect on evoked potential reinterpreted. Acta Psychologica, 1978, 42, 313-329.
Näätänen, R., Gaillard, A.W.K. & Mäntysalo, S., Brain-potential correlates of voluntary and involuntary attention. In: Kornhuber & Deecke, 1980b, 54, 343-348.
Näätänen, R., Gaillard, A.W.K. & Varey, C.A., Attention effects on Auditory EPs as a function of interstimulus interval. Bio.Psy., 1981, 13, 173-187.

Näätänen, R., Hukkanen, S. & Järvilehto, T., Magnitude of stimulus deviance and brain potentials. In: Kornhuber & Deecke, 1980a, 54, 337-342.
Näätänen, R. & Michie, P.T., Early selective attention effects on the evoked potential. A critical review and reinterpretation. Bio.Psy., 1979a, 8, 81-136.
Näätänen, R. & Michie, P.T., Different variants of endogenous negative brain potentials in performance situations: A review and classification. In: D. Lehmann & E. Callaway (Eds.), Human evoked potentials. New York: Plenum Press, 1979b, pp. 251-267.
Näätänen, R., Sams, M., Järvilehto, T. & Soininen, K., Probability of deviant stimulus and event-related brain potentials. In: R. Sinz & M. Rosenzweig (Eds.), Proceedings of Psychophysiology. Berlin, DDR: Gustav Fisher, in press.
Näätänen, R., Simpson, M. & Loveless, N.E., Stimulus deviance and evoked potentials. Bio.Psy., 1982, 14, in press.
Neisser, U., Cognitive Psychology. New York: Appleton Centry Crofts, 1967.
Neville, H.J., Electroencephalographic testing of cerebral specialization in normal and congenitally deaf children: A preliminary report. In: S.J. Segalowitz & F.A. Gruber (Eds.), Language Development and Neurological Theory. New York: Academic Press, 1977, pp. 121-131.
Neville, H.J., Electrographic correlates of lateral asymmetry in the processing of verbal and nonverbal auditory stimuli. Journal of Psycholinguistic Research, 1974, 3, 151-163.
Neville, H.J., Event-related potentials in neuropsychological studies of language. Brain and Language, 1980, 11, 300-318.
Newell, K.M., Some issues on action plans. In: G.E. Stelmach (Ed.), Information processing in motor control and learning. New York: Academic Press, 1978, pp. 41-54.
Nicholson, C. & Freeman, J.A., Theory of current source-density analysis and determination of conductivity tensor for anuran cerebellum. Journal of Neurophysiology, 1975, 38, 356-368.
Nieme, P. & Näätänen, R., Foreperiod and simple reaction time. Psychological Bulletin, 1981, 89, 133-162.
Novick, B., Kurtzberg, D. & Vaughan, H.G. Jr., An electrophysiological indication of defective information storage in childhood autism. Psychiatry Research, 1979, 1, 101-108.
Novick, B., Vaughan, H.G. Jr., Kurtzberg, D. & Simson, R., An electrophysiological indication of auditory processing defects in autism. Psychiatry Research, 1980, 3, 107-114.
Obrist, P.A., The cardiovascular-behavioral interaction -- as it appears today. Psy.Phys., 1976, 13, 95-107.
O'Connor, K.P., The intentional paradigm and cognitive psychophysiology. Psy.Phys., 1981, 18, 121-128.
O'Connor, K.P. & Shaw, J.C., Field dependence, laterality and the EEG. Bio.Psy., 1978, 6, 93-109.
O'Gorman, J.G., The orienting reflex: Novelty or significance detector? Psy.Phys., 1979, 16, 253-262.
Öhman, A., Differentiation of conditioned and orienting response components in electrodermal conditioning. Psy.Phys., 1971, 8, 7-21.
Öhman, A., The orienting response, attention, and learning: an information-processing perspective. In: Kimmel et al., 1979, pp. 443-472.
Öhman, A., Bjorkstrand, P. & Ellstrom, P., Effect of explicit trial-by-trial information about shock probability in long interstimulus GSR conditioning. Journal of Experimental Psychology, 1973, 98, 145-151.

Öhman, A. & Lader, M.H., Short-term changes of the human auditory evoked potentials during repetitive stimulation. In: J.E. Desmedt (Ed.), Auditory Evoked Potential in Man: Psychopharmacology Correlates of Evoked Potentials. Basel: Karger, 1977.

O'Keefe, J. & Nadel, L., The Hippocampus as a Cognitive Map. London: Oxford University Press, 1978.

Okita, T., Slow negative shifts of the human event-related potential associated with selective information processing. Bio.Psy., 1981, 12, 63-75.

Oldfield, R.C., The assessment and analysis of handedness: The Edinburgh Inventory. Neuropsychologia, 1970, 9, 97-113.

Oltman, P.K., Semple, C. & Goldstein, L., Cognitive style and interhemispheric differentiation in the EEG. Neuropsychologia, 1979, 17, 699-702.

Olton, D.S., Becker, J.T. & Handelman, G.E., Hippocampus, space and memory. The Behavioral and Brain Sciences, 1979, 2, 313-366.

Orgogozo, J.M., Larsen, B., Roland, P.E. & Lassen, N.A., Activation de l'aire motrice supplementaire au cours des mouvements volontaires chez l'homme. Revue Neurologique, 1979, 135, 705-717.

Orme-Johnson, D., EEG coherence during Transcendental Consciousness (abstract). EEG j., 1977, 43, 581-582.

Ornstein, R.E. & Galin, D., Physiological studies of consciousness. Institute for Cultural Research, Monograph 11, 1973.

Ornstein, R., Johnstone, J., Herron, J. & Swencionis, C., Differential right hemisphere engagement in visuospatial tasks. Neuropsychologia, 1980, 18, 49-64.

Orr, W.C. & Naitoh, P., The coherence spectrum: an extension of correlation analysis with application to chronobiology. Internation Journal of Chronobiology, 1976, 3, 171-192.

Osgood, C., Suci, G. & Tannenbaum, P., The Measurement of Meaning. Urbana: University of Illinois Press, 1957.

Otto, D., Multidisciplinary Perspectives in Event-related Brain Potential Research. Washington, D.C.: U.S. Government Printing Office, EPA-600/9-77-043, 1978.

Otto, D.A., Benignus, V.A., Ryan, L.G. & Leifer, L.J., Slow potential components of stimulus, response and preparatory processes in man. In: Desmedt, Vol. 1, 1977, pp. 211-230.

Otto, D.A., Houck, K., Finger, H. & Hart, S., Event-related slow potentials in aphasic, dyslexic and normal children during pictorial and letter-matching. In: W.C. McCallum & J.R. Knott (Eds.), The Responsive Brain. Bristol: Wright, 1976, pp. 172-177.

Otto, D.A. & Leifer, L.J., The effect of modifying response and performance feedback parameters on the CNV in humans. EEG j., 1973, Suppl. 33, 29-37.

Otto, D.A. & Reiter, L. (Eds.), Environmental neuro-toxicology. In: Otto, 1978, pp. 407-498.

Palermo, D.S., Psychology of Language. Glenview: Scott, Foresmand and Company, 1978.

Paul, D.D. & Sutton, S., Evoked potential correlates of response criterion in auditory signal detection. Science, 1972, 177, 362-364.

Papakostopoulos, D., The present state of brain macropotentials in motor control research - A summary of issues. In: Otto, 1978a, pp. 77-81.

Papakostopoulos, D., Electrical activity of the brain associated with skilled performance. In: Otto, 1978b, pp. 134-137.

Papakostopoulos, D., A no-stimulus, no-response event-related potential of the human cortex. EEG j., 1980, 48, 622-638.

Papakostopoulos, D., Cooper, R. & Crow, H.J., Inhibition of cortical evoked potentials and sensation by self-initiated movement in man. Nature, 1975, 258, 321-324.

Papanicolaou, A.C., Cerebral excitation profiles in language processing: The photic probe paradigm. Brain and Language, 1980, 9, 269-280.

Papanicolaou, A.C. & Molfese, D.L., Neuroelectrical correlates of hemisphere and handedness factors in cognitive tasks. Brain and Language, 1978, 5, 236-248.

Parasuraman, R., Auditory evoked potentials and divided attention. Psy.Phys., 1978, 15, 460-465.

Parasuraman, R. & Beatty, J., Brain events underlying detection and recognition of weak sensory signals. Science, 1980, 210, 80-83.

Parasuraman, R., Richer, F. & Beatty, J., Detection and recognition: Concurrent processes in perception. Perception and Psychophysics, 1982, 31, 1-12.

Pass, H.L., Klorman, R., Salzman, L.F., Klein, R.H. & Kaskey, G.B., The late positive component of the evoked response in acute schizophrenics during a test of sustained attention. Biological Psychiatry, 1980, 15, 9-20.

Pavlov, I.P., Conditioned reflexes. An investigation of the physiological activity of the cerebral cortex. Oxford: Oxford University press, 1927. Reprinted, New York: Dover, 1960.

Pendergrass, V.E. & Kimmel, H.D., UCR diminution in temporal conditioning and habituation. Journal of Experimental Psychology, 1968, 77, 1-6.

Perdok, J. & Gaillard, A.W.K., The terminal CNV and stimulus discriminability in motor and sensory tasks. Bio.Psy., 1979, 8, 213-223.

Peters, J.F., Knott, J.R., Miller, L.H., Van Veen, W.J. & Cohen, S.I., Response variables and magnitude of the contingent negative variation. EEG j., 1970, 29, 608-611.

Pfefferbaum, A., Horvath, T.B., Roth, W.T. & Kopell, B.S., Event-related potential changes in chronic alcoholics. EEG j., 1979, 47, 637-647.

Pfefferbaum, A., Ford, J.M., Roth, W.T. & Kopell, B.S., Age differences in P3-reaction time associations. EEG j., 1980a, 49, 257-265.

Pfefferbaum, A., Ford, J.M., Roth, W.T. & Kopell, B.S., Age-related changes in auditory event-related potentials. EEG j., 1980b, 49, 266-276.

Pfurtscheller, G., Maresch, H. & Schuy, S., Inter- and intrahemispheric differences in the peak frequency of rhythmic activity within the alpha band. EEG j., 1977, 42, 77-83.

Phillips, C.G. & Porter, R., Corticospinal neurones. Their role in movement. London: Academic Press, 1977.

Pick, A.D., Frankel, D.G. & Hess, V.L., Children's attention: The development of selectivity. In: E.M. Hetherington (Ed.), Review of Child Development Research. Chicago: The University of Chicago Press, 1975, Volume 5, pp. 325-383.

Picton, T.W., Campbell, K.B., Baribeau-Braun, J. & Proulx, G.B., The neurophysiology of human attention: A tutorial review. In: J. Requin (Ed.), Attention and Performance, VII. Hillsdale, N.J.: Erlbaum, 1978, pp. 429-467.

Picton, T.W. & Hillyard, S.A., Human auditory evoked potentials. II: Effects of attention. EEG j., 1974, 36, 191-199.

Picton, T.W., Hillyard, S.A., Krausz, H.I. & Galambos, R., Human auditory evoked potentials. I: Evaluation of components. EEG j., 1974, 36, 179-190.

Picton, T.W. & Stuss, D.T., The component structure of the human event-related potentials. In: Kornhuber & Deecke, 1980, pp. 17-49.

Picton, T.W., Woods, D.L., Baribeaum-Braun, J. & Healey, T.M.G., Evoked potential audiometry. Journal of Otolaryngology, 1977, 6, 90-119.

Picton, T.W., Woods, D.L. & Proulx, G.B., Human auditory sustained potentials. I. The nature of the response. EEG j., 1978a, 45, 186-197.

Picton, T.W., Woods, D.L. & Proulx, G.B., Human auditory sustained potentials. II. Stimulus relationships. EEG j., 1978b, 45, 198-210.

Pieper, C.F., Goldring, S., Jenny, A.B. & McMahon, J.P., Comparative study of cerebral cortical potentials associated with voluntary movements in monkey and man. EEG j., 1980, 48, 266-292.
Plooij-van Gorsel, P.C., Persoonlijkheid en arousal. Amsterdam: Swets & Zeitlinger, 1980.
Plooij-van Gorsel, P.C. & Janssen, R.H.C., Contingent negative variation (CNV) and extraversion in a psychiatric population. In: C. Barber (Ed.), Evoked Potential. Proceedings of an International Evoked Potentials Symposium. M.T.P. Press Ltd., 1980.
Poon, L.W., Thompson, L.W., Williams, R.B. & Marsh, G., Changes of anterio-posterior distribution of CNV and late positive component as a function of information processing demands. Psy.Phys., 1974, 11, 660-673.
Posner, M.I., Psychobiology of attention. In: M.S. Gazzaniga & C. Blakemore (Eds.), Handbook of Psychobiology. New York: Academic Press, 1975, pp. 441-480.
Posner, M.I., Chronometric explorations of mind. Hillsdale, N.J.: Erlbaum, 1978.
Posner, M.I., Klein, R., Summers, J. & Buggies, S., On the selection of signals. Memory and Cognition, 1973, 1, 2-12.
Posner, M.I. & Snyder, C.R.R., Attention and cognitive control. In: R.L. Solso (Ed.), Information Processing and Cognition. Hillsdale, N.J.: Erlbaum, 1975, pp. 55-85.
Porjesz, B., Begleiter, H. & Samuelly, I., Cognitive deficits in chronic alcoholics and elderly subjects assessed by evoked brain potentials. Acta Psychiatrica Scandinavia, Suppl. 286, 1980, 62, 15-30.
Pribram, K.H. & McGuinness, D., Arousal, activation, and effort in the control of attention. Psychological Review, 1975, 82, 116-149.
Prichep, L.S., Sutton, S. & Hakerem, G., Evoked potentials in hyperkinetic and normal children under certainty and uncertainty: A placebo and methylphenidate study. Psy.Phys., 1976, 13, 419-428.
Pritchard, W.S., Psychophysiology of P300. Psychological Bulletin, 1981, 89, 506-540.
Prokasy, W.F. & Kumpfer, K.L., Classical conditioning. In: W.F. Prokasy & D.C. Raskin (Eds.), Electrodermal Activity in Psychological Research. New York: Academic Press, 1973, pp. 157-202.
Rabbitt, P.M., Signal discriminability, S-R compatibility and choice reaction time. Psychonomic Science, 1967, 7, 419-420.
Ragot, R. & Renault, B., P300 as a function of S-R compatibility and motor programming. Bio.Psy., 1981, 13, 289-294.
Raskin, D.C., Kotses, H. & Bever, J., Autonomic indicators of orienting and defensive reflexes. Journal of Experimental Psychology, 1969, 79, 69-76.
Rasmussen, C.T., Allen, R. & Tarte, R.D., Hemispheric asymmetries in cortical evoked potential as a function of arithmetic computations. Bulletin of the Psychonomic Society, 1977, 10, 419-421.
Ravizza, R. & Belmore, S., Auditory forebrain. In: R.B. Masterton (Ed.), Handbook of Behavioural Neurobiology, Vol. 1. New York: Plenum Press, 1978.
Ray, R.L., The effect of stimulus intensity and intertrial interval on long-term retention of the OR. In: Kimmel et al., 1979, pp. 373-379.
Rebert, C.S., Slow potential changes in the monkey's brain during reaction time foreperiod. In: W.C. McCallum and J.R. Knott (Eds.), The responsive brain. Bristol: Wright, 1976, pp. 191-194.
Rebert, C.S., Functional cerebral asymmetry and performance I: Reaction time to words and dot patterns as a function of electroencephalographic alpha asymmetry. Behavioral Neuropsychiatry, 1977a, 8, 90-98.

Rebert, C.S., Functional cerebral asymmetry and performance II: Individual differences in reaction time to word and pattern stimuli triggered by assymetric alpha bursts. Behavioral Neuropsychiatry, 1977b, 8, 99-103.

Rebert, C.S. & Low, D.W., Differential hemispheric activation during complex visuomotor performance. EEG j., 1978, 44, 724-734.

Rebert, C.S. & Lowe, R.C., Task-related hemispheric asymmetry of contingent negative variation. In: Kornhuber & Deecke, 1980, pp. 776-781.

Rebert, C.S. & Mahoney, R.A., Functional cerebral asymmetry and performance III: Reaction time as a function of task, hand, sex and EEG asymmetry. Psy.Phys., 1978, 15, 9-16.

Rebert, C.S., McAdam, D.W., Knott, J.R. & Irvin, D.A., Slow potential change in human brain related to level of motivation. Journal of Comparative and Physiological Psychology, 1967, 63, 20-23.

Rebert, C.S. & Tecce, J.J., A summary of CNV and reaction time. In: W.C. McCallum & J.R. Knott (Eds.), Event-related slow potentials of the brain: Their relations to behavior. EEG j., 1973, Suppl. 33, 173-178.

Rémond, A., Integrated and topographical analysis of the EEG. EEG j., 1961, 20, 64-67.

Renault, B. & Lesèvre, N., Variations des potentiels évoquès par des patterns (P.E.P.) chez l'homme dans une situation de temps de reaction: influence de la vigilance et de la motricité. Revue d'EEG et de Neurophysiologie Clinique, 1975, 5, 360-366.

Renault, B. & Lesèvre, N., Topographical study of the emitted potential obtained after the omission of an expected visual stimulus. In: Otto, 1978, pp. 202-208.

Renault, B. & Lesèvre, N., A trial-by-trial study of the visual omission response in reaction time situations. In: D. Lehmann & E. Callaway (Eds.), Human evoked potentials. New York: Plenum Press, 1979, pp. 317-329.

Renault, B., Ragot, R. & Lesèvre, N., Correct and incorrect responses in a choice reaction time task and the endogenous components of the evoked potential. In. Kornhuber & Deecke, 1980b, pp. 647-654.

Renault, B., Ragot, R., Furet, J. & Lesèvre, N., Etude des relations entre le potentiel émis et les mécanismes de préparation perceptivo-motrice. In: J. Requin (Ed.), Anticipation et Comportement. Paris: édition du CNRS, 1980a, 168-182.

Renault, B., Ragot, R., Lesévre, N. & Remond, A., Onset and offset of brain events as indices of mental chronometry. Science, 1982, 215, 1413-1415.

Rhodes, L.E., Obitz, F.W. & Creel, D., Effect of alcohol and task on hemispheric asymmetry of visually evoked potentials in man. EEG j., 1975, 38, 561-568.

Ritter, W., Rotkin, L. & Vaughan, H.G. Jr., The modality specificity of the slow negative wave. Psy.Phys., 1980, 17, 222-227.

Ritter, W., Simson, R. & Vaughan, H.G. Jr., Association cortex potentials and reaction time in auditory discrimination. EEG j., 1972, 33, 547-555.

Ritter, W., Simson, R. & Vaughan, H.G. Jr., Event-related potential correlates of two stages of information processing in physical and semantic discrimination tasks. Psy.Phys., in press.

Ritter, W., Simson, R., Vaughan, H.G. Jr. & Friedman, D., A brain event related to the making of a sensory discrimination. Science, 1979, 203, 1358-1361.

Ritter, W., Simson, R., Vaughan, H.G. Jr. & Macht, M., Manipulation of event-related potential manifestations of information processing stages. Science, in press.

Ritter, W. & Vaughan, H.G. Jr., Averaged evoked responses in vigilance and discrimination: A reassessment. Science, 1969, 164, 326-328.

Ritter, W., Vaughan, H.G. Jr. & Costa, L.D., Orienting and habituation to auditory stimuli: a study of short term changes in averaged evoked responses. EEG j., 1968, 25, 550-556.

Rix, R. & Korth, M., Gleichspannungsanteile in visuell evozierten potentialen (VER, VEP) des menschlichen elektroenzephalogramms (EEG). Ber 74. Zusammenkunft des München, J.V. Bergmann, 1977.

Robbins, K.I. & McAdam, D.W., Interhemispheric alpha asymmetry and imagery mode. Brain and language, 1974, 1, 189-193.

Rockland, K.S. & Pandya, D.N., Laminar origins and terminations of cortical connections of the occipital lobe in the rhesus monkey. Brain Research, 1979, 179, 3-20.

Rohrbaugh, J.W., Syndulko, K. & Lindsley, D.B., Brain wave components of the contingent negative variation in humans. Science, 1976, 191, 1055-1057.

Rohrbaugh, J.W., Syndulko, K. & Lindsley, D.B., Cortical slow negative waves following nonpaired stimuli: Effects of task factors. EEG j., 1978, 45, 551-567.

Rohrbaugh, J.W., Syndulko, K. & Lindsley, D.B., Cortical slow negative waves following non-paired stimuli: Effects of modality, intensity and rate of stimulation. EEG j., 1979, 46, 416-427.

Rohrbaugh, J.W., Syndulko, K., Sanquist, T.F. & Lindsley, D.B., Synthesis of the contingent negative variation brain potential from non-contingent stimulus and motor elements. Science, 1980, 208, 1165-1168.

Rohrbaugh, J.W., Newlin, D.B., Varner, J.L. & Ellingson, R.J., Bilateral distribution of the O wave. Paper presented at Evoked Potential International Congress VI, Lake Forest, Illinois, 1981.

Roland, P.E., Somatotopical tuning of postcentral gyrus during focal attention in man: A regional cerebral blood flow study. Journal of Neurophysiology, 1981, 46, 744-754.

Roland, P.E., Larsen, B., Lassen, N.A. & Skinhøj, E., Supplementary motor area and other cortical areas in organization of voluntary movements in man. Journal of Neurophysiology, 1980, 43, 118-136.

Roland, P.E., Skinhøj, E. & Lassen, N.A., Focal activations of human cerebral cortex during auditory discrimination. Journal of Neurophysiology, 1981, 45, 1139-1151.

Rosch, E., Cognitive representation of semantic categories. Journal of Experimental Psychology, 1975, 104, 192-233.

Rosch, E., Mervis, C., Gray, W., Johnson, D. & Boyes-Braem, P., Basic objects in natural categories. Cognitive Psychology, 1976, 8, 382-439.

Rösler, F., Evozierte Hirnrindenpotentiale und Informationsverarbeitungsprozesse - "Ubersichtsreferat. In: W.H. Tack (Ed.), Bericht über den 30. Kongress der Deutschen Gesellschaft für Psychologie. Göttingen: Hogrefe, 1977, pp. 339-350.

Rösler, F., Cortical potential correlates of selective attention in multidimensional scaling. Bio.Psy., 1978, 7, 223-238.

Rösler, F., Event-related positivity and cognitive processes. In: M. Koukkou, D. Lehmann & J. Angst (Eds.), Functional states of the brain: Their determinants. Amsterdam: Elsevier, 1980, pp. 203-222.

Rösler, F., Statistische Verarbeitung von Biosignalen: Die Quantifizierung -irnelektrischer Signale. In: U. Baumann, H. Berbalk & G. Seidenstücker (Eds.), Klinische Psychologie: Trends in Forschung und Praxis: 3. Bern: Huber, 1980a.

Rösler, F., Event-related brain potentials in a stimulus-discrimination learning paradigm. Psy.Phys., 1981, 18, 447-455.

Rösler, F. & Manzey, D., Principal components and VARIMAX-rotated components in event-related potential research: Some remarks on their interpretation. Bio.Psy., 1981, 13, 3-26.

Roth, W.T., Auditory evoked responses to unpredictable stimuli. Psy.Phys., 1973, 10, 125-137.

Roth, W.T., Late event-related potentials and psychopathology. Schizophrenia Bulletin, 1977, 3, 105-120.

Roth, W.T., How many late positive waves are there? In: Otto, 1978, pp. 170-172.

Roth, W.T., Blowers, G.H., Doyle, C.M. & Kopell, B.S., Auditory stimulus intensity effects on components of the late positive complex. Submitted for publication, 1982.

Roth, W.T. & Cannon, E.H., Some features of the auditory evoked response in schizophrenics. Archives of General Psychiatry, 1972, 27, 466-471.

Roth, W.T., Doyle, C.M., Pfefferbaum, A. & Kopell, B.S., Effects of stimulus intensity on P300. In: Kornhuber & Deecke, 1980, pp. 296-300.

Roth, W.T., Ford, J.M. & Kopell, B.S., Long latency evoked potentials and reaction time. Psy.Phys., 1978, 15, 17-23.

Roth, W.T., Ford, J.M., Krainz, P.L. & Kopell, B.S., Auditory evoked potentials, skin conductance response, eye movement, and reaction time in an orienting response paradigm. In: Otto, 1978, pp. 209-214.

Roth, W.T., Ford, J., Pfefferbaum, A., Horvarth, T.B., Doyle, C.M. & Kopell, B.S., Event-related potential research in Psychiatry. In: D. Lehmann & E. Callaway (Eds.), Human Evoked Potentials: Applications and Problems. New York: Plenum Press, 1979, 331-346.

Roth, W.T., Ford, J.M., Lewis, S.J. & Kopell, B.S., Effects of stimulus probability and task-relevance on event-related potentials. Psy.Phys., 1976, 13, 311-317.

Roth, W.T., Horvath, T.B., Pfefferbaum, A. & Kopell, B.S., Event-related potentials in schizophrenics. EEG j., 1980a, 48, 127-139.

Roth, W.T. & Kopell, B.S., P300 - an orienting reaction in the human auditory evoked response. Perceptual and Motor Skills, 1973, 36, 219-225.

Roth, W.T., Kopell, B.S., Tinklenberg, J.R., Darley, C.F., Sikora, R. & Vesecky, T.B., The contingent negative variation during a memory retrieval task. EEG j., 1975, 38, 171-174.

Roth, W.T., Pfefferbaum, A., Horvath, T.B., Berger, P.A. & Kopell, B.S., P3 reduction in auditory evoked potentials of schizophrenics. EEG j., 1980b, 49, 497-505.

Rubens, A.B., Mahowald, M.W. & Hutton, J.T., Asymmetry of the lateral (sylvian) fissures in man. Neurology, 1976, 26, 620-624.

Ruchkin, D.S. & Sutton, S., Equivocation and P300 amplitude. In: Otto, 1978b, pp. 175-177.

Ruchkin, D.S. & Sutton, S., Emitted P300 potentials and temporal uncertainty. EEG j., 1978a, 45, 268-277.

Ruchkin, D.S. & Sutton, S., CNV and P300 relationships for emitted and for evoked cerebral potentials. In: Desmedt, Vol. 6, 1979, pp. 119-131.

Ruchkin, D.S., Sutton, S., Kietzman, M.L. & Silver, K., Slow wave and P300 in signal detection. EEG j., 1980a, 50, 35-47.

Ruchkin, D.S., Sutton, S., Munson, R., Silver, K. & Macar, F., P300 and feedback provided by absence of the stimulus. Psy.Phys., 1981, 18, 271-282.

Ruchkin, D.S., Sutton, S. & Stega, M., Emitted P300 and slow wave event-related potentials in guessing and detection tasks. EEG j., 1980b, 49, 1-14.

Ruchkin, D.S., Sutton, S. & Tueting, P., Emitted and evoked P300 potentials and variation in stimulus probability. Psy.Phys., 1975, 12, 591-595.

Rugg, M.D., Electrophysiological studies of divided visual field stimulation in man. In: J.G. Beaumont (Ed.), Divided visual field Studies of Cerebral organisation. London: Academic Press, 1982.

Rugg, M.D. & Beaumont, J.G., Late positive component correlates of verbal and visuospatial processing. Bio.Psy., 1979, 9, 1-11.
Rugg, M.D. & Beaumont, J.G., Visual evoked responses to visual-spatial and verbal stimuli: evidence of differences in cerebral processing. Physiology Psychology, 1978a, 6, 501-504.
Rugg, M.D. & Beaumont, J.G., Interhemispheric asymmetries in the visual evoked response: effects of stimulus lateralisation and task. Biological Psychology, 1978b, 6, 283-292.
Rugg, M.D. & Venables, P.H., EEG correlates of the acquisition of high- and low-imagery words. Neuroscience Letters, 1980, 16, 67-70.
Sams, M., Alho, K. & Näätänen, R., Sequential effects on brain-potential responses to deviant stimuli. In preparation.
Sanders, A.F., The selective process in the functional visual field. Assen: Van Gorcum, 1963.
Sanders, A.F., Stage analysis of reaction process. In: G.E. Stelmach & J. Requin (Eds.), Tutorials in motor behavior. Amsterdam: North-Holland, 1980, pp. 331-354.
Sanquist, T.F., Beatty, J.T. & Lindsley, D.B., Slow potential shifts of human brain during forewarned reaction. EEG j., 1981, 51, 639-649.
Sanquist, T.F., Rohrbaugh, J.W., Syndulko, K. & Lindsley, D.B., Electrocortical signs of levels of processing: perceptual analysis and recognition memory. Psy.Phys., 1980, 17, 568-576.
Sasanuma, S., Acquired dyslexia in Japanese: clinical features and underlying mechanisms. In: M. Coltheart, K. Patterson & J.C. Marshall (Eds.), Deep Dyslexia. London: Routledge and Kegan Paul, 1980.
Satterfield, J.H. & Braley, B.W., Evoked potentials and brain maturation in hyperactive and normal children. EEG j., 1977, 43, 43-51.
Schandry, R. & Hoefling, S., Interstimulus interval length and habituation of the P300. In: Kimmel et al., 1979, pp. 129-134.
Schaub, D., Event-related potential (ERP) and behavioral correlates of attention in reading disabled boys. Ph.D. dissertation, SUNY, Stony Brook, 1981, and manuscript in preparation.
Schell, A.M. & Grings, W.W., Judgments of UCS intensity and diminution of the unconditioned GSR. Psy.Phys., 1971, 8, 427-432.
Schmidt, R.A., A schema theory of discrete motor skill learning. Psychological Review, 1975, 82, 225-260.
Schneider, W. & Shiffrin, R.M., Controlled and automatic human information processing: I. Detection, search and attention. Psychological Review, 1977, 84, 1-66.
Schulman-Galambos, C. & Calambos, R., Cortical responses from adults and infants to complex visual stimuli. EEG j., 1978, 45, 425-435.
Schwartz, G.E., Davidson, R.J. & Maer, F., Right hemisphere lateralization for emotion in the human brain: Interactions with cognition. Science, 1975, 190, 286-288.
Schwent, V.L. & Hillyard, S.A., Evoked potential correlates of selective attention with multi-channel auditory inputs. EEG j., 1975, 38, 131-138.
Schwent, V., Hillyard, S.A. & Galambos, R., Selective attention and the auditory vertex potential. I. Effects of stimulus delivery rate. EEG j., 1976a, 40, 604-614.
Schwent, V., Hillyard, S.A. & Galambos, R., Selective attention and the auditory vertex potential. II. Effects of signal intensity and masking noise. EEG j., 1976b, 40, 615-622.
Shallice, T., Dual functions of consciousness. Psychological Review, 1972, 79, 383-393.

Shallice, T., The dominant action system: An information-processing approach to consciousness. In: K.S. Pope & J.L. Singer (Eds.), The Stream of Consciousness. New York: Plenum Press, 1978, pp. 117-157.

Shannon, C.E. & Weaver, W., The Mathematical Theory of Communication. Urbana: University of Illinois, 1949.

Shapiro, B.E., Grossman, M. & Gardner, H., Selective musical selective processing deficits in brain damaged populations. Neuropsychologia, 1981, 19, 161-169.

Sharrard, G., Further conclusions regarding the influence of word meaning on the cortical averaged evoked response in audiology. Audiology, 1973, 12, 103-115.

Shaw, J.C., An introduction to the coherence function and its use in signal analysis. Unpublished manuscript. MRC Clinical Psychiatry Unit, Graylingwell Hospital, Chichester, 1980.

Shaw, J.C. & Ongley, C., The measurement of synchronisation. In: H. Petsche & M.A.B. Brazier (Eds.), Synchronisation of EEG Activity in Epilepsies. New York: Springer, 1972, pp. 204-216.

Shaw, J.C., Brooks, S., Colter, N. & O'Connor, K.P., A comparison of schizophrenic and neurotic patients using EEG power and coherence spectra. In: J. Gruzelier & P. Flor-Hendry (Eds.), Hemisphere Asymmetries of Function and Psychopathology. Amsterdam: Elsevier/North Holland, 1979, pp. 257-284.

Shelburne, S.A., Visual evoked responses to word and nonsense syllable stimuli. EEG j., 1972, 32, 17-25.

Shelburne, S.A., Visual evoked responses to language stimuli in normal children. EEG j., 1973, 34, 135-143.

Shelburne, S.A. Jr., Visual evoked potentials to language stimuli in children with reading disabilities. In: Otto, 1978, pp. 271-274.

Shepherd, R. & Gale, A., EEG aysmmetries in sustained attention (abstract). Bulletin of the British Psychological Society, 1980, 33, 22.

Shibasaki, H., Barrett, G., Halliday, E. & Halliday, A.M., Components of the movement-related cortical potential and their scalp topography. EEG j., 1980a, 49, 213-226.

Shibasaki, H., Barrett, G., Halliday, E. & Halliday, A.M., Cortical potentials following voluntary and passive finger movements. EEG j., 1980b, 50, 201-213.

Shiffrin, R.M. & Schneider, W., Controlled and automatic human information processing: II. Perceptual learning, automatic attending, and a general theory. Psychological Review, 1977, 84, 127-190.

Shucard, D.W., Cummins, K.R., Thomas, D.G. & Shucard, J.L., Evoked potentials to auditory probes as indices of cerebral specialisation of function - replication and extension. EEG j., 1981, 52, 389-393.

Shucard, D.W., Shucard, J.L. & Thomas, D.G., Auditory evoked potentials as probes of hemispheric differences in cognitive processing. Science, 1977, 197, 1295-1297.

Shwartz, S.P., Pomerantz, J.R. & Eseth, H.E., State and process limitations in information processing: An additive factors analysis. Journal of Experimental Psychology: Human Perception and Performance, 1977, 3, 402-410.

Siddle, D.A.T., The orienting response and stimulus significance: Some comments. Bio.Psy., 1979, 8, 303-309.

Siddle, D.A.T. & Heron, P.A., Stimulus omission and recovery of the electrodermal and digital vasoconstrictive components of the orienting response. Bio.Psy., 1975, 3, 277-293.

Siddle, D.A.T. & Heron, P.A., Effects of length of training and amount of tone frequency change on amplitude of autonomic components of the orienting response. Psy.Phys., 1976, 13, 281-287.

Siddle, D.A.T., O'Gorman, J.G. & Wood, L., Effects of electrodermal lability and stimulus significance on electrodermal response amplitude to stimulus change. Psy.Phys., 1979a, 16, 520-527.
Siddle, D.A.T., Kyriacou, C., Heron, P.A. & Matthews, W.A., Effects of changes in verbal stimuli on the skin conductance response component of the orienting response. Psy.Phys., 1979b, 16, 34-40.
Siddle, D.A.T., Kuiack, M. & Kroese, B.S., In: A. Gale & J. Edwards (Eds.), Physiological correlates of human behavior. London: Academic Press, in press.
Simons, R.F. & Lang, P.J., Psychophysical judgment: Electro-cortical and heart rate correlates of accuracy and uncertainty. Bio.Psy., 1976, 4, 51-64.
Simons, R.F., Ohman, A. & Lang, P.J., Anticipation and response set: Cortical, cardiac, and electrodermal correlates. Psy.Phys., 1979, 16, 222-233.
Simson, R., Vaughan, H.G. & Ritter, W., The scalp topography of potentials associated with missing visual or auditory stimuli. EEG j., 1976, 40, 33-42.
Simson, R., Vaughan, H.G. Jr. & Ritter, W., The scalp topography of potentials in auditory and visual discrimination tasks. EEG j., 1977a, 42, 528-535.
Simson, R., Vaughan, H.G. Jr. & Ritter, W., The scalp topography of potentials in auditory and visual go/nogo tasks. EEG j., 1977b, 43, 864-875.
Skerchock, J.A. & Cohen, J., Alcoholism, organicity and event-related potentials. In: J. Cohen, R. Karrer & P. Tueting (Eds.), Annals of the New York Academy of Sciences, in press.
Skinner, J.E. & Yingling, C.D., Central gating mechanisms that regulate event-related potentials and behavior: A neural model for attention. In: Desmedt, Vol. 1, 1977, pp. 30-69.
Smith, E.E., Choice reaction time: An analysis of the major theoretical positions. Psychological Bulletin, 1968, 69, 77-110.
Smith, T.S., Neilson, B. & Thistle, A.B., Question of asymmetries in auditory evoked potentials to speech stimuli. Journal of the Acoustical Society of America, 1975, 58, 1, 557.
Smyk, K. & Darwaj, B., Dominance of one cerebral hemisphere in the electroencephalographic record. Acta Physiologica Polonica, 1972, 23, 359-367.
Snyder, E. & Hillyard, S.A., Long-latency evoked potentials to irrelevant, deviant stimuli. Behavioural Biology, 1976, 16, 319-331.
Snyder, E., Hillyard, S.A. & Galambos, R., Similarities and differences among the P3 waves to detected signals in three modalities. Psy.Phys., 1980, 17, 112-122.
Sokolov, E.N., Neural models and the orienting reflex. In: M.A. Brazier (Ed.), The central nervous system and behaviour. New York: Macy, 1960.
Sokolov, E.N., Perception and the conditioned reflex. New York: Pergamon Press, 1963.
Sokolov, E.N., Higher nervous functions: the orienting reflex. Annual Review of Physiology, 1963a, 25, 545-580.
Sokolov, E.N., Orienting reflex as information regulator. In: A. Leontiev, A. Luria & A. Smirnow (Eds.), Psyhcological Research in the UUSR, Moscow: Progress Publishers, 1966, pp. 334-360.
Sokolov, E.N., The modelling properties of the nervous system. In: M. Cole & I. Maltzman (Eds.), A handbook of contemporary Soviet Psychology. New York: Basic Books, 1969, pp. 671-704.
Sokolov, E.N., The neuronal mechanisms of the orienting reflex. In: E.N. Sokolov & O.S. Vinogradova (Eds.), Neuronal mechanisms of the orienting reflex. Hillsdale, N.J.: Erlbaum, 1975, pp. 217-235.

Sokolov, E.N. & Vinogradova, O.S. (Eds.), Neural Mechanisms of the orienting Reflex. Hillsdale, N.J.: Erlbaum, 1975.

Soltaire, G.B. & Lamarche, J.B., Alzheimer's disease and senile dementia as seen in mongoloids: neuropathological observations. American Journal of Mental Deficiency, 1966, 70, 840-849.

Sperry, R.W., Neurology and the mind-brain problems. American Scientist, 1952, 40, 291-312.

Spinks, J.A. & Siddle, D.A.T., Effects of stimulus information and stimulus duration on amplitude and habituation of the electrodermal orienting response. Bio.Psy., 1976, 4, 29-39.

Spydell, J.D., Ford, M.R. & Sheer, D.E., Task dependent cerebral lateralization of the 40 hertz EEG rhythm. Psy.Phys., 1979, 16, 347-350.

Squires, K.C., Chippendale, T.J., Wrege, K.S., Goodin, D.S. & Starr, A., Electrophysiological assessment of mental function in aging and dementia. In: L.W. Poon (Ed.), Aging in the 1980s. Washington, D.C.: APA, 1980, pp. 125-134.

Squires, K.C., Donchin, E., Herning, R.I. & McCarthy, G., On the influence of task relevance and stimulus probability on event-related potential components. EEG j., 1977, 42, 1-14.

Squires, K.C., Hillyard, S.A. & Lindsay, P.H., Cortical potential evoked by confirming and disconfirming feedback following an auditory discrimination. Perception and Psychophysics, 1973a, 13, 25-31.

Squires, K.C., Hillyard, S.A. & Lindsay, P.H., Vertex potentials evoked during auditory signal detection: Relation to decision criteria. Perception and Psychophysics, 1973b, 14, 265-272.

Squires, K.C., Petuchowski, S., Wickens, C. & Donchin, E., The effects of stimulus sequence on event-related potentials: A comparison of visual and auditory sequences. Perception and Psychophysics, 1977, 22, 31-40.

Squires, K.C., Squires, N.K. & Hillyard, S.A., Vertex evoked potentials in a rating scale detection task: Relation to signal probability. Behavioral Biology, 1975a, 13, 21-34.

Squires, K.C., Squires, N.K. & Hillyard, S.A., Decision-related cortical potentials during an auditory signal detection task with cued observation intervals. Journal of Experimental Psychology: Human Perception and Performance, 1975b, 1, 268-279.

Squires, K.C., Wickens, C., Squires, N.K. & Donchin, E., The effect of stimulus sequence on the waveform of the cortical event-related potential. Science, 1976, 193, 1142-1146.

Squires, N.K., Donchin, E., Squires, K.C. & Grossberg, S., Bisensory stimulation: Inferring decision-related processes from the P300 component. Journal of Experimental Psychology: Human Perception and Performance, 1977, 3, 299-315.

Squires, N.K., Galbraith, G.C. & Aine, C.J., Event related potential assessment of sensory and cognitive defects in the mentally retarded. In: D. Lehman & E. Callaway (Eds.), Evoked Potentials: Applications and Problems, New York: Plenum Press, 1979, pp. 397-413.

Squires, N.K., Squires, K.C. & Hillyard, S.A., Two varieties of long-latency positive waves evoked by unpredictable auditory stimuli in man. EEG j., 1975, 38, 387-401.

Stanovitch, K.E. & Pachella, R.G., Encoding, stimulus-response compatibility, and stages of processing. Journal of Experimental Psychology, Human Perception and Performance, 1977, 3, 411-421.

Steinberg, D. & Jakobovitz, L. (Eds.), Semantics. Cambridge: Cambridge University Press, 1971.

Steinhauer, S.R., Emitted and evoked pupillary responses and event-related potentials as a function of reward and task involvement. Unpublished doctoral dissertation, Queens College, CUNY, 1981.

Steinschneider, M., Arezzo, J.C. & Vaughan, H.G. Jr., Speech evoked activity in the auditory radiations and cortex of the awake monkey. Brain Research, 1982, in press.

Stephenson, W.A. & Gibbs, F.A., A balanced non-cephalic reference electrode. EEG j., 1951, 3, 237-240.

Stephenson, D. & Siddle, D.A.T., Effects of "below-zero" habituation on the electrodermal orienting response to a test stimulus. Psy.Phys., 1976, 13, 10-15.

Sternberg, S., The discovery of processing stages: Extensions of Donders' method. In: W.G. Koster (Ed.), Attention and Performance II. Amsterdam: North-Holland, Acta Psychologica, 1969a, 30, 276-315.

Sternberg, S., Memory scanning: Mental processes revealed by reaction-time experiments. American Scientist, 1969b, 57, 421-457.

Stevens, K. & Blumstein, S., Invariant cues for place of articulation in stop consonants. Journal of the Acoustical Society of America, 1978, 64, 1358-1368.

Stuss, D.T. & Picton, T.W., Neurophysiological correlates of human concept formation. Behavioral Biology, 1978, 23, 135-162.

Stuss, D.T., Toga, A., Hutchinson, J. & Picton, T.W., Feedback evoked potentials during an auditory concept formation task. In: Kornhuber & Deecke, 1980, pp. 403-409.

Surwillo, W.W., Interhemispheric EEG differences in relation to short term memory. Cortex, 1971, 7, 246-253.

Sutton, S., The specification of psychological variables in an average evoked potential experiment. In: E. Donchin & D.B. Lindsley (Eds.), Average Evoked Potentials -- Methods, Results and Evaluations. Washington, D.C.: NASA, SP-191, 1969, pp. 237-262.

Sutton, S., P300 -- thirteen years later. In: H. Begleiter (Ed.), Evoked Brain Potentials and Behavior. New York: Plenum Press, 1977, pp. 107-126.

Sutton, S., Braren, M., Zubin, J. & John, E.R., Evoked-potential correlates of stimulus uncertainty. Science, 1965, 150, 1187-1188.

Sutton, S., Braren, M., Zubin, J. & John, E.R., Information delivery and the sensory evoked potential. Science, 1967, 155, 1436-1439.

Sutton, S. & Ruchkin, D.S., The late positive complex -- Advances and new problems. In: J. Cohen, R. Karrer & P. Tueting (Eds.), Annals of the New York Academy of Sciences, in press.

Symmes, D. & Eisengart, M.A., Evoked response correlates of meaningful visual stimuli in children. Psy.Phys., 1971, 8, 769-778.

Syndulko, K. & Lindsley, D.B., Motor and sensory determinants of cortical slow potential shifts in man. In: Desmedt, Vol. 1, 1977, pp. 97-131.

Syndulko, K., Hansch, E.C., Cohen, S.N., Pearce, J.W., Goldberg, Z., Montan, B., Tourtellotte, W.W. & Potvin, A.R., Long-latency event related potentials in normal aging and dementia. In: J. Courjon, F. Mauguiere, M. Revol & F. Peronnet (Eds.), Clinical Applications of Evoked Potentials in Neurology. New York: Raven, in press.

Tanguay, P., Taub, J., Doubleday, C. & Clarkson, D., An interhemispheric comparison of auditory evoked responses to consonant-vowel stimuli. Neurophysiologia, 1977, 15, 123-131.

Taylor, D.A., Stage analysis of reaction time. Psychological Bulletin, 1976, 83, 161-191.

Taylor, M.J., Bereitschaftspotential during the acquisition of a skilled motor task. EEG j., 1978, 45, 568-576.

Tecce, J.J., Contingent negative variation (CNV) and psychological processes in man. Psychological Bulletin, 1972, 77, 73-108.

Tecce, J.J., Boehner, M.B. & Cattanach, L., CNV and myogenic functions: II. Divided attention produces a double dissociation of CNV and EMG. Paper presented at Evoked Potentials International Congress VI, Lake Forest Illinois, 1981a.
Tecce, J.J. & Cole, J.O., The distraction-arousal hypothesis, CNV and schizophrenia. In: D.I. Mostofsky (Eds.), Behavior Control and Modification of Physiological Activity. Englewood Cliffs, N.J.: Prentice-Hall, 1976, 162-219.
Tecce, J.J., Savignano-Bowman, J. & Dessonville, C.L., CNV and myogenic functions: I. Muscle tension produces a dissociation of CNV and EMG. Paper presented at Evoked Potential International Congress VI, Lake Forest, Illinois, 1981b.
Tecce, J., Savignano-Bowman, J. & Meinbresse, D., Contingent negative variation and the distraction-arousal hypothesis. EEG j., 1976, 41, 227-286.
Teyler, T., Roemer, R., Harrison, T. & Thompson, R., Human scalp-recorded evoked-potential correlates of linguistic stimuli. Bulletin of the Psychonomic Society, 1973, 1, 333-334.
Thatcher, R.W., Electrophysiological correlates of animal and human memory. In: R.D. Terry & S. Gershon (Eds.), Neurobiology of Aging. New York: Raven Press, 1976, pp. 43-102.
Thatcher, R.W., Evoked potential correlates of hemispheric lateralization during semantic information processing. In: S. Harnad, R.W. Doty, L. Goldstein, J.J. Jaynes, G. Krauthamer (Eds.), Lateralization in the nervous system. New York: Academic Press, 1977a.
Thatcher, R.W., Evoked potential correlates of delayed letter matching. Behavioural Biology, 1977b, 19, 1-23.
Thatcher, R.W. & April, R.S., Evoked potential correlates of semantic information processing in normals and aphasics. In: R.W. Reiber (Ed.), The Neuropsychology of Language. New York: Plenum Press, 1976.
Thompson, R.F. & Spencer, W.A., Habituation: a model for the study of neuronal substrates of behavior. Psychological Review, 1966, 173, 16-43.
Timsit-Berthier, M., Delaunoy, J., Koninckx, N. & Rousseau, J.C., Slow potential changes in psychiatry. I. Contingent negative variation. EEG j., 1973a, 35, 355-361.
Timsit-Berthier, M., Delaunoy, J. & Rousseau, J.C., Slow potential changes in psychiatry. II. Motor potential. EEG j., 1973b, 35, 363-367.
Tojo, Y., EEG asymmetry during mental arithmetic and recall of verbal (kanji, kana alphabet) and visual imageries. Japanese Journal of Psychology, 1978, 49, 288-291.
Towey, J., Rist, F., Hakerem, G., Ruchkin, D.S. & Sutton, S., N250 latency and decision time. Bulletin of the Psychonomic Society, 1980, 15, 365-368.
Treisman, A.M., Perception and recall of simultaneous speech stimuli. In: A.F. Sanders (Ed.), Attention and Performance III, Acta Psychologica, 1970, 33, 132-148.
Treisman, A.M. & Gelade, G., A feature-integration theory of attention. Cognitive Psychology, 1980, 12, 97-136.
Trotman, S.C.A. & Hammond, G.R., Sex differences in task-dependent EEG asymmetries. Psy.Phys., 1979, 16, 429-431.
Tucker, D.M., Sex differences in hemispheric specialization for synthetic visuospatial functions. Neuropsychologia, 1976, 14, 447-454.
Tueting, P., Event-related potentials, cognitive events, and information processing: A summary of issues and discussion. In: Otto, 1978, pp. 159-169.

Tueting, P. & Sutton, S., The relationship between pre-stimulus negative shifts and post-stimulus components of the averaged evoked potential. In: H.H. Kornblum (Ed.), Attention and Performance IV. New York: Academic Press, 1973, pp. 185-207.

Tueting, P., Sutton, S. & Zubin, J., Quantitative evoked potential correlates of the probability of events. Psy.Phys., 1970, 7, 385-394.

Turvey, M.T., On peripheral and central processes in vision: Inferences from an information processing analysis of masking with patterned stimuli. Psychological Review, 1973, 80, 1-52.

Uttal, W.R., Do compound evoked potentials reflect psychological codes? Psychological Bulletin, 1965, 64, 377-392.

Vallbo, A.B., Human muscle spindle discharge during isometric voluntary contractions. Amplitude relations between spindle frequency and torque. Acta Physiologica Scandinavica, 1974, 90, 319-336.

Vanderwolf, C.H. & Robinson, T.E., Reticulo-cortical activity and behavior: A critique of the arousal theory and a new syntheis. The Behavioral and Brain Sciences, 1981, 4, 459-514.

Van Voorhis, S. & Hillyard, S.A., Visual evoked potentials and selective attention to points in space. Perception and Psychophysics, 1977, 22, 54-62.

Van Olst, E.H., The orienting reflex. The Hague: Mouton, 1971.

Vaughan, H.G. Jr., The relationship of brain activity to scalp recordings of event-related potentials. In: E. Donchin & D.B. Lindsley (Eds.), Averaged evoked potentials: Methods, results, evaluations. Washington, D.C.: NASA, 1969, pp. 45-94.

Vaughan, H.G. Jr., The analysis of scalp-recorded brain potentials. In: R.F. Thompson & M.M. Patterson (Eds.), Bioelectric recording techniques. Part B. Electroencephalography and human brain potentials. New York: Academic Press, 1974, pp. 157-207.

Vaughan, H.G. Jr., The motor potentials. In: A. Remond (Ed.), Handbook of Electroencephalography and Clinical Neurophysiology. Vol. 8, Part A. Amsterdam: Elsevier, 1975, pp. 86-91.

Vaughan, H.G. Jr., The neural origins of human event related potentials. Annals of the New York Academy of Sciences, in press.

Vaughan, H.G. Jr., Costa, L.D. & Gilden, L., The functional relation of visual evoked response and reaction time to stimulus intensity. Vision Research, 1966, 6, 645-656.

Vaughan, H.G. Jr., Costa, L.D., Gilden, L. & Schimmel, H., Identification of sensory and motor components of cerebral activity in simple reaction-time tasks. Proceedings of the 73rd Conference of the American Psychological Association, 1965, 1, 179-180.

Vaughan, H.G. Jr., Costa, L.D. & Ritter, W., Topography of the human motor potential. EEG j., 1968, 8, 135-147.

Vaughan, H.G. Jr., Bossom, J. & Gross, E.G., Cortical motor potential in monkeys before and after upper limb deafferentiation. Experimental Neurology, 1970, 26, 253-262.

Vaughan, H.G. Jr. & Ritter, W., The sources of auditory evoked responses recorded from the human scalp. EEG j., 1970, 28, 360-367.

Vaughan, H.G. Jr., Ritter, W. & Simson, R., Topographic analysis of auditory event-related potentials. In: Kornhuber & Deecke, 1980, pp. 279-285.

Velden, M., An empirical test of Sokolov's entropy model of the orienting response. Psy.Phys., 1974, 11, 682-691.

Velden, M., Some necessary revisions of the neuronal model concept of the orienting response. Psy.Phys., 1978, 15, 181-185.

Verbaten, M.N., Relation between the visual orienting reaction (VOR) and phasic and tonic aspects of electrodermal activity. Psychological Laboratory Reports, 1976, 4.

Verbaten, M.N., Woestenburg, J.C. & Sjouw, W., The influence of visual information on habituation of the electrodermal and the visual orienting reaction. Bio.Psy., 1979, 8, 189-201.

Verbaten, M.N., Woestenburg, J.C. & Sjouw, W., The influence of task relevance and stimulus information on habituation of the visual and the skin conductance orienting reaction. Bio.Psy., 1980, 10, 7-19.

Verbaten, M.N., Woestenburg, J.C., Sjouw, W. & Slangen, J.L., The influence of uncertainty and visual complexity on habituation of the electrodermal and visual orienting reaction. Psy.Phys., 1982, in press.

Verleger, R. & Cohen, R., Effects of certainty, modality shift and guess outcome on evoked potentials and reaction times in chronic schizophrenics. Psychological Medicine, 1978, 8, 81-93.

Vinogradova, O.S., Functional organization of the limbic system in the process of registration of information: facts and hypotheses. In: R.L. Isaacson & K.H. Pribram (Eds.), The Hippocampus, Vol. 2: Neurophysiology and Behavior. New York: Plenum Press, 1975, pp. 3-67.

Vinogradova, O.S., The hippocampus and the orienting reflex. In: E.N. Sokolov & O.S. Vinogradova (Eds.), Neural Mechanisms of the Orienting Reflex. Hillsdale N.J.: Erlbaum, 1975, pp. 128-154.

Walter, W.G., Slow potential changes in the human brain associated with expectancy, decision and intention. EEG j., 1967, 26, 123-130.

Walter, W.G., Cooper, R., Aldridge, V., McCallum, W.C. & Winter, A.L., Contingent negative variation: An electric sign of sensorimotor association and expectancy in the human brain. Nature, 1964, 203, 380-384.

Warren, C.A., The contingent negative variation and late evoked potential as a function of task difficulty and short-term memory load. Doctoral dissertation. University of Illinois, 1974.

Warren, L.R. & Harris, L.J., Arousal and memory: phasic measures of arousal in a free recall task. Acta Psychologica, 1975, 39, 303-310.

Warren, C.A., Karrer, R. & Cone, Movement-related potentials in children: A replication of waveforms, and their relationships to age. performance and cognitive development. Paper presented at Evoked Potential International Congress VI, Lake Forest, Illinois, 1981.

Wastell, D.G., Statistical detection of individual evoked responses: An evaluation of Woody's adaptive filter. EEG j., 1977, 42, 835-839.

Wastell, D.G., Attention and the habituation of human brain potentials. Doctoral thesis, Durham University, 1978.

Wastell, D.G., On the independence of P300 and the CNV: A short critique of the principal components analysis of Donchin et al. (1975). Bio.Psy., 1979, 9, 171-176.

Weerts, T. & Lang, P.J., The effects of eye fixation and stimulus and response location on the contingent negative variation (CNV). Bio.Psy., 1973, 1, 1-19.

Weinberg, H. & Papakostopoulos, D., The frontal CNV: Its dissimilarity to CNVs recorded from other sites. EEG j., 1975, 39, 21-28.

Welch, J.C., On the measurement of mental activity through muscular activity and the determination of a constant of attention. American Journal of Physiology, 1898, 1, 253-306.

Weller, M. & Montagu, J.D., Electroencephalographic coherence in schizophrenics: a preliminary study. In J. Gruzelier & P. Flor-Henry (Eds.), Hemisphere Asymmetries of Function and Psychopathology. Amsterdam: Elsevier/North Holland, 1979, pp. 285-292.

Wieneke, G.H., Deinema, C.H.A., Spelstra, P., Storm van Leeuwen, W. & Versteeg, H., Normative spectral data on alpha rhythm in male adults. EEG j., 1980, 49, 636-645.

Wiet, S.G. & Goldstein, L., Successful and unsuccessful university students: Quantitative hemispheric EEG differences. Bio.Psy., 1979, 8, 273-284.

Wilke, J.T. & Lansing, R.W., Variations in the motor potential with force exerted during voluntary arm movements in man. EEG j., 1973, 35, 259-266.

Wilkinson, R.T. & Lee, M.V., Auditory evoked potentials and selective attention. EEG j., 1972, 33, 411-418.

Wilkinson, R.T. & Morlock, H.C., Evoked cortical response and performance. Proc. London Conference of British Psychological Society, 1965, Abstract in Bulletin of British Psychological Society, 1966, 19, 10a.

Williams, W.C. & Prokasy, W.F., Classical skin conductance response conditioning: effects of random intermittent reinformcement. Psy.Phys., 1977, 14, 401-407.

Winfield, D.A., Rivera-Dominguez, M. & Powell, T.P.S., The termination of geniculocortical fibres in Area 17 of the visual cortex in the macaque monkey. Brain Research, 1982, 231, 19-32.

Wing, A.M. & Kristofferson, A.B., The timing of inter-response intervals. Perception and Psychophysics, 1973, 13, 455-460.

Witwer, J.G., Trezek, G.J. & Jewett, D.L., The effect of media inhomogenities upon intracranial electrical fields. IEEE Transactions in Biomedical Engineering, 1972, 5, 352-362.

Woestenburg, J.C., Verbaten, M.N., Sjouw, W.P.B. & Slangen, J.L., A statistical Wiener filter using complex analyses of variance. Bio.Psy., 1981a, 13, 215-225.

Woestenburg, J.C., Verbaten, M.N. & Slangen, J.L., The influence of information on habituation of the "Wiener" filtered visual event-related potential and the skin conductance reaction. Bio.Psy., 1981b, 13, 189-201.

Woestenburg, J.C., Verbaten, M.N., Sjouw, W.P.B. & Slangen, J.L., The influence of task-relevance and stimulus information on habituation of "Wiener" filtered evoked potentials and the skin conductance reaction. 1981c, in preparation.

Wood, C.C., Auditory and phonetic levels of processing in speech perception: Neurophysiological and information-processing analyses. Journal of Experimental Psychology: Human Perception and Performance, 1975, 104, 3-20.

Wood, C.C., Allison, T., Goff, W.R., Williamson, P.D. & Spencer, D.B., On the neural origin of P300 in man. In: Kornhuber & Deecke, 1980, pp. 51-56.

Wood, C.C., Goff, W.R. & Day, R.S., Auditory evoked potentials during speech perception. Science, 1971, 173, 1248-1251.

Wood, C.C., McCarthy, G., Squires, N.K., Vaughan, W.G. Jr., Woods, D.L. & McCallum, W.C., Anatomical and physiological substrates of event-related potentials: Two case studies. In: J. Cohen, R. Karrer & P. Tueting (Eds.), Annals of the New York Academy of Sciences, in press.

Woods, D.L., Courchesne, E., Hillyard, S.A. & Galambos, R., Recovery cycles of event-related potentials in multiple detection tasks. EEG j., 1980, 50, 335-347.

Woods, D.L. & Hillyard, S.A., Attention at the cocktail party: Brainstem evoked responses reveal no peripheral gating. In: D. Otto (Ed.), Proceedings of the Fourth International Congress on Event-Related Slow Potentials of the Brain. Washington, D.C.: U.S. Government Printing Office, 1979, pp. 230-233.

Woods, D.L., Hillyard, S.A., Courchesne, E. & Galambos, R., Electrophysiological signs of split-second decision making. Science, 1980, 207, 655-657.

Woodworth, R.S., Experimental Psychology. New York: Holt, 1938.
Zaidel, E. & Peters, A.M., Phonological encoding and ideographic reading by the disconnected right hemispere: Tw case studies. Brain and Language, 1981, 14, 205-234.